T5-CQJ-669

Electronic Systems Quality Management Handbook

Electronic Packaging and Interconnection Series

Charles M. Harper, Series Advisor

ALVINO • *Plastics for Electronics*

CLASSON • *Surface Mount Technology for Concurrent Engineering and Manufacturing*

DI GIACOMO • *Reliability of Electronic Packages*

GINSBERG AND SCHNOOR • *Multichip Module and Related Technologies*

HARPER • *Electronic Packaging and Interconnection Handbook*

HARPER AND MILLER • *Electronic Packaging, Microelectronics, and Interconnection Dictionary*

HARPER AND SAMPSON • *Electronic Materials and Processes Handbook, 2/e*

HWANG • *Modern Solder Technology for Competitive Electronics Manufacturing*

LAU • *Ball Grid Array Technology*

LAU • *Flip Chip Technologies*

LICARI • *Multichip Module Design, Fabrication, and Testing*

SMITH • *Optimizing Quality in Electronic Assemblies*

Related Books of Interest

BOSWELL • *Subcontracting Electronics*

BOSWELL AND WICKAM • *Surface Mount Guidelines for Process Control, Quality, and Reliability*

BYERS • *Printed Circuit Board Design with Microcomputers*

CAPILLO • *Surface Mount Technology*

CHEN • *Computer Engineering Handbook*

COOMBS • *Electronic Instrument Handbook, 2/e*

COOMBS • *Printed Circuits Handbook, 4/e*

DI GIACOMO • *Digital Bus Handbook*

DI GIACOMO • *VLSI Handbook*

FINK AND CHRISTIANSEN • *Electronics Engineers' Handbook, 3/e*

GINSBERG • *Printed Circuits Design*

JURAN AND GRYNA • *Juran's Quality Control Handbook*

JURGEN • *Automotive Electronics Handbook*

MANKO • *Solders and Soldering, 3/e*

RAO • *Multilevel Interconnect Technology*

SZE • *VLSI Technology*

VAN ZANT • *Microchip Fabrication*

Electronic Systems Quality Management Handbook

Marsha Ludwig-Becker, Editor
Operations Project Manager
The Boeing Company
Anaheim, California

McGraw-Hill
New York San Francisco Washington, D.C. Auckland Bogotá
Caracas Lisbon London Madrid Mexico City Milan
Montreal New Delhi San Juan Singapore
Sydney Tokyo Toronto

Library of Congress Cataloging-in-Publication Data

Electronic quality management handbook / Marsha Ludwig-Becker, editor.
p. cm.
Includes index.
ISBN 0-07-039055-X
1. Electronic industries—Quality control. 2. Total quality management. I. Ludwig-Becker, Marsha.
TK7836.E4656 1997 621.381'068'5—dc21 97-20755
CIP

McGraw-Hill 

*A Division of The **McGraw-Hill** Companies*

1 2 3 4 5 6 7 8 9 0 DOC/DOC 9 0 2 1 0 9 8 7

ISBN 0-07-039055-X

The sponsoring editor for this book was Stephen S. Chapman, the editing supervisor was Caroline R. Levine, and the production supervisor was Tina Cameron. This book was set in Century Schoolbook by Dina John of McGraw-Hill's Professional Book Group composition unit.

Printed and bound by R. R. Donnelley & Sons Company.

This book is printed on recycled, acid-free paper containing a minimum of 50% recycled de-inked fiber.

To that small group of leaders who enlightened me to understand that quality management is management of a quality company, and who are leading the United States to be world class internationally. They are: Richard Schwartz, Jim Albaugh, Dwayne Weir, Ralph Moslener, Paul Smith, and Derek McLuckey. To all the contributors who worked so hard to put this book together. And to my husband for his patience, love, understanding, and all the excellent art work he did for this book.

Contents

Preface xiii
List of Contributors xix

Chapter 1. Introduction 1

1.1 Definition 1
1.2 Scope 9
1.3 Application 16

Chapter 2. The New Definition of Quality 27

2.1 The New Quality 27
2.1.1 New definitions 32
2.1.2 The total system life cycle 56
2.2 Performance-Based Contracting for Electronics Systems 62
2.2.1 The contracting process 64
2.2.2 Detailing the performance, specification, characteristics, and checklist 80
2.2.3 The new approach to test and evaluation—impact on quality assurance 88
2.3 The World Class Electronics Company 92
2.3.1 Overview 92
2.3.2 Schonberger's view of world class manufacturing 95
2.3.3 Competitions in quality: the Baldrige, Deming, and *Industry Week* programs 109
2.3.4 Implementing a world class electronics company 117
2.3.5 Performance measures for a world class company 119
2.3.6 A case study in world class manufacturing 121

Chapter 3. New and Advanced Quality Systems for the New World 131

3.1 Introduction 131
3.2 One Quality System 132
3.2.1 ISO 9000 international quality assurance requirements 136
3.3 TQM and Teams 140
3.4 Management and Leadership 142

3.5 Empowering the Workforce 144
3.6 Metrics 148
3.6.1 Metrics to get to world class 153
3.7 Advance Quality Concepts/Techniques 154
3.7.1 Quality function deployment 156
3.7.2 Robust designs with Taguchi's Method 165
3.7.3 Statistical process control and 6 sigma 176
3.8 Nothing Left but to Do It 198

Chapter 4. ISO 9000 International Quality Management Standards 203
4.1 Introduction 203
4.2 The Elements of ISO 9001 210
4.2.1 Management responsibility 210
4.2.2 Quality system 214
4.2.3 Contract review 220
4.2.4 Design control 220
4.2.5 Document and data control 224
4.2.6 Purchasing 227
4.2.7 Control of customer-supplied product 230
4.2.8 Product identification and traceability 230
4.2.9 Process control 231
4.2.10 Inspection and testing 236
4.2.11 Control of inspection, measuring, and test equipment 238
4.2.12 Inspection and test status 239
4.2.13 Control of nonconforming product 240
4.2.14 Corrective and preventive action 241
4.2.15 Handling, storage, packaging, preservation, and delivery 242
4.2.16 Control of quality records 243
4.2.17 Internal quality audits 244
4.2.18 Training 247
4.2.19 Servicing 249
4.2.20 Statistical techniques 250
4.2.21 One quality system for the automotive industry: QS-9000 250
4.3 Product Certification 252
4.3.1 In Europe 253
4.3.2 In the United States 253
4.4 Becoming Compliant: Getting Organized 256
4.4.1 Tasks to schedule to become compliant or get certified 265
4.4.2 Self-evaluation/assessment 269
4.5 Preparing for Certification/Registration 269
4.5.1 Selecting a registration company 271
4.5.2 Alternatives to the certification/registration process: 1996 278

Chapter 5. Managing the Deming Way 295
5.1 Introduction 296
5.1.1 Who was Deming? 299
5.1.2 What was the spark that ignited Japan? 300
5.1.3 Deming's aim 300
5.1.4 What is management? 301
5.1.5 Overview of rest of this chapter 301
5.2 What Was Deming's Theory? 301
5.2.1 Part 1—Appreciation for a system 302
5.2.2 Theory of variation 308

5.2.3 Theory of knowledge 314
5.2.4 Psychology 318
5.2.5 Deming's theory as a system 320
5.3 The Operating Principles for a Manager for Managing the Deming way 321
5.4 Translating the Theory into Action 324
5.4.1 Optimization thinking 325
5.4.2 Creating a work team focused on being the best 327
5.4.3 Educating, training, coaching, and counseling 327
5.4.4 Involving the workers in decision making and optimizing 328
5.4.5 Interacting with the other components of the system including suppliers and customers 329
5.4.6 Getting the job done 329
5.4.7 Constantly improving the system 330
5.5 A Review of Common Management Practices 331
5.5.1 Proliferation of unnecessary paperwork 332
5.5.2 Reactive management 332
5.5.3 Reliance on numerical goals 333
5.5.4 Performance appraisal and ranking 335
5.5.5 Pay for performance and other forms of incentive pay 335
5.5.6 Management by objectives (as practiced)—managing the pieces, not the whole 336
5.6 More about Leadership 337
5.6.1 Deming on leadership 337
5.6.2 Brightness, darkness, and frequency 338
5.7 Other Key Concepts for Success—Other Things Successful Managers Do 338
5.7.1 Always act with integrity 338
5.7.2 Respect others 338
5.7.3 Manage yourself 338
5.7.4 Embrace change 339
5.7.5 Embrace the paradox 339
5.7.6 Do what is important but not urgent 339
5.7.7 Take personal responsibility 340
5.7.8 Have a passion for learning 340
5.8 The Deming Prize 341
5.9 Conclusion: Nothing Left but to Do It 347

Chapter 6. Total Quality Management 351

6.1 Definition of Total Quality Management 351
6.2 Background 352
6.2.1 Dr. Deming, Japan, and TQM 353
6.2.2 The TQM gurus 355
6.3 TQM Model 356
6.3.1 Culture 357
6.3.2 Change processes 366
6.3.3 Implementation and measurement (I&M) processes 376
6.3.4 Recognition 380
6.4 Additional Thoughts 383

Chapter 7. Subcontractor/Supplier Quality 387

7.1 Introduction 387

7.2 Supplier Selection 388
7.3 Supplier Partnership/Teaming 392
7.3.1 Criteria for partner selection 394
7.3.2 Partnership tools 396
7.4 Supplier Quality Assessments 397
7.4.1 First party assessment 398
7.4.2 Second party assessment 399
7.4.3 Third party assessment 400
7.4.4 On-site assessment 402
7.4.5 Quality system survey 403
7.4.6 Quality system audit 404
7.4.7 Qualified manufacturers list (QML) 404
7.4.8 Process certification 406
7.4.9 Process survey 407
7.5 Flowing down Requirements 410
7.5.1 Feedback in procurement 412
7.5.2 Clear description 414
7.6 Assessing Product Compliance with Requirements 414
7.6.1 Importance of early detection 414
7.6.2 Receiving inspection and certification 416
7.6.3 Verification and validation 418
7.6.4 Destructive parts analysis 419
7.6.5 Assessment 419
7.6.6 Automated testing 420
7.6.7 Source inspection 421
7.7 Measurement and Rating Systems 421
7.7.1 Assigning Responsibility 424
7.7.2 Delivery 424
7.7.3 Supplier rating and incentive program 425
7.8 Supplier Certification 431
7.8.1 Delegated inspection 432
7.8.2 SPC/variability reduction 432
7.8.3 Combination programs 433
7.8.4 Additional elements in certification 433
7.9 World Class Purchasing 436

Chapter 8. Benchmarking 445

8.1 Definition 445
8.2 History 448
8.3 Using and Understanding Benchmarking 451
8.3.1 Benefits of benchmarking 457
8.3.2 Objections to benchmarking 460
8.3.3 Convincing the boss 460
8.4 The Benchmarking Process 461
8.4.1 Types of benchmarking 462
8.4.2 Benchmarking steps 463
8.5 Processes and Process Understanding 473
8.5.1 Process flow diagram 475
8.5.2 Expansion of blocks 478
8.5.3 Gathering data for blocks 481
8.6 Vital Signs 483

8.7 Ethical Data Gathering 487
8.7.1 The ethics of benchmarking 489
8.8 Data Analysis 490
8.9 Building a Team for Success 492
8.10 Defining the Team Goal in Terms of the Physical 495
8.11 World Class Benchmarking 497
8.12 Planning for Success and Continual Improvement 502
8.13 Xerox: A Case in Point 504

Chapter 9. The Malcolm Baldrige Award: Striving for World Class Quality and Total Customer Satisfaction 507

9.1 The Linkage between World Class Quality and Total Customer Satisfaction 507
9.1.1 Why it's important! 507
9.1.2 Quality recognition 508
9.1.3 Some world class quality companies 508
9.1.4 The Baldrige Award and ISO 9000 509
9.2 Baldrige Award History and Background: Establishing a National Quality Award 511
9.3 Award Values, Concepts, and Framework 512
9.3.1 Core Values and Key Concepts 512
9.4 Award Criteria Category and Item Description 518
9.4.1 Award criteria summary 518
9.4.2 Detailed award criteria 522
9.4.3 Highlights of 1997 criteria changes 522
9.5 Evaluation and Scoring Guidelines 541
9.6 Award Criteria Key Characteristics 543
9.7 Baldrige Award Winners and Highlights 544
9.8 Award Winners' Best Practices 553
9.9 Baldrige Total Quality Payback 553
9.9.1 Benefits derived from self-assessment 554
9.9.2 Value of implementing a Total Quality Management system 554
9.9.3 Stock investment performance of the award winners 555
9.10 Baldrige Self-Assessment 555
9.10.1 Implementing a self-assessment and improvement system 555
9.10.2 Using Deming's PDCA cycle 559
9.10.3 Following some useful hints 560
9.10.4 Barriers to self-assessment and ways to overcome them 561
9.10.5 A real-life example—Company X 562

Appendix A Roadmap for Quality in the Twenty-First Century 567

Appendix B Sample Electronics Company Quality System Manual 617

Appendix C The USAF R&M 2000 Variability Reduction Process 639

Index 693

Preface

The aim of this book is to transform quality management from a department to management of a quality organization. It focuses on electronics because electronics is the fastest-changing technology in the international market today. Quality systems and management have long been thought of as two separate entities. A quality system is a management system that operates on specific principles to lead an organization. When the ISO 9000 series of quality management standards appeared, there was (and is) much confusion about how this idea of quality management relates to management principles and not merely to the "quality department." This handbook intends to bridge the gap. It provides a means of interpreting the quality management system and its principles to management, and how the two are really one. It links the management of quality organization with attainment of world class manufacturing, and provides information about how to take an organization forward. Focused on the customer, incited through employee involvement, and managed by data are the three hallmarks of world class quality companies. Whether we call these hallmarks quality principles or management principles is not important. It is important that managers understand that this style of management is invaluable to them in the international electronics world, or they will likely find themselves on the losing side of business.

The five main points to be discussed in this text are:

1. Understanding and using this new management system to move a company to world class status.
2. Methods for your organization to save money, to develop better, less expensive products and services, and to develop mutually beneficial supplier relationships.
3. A complete, simple way to implement ISO 9000, with a sample quality manual.

4. A basic understanding of operating in the new style using Total Quality Management (TQM).
5. Information on the Deming Award, Industry Week's Best Plant Award, and the Baldrige Award.

The world has changed. It is much smaller. Trade laws and international specifications are required for world trade. There is more challenge in the electronics industry than ever before. As products and processes become more complex, innovative solutions are needed to provide the cost and reliability customers demand and expect. Electronic devices are becoming obsolete at an ever-increasing rate, requiring improved cycle times for new production introduction and lean manufacturing practices in order to stay competitive. As a consequence of obsolescense, companies are merged, taken over, or disappearing more than ever in the international electronics worlds. Company focuses have changed. Successful companies in the electronic industry are international concerns. They are agile and provide quality productions to their customers. Customers have changed. They demand best value and usually have a number of choices for a supplier. Technology has changed, and is continuing to change at an ever increasing rate. Competition is at an all time high in electronics as we hear, "make it faster, smaller, cheaper and more powerful." Where will it all end? Or will it? It is fitting that we hear W. Edwards Deming's name so often today, because he taught us that it was all about management, and not only about improving the process but also the product. He is known as the "quality guru," but he really is the "management" guru.

Quality management can be thought of as a science where empirical data can be used to make a company successful; however empirical data alone is no longer enough. Using the principles of new quality management, including empowerment, leadership, a concern for people, and continuous improvement is a key factor to making companies successful today. Understanding the role of a manager as leader is as important as data collection and analysis in today's quality organization. United States business is improving in the international world, but many companies are still struggling to understand the link between management and quality. This handbook aims to assist in that understanding.

The introduction, Chap. 1, provides an overview of management disciplines and quality in the rapidly changing electronics industry. Chapter 2 describes the new definitions, how they encompass the entire organization, and how they relate to becoming a world class electronics company. This chapter contains data from the latest books on world class performance. Mr. Richard Schonberger, renowned author of several books on world class manufacturing (WCM), and the publishers of Industry Week have both consented to put their checklists in this volume. Mr. Schonberger would like to track your company's progress on attaining world class by his form, which appears as Table 2.22. Chapter 3 is an overview of the management techniques that are being used in the quality management area. These approaches, called "new and advanced quality techniques," are techniques that affect the way a product is defined and manufactured throughout an entire organization. These techniques provide for customer-pleasing design and cost savings in production. This chapter includes a discussion on Six Sigma. Chapter 4 is a complete overview of ISO 9000 and how to implement it in an organization so that it assures continuous improvement when combined with the data in Chaps. 5 and 6. Chapter 4 includes a complete ISO 9000 checklist, exact methods for implementation, and a self-assessment matrix. Chapter 5 gives the manager an overview of Deming with ideas to put Deming's ideas to work in your organization. The Deming Award criteria complete this chapter. Chapter 6 speaks to implementing a real Total Quality Management and Continuous Improvement program. It provides illustrations and concrete ideas for changing your organization and includes a self-assessment matrix. Putting this chapter into action will make a significant difference in an organization; however that is not easily accomplished. Implementation represents a paradigm shift in the way many companies manage today and may require a major cultural change in the way we think, talk and act. There are hundreds of pages written on Quality and Management, but in this text, this chapter provides a systems look at the relationship between these two philosophies and why they are important to leaders and the future of their companies. Chapter 7 addresses the relationship of an organization with its suppliers and its strategic importance to international competitiveness. More and more companies are increasing the outsourcing of processes that are not core to their business. Production lines are demanding just-in-time material with increased quality and lower cost. Chapter 7 deals with these techniques and methods for dealing with suppliers, including the techniques to develop a world class purchasing organization. It includes a self-assessment at the end of the chapter. Chapters 8 and 9 provide ways to improve through two self-assessment techniques: Benchmarking and The Malcolm Baldrige Award Criteria. The benchmarking chapter, Chap. 8, examines in great detail, ways to determine best practices. It includes history, objectives, issues, and costs of benchmarking in the electronics industry. The Baldrige chapter ties it all together. It reinforces the importance of total quality management (TQM) which spans the entire organization, its products, services, and support structure. This chapter provides specific best practices from Malcolm Baldrige National Quality Award winners, data on their quality payback (earnings), and a model for self-assessment and methods to improve using the Baldrige framework and criteria. An update on the changes in 1997 is also included. The Appendix contains two information documents first produced by the U.S. government in association with U.S. industry. The first document, The Roadmap for Quality in the Twenty-First Century, published in April 1995, states for the first time in the United States, the policy for quality as it relates to industry and government. It is not widely distributed, but is a milestone because it is changing the way we view quality management and how important it is to world trade in the next century. The other document is an out-of-print copy of the USAF R&M Variability Reduction Process. Since excess variability costs money, variability reduction techniques can turn your organization into a profitable business that produces quality products. The third document in the Appendix is a sample quality manual in the style of the ISO 9000 standards. It is provided for companies to use and reference in implementing and improving quality management systems.

All chapters have data that an organization can easily adapt. Many diagrams, tables, and illustrations supplement the text for easy understanding. When systems are to be implemented there are examples and matrices which can be customized and used for self assessment. Many sources are provided for further reading. Throughout the handbook, the ideas and data from industry experts are provided to facilitate the electronic manager's ability to adapt to our rapidly changing industry.

There is no magic to anything in this handbook. It has been assembled by experts who have actual hands-on experience in implementing these principles in their fields of expertise. Mr. Primus Ridgeway is nationally known for leading the industry effort that unified industry and the U.S. government on the adoption of a single quality system, through the work of the Government and Industry Quality Liaison Panel (G&QLP). He is Director of Quality Assurance for Rockwell Collins, Inc., Avionics and Communication group, and serves as chairman of the Quality Committee of the Electronics Industry Association (EIA). Ms. Marsha Ludwig-Becker, CM, is well known for teaching the new quality management system throughout the United States and Latin and South America; she is an acknowledged expert in the implementation of ISO 9000, and in leading self-managed teams, and writing about the changeover from U.S. military standards to the industry/commercial standards. She is an Operations Project Manager for The Boeing Company where she participates on many teams for U.S. Acquisition Reform. Mr. Karl Haushalter, president of Optimization Works, lectures, consults, and teaches about W. Edwards Deming. He is also a very entertaining and inspiring speaker on Deming's Management Principles. Mr. Larry Coleman is Director of the Total Quality System at The Boeing Company and has implemented some of the most advanced TQM systems nationally. Mr. Bill Kirsanoff, a procurement quality engineer for The Boeing Company, is nationally known for his expertise of quality, and electronics quality. He works with suppliers daily, consults on the implementation of quality rating systems and speaks and writes for the American Society of Quality (ASQ). Mr. Gerald Borie implemented Benchmarking at Litton Guidance and Control Systems and then presented papers and spoke at several national conventions; now retired, he does consulting. Mr. Gene Carrubba is a past member (for four years) of the Baldrige Board of Examiners and a prolific writer of articles and books on quality. After his retirement from Motorola, where he was a Group Director of Quality, he formed his own consulting practice, the PALC Group. He is a senior member of the Institute of Electrical and Electronics Engineers (IEEE). A better overview of quality management principles could not have been produced by a more competent group of individuals.

It is the hope of the editor that this volume will provide a useful reference to the majority of questions about establishing, implementing, and maintaining a quality management system.

Acknowledgments

Thanks goes to the Quality Information Center (QIC) of the ASQ for its excellent assistance in the preparation of this book.

Thanks goes to all the companies/authors who provided their permission to make this text a meaningful and factual reference document on the relationship of Quality, Management, and World Class Organizations.

And lastly, a supreme thanks to the contributors who worked diligently to produce chapters on their expertise; and to my husband, Frederick J. Becker, for the creation of most of the illustrations, charts, and matrices.

Marsha Ludwig-Becker

List of Contributors

Marsha M. Ludwig-Becker *Operations Project Manager, the Boeing Company; seminar/trainer for Technology Training Corporation (TTC), Los Angeles, CA, and Systems Management and Development Center (SDMC), Springfield, VA. President of Beckers, Inc., Placentia, CA* (CHAPS. 2, 3, 4)

Gerald J. Borie *Consultant, and Past International President, Society of Reliability Engineers (SRE)* (CHAP. 8)

Eugene R. Carrubba *President, the PALC Group, Consultants for Operational Excellence, Wayland, MA; past member, Baldrige Board of Examiners* (CHAP. 9)

Larry Coleman *Director of Total Quality Management (TQM) for the Boeing Company* (CHAP. 6)

Karl Haushalter *President of Optimization Works, an organization based on Deming's philosophy, Palos Verdes (Los Angeles) CA.* (CHAP. 5)

William Kirsanoff *Procurement Quality Engineer, the Boeing Company; active member of American Society of Quality Control (ASQC) at local and national levels* (CHAP. 7)

Primus Ridgeway *Director of Quality Assurance, Rockwell Collins, Inc., Avionics and Communication Group; co-chairman and industry leader, Government and Industry Quality Liaison Panel (G&QLP), Washington, DC; and Chairman of the Quality Committee for the Electronics Industries Association (EIA)* (CHAP. 1)

Chapter

1

Introduction

Primus Ridgeway, Jr.

Director, Quality Assurance
Avionics and Communications
Rockwell Collins, Inc.

1.1 Definition

This is a book about quality management. It specifically addresses the quality management of electronics. All three key terms—quality, management, and electronics—are in a period of significant change. The subject of management has been studied, in depth, and it is well acknowledged that the manager who is responsible for the company makes the difference in what happens to that company. Management may be defined as the "*X*" and "*Y*" style (autocratic versus democratic). But in reality there are many autocratic managers still in the marketplace who, unless they understand the new theories of management and quality, will not succeed. Management leads and supervises; employees create the products and services. The manager must empower the workforce. The new management style is not autocratic; it is democratic. It may at times use autocratic techniques, but overall the new manager is leader, and more coach than boss. This handbook explains this management approach and puts its emphasis on electronic systems. It takes both styles, enlightened and on the right track, to create quality products and services, and to create an environment for continuous improvement where the company can succeed.

The new definition of quality is that quality is everybody's job. From the first person who gets a contract or a purchase order, or a product to

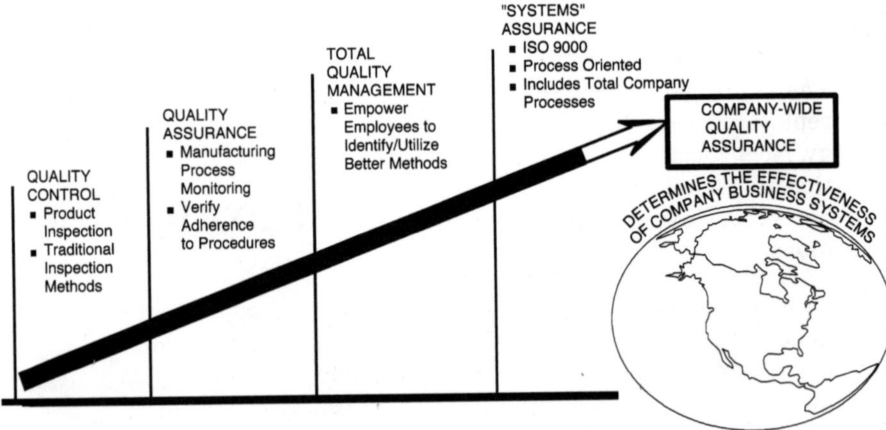

Figure 1.1 Evolution of quality assurance.

sell, to the last person who delivers the product or service—whether they are developing paper, software, or hardware, quality is part of their job. Each and every employee has an internal customer to whom he or she must deliver a quality product, and the company has an external customer who needs to be pleased. Quality is no longer a department or the "cop." It is no longer "conformance to requirements." It starts with the initial concept of a product or service and continues throughout development, delivery or installation, servicing, modification, and disposal. Quality has evolved from the inspection, over time, to the process-oriented, total company process that it is today (see Fig. 1.1). It is every one of us, from the marketing person to the creator of specifications or drawings, to the people who order parts and materials, to those who deliver and service the product to the customer.

An important challenge faces the development of quality products. Over the last 20 years, quality has been a major point of contention for companies. Better-quality products have an edge in the marketplace. Quality has become part of the strategic competitive position of the company. Some might think that advanced quality techniques may not be needed as the competitive position of companies increased. It seems this will not be the case for two reasons:

1. *The increasing complexity of products.* More and more components are being added to electronic products. They are becoming smaller and more powerful. The need for higher reliability in each component increases. As more intelligent capabilities are introduced to other products (where electronic chips are replacing mechanical systems), variation of complex products is unsatisfac-

tory in the eyes of customers. This is part of the reasoning behind the 6-sigma requirement of Motorola. [1]

2. *The rapid growth in technological support.* At first companies replaced simple repetitive, manual tasks with computer technology. Now companies are using computer technology to replace tasks that require a great deal of human intelligence and skill. Society is becoming very dependent on these technologies (some think too much so).

For example, railroad transportation systems now depend on sensing equipment installed along the tracks. No longer are there the signalers of the past. The need for highly reliable electronic equipment is great. Modern subway systems are totally run by electronic machines that "read" tickets and open gates. Just as calculators have eliminated the need to memorize arithmetic, so too will "smart cars" (with maps and directions) and programmed roadways eliminate the need to read maps. Again, advanced quality techniques must assure high quality, reliability, and durability.

The growth in technological support is not limited to products. In everyday service scenarios, from banks to hotels, customers find themselves dealing with a computer instead of a person. If the computer does not work, the service personnel stand by helplessly. Even the medical industry relies heavily on electronics to diagnose and treat illnesses. Modern hospitals and medical facilities are filled with electronic devices to treat and monitor everything from temperature to open heart surgery.

The consequences of poor quality in the future will be much worse than getting a defective product and throwing it out. Poor quality will lead to an inability to get to work, or to do one's job. There is no doubt that in this new world, there is a need for advanced quality techniques to fulfill customer's requirements and provide products reliable and durable enough to support the world market. [2]

During the American Society for Quality (ASQ) (formerly the American Society of Quality Control (ASQC) until July 1, 1997) fiftieth anniversary year, the society undertook a structured look at getting to the year 2010. A diverse group of seven professionals, with varied backgrounds and skills, met over 6 months. The team identified nine key forces that are powerful drivers of change for quality and the world. These forces are

1. Changing values are impacting both personal life and national priorities. Environmentalism, drug use, wealth distribution, ethnic nationalism, and employer-employee loyalty are among the value-changing forces perceived to be reshaping our lives.

2. The force of globalization includes the increasing role of world-spanning organizations and the enormous potential influence of the East, especially China and India, and the expansion of liberal democracy as a platform for change in other parts of the world.
3. The changing makeup of the work force reflects increased immigration, an age shift of working population, and a restructuring of the workplace, resulting in more individual agents and fewer lifetime employees.
4. The information revolution, of all forces, is having a great impact already, and will continue to increase. The Internet is changing the way data is managed. Bill Gates of Microsoft calls data "...the organizational challenge of the next century. Competitive success—even business survival—will depend on the ability to quickly turn it into comprehensible information." [3]
5. The velocity of change itself is seen as a force. In companies today, it seems the only constant is change.
6. Increased customer focus and individually tailored marketing will be a major determinant of commercial survival.
7. Leadership will remain a key force; it is speculated that one of its major thrusts will be promulgating values, and that it will likely influence people who are never seen in person.
8. Quality in new areas is the expansion of quality into new sectors. This is really the transition to the big "Q," where quality is involved in everything.
9. Change in quality practice will be reflected in the complete integration of quality practice into daily work. The average person will now know about quality practices, and understand how his or her job must be ensuring quality.

Overall, quality management will be integrated into the way things get done every day. The knowledge and application of quality tools and techniques will be mandatory. Communication, strategic thinking, information retrieval, and interpersonal skills will be essential to the work of advancing quality. [4]

Electronics is the science and technology of controlling the flow of electrons to produce useful results. It has had a tremendous influence on our lives. Television and radio receivers, stereo systems, computers, automatic controllers, and guidance systems for rockets all depend on electronics. Electronics is really only a means to an end. Electronic apparatuses are used to link input devices to output devices. Between those input and output devices, other electronics

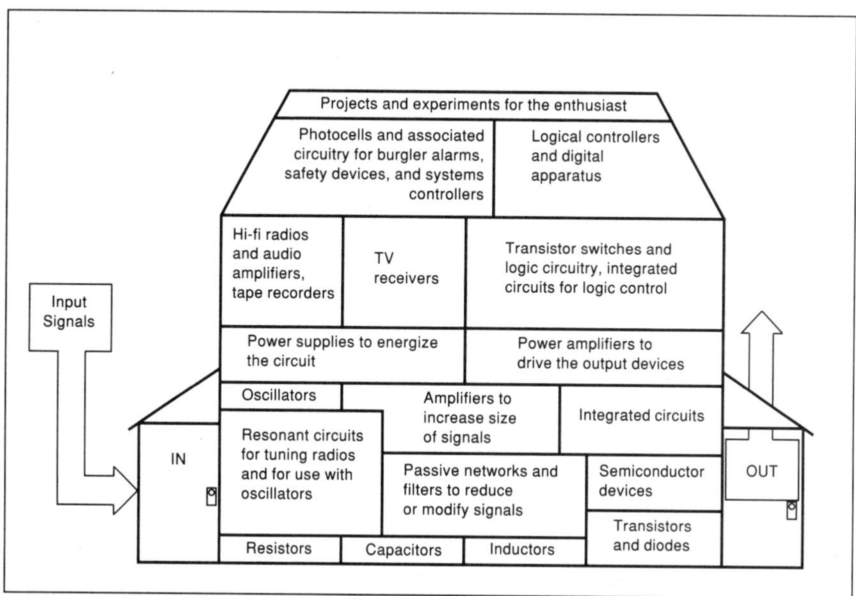

Figure 1.2 Electronic equipment "building blocks" produce desired circuits. (*Reprinted from Olsen, with permission.*)

apparatuses are required. A good example is a television camera: it cannot activate the picture tube in the television receiver directly because of differences in power requirements; therefore other electronics serve to output picture and sound. Figure 1.2 shows a prototype electronic "building" that illustrates how any type of electronic equipment may be built. [5]

Broadly speaking, electronics components can be divided into active and passive devices. Active devices, such as diodes, transistors, and integrated circuits, control the flow of energy in some way. Passive devices are resistors, conductors, capacitors, and inductors. [6] Active and passive electronic components are placed on printed wiring boards, in series, to produce electronic systems. Electronics is really a brand new field. Its entire history is limited to this century. It has grown so rapidly and become so important that two branches have developed: digital and analog. An analog circuit can be designed to produce an output for an infinite number of states. It measures voltage continuously from the very low to very high rather than in the one or two digits that a digital circuit shows. Things like height, weight, and speed are measured in an analog manner. Analog electronic systems were developed in the beginning and are well understood. Digital technology came into being with the integrated circuit and remarkable advances in the number of circuit functions that can be placed on a single silicon chip. A digital electronic device or circuit will recognize or produce an output of only several limited states. For example, most digital devices will respond to only two input conditions, low or high. Digital circuits are binary, based on 0 and 1 or "on" and "off." Digital electronics are more convenient to work with in electronic systems. It is common today to convert analog to digital so that information can be stored in a computer or on computer media. Digital media do not deteriorate with use as analog media do. This is well known by consumers, who are using more digital technology (compact disks) than analog (tapes) for recorded sound. The quality of sound is improved as well with digital media. More and more digital devices are being developed and fewer and fewer analog devices. The rapid growth of electronics in recent years is not likely to slow down. One of the key developments of the digital age is the microprocessor, a silicon chip the size of a watermelon seed containing the complete processing and control circuits of a small digital computer. [7] An entire new industry was born in electronics. Items like digital watches, microwave ovens, home appliances, and automated teller machines at banks are common in our daily life. Our work environment has changed dramatically because of the computer. NASA's space shuttle, now "old" electronic technology, could never have flown without the development of sophisticated electronics; and now even our automobiles and highways are employing even more sophisticated electronics for everyday uses. Almost every field is using electronics to make things work faster, more powerfully, and with greater automation.

It is interesting to understand how fast this "electronics" industry has grown. Two important developments at the beginning of the 1900s spurred the industry. The first was in 1901 when Marconi sent a message across the Atlantic Ocean using wireless telegraphy. Today we call this wireless communication *radio*. The second development came in 1906 when DeForest invented the audio vacuum tube. It was first used to make sounds louder, and was quickly adopted by the wireless inventors to improve their equipment. In 1906, Pickard used the first crystal radio detector. This was a great improvement that helped to make radio and electronics more popular. It also suggested the use of semiconductors (crystals) as materials with great promise for the future of the new field of radio and electronics. By 1920, commercial radio had changed the world by allowing nearly all homes to have one of the first electronic devices. Commercial television began around 1946 and was quite complicated compared to radios.

The first vacuum tube computer was build in 1943 at the University of Pennsylvania. It was large and expensive, but it became quite popular. Engineers were already working hard to replace vacuum tubes with semiconductor (crystal) devices. In 1947 at Bell Laboratories, the first working transistor was developed. This was such a major contribution to science and technology that the three inventors—Bardeen, Brittain, and Shockley—were awarded the Nobel Prize. Improvements came rapidly, and *solid state* is now a common term. Electronics in the early days was unreliable. Sometimes the electronic units, both for commercial and government use, were inoperable more than they were operable. By 1950, the U.S. Navy had begun studies of how to make their electronic vacuum tubes fail less frequently. The newly developed discipline was called *reliability.* The first reports on electronic reliability were issued by the U.S. Department of Defense (DOD) in 1951. [8] Reliability was a very important part of the new quality system, especially for electronics. Engineers were still trying to make faster, smaller, and more powerful electronics; and by 1958, Jack Kilby of Texas Instruments developed the integrated circuit (IC). Integrated circuits are complex combinations of several kinds of devices on a common base, a tiny piece of silicon. ICs offered lower cost, higher performance, and better reliability than an equivalent circuit built from separate parts. One IC could hold thousands of transistors. In 1971, Intel Corporation in California announced one of the most sophisticated circuits—the microprocessor, in which most of the circuitry of a computer is reduced to a single IC. The IC has produced an electronics explosion, and today electronics is an integral part of all technology. This is the electronic age. [9]

An integrated circuit, especially a sophisticated and complex one, when newly developed may have had a yield initially of less than 10 percent; nine out of ten devices would not pass tests and had be thrown away. This made the price of a new device very high. The electronics components industry, with established processes to increase yields and reliability, quickly raised the yield to 90 percent, then started measuring in parts per million the number that drop out of the process. The price then dropped drastically, and many new applications were found for the part because it had become economically attractive to use. The new parts were complex, powerful, and sophisticated. Their very complexity usually resulted in products that were easier to use. The term *user friendly* comes from this.

Not only is electronics growing rapidly, but it is also changing rapidly. Those who work in electronics must be willing to accept rapid change. Almost every technology now is using electronics to make products better, faster, and cheaper. The integrated circuit is the key to most electronic trends. The outlook is bright for the continued development of ICs. [10] In addition, the size of the IC is shrinking, while the power becomes greater. No one knows where it will end, or what the next major phase will be.

An electronic system is a set of electronic devices that make up a product. A block diagram defines an electronic system and shows all the individual functions of the system and how the signals flow through it. Schematic diagrams accompany the block diagram and show all of the individual parts of a circuit and how they are interconnected. Schematics are usually required for component-level troubleshooting of an electronic system. Component-level repair requires that the technician isolate and replace individual parts that are defective. System-level repair requires only a block diagram or knowledge of the block diagram; an entire module, panel, or circuit board can be replaced, and component level troubleshooting is not required. Component-level troubleshooting usually takes longer than system-level repair. Given that time is money, it may be better to replace an entire module or similar assembly. In a sophisticated, expensive, complex electronic system component-level troubleshooting is probably needed. [11] This involves a much more detailed set of schematics, and test equipment designed to a different level for troubleshooting and repair. In these areas, electronics works with its equally elusive partner, software, to create the product many consumers need, want, and desire. These things must be considered in the initial design.

Why is an electronic system different? It is different because many things electronic systems do when they go through design and manufacturing depend on the systems/processes that they use for development. The key word is *process.* Concurrent engineering, or integrated product development (IPD), must be used in designing electronic products and services. The design process must begin with understanding the customer requirements, support as well as performance. There is an excellent process for this known as quality function deployment (QFD).

An electronic system is also different because the first design and first packaging must be processed, and that requires an interface with the factory. The factory may or may not have the proper capability and systems in place. These processes are not easy to change because of the complex nature of the product and its interfaces with other products and services. The processes used in electronic systems, many of which are special processes, cannot be inspected by the usual inspection procedures. Electronic systems use "black magic," or propriety processes, which may be developed by small or large electronic manufacturers to produce an electronic product for a customer. Specifically, trained manufacturing engineers and production specialists, together with quality engineering personnel, must complete the design, including design of experiments and line proofing, with the people designing the product. First systems and processes must be developed as prototypes. Subcontractors and team members must be determined, and early in the process the entire team must work together to develop the quality product they wish to deliver. The company—whether large or small, defense or commercial—must develop a process to produce excellent quality, a big "Q" system, to repeatedly deliver a quality product or service.

1.2 Scope

This book is directed toward management of electronics design, development, and support. It covers both commercial and defense systems, because in today's world the U.S. government's goal is to integrate the two for their mutual benefit. Equally important to the U.S. government is to increase world trade and have world class manufacturers. The book is written for management, supervisors, leaders, and employees. It is written for all those who wish to understand the new quality and the new management of electronic systems. It does not contain all the data necessary to implement the complete process; this is readily available in the marketplace and is referred to in the book at the appropriate places. What is missing in the marketplace is the concept that the new quality manager is the chief executive officer (CEO) and quality is the company, not a department. This understanding is essential for the electronics supplier/company to survive in the entrepreneurial marketplace. Therefore, there is much emphasis on world class, because an electronics company that wants to succeed must be world class. CEOs must become the leaders of quality and make quality a company effort, not a department as in the old days. This is what this book is about.

This book covers electronic-systems-specific items that must be addressed in electronic development, manufacturing, and support. Wherever possible, the discussions are generic to systems, components, or printed wiring boards; at times, examples of one or the other are used. Nearly all systems/processes are applicable to all three, but are not implemented in quite the same way.

This book will identify and discuss elements of the essential integrated nature of a quality management system typically deployed by an enterprise. An effectively integrated quality management system will touch many different aspects of a business. These aspects include sales and marketing, procurement and order administration, engineering, manufacturing, enterprise management, and support. The single most important theme here is that the reader should recognize the need for each company to implement a quality management system that fits the specific requirements for the total business.

A standard on quality management is about to be released by the International Organization for Standardization (ISO) committee TC 176, which issues the ISO 9000 Quality Management Standards. In the document, eight quality management principles are defined. A quality management principle is defined as "a comprehensive and fundamental rule or belief, for leading and operating an organization, aimed at continually improving performance over the long term by focusing on customers while addressing the needs of all stakeholders. [12] Throughout this handbook, these eight principles are discussed. Table 1.1 lists and defines the principles and the chapters in which they are highlighted in the book.

Customer-focused organizations are the only kind of companies that will survive in the electronics world. It is well known that many organizations have gone out of business, and in many electronic industries it is extremely difficult to get certain electronic parts. Today, though, if the supplier is delivering parts, it must have almost perfect quality. Major suppliers, and original equipment manufacturers (OEMs) are demanding that they get 100 percent quality. With just-in-time initiatives, where the quality is left to the manufacturer, every component or subassembly that does not work causes more cost at the OEM. More cost means more dissatisfaction, and soon the OEM finds another supplier.

In the commercial marketplace, electronic companies that provide products that do not fail and have high reliability are taking over the marketplace. In addition, those companies are supplying improved warranties and 24-hour service. Some companies are even delivering loans or replacement products directly to consumer's homes or places of business.

Leadership has long been acknowledged as a positive trait. Good leaders seem to have no trouble in getting people to follow them. Although it has long been argued whether good leaders are born or taught, there are things that we all can learn to become good leaders. They are explained throughout this handbook. More important, in a period of immense change, it takes a leader with a vision to motivate, to endure—to change a culture. That is the task of leaders today. Leadership and management must go together in established organizations, but managers need to be leaders. The subject of leadership and management needs to be addressed just like the study of elec-

TABLE 1.1 Quality Management Principles and Quality Management Handbook

Principle	Definition	Location in this handbook
1. Customer-focused organizations	Organizations depend on their customers. They must strive to understand and exceed customer expectations.	Chap. 2: New Definitions Chap. 5: Deming Chap. 6: TQM Chap. 8: Baldrige
2. Leadership	Leaders establish unity of purpose, direction, and internal environment. They create an environment where people can succeed.	Chap. 3: New Quality Chap. 5: Deming Chap. 6: TQM Chap. 8: Baldrige
3. Involvement of people	People at all levels are the essence of an organization.	Chap. 3: New Quality Chap. 5: Deming Chap. 6: TQM Chap. 8: Baldrige
4. Process approach	A desired result is achieved more efficiently when related resources are managed as a process.	Chap. 3: New Quality Chap. 4: ISO 9000 Chap. 5: Deming Chap. 7: Baldrige Chap. 8: Benchmarking
5. System approach to management	Identifying, understanding, and managing a system of interrelated processes contributes to the effectiveness and efficiency of the organization.	Chap. 2: New Definitions Chap. 3: New Quality Chap. 5: Deming Chap. 6: TQM Chap. 8: Baldrige
6. Continual improvement	Continual improvement is a permanent objective of the organization.	Chap. 3: New Quality Chap. 5: Deming Chap. 6: TQM Chap. 8: Benchmarking Chap. 8: Baldrige
7. Factual approach to decision making	Effective decisions and actions are based on analysis of data and information.	Chap. 3: New Quality Chap. 5: Deming Chap. 8: Benchmarking Chap. 9: Baldrige
8. Mutually beneficial supplier relationships	Mutually beneficial relationships between organization and its supplier enhance the ability of both organizations to create value.	Primarily Chap. 7, but suppliers are talked about throughout the handbook

tronics or systems engineering. As part of leadership and management, good communication is extremely important. Techniques for communication must be agreed to for each company so that the employees are informed. Electronics again is dominating this world with electronic mail, video monitors, and electronic banners as well as electronic "face-to-face" town meetings.

People are what make the difference in the world. Even in the area of complete automation, there is a people connection. Tasks today are specialized at all levels of the organization. Other people need to have a generalized view, to implement the integration of systems and processes. Motivating people in the workforce has been long discussed. Also discussed has been the change in culture from the "old" workforce that felt they "owed" the company hard work, and the "new" culture that feels no commitment to the company.

The new term for people is *empowerment.* People want to be appreciated and part of the decision-making process. When they are, they are a tremendous force created to improve the entire company. It is people who make the difference. In the teams where people are empowered, the synergism makes a significant difference in productivity and quality.

Process, process, process. In the new world, at times "process" may seem to be an overused word. However, given the power of the word, the concept really is not overused. Everything is done by a process that includes people, equipment, materials, methods, and the environment. Through Total Quality Management (TQM) methods and statistical techniques these processes can be flowcharted, analyzed, and continuously improved.

The company is a set of interrelated processes that must be managed as a whole. That is part of the job of management—to make the processes work together. Each process interfaces with other processes. When the interfaces are defined on paper, through flowcharts, then analyzed for value-added work and increased productivity, cost reduction/avoidance will occur. Using this process approach, one soon sees that the entire company is interdependent, hence the new emphasis on integrated product/process teams, because that is the efficient way to do work.

In the electronics world, in particular, if a company does not continuously improve, it will not survive. There is continuous improvement in products and services in the 1990s like no other decade has seen. Everything from higher reliability of products to 24-hour call-in support to electronic banks and airline tickets is seen today. Process and product are on a continuous improvement program moving to the twenty-first century. By the time many electronic products are purchased, they are outdated and a new model is hitting store shelves. Electronics and software are at the heart of this improvement. The saying "better, faster, cheaper" is a fact of the 1990s.

Producing products and services must become more affordable, and that is why there is continuous improvement in both. Ten years ago getting to 100 parts per million (ppm) failure rate in the electronics component world was thought to be almost impossible; now 100 ppm is not even acceptable. Ten years ago Motorola was the only company talking about 6 sigma. Now many companies have equivalent quality systems producing excellent quality products.

In the service world, in the consumer area, the two best examples are banks and airlines. Remember how there used to be hundreds of tellers in banks? Think how many have been replaced by automatic teller machines (savings of people and manual accounting). Airlines are now moving to electronic tickets, saving not only paper and labor, but mailing and the disposal of all that paper into our environment. Both examples represent breakthroughs in the way business is done, and more of this is seen everyday, with the Internet. Continuous improvement using electronic equipment is the way of the twenty-first century.

Decision making should always be based on some sort of factual information. Even in the stock market world, there are reasons for market changes and data to back them up, even if the changes make no sense to the average person. In the electronics world, each company needs to collect data, and know what that data means. TQM brought the concept of data to the forefront of the quality management system. And not just data, but data on the process must be present so that the process can be evaluated. Every good TQM program has a phase where data is collected on the existing process so that when the change is made it can be quantified—either positively or negatively. First the existing process metrics must be known.

Competitive benchmarking is an eye-opening experience; it helps one to understand how competitors keep data. Some statisticians even argue about how Motorola calculates 6 sigma. The point is that data must be kept, and decisions need to be based on data. It is up to the CEO and the executive staff to enforce the rule that no decisions are made without data. People need to be trained that data makes the difference. Again, the TQM program is where this can be put into action.

Especially in the electronics world, most companies do not manufacture every item that they use in their products and services. That is not cost-effective. Suppliers are needed for many items, but the relationship must work both ways. Value must be created for both organizations. This is a significant change to the adversarial competitiveness of the last decade, but is becoming the norm with supplier partnerships and supplier chain management. The degree of trust changes supply chain management, with more internal communication between companies. This will take a culture change, but the outcome will be more beneficial supplier relationships. Currently, there are prototype systems set up for study in this area. This might be a good benchmarking activity for your company.

The CEO and the entire executive staff should understand that these quality management principles are the management theories for company management and a quality management style much like that which Deming lectured about.

The information provided in this book addresses the effective application of a quality management system to the applicable products, goods, and services provided by the electronics industry.

The scope, volume, and diversity of electronics manufacture in the world has traditionally lent itself to much experimentation in manufacturing techniques and focus. Traditional quality systems have typically focused on the identification and control of hardware that fails to meet specified requirements. In the traditional approach, the quality system functions properly when it prevents nonconforming hardware from getting into the hands of the customer. Identification of nonconforming material is usually accomplished through extensive inspection and testing. Once identified, the hardware is segregated and disposed of through the preliminary review or material review board process, in which a determination is made as to whether the hardware should be used as is, reworked, repaired, or scrapped.

Although preventing nonconforming material from reaching the hands of the customer is a critical function, the traditional quality assurance approach suffers from a number of drawbacks. Foremost among these drawbacks is that the identification and correction of defects have proved to be much more costly than preventing their occurrence in the first place. Such activities as inspection, test, segregation, nonconformance processing, and rework incur costs and yet add no value to the product. Secondly, inspection and test—even when performed on a 100 percent basis—often fail to identify all existing nonconformances. One hundred percent inspection has proved to be less than 100 percent effective in identifying defects. Lastly, the use of inspection and test as the principal means of determining product acceptability has frequently led to the perception that workers who perform such inspections and tests—rather than those who design, fabricate, and assemble the product—are responsible for product quality.

To effectively address these issues in the context of modern electronics manufacturing, quality assurance efforts must embrace the issue of controlling the effectiveness of the processes that create the product or service being provided. This technique is recognized by the term *advanced quality management practices.*

Advanced quality management practices contrast with traditional practices by emphasizing the prevention of defects versus after-the-fact detection of defects. Advanced quality management approaches emphasize quality in the development process to achieve producible designs and capable, controlled manufacturing processes. To achieve these objectives, advanced quality management practices also emphasize an integrated, multifunctional approach to quality throughout the product life cycle. Potential benefits directly attributable to the implementation of advanced process management practices include decreased cycle time as well as reductions in rework, engineering changes, and inspection and test. These benefits translate into improved affordability and reduced production transition risk.

In some cases an additional benefit to companies successfully implementing this approach occurs when their customers become aware of the focus and perceive it as a commitment to improve the quality of the goods or services being offered. This enables companies to successfully compete on the basis of the quality of their products. Companies such as Sony have long been leaders in their respective markets because they have managed to create in the mind of their customers the perception that their products are of higher quality than those of competitors. In some products that are not supplied as end items to the consumer, the process of convincing the customer of the inherent superior quality of one's product is more difficult. One company that has succeeded is Motorola's semiconductor business, with its now world-famous 6-sigma quality approach. In essence Motorola has convinced customers that its process control is so effective and superior to that of the rest of the industry that it makes good products by default. Motorola semiconductors are sought-after components in many applications, ranging from commercial to military to spaceflight.

When one examines elements of quality management, with specific emphasis on advanced quality processes, it becomes apparent that a company must incorporate such processes to receive full benefit. In addition, to be successful, a company must employ a good method to evaluate the effectiveness of the processes being managed.

An additional feature of advanced quality management versus traditional quality system approaches is the emphasis on reduction of program risk and prevention of defects, rather than on the identification and correction of defects after the fact. While both approaches share the objective of ensuring that only material that meets customer expectations is delivered to the customer, traditional systems tend to focus on the production phase and rely on inspection and test (i.e., defect detection) to sort out defective material. In contrast, advanced quality management approaches risk mitigation by matching the form, fit, and function requirements to the process limitations, and then controlling the process to ensure that only conforming product is produced. It is recognized that no system will be 100 percent effective in eliminating all risk. For this reason, risk mitigation, design, process, and material reviews, with associated corrective action systems, will still be required.

Since advanced quality management encompasses design, manufacturing, and support it is initially considered during the development phase (normally within an integrated product and process development framework) and focuses on achieving robust, producible designs, and ensuring that manufacturing processes are controlled and capable. The objectives of the advanced process management system are achieved in a systems engineering environment, utilizing a thorough knowledge of manufacturing and quality risks. While this early emphasis on quality necessitates the application of additional emphasis in the source-selection phases, the potential benefits (including decreased engineering changes, production cycle time, rework, and inspections) translate into improved life-cycle affordability and reduced program risk.

To achieve the goal of risk reduction, advanced process management emphasizes the optimization of the customer's desires, in terms of cost, schedule, and performance, and their impact on the design/manufacturing process interface. As stated earlier, implementation of advanced quality practices is ultimately viewed by the marketplace as the result of a contractor's initiative to provide "world class service" to its customers. [13]

1.3 Application

As our review and analysis of the quality assurance function progresses, we will examine several methodologies that represent effective applications of the concepts covered in the previous section. There are numerous other techniques, approaches, and applications of the quality discipline throughout the expanding universe of quality tools. However, those examined here have proved to be particularly effective when used in the design, development, manufacture, and support of electronic systems. Our approach is to present the new definitions and the data on what is important—including implementation information—and then suggest ways that your company can use these data to improve.

The definitions of quality assurance are discussed in Chap. 2. The chapter explains the transition from the little "q" to the big "Q" (see Table 1.2). Good systems engineering processes are defined, with discussions of performance specifications, and what they should contain. Overviews of the new voluntary standards that are replacing the military standards and specifications are provided. Performance-based contracting is discussed for four reasons: (1) the process serves as a model for commercial companies to internally define products in order to have good sales in the marketplace; (2) the discussion may provide the contract-electronics manufacturers with ideas to improve their contracting process; (3) because the government acquisition reform concept applies to all suppliers of electronics, there is opportunity for all electronic companies to expand their market share with the government; and (4) some of the processes may assist in making the government or large-customer contract, or supplier/vendor process, work more efficiently. Chapter 2 ends with guidelines for moving a company to world class status and a case study.

The new and advanced quality systems discussed in Sec. 1.2, are addressed in greater detail in Chap. 3. This chapter explains the magnificent strides that the U.S. government, with the help of a government/industry panel, has made in defining a single quality system for the United States. Work continues to further implement this

TABLE 1.2 Little "q" versus Big "Q"

Little "q"	Big "Q"
Manufacturing	All products and processes
Manufacturing	All support and business processes
Manufacturing	All services and customer support
Inspection	Process control
Serial approach	Concurrent processes
Defined by specification	Defined by customer
Management is boss	Management is coach/leader
Teams take direction	Teams are visible, apparent, work for continuous improvement
Employees do work	Employees empowered to solve problems, work for continuous improvement
Metrics—defects	Metrics cover all aspects of business

approach. This, in the style of what the Japanese did years ago, is the first move to a unified U.S. quality policy. This chapter discusses the new philosophies of quality assurance, including overviews of the main elements of a quality management approach: TQM and teams, management and leadership, and empowering the workforce. Views of quality found in the current literature that support the move to the world class company are presented. After a section on metrics, with lots of examples, and suggestions of ways to implement techniques, the chapter moves on to the advanced techniques of quality function deployment, to completely define customer's requirements; Taguchi's methods and robust design, to make the product producible by process and product characteristics; and statistical process control (SPC) and variability reduction (VR). These are techniques that focus on the evaluation and control of the processes used to create the hardware. The correct utilization of these techniques transcends traditional metrics, because the data they generate is used to control process parameters which affect the characteristics of the product (see Fig. 1.3).

No discussion of modern quality assurance practices for electronics would be complete without an examination of the International

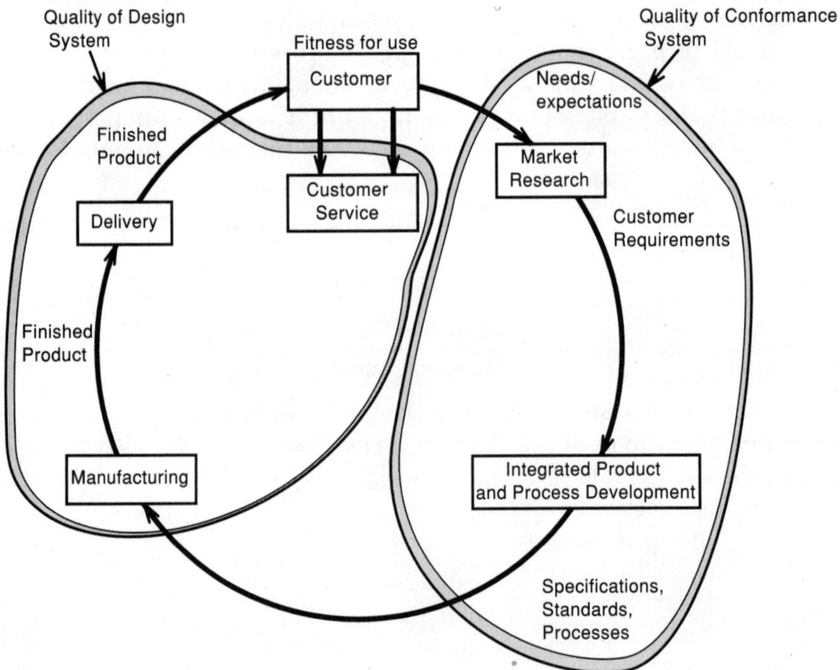

Figure 1.3 Quality system circle.

Figure 1.4 ISO unites quality throughout the world.

Organization for Standardization (ISO) 9000 series of documents and their impact on the development and application of quality management systems. In the international arena, quality assurance activities go primarily by the name *conformity assessment.* The ISO series of documents is designed with that intent.

The main impetus for the worldwide surge in the use of ISO 9000 comes from its adoption by the European Union (EU). EU has acted to accomplish what it calls the "harmonization" of standards among its various members, to facilitate commerce and ease of trade among member countries. Consequently, EU has mandated that most commercial products sold in any EU member companies must be produced under a quality management system that meets the requirements of ISO 9000. Several ministries of defense of various member countries have followed suit for their procurements. International attention is now on certification for the ISO 9000 processes as well as the specifications and standards that are used to design, manufacture, and test the products. This is being discussed internationally in great detail (see Fig. 1.4). Details are found in the "Product Certification" section of Chap. 4.

In the United States of America, the ISO series has been adapted by the American National Standards Institute (ANSI) and the American Society for Quality (ASQ). A system of registration, certification, and periodic reevaluations has been set in place with the idea of ensuring continuous conformity to the standard. Economics is the principal driver for ISO registration in the United States. Companies who have failed to adopt an ISO 9001–based quality system have found themselves faced with a "hidden trade barrier" to their products, depending on their industry or where they wish to sell the product in the world.

In Chap. 4, we will examine the ISO 9000 approach to quality assurance, its specific implications for electronic systems, the pros and cons, product certification and the world of international standards, and alternatives to the registration processes.

W. Edwards Deming has been referred to as the "father of quality management." Dr. Deming was not fond of the terms *quality management* or *TQM,* because he believed that quality was merely the by-product of good management, not the means with which to achieve it. His theory of management includes his system of profound knowledge (appreciation for a system, theory of variation, theory of knowledge, and psychology), his 14 points, and the learning cycle of plan–do–study–act (PDSA). His adamant position that the CEO must *personally* lead the transformation makes Deming's philosophy one that *must* be stressed in a book on quality management. Deming would have liked the big "Q" approach.

Figure 1.5 [14] illustrates Deming's vision of management's job to view the whole organization as a system and, using the four parts of profound knowledge, optimize the entire system over time. Managing for optimization of individual components will most likely lead to sub-optimization of the whole.

Chapter 5 lets the CEO and his/her staff know what they must do to move to this new style of management. Deming's philosophy emphasizes theory, leadership, and data. Each section contains guidelines for the electronics company to implement Deming's theory.

Deming influenced Japanese quality so much that the Japanese

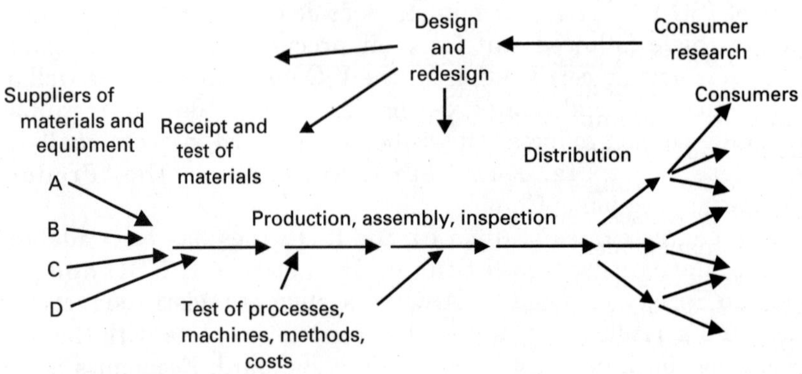

Figure 1.5 Deming's flow diagram.

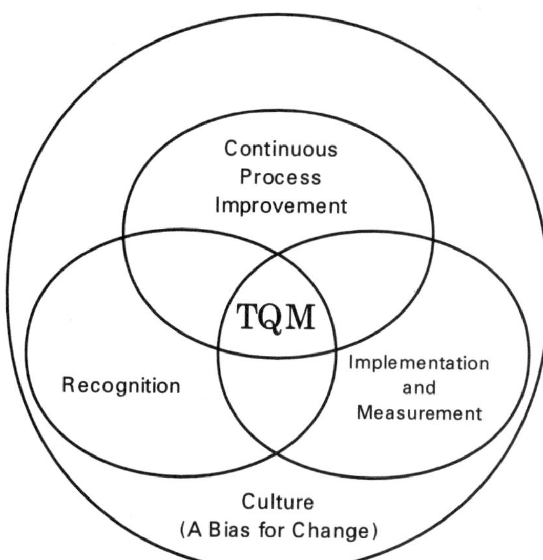

Figure 1.6 Total Quality Management (TQM).

Deming Prize is named for him; the Baldrige Award is modeled in part on the Deming Prize detailed in Chap. 5.

Our TQM chapter, Chap. 6, is not the detailed "how" of doing TQM, like many other books on the market. TQM is part of the culture change to get to the big "Q"; it is the organization, strategy, empowerment, problem-solving techniques, tools, models, and management strategy (see Fig. 1.6). These things are imperative in the new quality approach, and are in complete agreement with Deming's philosophy of giving the processes to the people. Deming did not like the way lip service was paid to TQM instead of instituting real change. This is where TQM is today. Lots of work has begun, but only a minority of companies have achieved a complete TQM philosophy of life where one lives with continuous change and data. Chapter 6 gives the big picture and indicates what is imperative to change in the culture. It provides the tools, philosophy, and background. Then it discusses the four key elements: culture change, change processes, implementation and measurement processes, and recognition processes. TQM concentrates on customer, management, people, change, improvement, and data. These are the keys to success in the electronics business: to please the customer, to keep improving, and to grow the marketplace. Electronics companies need to embrace these techniques, and it takes time.

With the increase in outsourcing of labor and subcontracting, there is also increased recognition of the importance of identifying and

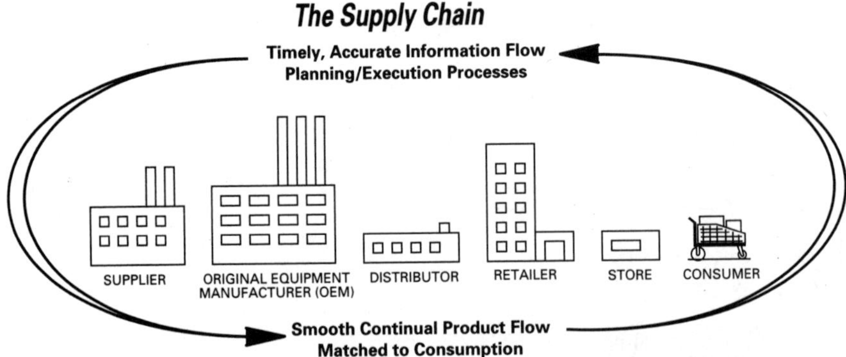

Figure 1.7 Linking the virtual enterprise. (*Adapted from Managing Automation, October 1996; used with permission.*)

selecting vendors and subcontractors capable of meeting requirements. A new term has arisen that captures this overall concept: *supply chain management* (see Fig. 1.7). An outgrowth of supplier-based management, that system recognizes the fact that an increasingly large percentage of equipment supplied by an original equipment manufacturer contains a substantial amount of parts manufactured by subcontractors. Supply chain management just goes the next step by establishing complete automation between supplier and customer, even into each other's scheduling system. This involves an increase in trust, and knowing that quality products will be delivered. In Chapter 7 we will explore the key attributes of attaining subcontractor/vendor quality. Supplier selection, supplier partnerships (including certification approaches), requirements flowdown, rating systems, and performance measurements are critical in the improvement of electronic systems. In addition, the items needed to comply with ISO 9001 purchasing requirements are pointed out. Supplier quality assurance is fast becoming part of the big "Q" as it changes from a stand-alone function to one intimately involved with purchasing, engineering, requirements planning, and manufacturing in a way that it never was in the old-style quality.

Many new techniques that have an impact on the effectiveness of quality have also surfaced in recent years. Others have been revamped and given a fresh look for modern times. One of the oldest and best known is benchmarking. The key to effective benchmarking lies in its effective execution. Benchmarking is discussed in Chap. 8. Effective execution is the only way to achieve the goal, which is a comprehensive and objective review of the performance of a business process. We will explore a disciplined approach to the benchmarking

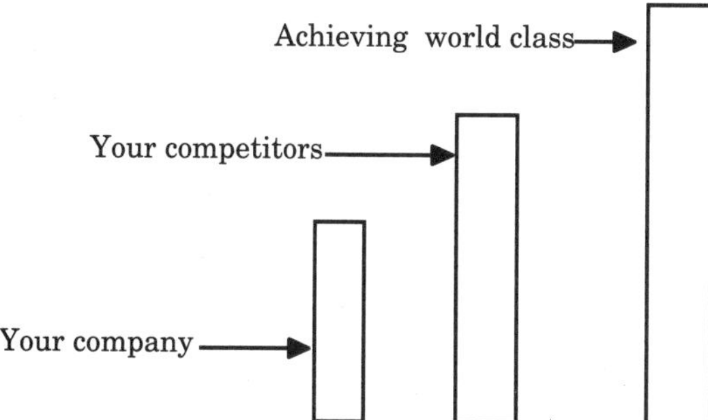

Figure 1.8 Benchmarking to achieve world class.

process with particular focus on planning and execution, as well as the roles and responsibilities of personnel involved. Without effective implementation of a disciplined approach, benchmarking activities become "industrial tourism." While this may be fun, it will not prove effective and the information gathered will be of little value. Benchmarking requires knowing your own processes and metrics so that you can truly use benchmarking to achieve world class (see Fig. 1.8).

Last, we will address the Malcolm Baldrige National Quality Award (MBNQA). The Baldrige Award is given in several categories by the National Institute of Standards and Technology (NIST). In the United States, the Baldrige Award is one of the top honors a company can achieve—it is synonymous with being a world class company. In Chap. 9, we will review the Baldrige Award values, concepts, key characteristics, and framework (see Table 1.3), and will examine the criteria. We will also look at some of the attributes of award-winning electronics companies. Finally, we will highlight the benefits of the award, and provide a case study of a company applying the award criteria for self-assessment and improvement.

Finally, a word about the future. Quality has changed and will not return to the practices of the past. It is important to remember that

TABLE 1.3 Baldrige Award Criteria (1996)

Leadership
Information and analysis
Strategic planning
Human resource development and management
Process management
Business results
Customer focus and satisfaction

TABLE 1.4 Quality is the strategic business discriminator

New: perceived quality, best value	Old paradigm: cost, technical
Cost/price	Cost/price
Technical performance	Technical performance
Quality	Schedules
Reliability and maintainability	Supporting functions
Configuration management	Quality assurance
Supporting functions	Configuration management
	Reliability and maintainability
	Manufacturing, etc.
Everything is important to the customer: best value	Only cost and technical performance interested the customer

the quality assurance approach applied by any enterprise has one goal that never really changes: to ensure that the goods or services supplied to the customer perform as they are intended. As the quality assurance discipline has matured, the advantages of the big "Q" approach to quality become more obvious. The concept of quality continues to mature from an inspection-based function done at the end of the production line, to a fully integrated management tool for evaluating the effectiveness of all business processes, not just in plant manufacturing–related activities. Any suboptimized business processes are too costly to accept in a world of shrinking profit margins.

Quality has been redefined. It is a strategic business discriminator (see Table 1.4). Quality was defined by Hagan, from a customer's point of view, relative to competitors [15]:

- It is not those who offer the product but those whom it serves—the customers, users, and those who influence or represent them—who have the final word on how well a product fulfills needs and expectations.
- Satisfaction is related to competitive offerings.
- Satisfaction, as related to competitive offerings, is formed over the product lifetime, not just at the time of purchase.
- A composite of attributes is needed to provide the greatest satisfaction to those whom the product serves.

The disciplined approach to process evaluation inherent in an effective quality management system is an effective tool that already exists in most mature successful enterprises. The challenge is to effectively increase the scope of that tool to encompass other key business systems that may be nontraditional areas for application of the quality function.

The tough business realities of global competition and the ever-increasing pace of changes in the business environment are at the core of the pressures driving companies to adopt practices that are prevention-based.

The quality assurance processes can be an effective management tool to address these uncertainties. Companies that can recognize and properly apply the quality discipline in this expanded role will find a ready-made resource already tailored to the particulars of their business.

References

1. Harry, Mikel J., *The Nature of Six Sigma Quality,* Motorola University Press, Rolling Meadows, IL., 1988, p. 1.
2. Gershon, Mark, "A Look at the Past to Predict the Future," *Quality Progress,* July 1996, p. 31.
3. Teresko, John, "Too Much Data, Too Little Information," *Industry Week,* vol. 245, no. 15, August 19, 1996, pp. 66–67.
4. Luther, David B., "Quality, the Future, and You," *Quality Progress,* July 1996, pp. 68–69, paraphrased.
5. Olsen, George H., revised by Forrest M. Mims III, *The Beginner's Handbook of Electronics,* Prentice-Hall, 1980, pp. 3–7.
6. Olsen, pp. 9–10.
7. Olsen, p. 269.
8. Garvin, David A., *Managing Quality, The Strategy and Competitive Edge,* The Free Press, New York, 1988, p. 15.
9. Schuler, Charles A., *Electronics: Principles and Applications,* 3d ed., Glencoe, Macmillan/McGraw-Hill, New York, 1993, pp. 1–3.
10. Schuler, p. 8.
11. Schuler, p. 6.
12. From the Internet, Quality Management Principles (QMP) forum, address: http://www.wineasy.se/qmp/about.html, November 24, 1996. Source document cited as ISO TC 176/SC2/WG15/N125. It will be available from American National Standards Institute (ANSI) when printed.
13. Findings of the Government and Industry Quality Liaison Panel, 1996.
14. Deming, W. Edwards, *Out of the Crisis.* (Reprinted by permission of MIT and The W. Edwards Deming Institute. Published by MIT, Center for Advanced Educational Services, Cambridge, MA 02139. Copyright 1986 by The W. Edwards Deming Institute)
15. Hagan, John T., "The Management of Quality: Preparing for a Competitive Future," *Quality Progress,* December 1984, p. 21.

Chapter

2

The New Definition of Quality

Marsha Ludwig-Becker, C. M.

Operations Project Manager
The Boeing Company
Anaheim, California

2.1 The New Quality

Consumers of electronic systems worldwide are looking for a new definition of quality. They want a reliable and realistically priced product that works the way they want it to work, in the place they want it to work. In the past, the quality of a product was ensured by thorough inspection. We still find little tags on our consumer items, "inspected by TBD." This process of inspection, the serial process, involved one person building a product and another person inspecting it, either "stamping" the product if passed or rejecting it if failed. If the product did not pass the inspection, the workers were the problem; they had not built the system per specifications. This approach/ideology assumed that everything could be inspected (which is not the case with electronics or electronic processing). This old approach is known as quality with a little "q." It is the type of quality that is controlled by one department, generally the "Quality Department," whose quality personnel, serving as the "police" of the company, conduct inspections. This Quality Department is *not* part of the process, and their experience on many problems in the past is not used as lessons learned on newly developed products.

Quality in the early 1970s was equivalent to a cop patrolling a beat—a person or department checking, inspecting, or verifying the work of another department or process. "Unfortunately, quality control departments have taken the job of quality away from the people who can contribute most to quality—management, supervisors, managers of purchasing and production workers." [1] Quality assurance was almost the same as inspection, or inspectors. This is the quality of the past. Quality today is the center of a progressive electronics company. Quality is the very heart of a company which focuses on producing a reliable, customer-pleasing, quality product. It is no longer an inspection process, but a total process designed to meet customer requirements. It has long been acknowledged that quality could not be "inspected" into a product. The day has finally come, this is acknowledged in the literature, and now we must start implementing it in the world.

The new quality, the big "Q," is defined by the customer. The big "Q" approach to developing, manufacturing, and servicing must be part of the company's culture. All employees are part of the quality system. There are "quality engineers," but they are on the team, in the process, and work to eliminate problems with the team. In most cases, the only "inspection" is a final inspection to ensure that processes and tests have been followed, and it is not necessarily done by an "inspector"; it can be done by another member of the manufacturing or design team. The big "Q" is characterized by active leadership and the empowerment of teams. The big "Q" is linked to competitive strategy of the company. [2] These are the essentials of the big "Q" system. Customers, not specifications, define the big "Q" in every aspect of business relevant to customers—price, service reliability, safety, delivery, and courtesy.

The new quality is performance-based, and concentrates on processes which begin early in the product life and become established and auditable. Metrics must be collected and used to manage the quality of the product.

John J. Hudiburg, Chairman of Florida Power and Light (FPL), developed the big "Q" and little "q" concept. In 1989, FPL was the first U.S. company to win the Deming Prize, given out by the Japanese Union of Scientists and Engineers for outstanding productive and quality accomplishments. Mr. Hudiburg stated [3]:

> There are basically two definitions of quality. The first is what we call quality. Little q means conforming to product specifications. Most companies think about quality and get stuck on the little q. They get so bogged down in the areas of specification that they can't even see the forest of other quality issues. Little q can have some impact on your general level of quality, but it's too limited to make your company really successful.

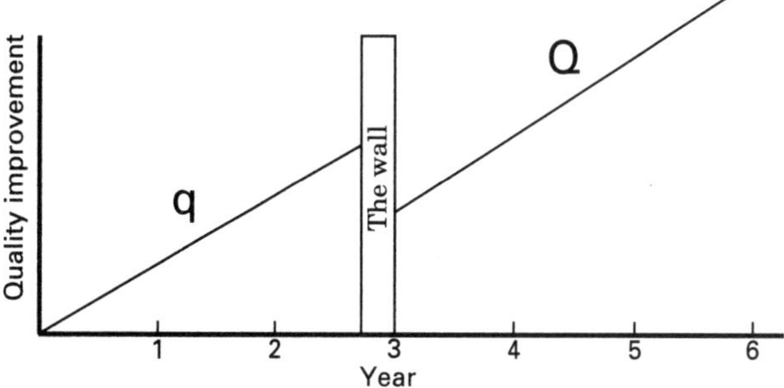
Quality assumptions:
• Quality is goodness
• Use lockstep approach
• Workers are the problem
Quality understanding:
• Quality is defined by customers
• Approach must be part of cultu
• Active leadership is essential
Quality improvement
q
The wall
Q
1
2
3
4
5
6
Year

Figure 2.1 The promise and shortcomings of total quality. (*From Ref. 3, with permission.*)

The challenge is learning exactly what the big "Q" means for your company and acting on it by implementing a quality program that makes your company the most successful in the electronics industry. It is a paradigm shift to go from the little "q" to the big "Q" (see Fig. 2.1) [4] and takes several years of enlightened leadership, training, and new principles from which to manage the electronics factory. This same management would begin establishing a Total Quality Management (TQM) centered approach where empowered teams begin to learn about managing the quality of their product or service. When all management and employees embrace the efforts of the big "Q," it happens. It takes a lot of retraining, replacing inspection with manufacturing verification, and proactive prevention teams, along with management that is willing to "walk what they talk" and respect all individuals.

Juran and Gyrna depict this change as a change in scope (Table 2.1) [5]. Formerly only manufacturing processes were considered

TABLE 2.1 Little "q" and Big "Q"

Topic	Contents of little "q"	Contents of big "Q"
Products	Manufactured goods	All products, goods, and services whether for sale or not
Processes	Processes directly related to manufacture of goods	All processes: manufacturing support, business, etc.
Industries	Manufacturing	All industries: manufacturing service, government, etc., whether for profit or not

SOURCE: Reference 5, p. 7, used by permission.

"quality controlled." Now it is the whole company. The contrasts of the traditional versus the quality-oriented organization are shown in Table 2.2 [6].

The traditional scope of quality activities is undergoing a radical and exciting change from the historical emphasis on quality of physical products in manufacturing industries (little "q") to what is now emerging as the application of quality concepts to all products, all functional activities, and all industries (big "Q").

Further definition of the big "Q" is evidenced in the concept of strategic quality planning, and the executive as the leader of quality reviewing the system. This emerged in the 1980s but was used centuries ago. In a discussion of managing for quality in ancient Rome, Ref. 7 states that the application to quality involves the following:

- The business plan is enlarged to include goals for quality.
- These goals are "deployed" to lower levels in order to determine the resources needed, agree on the actions to be taken, and fix responsibility for taking the actions.
- Measures are developed to permit evaluation of progress against the goals.
- Managers, including upper managers, review progress regularly.
- The reward system is revised to give appropriate weight to meeting the quality goals.

The traditional independent quality organization did not always produce the intended results. Even though they were independent, the transfer of responsibility and the lack of perceived ownership for achieved quality are major problems of the model. The newer continuous improvement programs, such as Total Quality Management, have much broader dimensions than the quality of the past. The new TQM approach includes [8]:

- Defining customer expectations and translating them into realistic organizational goals and objectives
- Managing projects and processes
- Obtaining appropriate quantitative and qualitative measurements and feedback, including real-time status reports and trend analyses
- Conducting effective performance-based assessments and balanced overview activities
- Carrying out realistic corrective, adaptive, and preventive actions

TABLE 2.2 Contrasts in Traditional and Quality-Oriented Organizations

The Traditional Organization	The Quality-Oriented Organization
▪ Defines quality as it sees fit	▪ Focuses on the customer and how quality is defined by the customer
▪ Views the end user as the customer	▪ Sees the next person in the process as a customer and addresses both internal and external customers
▪ Is focused on maintaining the status quo and wants to do things the way it has always done	▪ Is focused on daily improvement (however small) involving everyone
▪ Is impatient and short-sighted	▪ Is patient and long-term–oriented
▪ Identifies acceptable quality levels	▪ Targets zero defects in products and services
▪ Is results-oriented	▪ Is process-oriented
▪ Trains technical and manager personnel	▪ Provides extensive training to all employees
▪ Relies on managers as the problem solvers	▪ Relies on teamwork and group problem solving involving all employees
▪ Makes quality the responsibility of managers or the quality control department	▪ Involves everyone in quality and process improvement
▪ Is focused on traditional accounting methods of cost reduction	▪ Is focused on quality and service improvement and maximizing the ratio of value-added to non-value-added work
▪ Organizes around a cycle of inspection and rework	▪ Targets reduction of variability and the prevention of errors
▪ Maintains a simplistic view of rework	▪ Recognizes the real cost of nonconformance and rework
▪ Relies on a few systematic procedures for quality and service improvement	▪ Relies on multiple methods and resources for quality and service improvement
▪ Tampers with common variations in product and quality	▪ Attacks real problems (special causes of variation) and improves systems and processes
▪ Awards supplier contracts on the basis of cost and places responsibility for quality on the supplier	▪ Awards contracts on the basis of quality and assumes joint responsibility for supplier quality
▪ Blames workers for poor effort and performance	▪ Understands that management is responsible for the process and thus performance

SOURCE: Reference 6; used with permission.

- Conducting root-cause analyses to identify the real cause of organizational problems and the most obvious opportunities for improvement
- Planning and realizing continuous improvement
- Using, developing, and empowering human resources and treating them as capital.

This is an even more extensive definition of the big "Q."

Japanese-style total quality control is different from traditional U.S. quality control. Japanese managers have coined the phrase *company-wide quality control* (CWQC) to distinguish the difference. CWQC is a means to "provide good and low cost products, dividing the benefits among consumers, employers and stockholders while improving the quality of people's lives. The Japanese even created a specification for CWQC. It is Japan Industrial Standard Z8101-1981. The standard states that 'implementing quality control effectively necessitates the cooperation of all people in the company, involving top management, managers, supervisors, and workers in all areas of corporate activities such as market research, research and development, product planning, design, preparations for production, purchasing, vendor management, manufacturing, inspection, sales and afterservice, as well as financial control, personnel administration, and training and education. This is called company wide quality control.'" [9] For more information on CWQC, see Chap. 5 (Table 5.24).

2.1.1 New definitions

Even with all the emphasis in the press on quality assurance in the 1970s and 1980s, and the movement to statistical process control (SPC), and Total Quality Management, the first time the new definition of quality came to the forefront in the United States was with the publication of the definition of quality in the Department of Defense Instruction 5000.2, 1991 edition. The instruction put the following note up front: "Note: Quality as discussed in this section is far more than the determination that the as-built system conforms to its manufacturing specifications. As such, its breadth is greater than the historical application of the referenced documents." [10] Previously quality experts had talked of quality of design and quality of conformance. Juran even wrote, in 1988, to define quality as "fitness for use," but agreed that it had not been universally accepted. [11]

In 1988, David A. Garvin of the Harvard Business School published a book entitled *Managing Quality: The Strategic and Competitive Edge,* which is highly respected in the world class manufacturing

(WCM) literature. Many WCM authors use his "new" definition of quality.

In Garvin's definition, it is essential for quality to assume a strategic role. He arrives at this conclusion because the scholars in many disciplines (philosophy, economics, marketing, and operations/manufacturing) defined quality differently. As Garvin researched the literature, five themes became apparent: transcendent, product-based, user-based, manufacturing-based, and value-based. Table 2.3 presents representative examples of each approach. [12]

Garvin says that "complex concepts like quality are difficult to penetrate," but then goes on to define eight dimensions or categories of quality that can be used for analysis [13]:

- *Performance.* The primary operating characteristics
- *Features.* The "bells and whistles" of products, secondary characteristics that supplement the basic functioning operation
- *Reliability.* The probability of failing over time
- *Conformance.* The degree to which a product's design and operations meet established standards
- *Durability.* A measure of product life, with both economics and technical dimension; the amount of use one gets from a product before it physically deteriorates
- *Serviceability.* The speed, courtesy, competence, and ease of repair
- *Aesthetics.* Subjective, but directly relates to what a customer's preference is, the fitness for use
- *Perceived quality.* Again subjective, relates to reputation, currently things like "Boeing makes good airplanes," or "Good electronics come from Japan"

Garvin's eight dimensions can be tracked (see Fig. 2.2) to the three interconnecting subefforts originally defined in the DOD Instruction 5000:

- Quality of the design
- Quality of conformance
- Quality of the fitness for use

It is apparent that all three subefforts make a thorough definition of quality. Customers are demanding that all three of these characteristics must be met—these make up the "new definition of quality."

TABLE 2.3 Five Definitions of Quality

I. Transcendent

"Quality is neither mind nor matter, but a third entity independent of the two...even though Quality cannot be defined, you know what it is." (Robert M. Pirsig, *Zen and the Art of Motorcycle Maintenance,* Bantam Books, New York, 1974, pp. 185, 213.)

"...a condition of excellence implying fine quality as distinct from poor quality....Quality is achieving or reaching for the highest standard as against being satisfied with the sloppy or fraudulent." (Barbara W. Tuchman, "The Decline of Quality," *New York Times Magazine,* November 2, 1980, p. 38.)

II. Product-based

"Differences in quality amount to differences in the quantity of some desired ingredient or attribute." (Lawrence Abbott, *Quality and Competition,* Columbia University Press, New York, 1955, pp. 126–127.)

"Quality refers to the amounts of the unpriced attributes contained in each unit of the priced attribute." (Keith B. Leffler, "Ambiguous Changes in Product Quality," *American Economic Review,* December 1982, p. 956.)

III. User-based

"Quality consists of the capacity to satisfy wants..." (Corwin D. Edwards, "The Meaning of Quality," *Quality Progress,* October 1968, p. 37.)

"In the final analysis of the marketplace, the quality of a product depends on how well it fits patterns of consumer preferences." (Alfred A. Kuehn and Ralph L. Day, "Strategy of Product Quality," *Harvard Business Review,* November–December 1962, p. 101.)

"Quality is fitness for use." (J. M. Juran (ed.), *Quality Control Handbook,* 3d ed., McGraw-Hill, New York, 1974, p. 202.)

IV. Manufacturing-based

"Quality (means) conformance to requirements." (Philip B. Crosby, *Quality is Free,* New American Library, New York, 1979, p. 15.)

"Quality is the degree to which a specific product conforms to a design or specification." (Harold L. Gilmore, "Product Conformance Cost," *Quality Progress,* June, 1974, p. 16.)

V. Value-based

"Quality is the degree of excellence at an acceptable price and the control of variability at an acceptable cost." (Robert A. Broh, *Managing Quality for Higher Profits* (McGraw-Hill, New York, 1982, p. 3.)

"Quality means best for certain customer conditions. These conditions are (a) the actual use and (b) the selling price of the product." (Armand V. Feigenbaum, *Total Quality Control,* McGraw-Hill, New York, 1961, p. 1.)

SOURCE: Reference 12; used with permission.

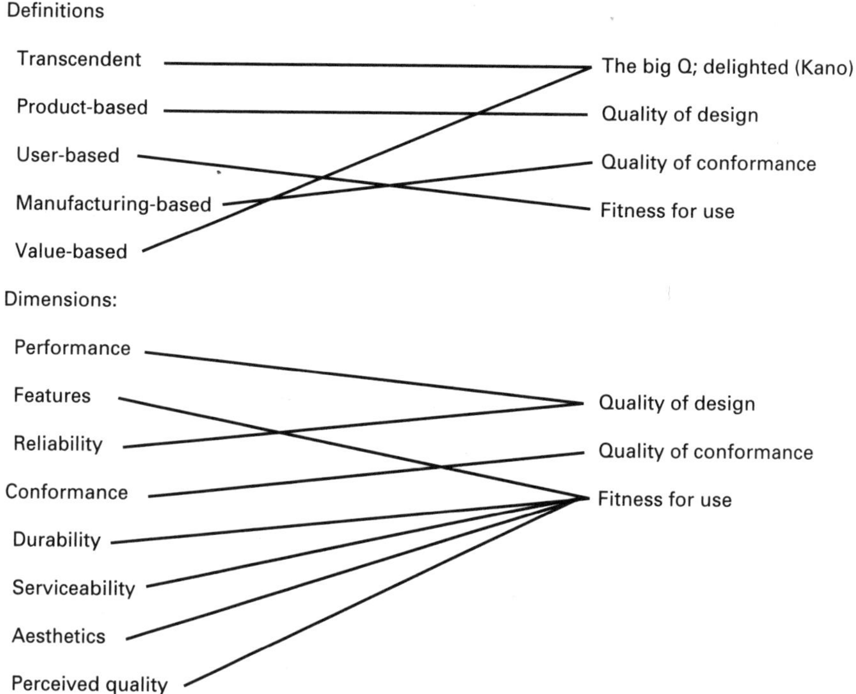

Figure 2.2 Garvin's definitions and dimension map to three categories of quality.

Quality of design

> The effectiveness of the design process in capturing the operational requirements and translating them into detailed design requirements that can be manufactured (or coded) in a consistent manner. [14]

To be successful in a highly competitive market, businesses must be more precise in understanding customer needs and in designing and manufacturing products/services that meet those needs. This is important in all systems, but it is even more critical for electronics and electronic systems. Electronic design is applied to mainly two kinds of products: electronic components and electronics systems. They are designed by the same approach with exceptions only for specialized items because of the unique nature of the product. For example, an integrated circuit has a complicated set of gates that must be designed as part of the total design. Electronic system designers are at the mercy of component designers for much of the expected performance. The discussion of design here will apply to both components and electronic systems, with special emphasis in some cases to one or the other.

The chief goal of the design process is to translate customer requirements into technical specifications for output. These specifications/drawings should provide “clear and definitive” instructions for:

- Quality aspects of the design
- Characteristics important to quality
- Instructions for performance of work
- Guidelines for verifying conformance to requirements
- Contingency plans

Electronic components form a significant part of the design of the electronic system, and are therefore often the critical determinants of the quality and reliability of the final system. Because components are purchased from outside suppliers and manufactured in different manufacturing plants, supplier quality control is essential for component quality and reliability. Specific procedures for quality reporting, test and inspection results sharing, quality tracing, and reliability performance must be developed.

For the most part, electronic components are produced in highly automated plants or sections of plants, in very large quantities. They are measured, handled, tested, and packaged in highly automated processes. They are often 100 percent tested, sometimes several times during the manufacturing processes. Because many components constitute the critical building blocks of complex systems (e.g., microprocessors used in personal computers), samples of components are often selected for life tests, stress tests, humidity tests, vibration tests, and accelerated life tests.

As the complexity of products and system increases, the need for high-quality, high-reliability components increases dramatically. Two distinct trends in quality and reliability of components have been apparent in the 1980s. The quality levels of components have been improved dramatically. Component quality is now commonly measured in parts per million (ppm) defective, whereas only a few years ago percent defective was the accepted norm. Average ppm ratings in 1983 were from 500 to 800 ppm; in 1985 they came in close to 100 ppm. World class quality standards show less than 50 ppm as the standard. [15] Complete data showing the history of the improvement can be found in the *Juran Quality Control Handbook,* pp. 29.2 and 29.3. [16] The Electronic Industries Association (EIA) publishes standards on ppm definition for components (EIA 554-A and ANSI/EIA 555-8). [17] These standards can monitor the acceptance of the parts, but the design too must be conceived to arrive at the highest of quality standards.

The key to good design definition is understanding the operational requirements so that a customer will be pleased. How should this process occur? What is the customer expecting? If the product is consumer goods, what will the customers buy? What is the company trying to sell? An idea may transform itself into a product brochure for prospective customers; it must specify the electronics design parameters. Electronics is being used to make every industry operate better, faster, and cheaper, and it is consuming more and more market share.

The successful electronics company, for either components or systems, must have a good systems engineering process for the design/development of the electronic product. This is the heart of an electronics company's success, for electronics is changing constantly. To keep market share and stay state-of-the-art in electronics, system engineering procedures/processes must be constantly applied, and the improvement of the product must be continually worked on. Good systems engineering in the electronics world accepts that quality is an integral part of all technical elements and should be proactive in preventing defects rather than correcting them (see Fig. 2.3). [18] Design verification and validation of electronic products is vital and a most important part of the new quality management requirements.

Systems engineering is well known as a standard process of requirements analysis, functional analysis, synthesis, systems analysis, and verification and validation. In order to ensure that the design process is always concentrating on the big "Q," progressive electronics design companies must have a complete set of engineering design processes which include the manufacturing and support of the product. A guide for the top-level processes is provided by ANSI/ASQC or ISO 9001:

- *Design and developmental planning*—to plan the design (see Sec. 4.2.4)
- *Organizational and technical interfaces*—to ensure all who should be involved in the design are involved
- *Design inputs*—to ensure that all requirements, customer and statutory, are included
- *Design outputs*—to ensure that the design will be documented
- *Design reviews*—to ensure that peer groups and customers can review and provide input to ideas or issues, or even contractual issues
- *Design verification and validation*—to test to requirements either internal or external
- *Design changes*—an inevitable part of the electronics design process

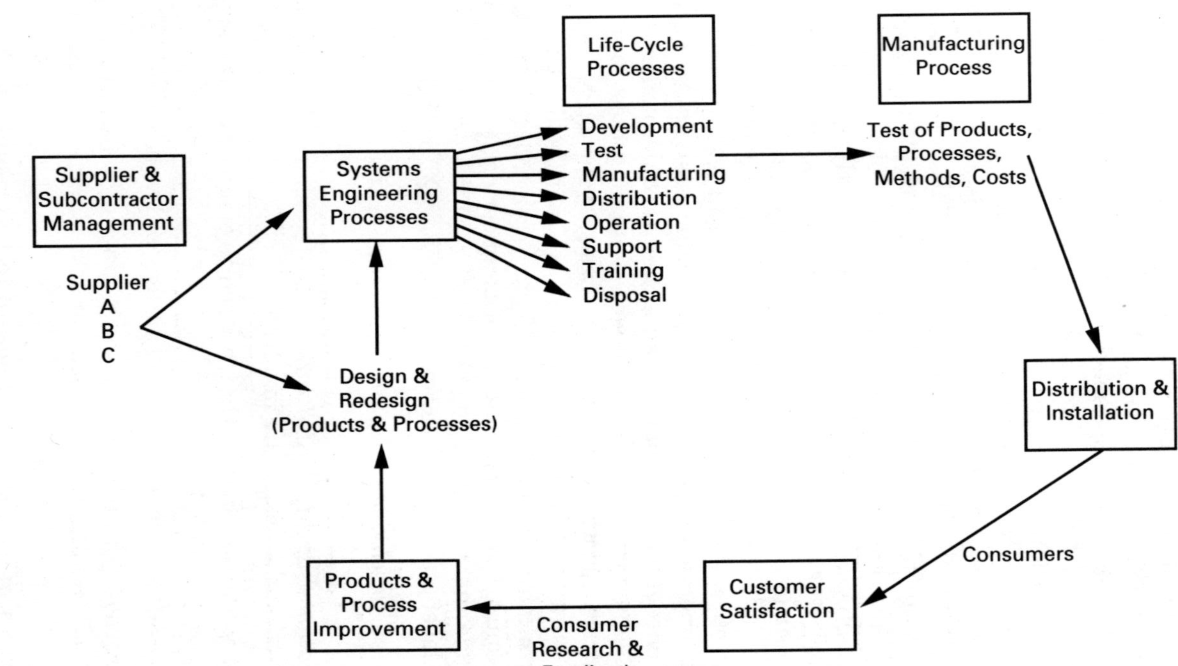

Figure 2.3 Elements of quality management. (*From Ref. 18; copyright IEEE, all rights reserved.*)

Design specifications, at times down to the individual component level, may be necessary for a complex design. Semiconductors and integrated circuits demand that detail designs be accomplished to ensure that functional requirements are met. In the new commercial world of world class quality, the IEEE P1220 systems engineering process extends the verification and validation phase between every traditional activity (see Fig. 2.4). [19] This ensures that the system engineering process keeps the customer's requirements in view

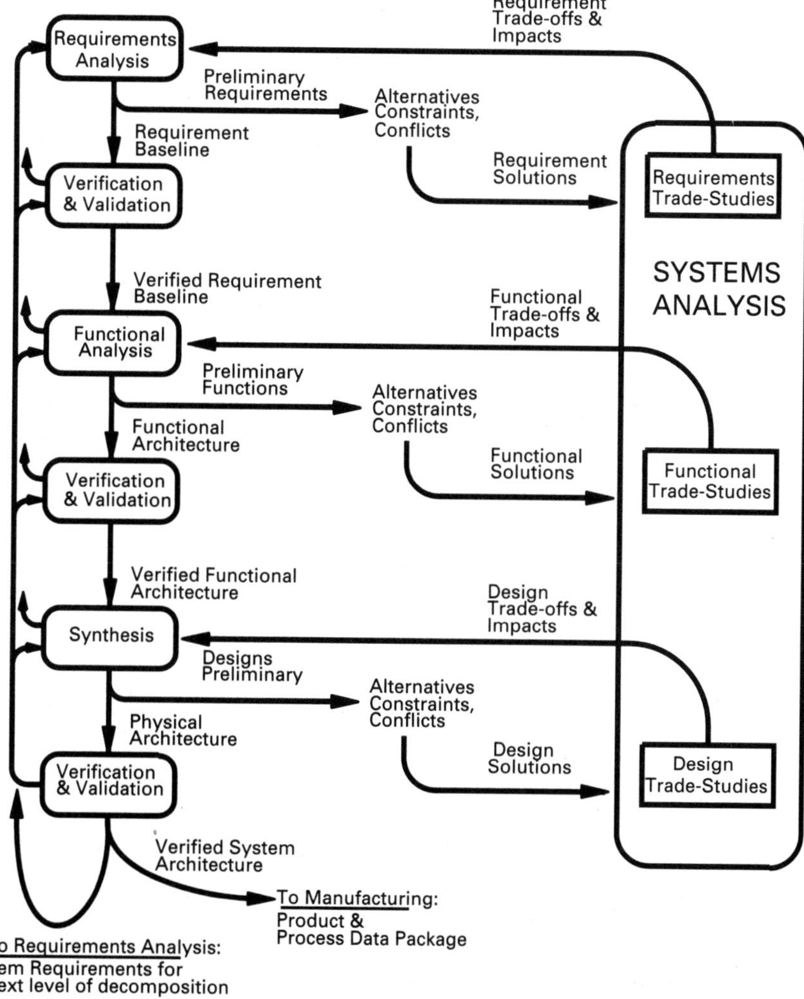

Figure 2.4 The systems engineering process. (*From Ref. 19; copyright IEEE 1993, all rights reserved.*)

throughout the design process. It helps to resolve conflicts in requirements and design solutions. It also focuses the process when the design is improved over time. Verification includes alternative calculations and similarity analysis with other equipment. It also includes tests, demonstrations, and constant reviews, including reviews of documentation prior to release.

Validation follows verification, and is performed under the operating conditions defined by the customer. It is normally performed on the final product—not the engineering model, but the first product unit. If there are multiple uses, multiple validations may be made. Functional configuration audits (FCAs) and physical configuration audits (PCAs), along with the first article inspection and the factory acceptance test, are considered validation.

Operational scenarios for the product under design should be known. The customer should divulge these to the contractor in the performance specification. These operational scenarios should be considered in the requirements phase of the system engineering process and allow for the system to be operational and reliable in that setting, and plan for maintenance if it cannot happen in that area. The operational scenarios should identify the exception interactions with the environment, other systems, and physical interconnections with external systems or products.

In its series, the American National Standards Institute/Institute for Interconnecting and Packaging Electronic Circuits (ANSI/IPC) defines operational classes as *performance classes.* Three general classes have been established to reflect progressive increases in sophistication, functional performance requirements, and testing/inspection frequency. It should be recognized that there may be an overlap of equipment categories in equipment classes. The user has the responsibility to specify in the contract or purchase order the performance classes required for each product and shall indicate any exceptions to specific parameters where appropriate. There are excellent top-level fitness-for-use criteria that can be used in the front end of the design process. At the beginning of the design phase, one of these three classes must be defined as part of the design/development plan for electronics:

- *Class 1, general electronic products.* Includes consumer products, some computer and computer peripherals, as well as general military hardware suitable for applications where cosmetic imperfections are not important and the major requirement is function of the completed printed circuit board.
- *Class 2, dedicated service electronic products.* Includes communications equipment, sophisticated business machines, instruments,

and military equipment where high performance and extended life are required and for which uninterrupted service is desired but not critical. Certain cosmetic imperfections are allowed.

- *Class 3, high-reliability electronic products.* Includes the equipment for commercial and military products where continued performance or performance on demand is critical. Equipment downtime cannot be tolerated; equipment must function when required, such as in life support items or flight control systems. Printed boards in these classes are suitable for applications where high levels of assurance are required and service is essential.

The ANSI/IPC specifications separate requirements for most electronics products by these three classes. The use of one class for a specified attribute does not mean that all other attributes must be at the same class. Selection should be based on minimum need; however, crossover between classes or other specification requirements requires a complete definition of test requirements in the procurement documentation. [20] IPC also prints a *Design Guide Manual* as IPC-D-330. [21]

These link to how thoroughly the manufacturing requirements are defined. A Class 3 high-reliability product where only two or three items are produced may not define the same set of variability reduction (VR) and statistical process control methods as does a Class 1 or 2 where large amounts of product may be produced.

Documenting design trade studies is necessary in this new world of understanding the design parameters and ensuring that there is solid logic for design decisions that are made in the many available options.

As part of design and developmental planning, quality function deployment (QFD) is a good tool for defining the product, including its functional characteristics, design requirements, and requirements for fitness for use or operational requirements for field use. QFD is not a tool of the Quality Department but a planning process used to transform customer needs into prioritized lists of design, manufacturing, and operational requirements critical to achieving customer satisfaction. It is a process that facilitates designing quality customer requirements into a product or service.

There are different approaches to QFD, but the purpose is the same: a team focuses on designing to satisfy customer needs, to build in rather than inspect for product quality. Approaches begin with identification of internal and external customers with clear statements of need and end with customer satisfaction. These activities correspond to QFD's voice of the customer. This analysis and quantification is an important element in QFD's ultimate application, driving the voice of the customer into prioritization of the product features. That capability provides a means to obtain direct flowdown of customer needs into product features, manufacturing processes, and process characteristics. It is the feature that accounts for QFD's ability to design quality into the final product. Details of the QFD process are found in Chap. 3.

Equally important is reliability. An unreliable electronics product is useless.

Reliability. Customers know, and designers agree, that a product should have a stated service life. Reliability is defined as "the ability of a product to perform a required function under stated conditions for a stated period of time." [22] A collection of tools known as *reliability engineering* has been developed to handle the complexity of the calculation of reliability information and the problem of unreliable electronics. In 1950, only one-third of the Navy's electronic devices were working properly at any given time. A study by the Rand Corporation at the time estimated that every vacuum tube that the military had plugged in and working was backed by nine others in a warehouse or on order. Serious equipment problems were encountered with missiles and other aerospace equipment. [23] A table of reliability figures of merit is found in Table 2.4. [24]

System reliability is determined in large part by the components' reliability. Performance and life testing of systems and completed products is difficult to verify, time-consuming, and expensive. For these reasons, component reliability information is critical in modern quality systems. Components are functionally tested, accelerated-life tested, operational-life tested, and burned in at many stress levels. Reliability estimates for components are carefully calculated and catalogued in reference books and/or databases. Designers use these estimates with system reliability models to predict system reliability early in the design process. When these predictions indicate that the system may not meet its reliability estimates, the designers may choose different components, add redundancy to the system, use components at less than full rated values, burn in critical components to improve reliability, or use other approaches. Figure 2.5 shows a typical electronic system reliability assurance program. The importance of component reliability is evident. As indicated in the figure, the systems designer has several options available to assure that components have the required reliability. When necessary, the designer may require burn-in of certain components to eliminate early-life reliability failures. During prototype testing, the designers are able to estimate system reliability and evaluate system reliability predictions

TABLE 2.4 Reliability Figures of Merit

Figure of Merit	Meaning
Mean time between failure (MTBF)	Mean time between successive failures of a repairable product
Failure rate	Number of failures per unit time
Mean time to failure (MTTF)	Mean time to failure of a nonrepairable product
Mean life	Mean failure life ("life" may be related to major overhauls, wear-out time, etc.)
Mean time to first failure (MTFF)	Mean time to first failure of a repairable product
Longevity	Wear-out time for a product
Availability	Operating time expressed as a percentage of operating and repair time
System effectiveness	Extent to which a product achieves the requirements of the user
Probability of success	Same as reliability (but often used for "one-shot" or non-time-oriented products)
b_{10} life	Life during which 10% of the population would have failed
b_{50} life	Median life, or life during which 50% of the population would have failed
Repairs/100	Number of repairs per 100 operating hours

SOURCE: Reprinted from Ref. 24, with permission.

and the reliability assurance programs. At this time, special supplier requirements for burn-in or reliability testing may be designed. During mass production, the reliability assurance program keeps track of failed components and sometimes performs failure mode analysis to determine causes and redesigns the parts to remove those causes from the manufacturing process. A reliability information system is necessary to provide feedback of reliability problems and improved component reliability estimates for use in new designs. The performance of the system in use in the customer's environment provides extremely useful reliability information. Component reliability information is collected through field performance tracking studies and through repair studies. [25]

Models are necessary for estimating component failure rates; no database will contain information on the reliability of every component that may be selected, since leading-edge designs will utilize unproved devices. Interpolation using the models permits prediction of failure

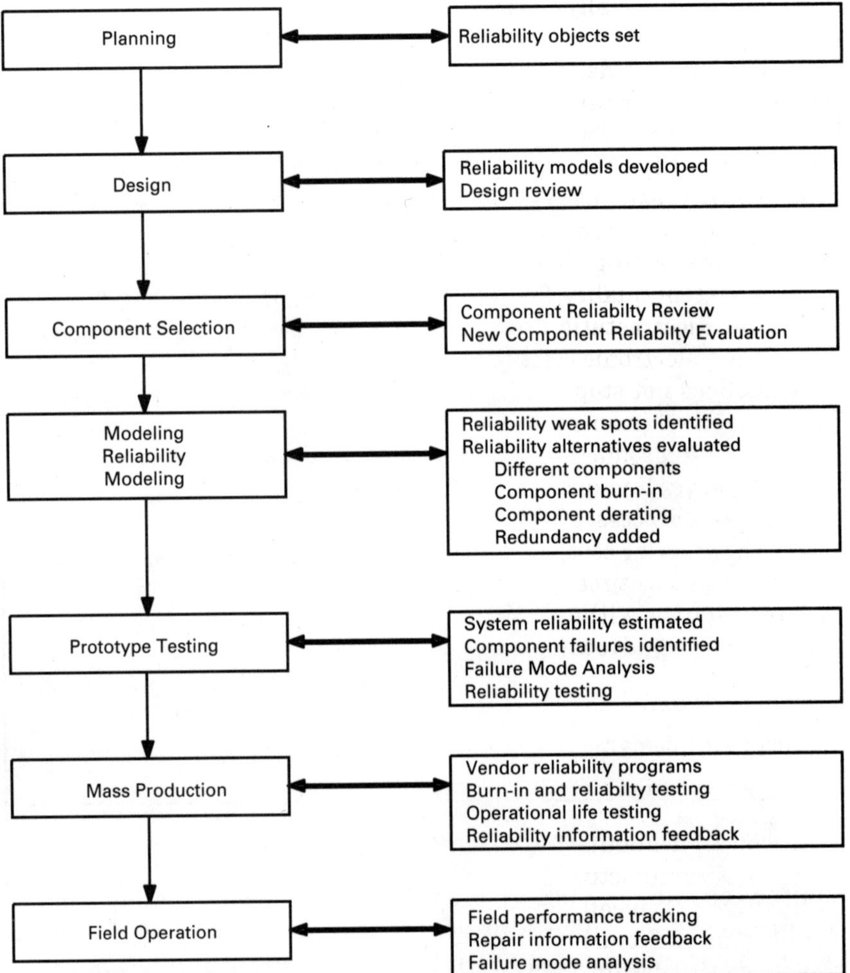

Figure 2.5 Reliability assurance program. (*Reprinted from J. M. Juran, Juran's Quality Control Handbook, 4th ed., McGraw-Hill, New York, 1988, with permission.*)

rates not given in the tables and allows extrapolations to components beyond the ranges of the tables. The most frequently used models for component reliability are contained in *Military Handbook 217D* (Department of Defense). Although these models have been used for years, recent studies show that they far underestimate reliability for integrated circuits now being produced for applications such as telecommunications. This has caused even the DOD to sometimes avoid using predictions, but instead to ask for actual data on former electronic designs and then ask for extrapolations based on those figures.

Reliability is usually determined by the design to a great extent, but many manufacturing steps affect reliability. The integrity of solder joints and welds, electrostatic discharge (ESD) latent effects, rework, and repair can all have significant impact on component reliability. Des Plas (1986) discusses several ways companies are building in reliability. For example, solder joints can be automatically inspected by means of laser inspection systems. Studies have shown that this system can improve quality and long-term reliability. Temperature testing (both hot and cold) of integrated circuits can improve system quality and reliability and significantly reduce total test costs. Electrostatic discharge controls are extremely important for assuring electronic component reliability. [26]

Design does not stop with the product in the electronics world. The Taguchi method is an engineering tool used prior to putting a product or process into production. The Taguchi method seeks to minimize the necessity for on-line quality control activity by reducing variability in products and processes. When this method is used, there will be less prototype material cost, less time for running experiments, less testing time, and less time for analyzing results. The primary intent is to improve quality without incurring capital and material cost. The Taguchi method advocates a three-step design process [27]:

1. System design
2. Parameter design
3. Tolerance design

Information about the Taguchi method is found in Chap. 3. A process for manufacturing the product must be designed, prototyped, and qualified. Typically, many electronic designs start out with state-of-the-art manufacturing, so linking the manufacturing process development to the design is even more important. Figure 2.6 shows the relationship between the two processes. [28] This is the crossover with quality of conformance.

Metrics for the design phase. Quality in the design phase is measured by the following:

- Design reviews
- Peer reviews
- Trade studies
- Drawing/specification checking (counting errors)
- Design changes after first (baseline) design, internally and externally driven

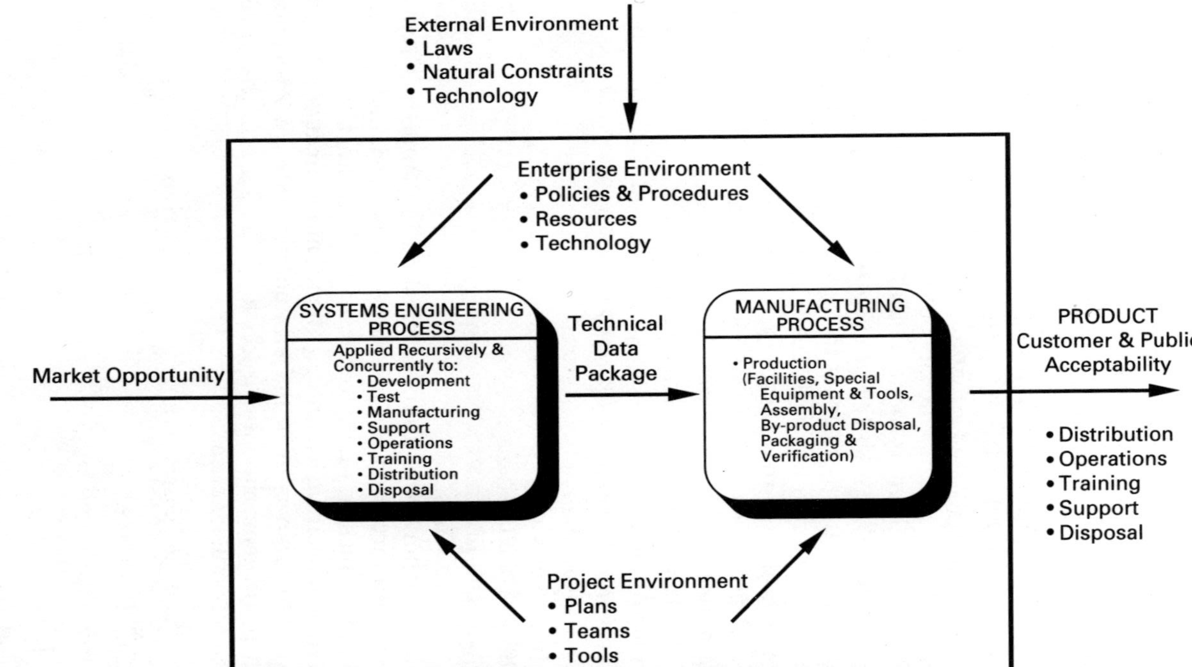

Figure 2.6 Systems engineering within an enterprise. (*From Ref. 28; copyright IEEE, all rights reserved.*)

Number of specifications

Number of drawings

Number of military versus commercial/industrial standards in the design

Number of months from first design to steady-state product

Ratio of predicted reliability to actual reliability

Total assembly time

Total number of operations

Ratio of number of parts to theoretic minimums

Quality of conformance

> The effectiveness of the design and manufacturing functions in executing the product manufacturing requirements and process specifications while meeting tolerance, process control limits, and target yields for a given product group. [29]

As broken down, this requirement assesses the effectiveness of the design and manufacturing requirements in product manufacturing. Process specifications should be used, and tolerances, process control limits, and target yields should be met. The first section assesses the effectiveness of the design and manufacturing requirements.

Where are the product design requirements delineated? They are found in the drawings, schematics, and specifications. A prime-item product fabrication specification defines the requirement for manufacture and acceptance of the prime item being manufactured. If the item is critical, the specification may be called a *critical product specification.* These specifications define physical and performance characteristics, and they specify the drawings, standards, and workmanship levels to which the items are manufactured, as well as the acceptance criteria. See Sec. 2.2.2 for complete definition of the specifications.

While manufacturing requirements are often defined in product specifications, it is critical that the manufacturing capability be ready to meet the requirements to effectively produce the design.

Line proofing is a good activity to ensure that the product can be manufactured. As part of the transition to production, the factory's capability in electronic processing must be demonstrated. Line proofing is the means to achieve this demonstration and serves a number of purposes: to verify the final "build to" package, to verify the capability of the test equipment, to exercise factory operations, and to provide systems integration and test experience required to produce the end product. The line-proofing effort may also be used to verify fault detection capabilities and other off-nominal operations.

The magnitude of the line-proofing effort will also depend on the availability of resources, the maturity of processes, and the extent to which production processes were employed to build test articles during the prototype phase. This effort should be considered for both the system electronics and component electronic manufacturers. The line-proofing effort should be conducted in conjunction with a physical configuration audit (PCA) of the design documentation such that the final line-proofing article is the PCA article.

The effort provides risk identification opportunities through the exercise of systems components. It also provides a means of problem resolution through iterative cycles of building and analyzing in a production line environment. Line-proofing activities at the prime and supplier levels can be performed incrementally, corresponding to the usual production-line process increments. When accomplished in conjunction with PCA, it can provide for more stable transition to production and minimize design changes to both products and processes.

In the past, production has frequently been attempted without a systematic and thorough verification of the "build to" documentation, and many problems have thus been encountered. It is important that, the first time, all of the system elements are brought together, assembled, tested, and integrated using production-type tooling, test equipment, and personnel. These integration problems point out areas for improvement in documentation and interchangeability of hardware and tooling. This results in the introduction of changes through experimentation as problems are encountered, and an unstable product configuration during early builds. By planning and executing a structured approach to check and verify the complete production process and the supporting infrastructure, it may be possible to eliminate a number of early production or rate buildup problems. A structured line-proofing approach can also allow iterative build, test, analysis, and cycle improvement to affect design and build processes. Also, because of varying lead times and other factors, there is frequently a gap between development and production efforts which can lead to the loss of key personnel at the major subcontractors. The line-proofing effort, although this is not its primary intent, may offer a solution to this problem.

Line proofing shall be conducted in a manner which supports risk management and abatement activities. The final line-proofing article shall be used as the PCA article. The electronics company should have the line-proofing effort fully integrated into the overall risk management effort. Suppliers should be involved and expected to do line proofing on new electronics processing as well. [30]

The process specification defines a process used to manufacture a product. It covers the equipment, materials, procedures, and operations, as well as how the process and its operators are certified. Process specifications and special process control are discussed in Chap. 4.

Tolerances are generally defined in the specifications and on the drawings. Tolerances should be closely reviewed during the transition to production. This review allows the widest possible tolerances the manufacturing processes can use in order to yield the best productivity while meeting the specification. Critical tolerances should be noted and manufacturing processes monitored closely for those tolerances.

Statistical process control is the technique used to control the quality of a process. It requires monitoring a process, detecting process changes, and correcting problems. As a result, problems are not passed on to the next part of the process.

SPC is called *statistical* process control because it uses information (data) taken directly from the process itself. Charts are used to record, calculate, plan, and find patterns. They allow you to recognize quickly whether the process is going to produce defects or errors and take corrective action.

SPC had its beginnings in 1931, when W. A. Shewhart published *Economic Control of Quality of Manufactured Product* (Van Nostrand, New York). During World War II, the United States used SPC. Since World War II, Japan has used SPC and other quality improvement techniques to a significant advantage. Around 1980, U.S. business became aware of the need for quality improvement and SPC. SPC is successful because it is used by the person who is most knowledgeable about doing the job and managing the process: the employee, not a quality control checker.

Once SPC is set up, variability reduction can be used as a system approach to reduce product/process variation. Emphasis shall be placed on developing manufacturing processes whose variability around target product critical attributes is minimized rather than simply held within the product tolerance.

VR is intended to improve life-cycle costs as well as operational availability and capability. For the supplier, the benefits are many: improved yields, reliable scheduling, reduced cycle times, reduced scrap and rework, better cost control, potentially improved profitability, and improved competitiveness in addition to the intangible benefits of improved customer satisfaction. Key elements of VR are a culture, an enabling structure, and a well-defined process that encourage and empower the factory workers, and a supporting management infrastructure. The techniques associated with process control and ongoing evaluations of process capability will identify candidates for process variability reduction. Additional information on SPC/VR is found in Chap. 3.

Quality control. Quality control is often associated with conformance. By the definition ISO 8402, "quality control is operational techniques and activities that are used to fulfill requirements for quality." In the notes section, ISO 8402 states that "quality control involves operational techniques and activities aimed at monitoring a process and at eliminating the cause of unsatisfactory performance in order to achieve economic effectiveness." [31]

Often the culture links "quality control" with inspection. This is not the case. Since we are talking about the big "Q," which involves the whole company, quality control would be monitoring of a process, any process, to eliminate unsatisfactory performance and achieve economic effectiveness. Therefore, quality control under the big "Q" is nothing more than keeping metrics on processes.

Juran and Gyrna, in *Quality Planning and Analysis,* use this definition of quality control and define the following principles to quality control subjects. [32] They should:

- Be customer-oriented; Table 2.5 is a good start for control subject categories in an electronics company.
- Be inclusive enough to evaluate current organizational performance and provide early warning of potential problems.
- Recognize both components of the definition of quality. Not just negative metrics like defects or errors represent quality; even if both of these measures are zero does not necessarily mean that the product is a "quality product."
- Recognize that all subjects, both customers internal and external, are sources of ideas for quality control subjects; especially those employees who deal with external customers should be used as

TABLE 2.5 Control Subject Categories and Examples

Category	Example
Document quality	Defects per thousand formatted output pages
Software quality	Defects corrected per thousand noncomment source statements
Hardware quality	Field removal rate
Process quality	Functional yields
System quality	Total outages

SOURCE: Reference 32; used with permission.

sources of quality control from the external customer's point of view.

- Be viewed by those that are measured as valid, appropriate, and easy to understand when translated into numbers. Of course this can be elusive in the real world.

Metrics for conformance. The following metrics show whether tolerances are met and process control limits are met or exceeded, and help to target yields:

Rework

Reinspection

Scrap

Material review boards—internal and external, if applicable

Design changes

Audits

Additional metrics are suggested in Sec. 3.6.

Fitness for use

> The effectiveness of the design, manufacturing, and support processes in delivering a system that meets the operational requirements under all anticipated operational conditions. [33]

In one word, this requirement could be stated as "supportability," or even better, meeting all the customers' requirements in a cost-effective manner so they feel they have received the best value.

In *Juran's Handbook of Quality Control,* "fitness for use" is defined as the customer's explicit or implicit quality requirements. [34] In Garvin's dimensions of quality, fitness for use is concerned with aesthetics and perceived quality. The totality of traits that a customer experiences will come together as the "quality image" [35] of the product. How the product looks, feels, sounds, tastes, or smells is clearly a matter of personal judgment, but it will impact the product's aesthetics. Reputation is one of the primary contributors of perceived quality. It comes from the unstated analogy that the quality of products manufactured by a company today is similar to that which was developed in earlier periods or in specific locations. [36]

Fitness for use must be discussed in design reviews from the first phases of design. Addressing the fitness for use, or operational and support criteria, as hard requirements in the QFD process is a first step. This assures getting the customers' fitness for use in the design. The fitness-for-use requirements should be detailed in the performance specification. With the requirements defined, the design verification and validation provides the data that show that operational requirements are met. The ANSI/ASQC Q9001 specification, Paragraph 4.4.8, "Design Validation," even points the supplier to fitness-for-use requirements when it states, "validation is normally performed under defined operating conditions," and "multiple validations may be performed if there are different intended uses." [37]

There are many measures of fitness for use. Two of special interest to most customers are documentation and the ability to test and troubleshoot in the field.

Documentation. Documentation for electronic systems comes in three main flavors: specifications, drawings, and user manuals. Schematics are important as a type of drawing that must be completed for electronic circuits.

The preference in government and industry today is have performance specifications that define the top-level customer requirements. Then industry builds specifications. For the government, they are now entitled commercial item descriptions (CIDs). These are simply specifications that follow a common specification table of contents depending on the items. The system requirements are decomposed and allocated to the system, segment, or configuration item specifications. Electronic parameters must be defined in a requirements specification. The level of the requirements specification depends on the product and the industry. Component specifications are quite detailed; electronic system specifications may allocate functions to subassemblies and be written at a higher level, referring to lower-level specifications. Requirements specifications precede functional specifications, which again are written at a level commensurate with product and industry. The functional specification is crucial in the electronics product, as it describes what the electronics product will do. A complete description of specifications and their contents is found in Sec. 2.2.2.

Drawings are created so that the product can be built. For electronics a logic diagram (also called a *block diagram*) and schematics are the bottom layer of the engineering documents that define the electronics system so that the customer and the team understand the electronics process. The Institute of Electrical and Electronics Engineers *Standard for Systems Engineering,* P1220, shows a list of data that should be generated for system architecture and the design information (see Table 2.6). User manuals are created so that the customer can know how to use, test, and interface with the electronics if there is a problem. The user documentation should contain schemat-

TABLE 2.6 Elements of Product and Process Data Packages

1. Arrangement drawings—show the relationship of the major subsystems of components of the system.
2. Assembly drawings—depict the relationship of a combination of parts and subassemblies required to form the next higher indenture level of equipment or system.
3. Connection drawing—shows the electrical connections of an installation or of its component devices or parts.
4. Construction drawing—shows the design of buildings or structures.
5. Product drawing—an engineering drawing that discloses configuration and configuration limitations, performance and test requirements, weight and space requirements, access clearances, pipe and cable attachments, support requirements, etc., to the extent necessary that an item may be developed, or procured on the commercial market to meet the stated requirements.
6. Detail drawing—depicts completed item requirements for the parts delineated on the drawing.
7. Elevation drawing—depicts vertical projections of building and structure or profiles of equipment.
8. Engineering drawings—an engineering document which discloses by means of pictorial or textual presentations, or a combination of both, the physical and functional end product requirements of an item.
9. Installation drawing—shows general configuration and complete information necessary to install an item relative to its supporting structure or to associated items.
10. Logic diagram—shows by means of graphic symbols or notations the sequence and function of logic circuitry.
11. Numerical control drawing—depicts complete physical and functional engineering and product requirements of an item to facilitate production by tape control means.
12. Piping diagram—depicts the interconnection of components by piping, tubing, or hose, and when desired, the sequential flow of hydraulic fluids or pneumatic air in the system.
13. Wire list—a book form drawing consisting of tabular data and instructions required to establish wiring connections within or between items.
14. Schematic diagram—shows, by means of graphical systems, the electrical connections and functions of a specific circuit arrangement.
15. Wiring and cable harness drawing—shows the path of a group of wires laced together in a specific configuration, so formed to simplify installation.
16. Software design documentation—shows the software item's architecture, design requirements, implementation logic, and source code to provide a means of support.
17. Models, simulations, or design databases—provide a view of any of the items listed above.

SOURCE: Reference 38;

ics if used for testing. Customer requirements should be met, and may depend on the user and the maintenance concept.

To evaluate the documentation on a given project, use the sample fitness-for-use matrix for documentation in Table 2.7.

Testing/troubleshooting in the field. Good-quality electronic products must be reliable, but at some point they must be serviced or repaired. This must be considered in the original design. How are the components organized? Will a printed wiring board be replaced or repaired? Can integrated circuits (ICs) be changed out? Has the product been designed modularly so that, when electronic components are obsolete, the unit can be upgraded?

Component-level repair requires that the technician isolate and replace defective parts. System-level repair requires only a block diagram or knowledge of the block diagram. An entire module, panel, or circuit board can be replaced, and component-level troubleshooting is not required. Component-level troubleshooting usually takes longer than system-level; given that time is money, it may be better to replace a panel, etc. In the new complicated, complex electronics, however, component-level troubleshooting is needed. [39] It takes significantly more sophisticated test equipment, programming, and documentation to set up a system to fault-isolate at the component level. This needs to be discussed with the customer and integrated in the total plan.

Reliability, maintainability, and availability (RMA) should be specified in the performance specifications. Then RMA is one of the contract requirements that must be met, regardless of the type of hardware being produced. RMA activities must be established early in the design cycle to assure meeting operations requirements and reducing life-cycle costs. RMA requirements shall be based on operational requirements and life-cycle cost considerations shall be quantified, in operational terms, and measured during the developmental and operational test and evaluation. They should be defined for all elements of the system, including support and training equipment. Reliability requirements shall address both mission reliability and overall product reliability. Maintainability shall address servicing, preventive, and corrective maintenance. Availability requirements shall address the readiness of the system. Often, when specifying nondevelopmental items (NDI), or commercial items, customers will specify the requirement as Availability (A_o). Reliability is specified as mean time between failure (MTBF). Maintainability is specified as mean time to repair (MTTR), so A_o must be calculated:

TABLE 2.7 Fitness-for-Use Compliance: Documentation

Documentation Attributes and Ratings					
Attribute	Rating 1 (20%)	Rating 2 (40%)	Rating 3 (60%)	Rating 4 (80%)	Rating 5 (100%)
Clarity	Poor format, hard to read. No figures or tables.	Fair format, descriptions not clear.	Good format, half the section is hard to read.	Most sections are clear except a few selections.	Excellent format; all sections easy to read.
Completeness	Incomplete. Major sections are missing.	All sections are identified but are very incomplete.	All sections identified and half the sections are complete.	Most sections are complete but a few sections need work.	All sections identified and complete.
Accuracy	Document does not reflect correct version of the product.	40% of the parameters of data identified is correct.	60% of the parameters or data identified is correct.	80% of the parameters or data identified is correct.	100% of the parameters or data identified is correct.
Configuration control	Not released, no control.	Document is revised and controlled by the originator only. Fair format, descriptions not clear.	Document is controlled by the originator but not ready for release.	Document ready for release; it is in the right format for document control but not released.	Document has been signed off by other functional groups and released by document control.

$$A_o = \frac{\text{MTBF}}{\text{MTTR} + \text{MTBF}}$$

Good guidelines are available on RMA from the ISO standard ISO 9000-4, *Quality Management and Quality Assurance Standards,* Part 4, *Guidelines on Dependability Programme Management.*

Demonstrations are the general method of verifying the RMA of the design on production-representative systems and actual operational procedures (technical orders, spare parts, tools, support equipment, and personnel who will use).

In the electronic world, the RMA requirements flow down to subcontractors and vendors, especially on electronic components. Most customers today do not want predictions of MTBF or MTTR; they want proof of the system, or the similar system upon which the new system is built.

A system should be set up as part of the program: when the system goes to the field or to customers, there needs to be a method of collecting the components that fail in order to verify the reliability, maintainability, and performance. This is not an easy task, but it needs to be done, especially in electronics, for the "health of a business." The level of repair, testing, and troubleshooting must be clearly defined in the beginning of a design.

2.1.2 The total system life

> Life cycle is defined as the scope of the system or product evolution, beginning with the identification of a perceived customer need, addressing development, test, manufacturing, operation, support and training, and continuing through various upgrades or evolution until the product and its related processes are disposed off. [40]

There is a life cycle in the electronics world (see Fig. 2.7). It is not restricted to the type of business that is being done, and in this fast paced life cycle of electronic products it can be modified. Sometimes prime contractors and/or the government defines part of the life cycle; at other times, the company may decide to define a new product. The new Department of Defense (DOD) 5000 defines acquisition phases which are applicable to all electronics.

Phase I: Concept exploration. This phase consists of competitive, parallel, short-term concept studies. The focus of these efforts is to define and evaluate the feasibility of concepts to provide a basis for design. The most promising system concepts shall be defined in terms of broad objectives for cost, schedule, performance, trade-offs, and test and evaluation. A prime contractor may ask an electronics company

Definition	Develop concepts to meet need; define alternatives; broad definition of best alternative strategy	Alternate concepts Cost drivers, life cycle cost estimates, performance trades, strategy decisions	Translate design into producible, supportable, cost effective design	Achieve operational capability, or full up production to serve needs ; support/ modification/ improvement	The effects of waste disposal to the environment–cost effective?
Life Cycle Phase	Concept Exploration	Program Definition and Risk Reduction	Engineering and Manufacturing Development	Production, Fielding, Operational, Support, Modification	Disposal; for military: demilitarization
Systems Engineering Phases	Requirements Analysis (QFD)	Functional Analysis	Synthesis		

Systems Analysis →

Verification and Validation →

Figure 2.7 The life cycle of an electronics system.

to define concept exploration; a commercial product company will do exploration as part of its marketing and future R&D program.

Phase II: Program definition and risk reduction. During this phase the program becomes more defined, and design approaches and/or parallel technologies are pursued as needed. Assessments of advantages and disadvantages of alternative concepts are refined. Prototyping, demonstrations, and marketing analysis is done. Included is risk to technology, manufacturing, and support. The government will look at life-cycle estimate, cost performance trades, interoperability, and acquisition strategy alternatives. A company in this phase primarily looks at cost to produce versus market potential.

Phase III: Engineering and manufacturing development (EMD). In this phase, the most promising design is translated into one or more products which can be manufactured and tested. Low-rate initial production occurs while the EMD continues to do line proofing, collect test results, design fixed, and upgrades. Low-rate initial production articles are generally used to provide representative articles for test and to establish the manufacturing requirements. The commercial company is finding ways to produce quicker, faster, and cheaper.

Phase IV: Production, fielding/deployment (distribution), and operational support. The objectives of this phase differ in commercial and defense manufacturing. The defense customer is concerned with achieving an operational capability to satisfy a mission. A commercial business has put out advanced planning on a new product, but now the advertising campaign accelerates; this is the "fielding/ deployment"/distribution phase.

> *Operational support.* For the defense agency, this is the time that the life-cycle benefits of support come into being. For the commercial house, this is when the customers buy the product and either like it and product sales increase, or do not like it and the product fails. The defense contractor worries about support, but the government may decide to support it itself. In the commercial electronics world, the company must support the product.
>
> *Modifications.* Both worlds use modification; some electronics products, such as computers, are modified (e.g., by adding more memory) before they even are taken home by the customer.

This phase of the traditional life cycle is the fielding, the deployment, the selling of the consumer product and shipping it to customers. This phase contains operational support, or customer service. The way that electronics must be supported has changed rapidly in the consumer and commercial worlds (e.g., airplanes, computers), and must in the high-reliability world. For customers, from the consumer who buys a computer to NASA buying a space station, electronics is being packaged so that it can easily be replaced. For the first time on a large scale, the 1990s are throwing away, and buying new, because it is more efficient and a better value to the customer. Computers are again the prime example and leading the market. But even in the new DOD programs for high-reliability products, if the components are inexpensive and getting cheaper, throwing them away has become a real option.

Phase V: Demilitarization and disposal. Demilitarization applies only to the government, but disposal applies also to electronics companies. What happens to all the "outdated product" thrown away? Is it recycled? Environmental laws and regulations control how the environment is affected in closing facilities. Closing a printed wiring board processing facility, where many hazardous materials are used, provides many challenges to the environment and the owner. What is happening to the disposal of electronics in the commercial world? Some electronics are being reused, others are collectors' items in people's garages.

The supplier needs to look at the entire life cycle of the electronics product/service or system. How specialized is this product? How long will it be supported? Can it be upgraded? Electronics is not the traditional life cycle. Again, as one examines the life cycle of specific devices this can change, and behave individually, but the generic approach is the same.

A concept is generated by a supplier, or a customer, and there is a great excitement. It may be in the constant development process of electronics, to make it operate faster, with greater capacity. It generally does not take the customer or the supplier long to agree that that concept is a "good" thing.

The next phase of program definition moves quickly also, whether the supplier creates specification requirements, or receives them from a customer. Typically, this also is fast, yet there are many issues as to whether this program definition is complete. This will be discussed in Sec. 2.2. Notwithstanding this, the program moves into the first-design/first-prototype build. The DOD calls this *engineering and manufacturing development.* It is when the first product actually can be built, and therefore tested.

This is where the life-cycle process slows down. Since their relationship began, suppliers, customers, and the DOD have been trying to find ways to speed up this phase of the life cycle. Generally, the concept is based on a need (or, in the consumer world, the concept is based on increasing a market share or making additional profits), so there is an urgency to get through the prototype development phase. This is where electronics as a discipline slows the process down. If a new electronic component is developed, extensive testing must occur. Commonly called *beta testing,* it is used to characterize the electronic component. The company must know what the electronic component can do.

Two things are happening in the 1990s: (1) Sometimes, suppliers and customers listen to component suppliers and believe that a new electronic product can be done quickly; schedules are compressed and risk goes up. (2) Alternatively, they seek an electronic product that is already designed, or a nondevelopmental item (NDI), or commercial off-the-shelf item (COTS). Many times, this NDI or COTS can then be used to move through the developmental/prototype phase, with an update to the system planned after the development of the next-generation electronics. This approach uses the concept of open architecture in the electronics world.

We see this in the consumer computer market constantly. No sooner does a computer hit the stores than the next generation or model, which is faster and more capable, is on the manufacturer's shipping dock. Now, this directly affects the next phase of life-cycle development, which is again different for electronics. The traditional model is that once a device is defined, it could be produced for as long as the customers want it. Because the electronics market is changing so rapidly, it affects all phases of production, whether for consumer, industrial, or high-reliability systems. Whether the product is a kitchen stove with digital controls or a rocket ship with digital controls, chances are the electronics supplier will guarantee that the electronic product that the prime supplier is buying will be produced for a few years (the common number seems to be 6 years). Now a new set of issues arises: What does the prime do after that 6 years, if the life cycle of the product is 4 to 5 times longer, to support or repair that product? That brings us to the reality in the 1990s.

Metrics for fitness for use. These metrics are not easy to get most of the time, because they do not come from within the enterprise. Metrics must be developed to find out the customer's view. Part of the ISO 9000 quality management premise is that a customer survey must be a part of the quality management system. If the contract provides for repair services or field support, metrics can be kept on those activities. A part failure system must be in place to collect the reliability/maintainability of parts that fail, and collect reliability measurement data.

How do we do customer surveys? Put them in box to be picked up by customers? Send them out? Call customers on the phone?

These methods are fine, and the customer may respond at first. But a method must be set up that follows up many months or perhaps years later.

The consumer electronics world, because of a need to provide customer service for failed electronics, has come up with service contracts; a consumer buys a contract and then the supplier repairs or replaces products at no additional cost. Another option used extensively in the computer world today is a toll-free phone number, a "1-800-my company" number that the consumer can call with a problem. Companies can keep a record of the type of problems that are called in. Microsoft Inc. uses this approach to find out what consumers do not like about their software. In the TQM fashion, the data can be kept.

It costs money to set up metrics with the field, but if supplier is worried about the big picture, of pleasing the customer and getting more business, then everyone is happier in the end. Because electronics is really "black magic" to the average consumer, this is good, and good business.

Some accounts of what customers are doing to monitor field performance and customer service are provided by Juran and Gyrna. In a survey of 267 companies done by the American Management Association in 1978 (see Table 2.8), 267 companies were asked if they conducted certain activities related to complaints. Note that many companies did not regularly count, report, or track the response time on complaints. [41]

A technique called *Weibull probability analysis* can be used to analyze early field data on warranty claims. A Weibull probability distribution is applicable for describing a wide variety of patterns in variation, including departures from the normal and exponential distributions. [42] Through such an analysis, it is possible to predict

TABLE 2.8 Collection and Use of Customer Comments and Complaints

Activity	Companies utilizing, %
Regular counts kept	72.6
Reports prepared and circulated	66.3
Response time tracked	64.8
Competition evaluated	34.1
Included in strategic planning	60.7
Part of performance appraisal	42.0

SOURCE: American Management Association (1978); from Ref. 41, with permission.

TABLE 2.9 Repair Information on Electrical Subassembly

Time in service in months	Repairs per 100 units (R/100)	Cumulative R/100
1	0.49	0.49
2	0.32	0.81
3	0.24	1.05
4	0.24	1.29
5	0.21	1.50
6	0.19	1.69
7	0.19	1.88
8	0.23	2.11

SOURCE: Reference 43, with permission.

what the cumulative number of claims will be at the end of the warranty period. Table 2.9 shows repair information collected by Ford in 1972. The cumulative repairs per 1000 units are interpreted as a cumulative rate in percentage. These data are summarized from a large number of warranty reports, and thus the data can be plotted directly. [43]

The reason this study is mentioned here is that, in the new world of quality, there needs to be more than intuitive ideas; because of the nature of Total Quality Management, real data are needed.

2.2 Performance-Based Contracting for Electronics Systems

This section is most applicable to the electronics company that does business with large customers, original equipment manufacturers, or government. This includes electronic contract manufacturing firms. It also talks to companies in the consumer market in regard to defining requirements. It is provided here to enable all electronics manufacturers to expand their electronics product line in this market. There is no "corner on the market" when it comes to a new idea in electronics. Think about it. Personal computers came along in the late 1970s and by the mid-1980s had killed the office typewriter, made secretaries as we knew them obsolete, created the largest continuous peacetime economic expansion in U.S. history, and made several people rich beyond belief. [44] Because of the rapid expansion of electronics in all phases of life and the U.S. government's desire to work with the commercial companies for reasons of economics and best value, it is necessary that especially electronic systems companies understand the process. Perhaps companies and the government can use suggestions in this chapter to make the process better. The opportunity is present; the understanding is what is needed and offered here.

Performance-based contracting can be used by any company for any purpose. It is especially applicable to electronics because the end goal of electronics is a product that performs as intended, and performance-based requirements can be met in many different ways. This allows competition and a leveling of the industry base for the best-value product and service.

Performance-based contracting uses performance specifications and evaluates proposals on the basis of demonstrated performance. Elements of performance-based contracting include:

- Best value
- Customer on integrated product development (IPD) teams
- Emphasis on nondevelopmental items/commercial off-the-shelf items
- Emphasis on risk management
- Emphasis on process
- Emphasis on metrics and
- Elimination of nonvalue work.

Performance-based contracting means structuring all aspects of an acquisition around the purpose of the work to be performed as opposed to either the manner in which the work is to performed or broad and imprecise statements of work. [45]

Performance-based contracting must concentrate on the big "Q," the quality of the design, conformance, and fitness for use. Customer objectives must be met. Capable processes must develop good electronics, variation must be reduced to ensure profitability and drive out defects, and there must be a system of feedback from the customer or the field.

Performance specifications state functional requirements in terms of results which are criteria for verifying compliance, without stating methods for achieving results (see Table 2.10). [46] The performance specification defines the environment in which it must operate, interface, and be interchangeable. It defines the functional elements and operating environments. Detailed specifications are then written by the supplier for a customer, or for a supplier wishing to sell to the consumer market. The design specification details the design solutions: how the requirements of the performance specification are

TABLE 2.10 Definition of Performance-Based Specifications

Performance-based specifications:
Define purpose of work
Define results required
Define criteria to verify compliance
Do not state methods for achieving results
Define functions to be performed
Define performance expected
Define essential physical characteristics
Define form/fit/function and interfaces
Define environment to be operated in

achieved, and how the product is constructed and tested. Process specifications define the control of processes critical to manufacturing of the electronic device or system.

In this section, we will to talk about the *process* of performance-based contracting. Performance-based contracting is new to the late twentieth century and needs much improvement; Peter Hybert wrote an article titled "Five ways to improve the contracting process" (*Quality Progress,* February 1996) that provides some easy solutions based on QFD and in a few pages lays out a good plan to improve the entire contracting process, whether defense or commercial. [61]

Almost every technology now is using electronics to make it better, faster, and cheaper. Electronics technology is also changing rapidly. Those who work in electronics must be willing to accept rapid change. Change is so rapid that good electronics systems must be designed to be "upgraded" as the electronics change. This is often called *open architecture.* Open systems architecture (OSA) is a physical and logical organization of system functions, structures, and operations based on open system interface standards to meet requirements. [47]

The performance specification defines the needs of the customer without directing the supplier in how the electronic system is detailed. For the U.S. government, the Federal Acquisition Streamlining Act (FASA) of 1994 states that the government will only provide objectives, and the supplier will detail the "how." The emphasis is on commercial and nondevelopmental items that can serve the U.S. government's needs.

2.2.1 The contracting process

Figure 2.8 shows the typical model of the contracting process. As it is easy to see, it is rather generic. This same process can apply internally to a commercial supplier that is developing a new product, or to an

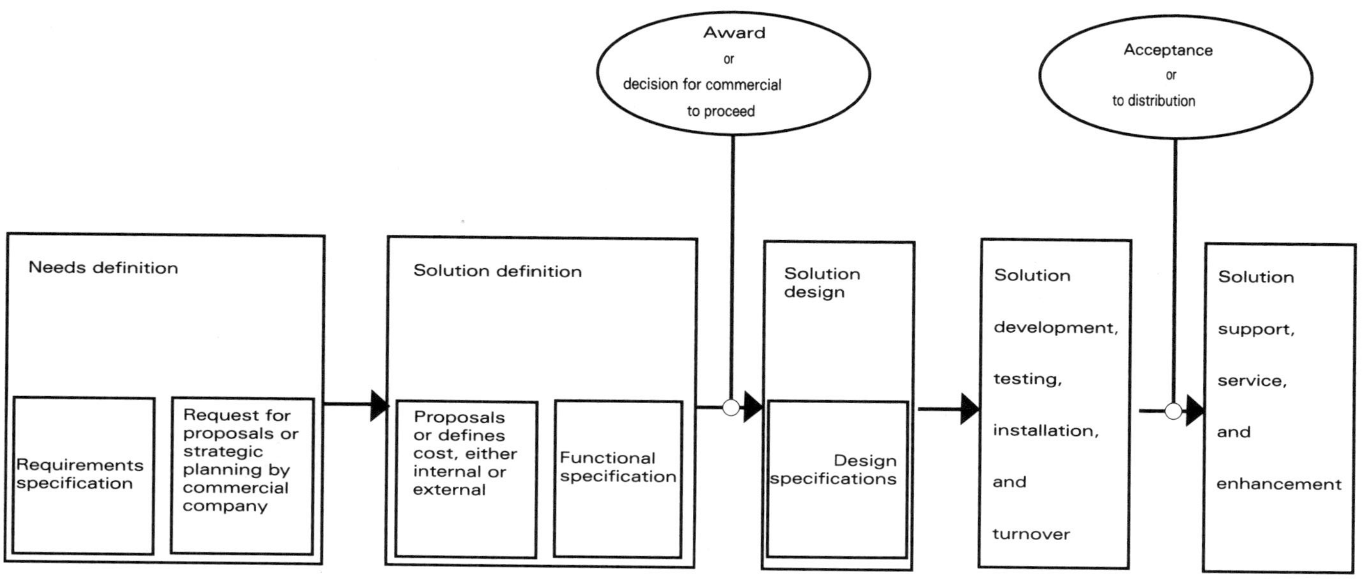

Figure 2.8 A generic model of the contracting or commercial internal business process. (*Adapted from Ref. 61, with permission.*)

upgraded product for the enterprise. It can apply to electronics contract manufacturing. The customers may differ—the marketplace or the U.S. or a foreign government—but now with the ISO 9000 series of quality management standards, the requirements to deliver a customer product are standardized across defense and commercial arenas. About the only difference is that the marketplace decides the success of the commercial product, while the taxpayer sometimes ends up with the cost if the U.S. government does not get the product that it needs, as more money is then spent to define the correct product to meet the need. The process has often failed to meet customer satisfaction, creates stress for both customer and contractor, and sometimes requires more resources than the customer expected. Out of a number of potential problems, Hybert identified the following five characteristics as primary causes that a good quality-based management approach can eliminate:

1. Poor up-front definition of customer needs
2. Incomplete evaluation criteria for awarding the contract (an overemphasis on price)
3. Poor planning of the project activities
4. Poor assimilation of necessary midstream project changes (driven by problems or improvements discussed during the project)
5. Rewards driving the wrong performance [61]

As we talk about the performance-based contracting process, we will examine his recommendations as part of the process of performance-based contracting.

The first two of these issues, poor up-front definition of customer needs and incomplete criteria for awarding the contract, are directly related to best-value procurements and the use of NDI/COTS. Also related is the development of specifications (covered in the next section). First we consider best value and NDI/COTS; then we look at Hybert's recommendations for improving these processes.

Best value. Best-value contracting is part of performance-based contracting. It is a process included in competitive contracting to select the most advantageous offer by evaluating and comparing factors in addition to cost/price. It seeks out quality which cannot be determined by price. It fits the fourth of Deming's 14 management principles, part of which states, "end the practice of awarding business solely on the basis of price." [48] It has an emphasis on past performance: records, deliveries, quality, cost control. It meets the customer's expectation of product quality—meeting the specified requirements and fitness for use. Sometimes comparing only the cost is an acceptable approach, especially in the commercial world but, according to Deming, not usually on a technically complex item. Even in the commercial world, if customers do not feel they are getting the best value in the market, the product will not sell and the supplier will lose the business. The best-value approach is used when it is likely that variations in industry solutions could result in beneficial differences in achieving customer objectives. Best value is a sophisticated process, takes more in-depth planning, more documentation, and more time. It defines requirements to a lower degree, and therefore provides for better meeting the customer needs. Best-value activities should be commensurate with acquisitions needs. In the commercial market, the product must meet market standards for best value. A best-value approach gives the customer freedom to consider innovative approaches, which is what the electronics industry is doing today. It drives cost and technical trade-offs, which are the heart of business today. Very highly complex technical capabilities and qualifications drive best value. The customer's view of quality must be examined. The customer needs to be excited or delighted about the product/service. To achieve a unique competitive advantage, we must focus on understanding the customer and the customer's implied requirements (see Fig. 2.9.) [49]

Solicitations [requests for proposal (RFPs)] must clearly state what costs are being evaluated: cost for basic effort, basic plus options, life

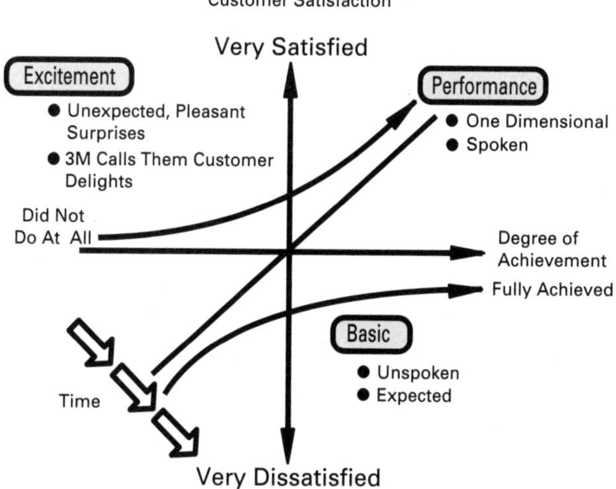

Figure 2.9 KANO model (of quality/features) (*From Ref. 49, with permission.*)

cycle, lowest production. If not clear, the company needs to talk to the customer to find out. In a commercial environment, the company needs to know what will sell.

The U.S. government often uses the best-value technique to ensure the best-value decision. Large commercial companies use the same process, with specific procedures that benefit the specific industry. When the U.S. government uses it, (1) it is based on comparative analysis of the proposals, (2) it is consistent with stated evaluation criteria, and (3) it considers whether perceived benefits are worth any price premium. It makes the decision on a rational basis and sets it forth in an independent, stand-alone, defensible document. [50] Large companies and OEM's generally mirror these kinds of methods for contract selection.

Nondevelopmental items/commercial off-the-shelf: the world of today. NDI/COTS offers four major benefits: quick response to operational needs, elimination/reduction of research and development costs, application of state-of-the-art technology to current requirements, and reduction of technical, cost, and schedule risks.

We begin with some definitions:

NDI means nondevelopmental item. Nondevelopmental means "not requiring development." Nondevelopmental items are defined by the DOD as: "(1) any previously developed item of supply used exclusively for governmental purposes by a Federal Agency, a State or local government, or a foreign government with which the United States has a mutual defense cooperation agreement; (2) any item described in (1) that requires only minor modification or modifications of the type customarily available in the commercial marketplace in order to meet the requirements of the procuring department or agency, or (3) an item described in (1) or (2) solely because the item is not yet in use (FAR 2.101)." [51]

A commercial product is defined as an NDI that has been produced for sale in the commercial marketplace. [52] In the new 1996 DOD regulations, the definitions of commercial items were expanded to allow more government agencies to buy commercial products that will fit government uses, including commercial products that are just being developed.

Often, the benefits of getting something done quickly or by the commercial market may be offset by performance trade-offs, and other compromises which are not supportable. Using NDI/COTS to build up an electronics system is significantly different from development of the same system. Sometimes it is more difficult to assemble several different electronic systems together than it is to build a new one. An

NDI may have to go through increased testing in the operational environment in which it must perform.

An NDI that is to operate in a different environment than originally defined commercially should have:

- Performance specifications to define the operational and functional requirements
- Altered item drawings to define the modifications of the NDI/COT to meet the requirements
- Specification control drawings to control the design
- Commercial format drawings so as not to incur the cost of doing military drawings
- Full disclosure of sources of supply so the government, prime contractor, or OEM knows where to buy parts or implement service contracts to support electronics repair and/or replacements
- Unique or special processes required to achieve the level of performance needed
- Identification of critical parts and critical characteristics

Quality assurance is very important to NDI/COTS. If the equipment was built where there was a good quality system and quality data, it can benefit the customer. Testing where the electronics is being used on other applications should be known as well as the way that equipment reacts in a given situation. [53]

Electronics is often the heart of a system. From telephones to medical equipment, from airplanes to cars to ships, electronics must be performance-based.

The contractor/supplier must interface with the customer to get the real requirements or, in a commercial world, thoroughly understand the market to understand what the customer needs and wants. The supplier must understand the customer's implied requirements—the customer's view of world class quality. Customers come in different flavors and have personal likes/dislikes that affect the perceived quality of the product. Two good and readily available sources for researching NDI/COTS for the U.S. government procurements are

- SD-2, *Buying NDI,* U.S. Government Printing Office
- American Defense Preparedness Association, *Commercial-Off-the-Shelf (COTS) Supportability Study,* 22 August 1994; Undersea Warfare Systems, American Defense Preparedness Association

It is essential when dealing with NDI and COTS to know the customer's requirements. They are defined in four main areas:

1. *Operational.* Desired operational (or performance) requirements; what the system does.
2. *Functional.* The deliverable in terms of form, fit, and function.
3. *Interfaces.* All the components that the system must interface with, and the exact interface requirements, in terms of watts, ohms, voltages, connectors, pins, etc.
4. *Environment.* The environmental conditions that the system must operate in; the amount of heat, moisture, salt, movements, etc.

Interface with customer. A good interface with the customer is required to understand the performance requirements. In the commercial world, a good understanding of the customer buying the product is required. One way to accomplish this in a complex new design is to have the customer on the design team. With the technology as it is now, with the Internet and other groupware and local-area networks (LANs), it is quite possible for the customer to be part of the design team and the decision-making process to make sure that the customer's requirements are being met. The cost of travel and meetings can be significantly cut down with the right communication media set up in the beginning of the program.

Hybert points out that there is poor up-front definition of customer needs. [61] The customer may have a good top-level concept, but deficiencies in the needs-definition process snowball as they roll further into the project, and many other issues arise. Improving this process is a critical step toward more effective projects. The core of the needs-definition process is the series of specifications that are developed by the customer and possibly by competing suppliers. If contractors were to complete an application for the Malcolm Baldrige National Quality Award, they would probably view their specification process as a plus. After all, are not specifications the result of working with customers to define their requirements? Ideally, yes. In actuality, however, specifications usually define only the technical aspects of the solution. They might not address quality requirements, customer satisfaction requirements, or cost considerations (such as development or life-cycle cost targets). So, while the specifications do describe some customer requirements, they are incomplete because they are directed to only one segment of the customer population: the technical *evaluators*. More important, because the specifications are developed up-front rather than throughout the process, they are often inaccurate or incomplete in retrospect.

Good electronics design must list the important quality characteristics, and control parameters, along with a plan of how the parameters will be verified. This includes key and critical control parameters. [54]

Hybert recommends using a quality function deployment approach [61]. Customers and contractors have a common interest in clearly defining needs in the early stages of a project. A QFD approach for defining functionality, quality, and cost requirements can reduce time and errors in this part of the process. If the contractor's goal is to deliver a system or product that meets long-term customer requirements, its mind-set has to shift from deliverables and technical specifications (i.e., "What is the minimum I owe you?") to customer functionality and fitness for use (i.e., "What work will the customer be doing, and how will this system or product improve the ease, cost, and quality of that work?"). For this to happen, the contractor must incorporate the needs of all stakeholders in the customer's organization. Again, this is a reflection of the big "Q."

For example, suppose the customer is a chemical processing plant looking for a computer-based control system for a new process. The plant's executives are concerned about return on investment, cash flow, and life-cycle costs. Its operations managers are interested in reliability, maintainability, and resources needed to support the system. The operators, or users, of the system are concerned with ease of learning, ease of use, and ease of access to reference information. (Too often, the users, including those who support and maintain the system, have the least input on its design.) Finally, technical experts are looking for system data for quality assurance tracking, the ability to adjust the system for varying situations, and so forth. There might even be regulatory needs as well. The needs of all these stakeholders must be met for the control system to improve the ease, cost, and quality of their work.

This is where the QFD matrix (often referred to as the *House of Quality*) and the group consensus process can help. A team consisting of key personnel from both the customer and contractor organizations should use a consensus approach to develop the customer requirements portion of the QFD matrix (i.e., collect, sort, and prioritize customer requirements). Many approaches for gathering information can be used (e.g., focus groups or individual interviews), as long as the primary goal is to understand what the customer needs and expects—and not to create a perfect QFD matrix.

The value of the QFD approach is that the identified needs are clear, specific, and easily understood. They also truly reflect the customer's requirements because generating the matrix takes dialogue, clarification, and more than a single iteration (the needs can be reviewed and revised by all of the stakeholders before awarding the contract). Although documenting customer needs with the QFD approach will increase the time from initial project concept to bidding, it will result in time and dollar savings downstream because changes and conflicts will be reduced.

Figure 2.10 illustrates how other QFD elements can be integrated into the contracting process. Once the customer needs are defined, they can be translated into technical requirements and linked to elements of the solution, completing the QFD matrix. Upon award of the contract, the key result measures or solution elements can be deployed to various team members to ensure that they are not later forgotten. Then, part of each team meeting can be spent focusing on progress toward meeting customer needs. Details on QFD may be found in Chap. 3.

Having the customer on the IPD team keeps the customer involved in the process. Change is inevitable over time, especially with electronics; before you put it on the shelf, it is out of date. The customer must agree on what changes occur, big or little. Having a good process for design, which examines changes and the risk to the program, is essential. It is a complicated formula, the changing technology, the changing customer, and the risk to the program, but there are good design/development processes available. With the customer involved, this can be overcome, but there is lots of work to do.

Part of the integrated product team approach is to team with customers and suppliers. Teams with suppliers have been used in product development for years. The concept of teaming (or partnering) with suppliers stresses having fewer suppliers and working closely with them so they understand the customer's needs well. In this way, both the customer and the suppliers have a stake in each other's success.

In the contracting process, the basic idea is for a customer to select a supplier (contractor) based on its ability to provide a solution. Then the customer and the contractor can develop the plan for the solution together, and can concurrently develop the plan with pricing the project. This becomes the project plan. They can share what they both know and do best for the benefit of the customer. The process of performance-based contracting needs to be improved. Especially where electronics products are concerned, the traditional views of the contracting process do not work. The customer needs to be involved in the planning of the product. It is necessary to acknowledge up front that the electronics process is going to change. So put that in the plan. Risk must be managed as part of the process, not as an add-on, and nonvalue work must be eliminated.

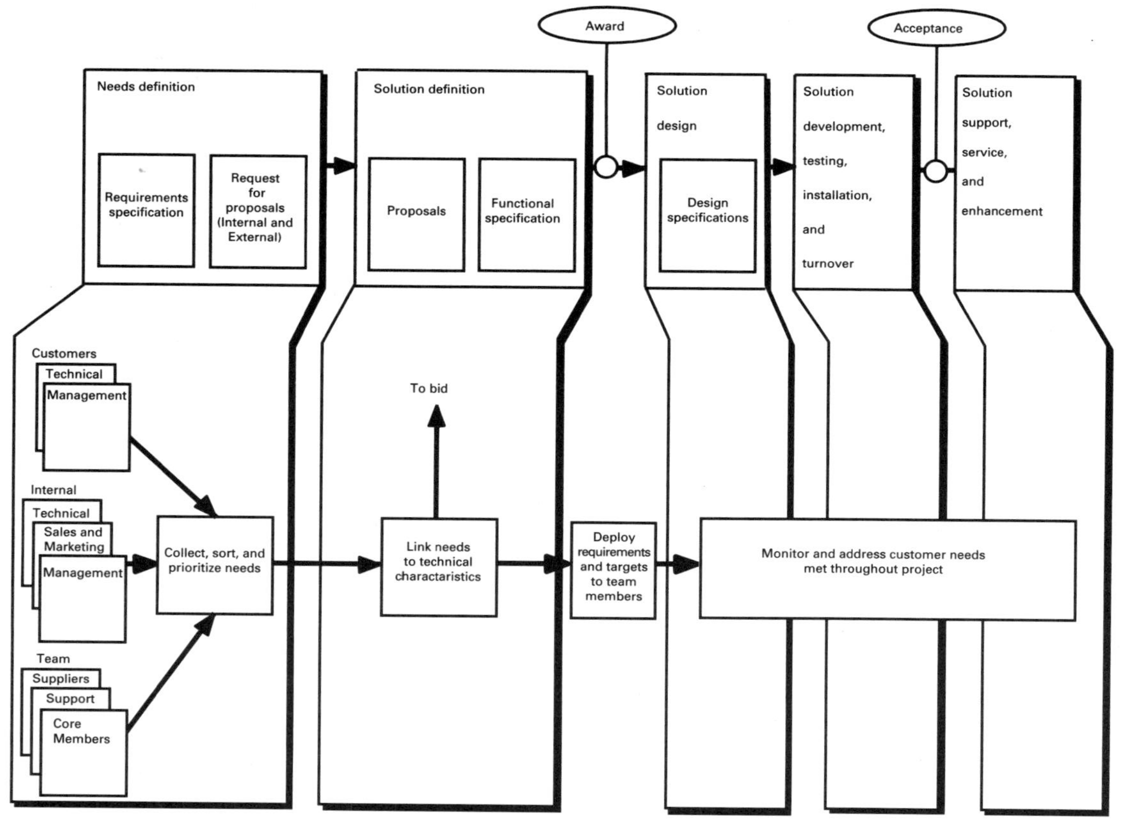

Figure 2.10 How QFD can be integrated into the contracting process. (*From Ref. 61, with permission.*)

Planning the project. Hybert's recommendations on planning the project and assimilating project changes are directly applicable for improvement in these processes. Poor planning of the project activities is well known in industry at this time. Defects in a project plan can often be traced to faulty performance in early process steps. If the needs are poorly defined, the tasks required to meet those needs will be vague or wrong. Typically, project plans either lack the detail to prove that customer requirements are being met or are so detailed that nobody checks their validity or buys into them. A worst-case example involved a project to develop a graphical interface for a control system. The team leader asked the functional subteam leaders to put a plan together for their individual parts of the project. Once all of the plans were turned in, the team leader simply compiled them on his computer and distributed them to all key players. Unfortunately, everyone was too busy to review the compiled plan for the hand-offs (i.e., the passing of project information and deliverables from one subteam to another) that had to occur. The result was that the project was more than 12 months late, due to the stacking of underestimated time frames for testing and approvals. There were misperceptions regarding the development of the project's design, prototype, training, and documentation requirements and poor assumptions about when people would be available to work on the project. In a related case, a company decided to develop a computerized control system device. The project manager prepared a project plan based on subteam requirements. Management, however, had already promised a delivery date much earlier than that given in the team's plan. Not surprisingly, the device was delivered 6 months after the promised delivery date. (In fact, many of the previous product launches at this company were either late or on time but with reduced product functionality.) Based on anecdotal evidence from many industries, these types of scenarios are much more common than one would expect.

Poor planning and non-value-added activities lead to increased costs, development schedules being missed, and nonquality products getting to market. In the 1990s, the emphasis on the elimination of nonvalue work went hand in hand with streamlining processes and empowering employees to deliver a quality product at best cost under the TQM umbrella. [61]

Value-added work. First, we define *value-added:* it pertains to work that increases the worth of a product or service to a customer. It is the work that a company is paid to do. [55] *Non-valued-added* pertains to a process, activity, or task that does not provide any value to the product or service. [56] By knowing the five categories of work, according to Conway—value-added, necessary non-value-added, rework, other unnecessary work, and not working—people can learn to look for wasted time and to maximize the amount of time spent on value-added work. [57] Non-value work to be eliminated would include rework, other unnecessary work, and not working, by Conway's definition. Non-value work can apply to paper and soft processes, such as design and procurement, as much as it can apply to the production floor. By the second or third iteration of the continuous improvement, we should look above waste elimination and concentrate on how the business can be improved. [58]

Conway, in *The Right Way to Manage,* developed five questions to improve on work processes. The questions are intended to get people to challenge the status quo and to think about necessary changes. These questions are asked in the order presented and are asked only after determining that the work being improved should be done at all. These five questions must be asked constantly about any process:

1. Can steps be eliminated?
2. Should the process be changed?
3. Should steps be defined?
4. Should steps be rearranged?
5. Can work be simplified?

How would one know if work should be done at all? Every organization is both a customer and supplier and has both customers and suppliers. (For a definition of customers/suppliers, see Chap. 6.) To understand the customer's requirements is to work on the "right" things. Customer requirements determine what constitutes value-added or real work. By understanding customer requirements, both internal and external, an organization can know what work to do.

Hybert's recommendation, now echoed by the 1996 edition of DODI 5000 for all U.S. government programs [59], as well as ISO 9001 for design (paragraph 4.4) [60], is to use a group-based integrated planning approach. [61]

If the team members, including customer and contractor personnel from the key functions involved in the project, work as a group team to develop the plan, they are more likely to develop an integrated project plan than if the project manager works independently. An integrated plan includes, at the minimum, sufficiently defined team deliverables and hand-offs to allow each subteam a reasonable comfort level. Once the baseline project plan is developed, team members are in a better position to work the plan (i.e., make local decisions that fit the overall team direction) because they understand the details and trade-offs within the plan, they have committed to it in front of others, and they have developed the working relationships within the team that promote extra effort to make sure deadlines are met. This is probably the first point at which the plan should be entered into a computer program. It could be argued that the group-based integrated planning approach cannot be used for complex, large-scale projects. But it could also be argued that, if the plan is so complicated that it cannot be comprehended, it cannot be effectively executed either. This approach will work on large-scale projects; they just might require several levels of planning.

Working together in meetings to determine the milestones and criteria for hand-offs will bring conflicts and difficult decisions, but the payoff is immense. For example, one contractor asked a group of five different suppliers (many of which had overlapping services and interests) to join the team to help prepare a bid for a project. Using the group-based integrated planning approach, the team developed a complex bid for more than $1 million of work during a 1-day meeting. In addition, the group accomplished some initial team building.

The management of risk is vital for keeping a customer happy. Risk can be reduced; we need to change the way we are contracting, because we are in a new-paradigm world. As stated previously, changes are not assembled in a good manner as the project needs change, especially with electronics, which is changing so fast.

When using a group-based integrated planning approach, the contracting process must allow for the extra dialogue and for modifications to the original plan. Instead of thinking that the defined and documented requirements are set in stone or forcing finalization earlier in the process than is reasonable, the project plan team should include intermediate milestones and reassessments to better fit the reality of how people come to a common understanding of needs, contract deliverables, and so forth.

Most plans display the bias that change is an anomaly when, in fact, change is inevitable. Customers might change their business strategies during the project and, hence, their requirements for the solution. Technology might advance, or the new technology planned for the solution might hit a snag in development. If the project plan has no room for adjustments to deal with unforeseen needs or problems, there is no room to work around those needs or problems.

Several words of caution, however, must be given in regard to making changes to the project plan: While the cost of small in-process changes can be absorbed, big ones can significantly increase the project cost. The team leader must be alert for all changes and not let big ones slip through unnoticed (many big ones look small at first). Although many contractors consider change orders to be "good business," they are a major source of potential customer dissatisfaction. This definitely leads to a company in the commercial world losing market share. From the customers' perspective, a change order means they have to go back to their managers to sell them on why more funds are needed for the project. Customers often feel taken advantage of because they know that margins are typically higher on change-order work than the original-project work. They might blame the contractors, thinking that the contractors did not listen closely enough or did not understand their businesses well enough to provide what they *needed,* instead of what they *asked for.* From the contractors' perspective, changes are very expensive (and frustrating) to incorporate in the process. There is a real cost in productivity and energy associated with documenting a change, identifying its effect on other project areas, and communicating the information to everyone who needs to know. Every change is another opportunity to make a mistake; it introduces potential errors or discrepancies in the project information. Sometimes, the customer wants to change the solution back to what the contractor originally proposed, even though the customer initially refused it because of its cost. [61]

Hybert recommends that the model be changed. [61] There is a consensus that this is a good approach, but in reality it has not yet happened. As Fig. 2.8 shows, the typical model used to illustrate the contracting process is linear. A better representation, however, would be a communication spiral (see Fig. 2.11). The spiral shows the incremental progress in all areas of the solution and ensures communication among all players in the process. In the product development world, this approach is sometimes called *concurrent engineering, simultaneous engineering,* or *integrated product development* because the people who produce, maintain, and support the solution begin developing their processes while the product is being designed. A good system engineering process works hand in hand with changing the model to create a producible quality product that excites or delights the customer. Actually, the contracting process *can* be accurately represented by the typical linear model if a subtle point is well understood. The phases refer to activities, not individual functions, with team involvement in all phases. For example, in designing a software system, the design specifications should still be created before the final software code, but the software developers should be able to review and provide input on the specifications as they are written to ensure that the coding can be written efficiently. The project plan should incorporate several features to minimize the pain of change

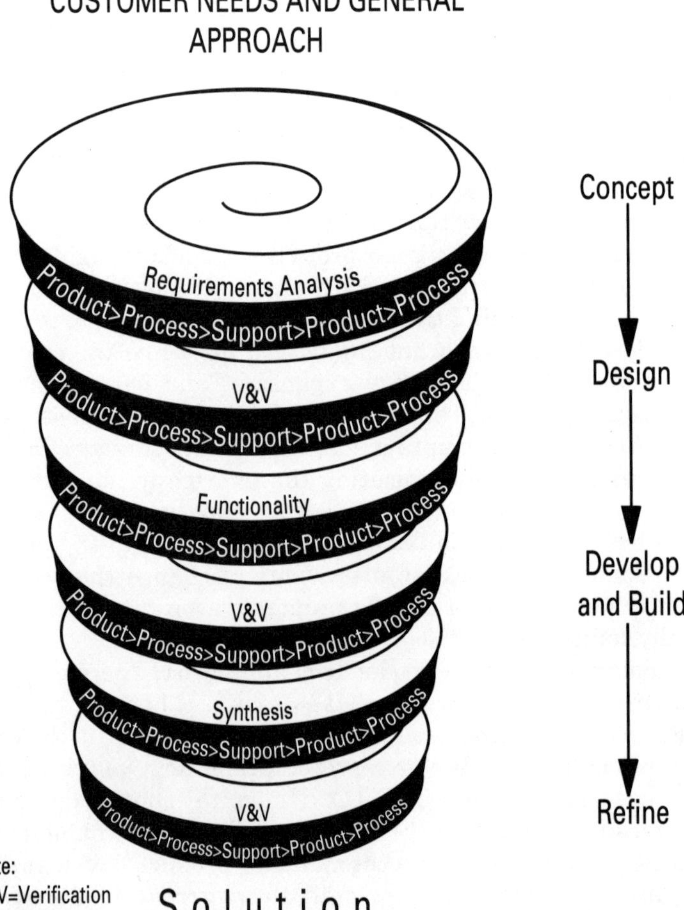

Figure 2.11 The spiral model of the contracting process. (*From Ref. 61, with permission.*)

and to account for in-process learning. "Cushions" should be allowed in schedules and budgets, but these cushions must be managed by the team and not buried in every task so that milestones are not taken seriously. Risks pertaining to potential changes should be systematically analyzed and managed. Accelerated prototypes or demos of critical elements should be prepared to get feedback on the solution earlier in the project (e.g., "rapid prototyping" as used in software development). Of course, traditional change control and management procedures are also important.

Metrics must be correct. The old cliché, "You can't manage what you cannot measure," is directly applicable. Underlying performance-based contracting, the new internal quality systems, and the U.S. government's and large contractors' emphasis on improving U.S. industry is the emphasis on metrics. Hybert points out that so many of the current metrics and rewards are driving the wrong performance. Last in the set of Hybert's recommendations is that metrics be focused on customer satisfaction and quality. Measurement is a key part of the quality discipline. Project metrics can be determined up front if the project is geared toward a desired business result for the customer, such as reduced cost, more productivity, or improved quality. The people involved early in the sale of the solution are often involved in these types of issues, but the people doing its postsale implementation are not. This results in project metrics based on schedule and cost plan versus actual. At its worst, this can result in a project manager completing a project on time at the planned margin but without it meeting the customer's needs.

To the customer, the most important consideration is the suitability of the end product over the long term. The customer must have a solution that serves its operational purposes and a system that provides the anticipated payback, or the project will be deemed a poor business decision. The right metrics need to be identified and made visible to the team, the customer, and the contractor.

Metrics for project outputs are typically assessed during acceptance testing. Metrics on the contracting process itself such as customer satisfaction, disruption of customer operations, and decision-making cycle are more often overlooked. Yet, these types of data could serve as an early warning system for potential downtime problems. The QFD format can be used to help link metrics to requirements. Metrics are addressed in many areas including the main discussion in Sec. 3.6.

A change in mind-set is needed, or, stated another way, a change in culture is needed. The recommendations described here are neither costly nor difficult to implement. An increasing number of contractors are already using QFD, advanced quality practices, and integrated product teams to define the product and service features and quality requirements. The DOD acquisition process is changing to become a model for U.S. industry cooperation according to the principles outlined in Chap. 3. Their new model includes a number of quality-related provisions (especially risk assessment and management provisions and provisions for evolving requirements to improve the effectiveness of its contracting process). Throughout the business world, experienced project managers are learning to create sound, well-integrated plans.

In changing mind-set, an issue is the trust among the customer, contractor, and suppliers. Since contracting has historically been adversarial, it is unrealistic to expect companies to trust each other without first establishing a relationship. This relationship can be created by:

- Paying the contractor for the end deliverable and the consulting work needed to effectively define the customer's needs up front
- Creating a mutually beneficial partnership between the contractor and customer
- Developing well-integrated project plans that are aligned with the business goals of the contractor, customer, and suppliers
- Developing project plans that minimize the pain of change and account for in-process learning
- Setting up project and process metrics that focus on customer satisfaction and quality

Although establishing trust and adapting quality principles to fit a contracting environment are not easy tasks, there are certainly rewards to those who invest in the effort. [61]

2.2.2 Detailing the performance specifications, characteristics, and checklist

It is important to understand what performance specifications are and how they are related to performance characteristics. Performance characteristics may be defined in the performance specification. Performance characteristics define what the product must do, and how it must act. SD-15, *Performance Specification Guide,* published by the Defense Standardization Program in June 1995, is a good guide for performance specification writing. Things that are important to the electronics world would be

- Dimensions: weight, height, and depth
- Power consumption
- Speed of operation
- Capability to store information
- Output: digital or analog
- Reliability, maintainability, and/or availability

In addition, a customer may want a system to do certain things, ranging from inventory management to complex calculations of speeds or requirements.

The new performance specification defines what is required versus how to perform the actions. It defines functional requirements, operating environment, interfaces, and interchangeability requirements. It states requirements in terms of results with criteria for verifying compliance, without stating methods for achieving results.

To make sure that you are pleasing the customer, define the specification values in the specifications. The qualified product-expectations specifications include both the target values (values that provide ideal customer satisfaction) and the customer's tolerances around the target values. More completely, the product expectations can specify the values that will give the customer complete satisfaction, and specify some measurement of degradation in customer satisfaction as the values actually achieved deviate from target values. To be realistic, the degradation should be related to cost. This clarifies to all how design and development is tied to cost, much as is the manufacturing process. This way the entire cost of the perfect quality product (in terms of customer desires) is defined. We can make a determination based on expectations and customer satisfaction. For example, if the customer satisfaction target is 5.2 volts, with a tolerance of ± 0.1 volt, we can then actually calculate the loss of customer satisfaction by stating the quality loss in terms of tolerance limits. [62] This gets us back to quality function deployment and the House of Quality. The House of Quality can specify customer expectations, set the values for each characteristic based on benchmarking, and then extrapolate for new developments, new products, and continuous improvement. (See more information on QFD and the House of Quality in Chap. 3.)

In the traditional process, the specification values were usually known, but often unrealistic, and were not compared with the competitive marketplace, or even past performance; even worse, the characteristics themselves were weakly defined relative to customer perceptions, and often there was a lack of consensus within the design team. The specifications should be responsive to customers, competitive, and commonly understood (agreed to by consensus).

The clear specifications provide the foundation for the remainder of the development work. [63]

Specifications sometimes are written with the intent to influence the customer's requirements. Successful contractors try to influence competitive-bid pressure by working up front with customers to position themselves as the supplier of choice. This can result in specifications that indicate not what the customers need but rather what the contractor wants to give them. Once customers discover that their needs are not being developed, customer dissatisfaction grows. It is not good for a supplier to have a reputation of always trying to "tell" customers what they need. This flies in the face of the old adage, "The customer is always right." So it is better to let the customer, or the market, define the requirements with the developer. [61]

It is important to understand some basic principles that apply to performance specifications and the detail specifications that may be written in response to the top-level performance specification.

Principles for contents of specification [64]:

1. If something is not specified, it will not be provided.
2. Every requirement increases price.
3. The shorter the specification, the less time it takes to prepare it.
4. The specification is equally binding on both purchaser and vendor.

Performance-based specifications specify the purpose of the work, the results required in the environment that the equipment is to be operated in, and the interface and interchangeability requirements. They also include the criteria to verify compliance to these "specified requirements." If specific functions, performance, and/or physical characteristics must be met, they must be defined in the performance specification.

From a performance specification that may be written by a customer or for a commercial business that is defining a market need for a new product, a company may write a performance-based specification, say, by the marketing organization to define a new or next-generation product. From that performance specification, basically a set of detailed specifications is written. Detailed specifications can cover everything from detail design, including separate specifications for software, fabrication, processes, and test. Even repair manuals fit into a specification environment if they define the process for repair of the product. Specifications are extremely important to detail the design, and proper documentation is important in an electronics system to ensure that it meets customer requirements to use, maintain, and repair.

Over- or underspecification can lead to non-value-added work. When inadequate tolerances for parts are established, when operational terms are not defined, when the exact detail and measurements for products and services are not specified, non-value-added work is created. Over- and underspecification brings into the system the variation that spawns waste and reduces quality and productivity. Even variation can be reduced by creating the specification without over- or underspecifying. [65]

For a good electronic system design, several layers of specifications need to be written. One of the first and most important is a specification that defines the product. Juran suggests a good outline. Here are the contents of a product specification according to *Juran's Handbook* [66]:

1. Applicable documents
   a. Internal: main drawing, packaging drawings, component specifications
   b. External: rules and standards: Underwriters Laboratories etc.

2. Product description
   a. General features
   b. Variants

3. Product provisions
   a. Function: general data, operating characteristics, acceptable noise level, reliability
   b. Materials and workmanship
   c. Grounding and insulation (safety requirements)
   d. Dimensions
   e. Finish, appearance, corrosion resistance
   f. Marking: contents and location of data plate

4. Manufacture
   a. Fabrication
   b. Painting
   c. Assembly

5. Shipping
   a. Packaging: requirements, tests
   b. Marking: contents and location of labels

6. Inspection

Commercial item descriptions, as defined by the U.S. government in DOD 4120, are simplified product descriptions managed by the General Services Administration (GSA) that define the functional and performance characteristics that will satisfy the Government needs. Since it has become so popular for the U.S. government to buy commercial items, this term has been used throughout the government and no longer describes only a GSA item. [67]

As part of the development of specifications, other items are important to defining an electronic product. An electronic system is a set of electronic devices that make up a product. A block diagram defines an electronic system and shows all the individual functions of the system and how the signals flow through the system. Schematic diagrams accompany the block diagram and show all of the individual parts of a circuit and how they are interconnected. Schematics are usually required for component-level troubleshooting of an electronic system.

From a product specification, detail specifications need to be developed for most electronic products and the testing thereof. Detail specifications can begin with components, define software requirements, and/or define fabrication or processes. For electronic components, specifications are essential to define the component, including testing and the process by which the component is built. Two specifications are extremely important to electronic assemblies. One is the specification to assemble and test as the product is built up, especially in critical processes, and the other is the test specification.

There are basically two kinds of specifications: performance and detail. Performance specifications state requirements in terms of required results with criteria for verifying compliance, but without stating the methods for achieving the required results. A performance specification defines the functional requirements for the item, the environment in which it must operate and interface, and interchangeability characteristics. Detail specifications are documents that specify design requirements such as materials to be used, how a requirement is to be achieved, or how an item is to be fabricated or constructed. A specification that contains both performance and detail requirements is considered a detail specification.

Specific types of specifications are

1. *System specifications* define the system requirements and then detail design.
2. *Item specifications* define items that may be either critical or key items to the design. This category would include the product specification.
3. *Software specifications* define software requirements and design.
4. *Material specifications* apply to raw materials, mixtures, or semifabricated material.
5. *Process specifications* define processes. [68] Process specifications cover manufacturing techniques and special processes. For more information on process specifications see "Process control" in Sec. 4.2.

Two more aspects of specifications need to be addressed: writing them and determining what they should contain. First we cover some basics in writing a specification (see Table 2.11). To prepare a specification takes a team—many different specialties must be involved, such as engineering, production, quality assurance, and test, just to name a few. In the new ISO 9001, para. 4.4.3, as part of the design control section, specifically addresses the organizational and technical interfaces that shall be assigned to the design team. [69] A team leader should be assigned who has the technical and managerial ability to lead the team to complete the development. A budget and a schedule must be known and/or developed and agreed to by manage-

TABLE 2.11 Writing a Specification

1. Accept the assignment: know budget, schedule, and technical needs, personnel needs
2. Define the project—ground rules, scope, specification types, format, procedures for estimating cost
3. Gather information, design parameters, features, and constraints (including statutory)
4. Choose a writing technique (software, groupware); establish configuration management technique
5. Prepare an outline
6. Write
7. Check
8. Secure appropriate approvals
9. Coordinate with team
10. Release

Source: Adapted from Ref. 64 with permission.

ment. The project must be defined with ground rules. Security of a new product in the commercial world can be just as serious as security with the U.S. government. Setting ground rules early must be done in the current environment: what are the requirements (commercial, military), what specifications are to be used, and where will the items be produced seem like nits in the beginning; but when detailed specifications are put together, inattention to those very details holds up progress, and they often are not easily resolved. A team must be assigned and must accept the assignments, the budget, and the schedule. After that, the details of requirements and the solutions of the enterprise putting the specification together are developed by the next level down.

A good resource library needs to be set up to gather information and catalog it properly. A configuration management system must be set up to ensure that all team members are writing to the latest revision, and that back-up copies are being made. Clear assignments of responsibility and authority, along with a plan for review and resolution of issues, must be developed up front.

A template for contents of a specification is in Table 2.12. [64] It includes almost all things that a company might have to do, whether commercial or defense, and can be tailored for the specific operation under consideration. Generally, one company does not manufacture an entire electronic system, but designs and manufactures some of it and has other suppliers for different segments. It is important that the prime contractor flow down the same performance standards to the subcontractor and work closely for customer needs. Again, it is good for the suppliers to form a team to work together for the benefit of the customers. If subcontractors are doing any part of electronic

TABLE 2.12 Template for Specification Development

Specification contents	
1. Introduction	
Name/identifying number	
Name of purchaser/customer	
Short description of project	
Origin - where came from, who, etc.	
Functions/Operational scenarios	
Applications - part of another system interface	
Action required - preparation of proposal, preparation of specifications, block diagrams, schematics, detail drawings, manufacture of prototypes.	
2. Scope-cover these items	
Customer specification or market survey	
Environment to operate in	
Type of specfications and standards required	
List of equipment	
List of services	
Power requirements/constraints	
Documentation	
Furnish, install, delivery	
Spare parts	
Safety	
Tests	
Customer support/repair/field support	
3. Work by others	
Interfaces with vendors	
Role of all functions on team	
4. Delivery Program/Cost	
What is required?	
Is there a target price?	

5. Specification, Standards, Codes	
Customer specification/market survey-requirements	
Industry standards/specifications	
Statutory codes, regulations	
Other requirement documents	
6. Functional requirements	
Customer or internal performance requirements	
Interfaces	
Interchangeability	
Environmental factors	
Acceptable tolerances	
Requirenments to operate, store and repair	
Constraints to deliver, install, put on the market	
Other functional requirements	
7. Technical requirements	
Design	
Tolerances	
Operational market scenarios	
Construction/fabrication features/constraints	
Reliability, maintainability, availability	
Special/other requirements/issues/constraints	
Performance	
State of the art	
New or used	
Size, weight, volume constraints	
Unacceptable modes of failure	
8. Documentation	
Specifications	
Block digrams	
Schematics	
Drawings	
Design instructions	
Operating manuals	
Maintenance manuals	
Code certiciates	
Quality control records	
Engineer's certicates	
Progress reports	
Coordination with customer	

TABLE 2.12 Template for Specification Development (*Continued*)

9. Materials	
Construction materials	
Paints and coatings	
Lubricants/fluids/adhesives	
10. Test and Inspection	
Required tests and inspections	
Qualification tests - see test for options	
Conformance tests - for every acceptance	
Scheduling	
Witnessing	
Documentation	
Customer supplied equipment	
Customer inspections/witnessing	
Life/stress testing	
11. Safety requirements	
System	
Statutory federal, state and local	
Personnel	
12. Quality Assurace Requirements	
Quality program	
Quality control	
Parts control	
Product quality requirements	
13. Support	
Spare parts	
Repair capability	
Field support	
Servicing manual	
14. Service by Vendor	
Erection supervision	
Startup and test supervision	
Technical support	
Training	
Obtaining permits	
15. Services by Purchaser	
Electricity	
Water	
Other	

16. Project Procedures	
Correspondence	
Document approval	
Change to contract	
Data bank	
17. Schedule	
18. Shipment and Packaging	
Location for passage of title	
Packaging requirements	
19. Warranties, Bonuses, Penalties	
Schedule	
Performance	
20. Instruction to Bidders	
Time/Place of response	
Information to be submitted	
Questions, answers, addendas	
Scheduled award date	
Bidder's meeting	
Site visit	
21. Evaluation Criteria	
Best value	
Price	
Payment schedule	
Discount for future payments	
Escalation	
Schedule	
Qualifications	
Efficiency	
Reliability	
Local manufacture	
22. Attachments	
Drawings	
Data sheets	
Subcontractor/vendor information	
Standards specifications	
Block digrams	
Schematics	
23. Other Information	
In program office	
Models	
Site visits	
Bidders meeting	

Source: Adapted from Ref. 64 with permission.

design, it is important that they complete or prepare the documentation about their part of the electronic project.

2.2.3 The new approach to test and evaluation—impact on quality assurance

Testing and evaluation is the proof of the pudding. If the customer provides good requirements, the product is made traceable, and the requirements are mechanized, the testing and validation process proves an efficient process which shows the customer that the supplier has met the requirements. Test and evaluation is not a one-time event. It should be done in every phase of the design/development process to verify the design. The U.S. government and large companies may ask for a test and demonstration plan or data about how the product performance was tested and demonstrated. A good set of tests to evaluate a design is found in Table 2.13. [69*a*]

Each company should have established test processes in accordance with ANSI/ASQC Q9001-1994, para. 4.10, which states "The supplier shall establish and maintain documented procedures for inspection and testing activities in order to verify that the specified requirements for the product are met." [70]

TABLE 2.13 Summary of Tests Used to Evaluate a Design

Type of test	Purpose
Performance	Determine ability of product to meet basic performance requirements
Environmental	Evaluate ability of product to withstand defined environmental levels; determine internal environments generated by product operations; verify environmental levels specified
Stress	Determine levels of stress that a product can withstand in order to determine the safety margin inherent in the design; determine modes of failure that are not associated with time
Reliability	Determine product reliability and compare to requirements; monitor for trends
Maintainability	Determine time required to make repairs and compare to requirements
Life	Determine wear-out time for a product and associated failure modes
Pilot run	Determine if fabrication and assembly processes are capable of meeting design requirements; determine if reliability will be degraded

SOURCE: From Ref. 69*a*, p. 269; used with permission.

Companies should start with industry standards, but then tailor them if they do not fit the business of the company. The standards found in the Electronic Industries Association (EIA) catalog or the Institute for Interconnecting and Packaging Electronic Components (IPC) catalog are a good place for the electronic industry to start. The company must know the business and write the test processes around it. Some companies deliver multiple kinds of product and therefore have more than one set of testing standards. Testing can be different for the unique high-reliability system, the midlevel or industrial system, and commercial equipment. Chapter 3 outlines the three different quality systems used for these purposes; the test processes would follow along with the quality system selected.

The most difficult part of creating a test plan is determining what to test. We must decide not only whether to sample or to 100% test and in what order to test characteristics, but we must also decide whether to test each characteristic once or several times, and whether to check for intermittent failures or test over a period of time to measure reliability. It is almost impossible to completely characterize all the failure modes, but it is important that we understand the failure modes and characteristics of components and systems.

For electronic components it is a bit more complicated. For each type of component there are distinctive characteristics and recommended tests.

Figure 2.12 gives an example of Takanashi's qualification tests for semiconductors at Toshiba. In Fig. 2.13, Takanashi maps the observed failure modes to failure mechanisms and corrective actions. The use of the historical review concept is recommended here to capture this information in rules for future design improvements.[70*a*]

Once the developmental tests are completed on components, the testing should be set up with standard requirements of the EIA, ISO 9001, the Qualified Manufacturer's List (QML), International Electrotechnical Commission Quality Assessment System (IECQ), or IPC.

Automated testing. Today much testing and inspection in component and system manufacture is automated. Automated test equipment (ATE) has the ability to perform a sequence of preprogrammed tests without operator intervention. Companies like Intel manufacture and test almost all their products by computer control. [71] ATE is usually controlled by stored computer programs, digital switches, program cards, and paper tapes. New ATE includes visual test capabilities, sound sensing, and test functionality integrated with robotics. This has changed the level of training that is necessary for people in the electronics testing environment.

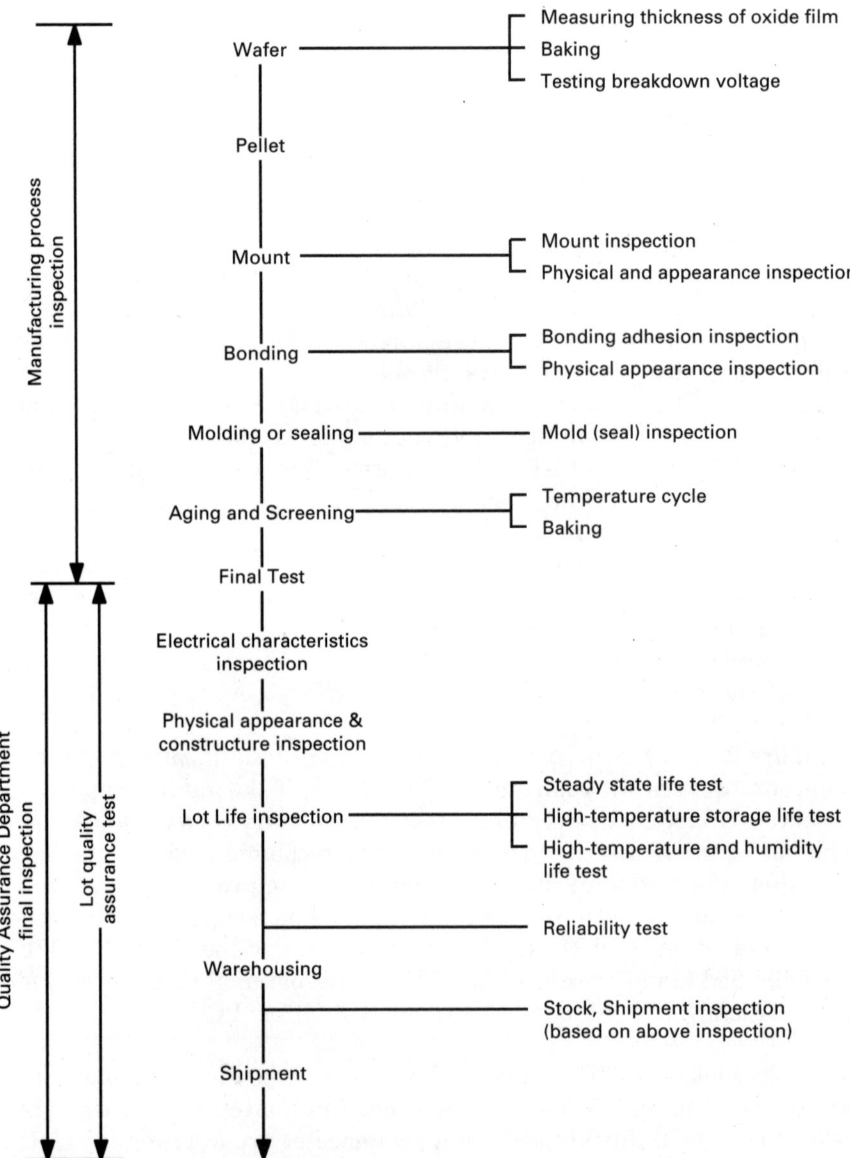

Figure 2.12 Toshiba's qualification test system (from Takanashi and Masahide, "Quality Assurance System for the Integrated Circuit," ILQC 78–Tokoyo. Japanese Union of Scientists and Engineers, Tokyo, pp. CI-SI to CI-56, 1978). (*From Ref. 70a p. 29. 12, with permission*)

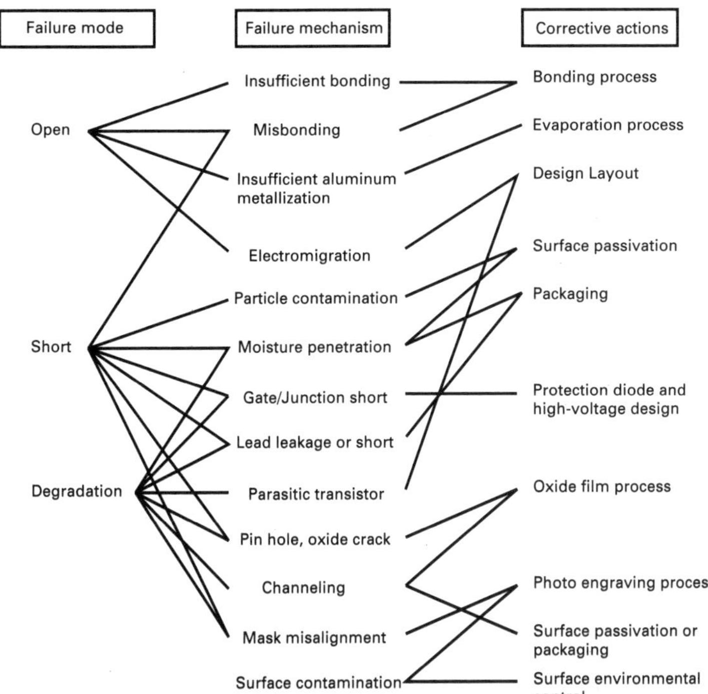

Figure 2.13 Mapping of failure modes to corrective actions (from Takanashi and Masahide, "Quality Assurance System for the Integrated Circuit," ILQC 78–Tokoyo. Japanese Union of Scientists and Engineers, Tokyo, pp. CI-SI to CI-56, 1978). (*From Ref. 70a p. 29. 12, with permission*)

Militarized items are items that are designed, manufactured, and tested to military specifications. Commercial items are those that are designed, manufactured, and tested to commercial, industrial, or company-unique specifications. Between those two extremes is a category called "ruggedized." *Ruggedized* generally refers to commercial equipment that is modified for military use. The modifications may be in the form of added parts, as a direct modification of COTS equipment. Often COTS equipment can be tested to ensure that it is ruggedized. The government must decide for each application if it needs fully militarized, ruggedized, or COTS products. This definition should be in the performance specification.

NDI testing may be different. Manufacturer-supplied data can be used, and testing that has been done can be acknowledged by the customer. However, NDI still requires verification of performance in its intended environments. Guidelines on the exact process follow.

From the performance specification, create a matrix to show how each requirement will be verified. Generally, they are verified by analysis, demonstration, inspection, and test. Some requirements are verified in the design/development process; others are verified each time the item is delivered. Sometimes these are referred to as qualification versus conformance tests. Qualification tests are one-time tests, whereas conformance tests are conducted each time the unit is shipped.

Qualification testing generally includes environmental testing (see Table 2.14), and performance verification testing. This includes vibration, stress, and specialized tests. It may include life testing to see what the life of the electronics is in a test system.

In a world class factory, environment stress screening (ESS) is a normal part of the process. ESS checks the product as it goes through the electronics factory. Sometimes the ESS process is called *burn-in.* Electronics is built and checked at various points for continuity and temperature through a series of in-process tests that include vibration and temperature cycling. If this process is used, when the product gets to final acceptance test, you know you have a good product. More data on burn-in may be found in this chapter under ISO Process Control, Sec. 4.2.9.

A good way to grade your test process would be to construct a matrix similar to that for fitness for use: verification testing, test proofing, and characteristics (see Table 2.15). In this way you find and improve weak processes.

2.3 The World Class Electronics Company

2.3.1 Overview

A world class electronics company is a world class manufacturing (WCM) company, according to the literature on today's shelves. WCM includes all types of manufacturing, but an extremely large subset is electronics. This section is not intended to give the reader all the information about WCM; its intent is to provide an overview of WCM as it relates to the electronics company. This is applicable in a chapter on the "New Definition of Quality," because the top-level tenets of a world class electronics manufacturing company are the embodiment of the new definition of quality. Specific references on quality systems for the new world are covered in Secs. 2.1 and 2.1.1. Here are highlights of what an electronics company must understand to get to world class.

Garvin, after finishing his study of multiple air-conditioning plants, summarized that all the plants together conveyed an important message: superior quality is associated with well-defined management practices and is not simply a supportive corporate culture. Design,

TABLE 2.14 Examples of Environmental Tests

Acceleration	Power inputs
Altitude Fluids susceptibility	Emission of radio-frequency energy (radiated and conducted)
Humidity	Induced signal susceptibility
Icing	Lightning-induced transient susceptibility
Leakage (immersion)	Lightning direct transient effects
Rain (waterproofness)	Magnetic effects
Salt fog	Pressure—ambient
Solar radiation Temperature (high and low)	Radio-frequency susceptibility (radiated and conducted)
Temperature (ambient)	Voltage spike
Temperature and altitude	Bonding
Temperature shock	Conducted emissions
Temperature variation	Conducted susceptibility
Bench handling	Radiated emissions
Shock (half-sine, peak, trapezoidal)	Radiation susceptibility
Crash safety (impulse) Hammer blow	High-intensity radiated field (10 kHz–40 GHz at 2000 V/m or greater)
Transit drop	Electromagnetic pulse
Vibration (gunfire, random, sine, sine on random)	Acoustic noises
Contaminants	Fungus resistance
Climate	Pyrotechnic shock
Audio-frequency conducted susceptibility	Sand and dust

purchasing, and manufacturing activities all play a role, but they must be accomplished by the right policies and attitudes. The plant's perspective on quality is intimately related to its overall quality performance. [72]

What is world class manufacturing? It has been defined as leading in the business of manufacturing, globally [73]; as improving performance and getting the competitive edge [74]; as total customer satisfaction, superior product, values and availability [75]; and as continual and rapid improvement in manufacturing [76]. It is all of these things and more. In order for an electronics business to be successful in the twenty-first century, it will have to approach world class because of the competition, the number of capable entrepreneurs, and

TABLE 2.15 Fitness-for-Use Matrix for Verification Testing, Test Proofing, and Characterization Efforts

Attribute	Rating 1 (20%)	Rating 2 (40%)	Rating 3 (60%)	Rating 4 (80%)	Rating 5 (100%)
Adequacy of plan	No formal plan exists	Handwritten plan exists but very incomplete	Test plan only 60% complete	Test plan is ready for release but a few sections are incomplete	Test plan is released through documented control and all sections are included
Thoroughness	Less than 10% of the features identified	Less than 40% of the features identified	Less than 60% of the features identified	Less than 80% of the features identified	All 100% of the features listed in the data sheet are identified in the plan
Effectiveness of best cases	No formal test cases written	Test cases only verify normal operation of features identified	Test cases verify normal, full range of operation on features identified	Test cases verify normal, full range, and worst case of features identified	Test cases verify the features completely as well as in many combinations of the parameters
Timeliness	Plan is conceived at the last minute before beta shipment	Plan developed after product design is complete	Plan developed after middle of product design stage	Detail plan was developed at the same time as the design was started	Plan developed at the beginning of product cycle development
Conformance to the plan	None	Less than 40% of the plan is actually done	Less than 60% of the plan is actually done	Less than 80% of the plan is actually done	All the test cases in the test plan were actually run

the demand for quality that customers want. In order for the electronics enterprise to achieve a world class status, it will need to be focused and trained. It will need to know the tenets of what has been proved to create world class and then to work according to those tenets.

WCM performance to a U.S. company today means that the following quantum leaps need to be made relative to today's averages:

Quality—increase by a factor of 100 to 1000

Costs—decrease by 30 to 50%

Inventory turns—increase work-in-process (WIP) turn by a factor of 4 to 10; increase overall turns by a factor of 3 to 5

Productivity—increase by a factor of 2 to 4

Order-to-delivery lead times—reduce by a factor of 5 to 10

New product and process design lead times—decrease by a factor of 30 to 60% [77]

2.3.2 Schonberger's view of world class manufacturing

The term *world class manufacturing* is synonymous with Richard Schonberger. In 1980, Schonberger, who had studied Japanese manufacturing and WCM as a process, defined *world class factory.* His view was that the Japanese had better quality and much tighter controls on waste. The message he learned and then told was that Japanese success is not culture-based. Its basis is quite a different set of concepts, principles, policies, and techniques for managing and operating a manufacturing enterprise. All of it is easy to understand and accept, teach, and apply. He detailed these manufacturing techniques, under the title *Japanese Manufacturing Techniques: Nine Lessons in Simplicity,* in 1982. In 1986, a follow-up book, *World Class Manufacturing: The Lessons of Simplicity Applied,* outlined how Americans could use the lessons of simplicity that he had previously described.

Schonberger defined WCM as equivalent to continual and rapid improvement. [78] From a new quality point of view, this is perfect. The task is to set up the means for continuous improvement that allows rapid improvement. The problem is that it is the nature of the current management style to achieve something, and then *not* do rapid improvement. The culture needs to advance more toward rapid improvement to break the paradigm.

According to Schonberger, the improvement journey follows a well-defined path. It requires clearing away obstacles so that product can be simplified. All of the lessons of simplicity can and do apply to electronics companies in the business of manufacturing.

At a summary level, these are the main ideas that Schonberger presented in his books, with special emphasis for the electronic company that wishes to become world class: Overall, the Japanese develop specialized manufacturing competency in a more narrowly focused area. They resist vertical integration, or expanding into manufacture of parts previously supplied by other plants. Local initiative and pride are keynotes, in an organization not in a huge vertical chain—one of fewer than 1000 employees, all of whom can have clearly identifiable work groups. [79]

Key ideas are

1. *Faster, higher, stronger.* This Olympic goal was changed to the WCM equivalent of continual and rapid improvement as the first goal of WCM. Do it, judge it, measure it, diagnose it, fix it on the factory floor. Produce immediate reports. The subject of variability reduction was written about as the second universal goal. "Variability is a universal enemy," Schonberger wrote in 1986, and he said the agreement on this was spreading. Schonberger suggests "papering the wall with charts showing the measured results"—and he is convinced this will lead the company to world class status. [80]

2. *Empower the people.* This key means getting the people involved, charting their work, and keeping results. The western world tends to go to overspecialization and therefore limits flexibility. In the Japanese system, if one worker is having difficulty, others will assist. This gives flexibility to the production line. More-flexible employees can expose problems across a line in concert with TQM. The flexible workers will be able to adapt more quickly, as the electronic system changes models. This is group technology, in which one worker handles several tasks. The emphasis is on flexibility, not stability as in the western companies. [81] Less reporting is required under this scheme. Again, this would save time and money.

3. *Focus on production.* Japanese experience shows that operators on the floor have impact on about 15 percent of all the problems; the rest is up to staff. Schonberger feels this takes the importance away from the operator on the floor and is counter to world class operations. The solution is to put staff people on the floor, supporting the line. This also creates more involvement with industrial engineering, purchasing, manufacturing engineering, and design engineering. It means better support and fewer people for maintenance, accounting, quality assurance, production control, materials management,

and data processing—all linked to just in time (JIT). This is where Schonberger's famous "count it, move it, store it, expedite it, search for it, taking from one container to another, and inspection in" was originally written with the question, "What adds value?" [82]

Parts count must be reduced. Why? Large numbers cause more stock locations, more suppliers, more buyers, just plain higher costs. More parts variability comes with more parts, therefore creating more scrap, rework, tests, technicians, production engineers, higher costs, and lower quality. More parts require more process steps, more setups, and more handling, and therefore cause more defects, more rework, and again higher cost and lower quality. Finally, more parts mean more parts to fail and therefore more scrap, problems to solve, rework, service and repair costs, tests, engineers, and—once more—higher costs and lower quality. [83]

4. *Economy of multiples.* In a world class supplier, matching capacity growth to demand is a key concept of capital investment. It is not good to spend huge amounts of money on "huge" machines; better to buy capacity in small increments as demand grows. Two lessons are cited:

- *More than one team, cell, line, or machine is better than one.* Two teams and sets of equipment making the same product or product facility are in friendly competition for results when things are going well. They back each other up so that sales need not be lost when things are not going well.
- *Add fixed capacity the way we add people: in small increments as demand grows.* The lesson applies not only to single machine types and cells but also to whole product flow lines, manual or automated, or even whole factories.

As sales of an item rise, its unit costs drop. This is especially true in electronics. Today, product life cycles are being depressed and movability of equipment is even more important if we intend to stay a WCM because we must achieve continual and rapid improvement. There must be a plan for moving old equipment out and replacing it with newer, flexible, more movable equipment. When measured against WCM standards, a 2 or 3 percent annual improvement is reported. In 1986, for his book, Schonberger's companies cut their lead times at least fivefold, typically in a year or two.

5. *Responsibility centers.* In 1982, Schonberger wrote, that just-in-time (JIT) manufacturing is ideal, with all materials in active use as elements of work in process, never at rest collecting carrying charges. It is "stockless production." JIT leads to significantly higher quality

and productivity and provides visibility for results so that work responsibility and commitment are improved. The combination of JIT and total quality control (TQC) marshals a rate of productivity and quality improvement that demoralizes foreign competitors or Japanese manufacturers. See Table 2.16 for benefits of the JIT approach.

JIT and TQC offer three ways to run responsibility centers:

a. Overlap production. Overlapping production is normal for JIT; a job order can be spread out so it can be in several stages of manufacturing at the same time. This means having many flexible people.

b. Slow down for problems. A companion to overlapping production is slowing down or stopping production if the user down-

TABLE 2.16 Just-in-Time Benefits

Cost of parts
Low inventory-carrying cost
Decreasing cost of parts, because of long-term learning in use at suppliers
Low scrap cost, since defects are detected early
Quality
Fast detection of defects, since deliveries are frequent
Fast correction of defects, since supplier setups are frequent and lots are small
Less need for inspection (of lots), since process control is encouraged
Higher quality of parts purchased—and of products they go into
Design
Fast response to engineering changes
Design innovativeness, since suppliers are expert and not hamstrung by restrictive specifications
Administrative efficiency
Fewer requests for bids
Few suppliers to contract with
Contracts negotiated infrequently
Minimal release paperwork
Little expediting
Short travel and telephone distances and costs
Simple accounting for parts received, if suppliers use standard containers
Reliable identification of incoming orders, if suppliers use the container labeling
Productivity
Reduced rework
Reduced inspection
Reduced delay because of off-spec parts, late deliveries, other underages
Reduced purchasing, production control, inventory control, supervision, with more reliable parts provisioning and quantities carried

SOURCE: From Ref: 85, p. 161, with permission.

stream is having other work problems. This is an excellent practice for electronics so that bad or problem product is not produced.

c. *Make only what is used.* Again, electronics companies should produce only what they know is selling. The change is too rapid. The Japanese tend to buy from the same few suppliers year after year, so that the suppliers develop a competency that is particularly attuned to the delivery and quality needs of the buying firm. Confidence in the supplier reduces buffer inventories, and delivery frequency from the suppliers is sometimes more than once a day. Some suppliers achieve quality levels high enough to bypass receiving inspection. [84] Make the supplier like family; Remember, in the Japanese environment the entire plant is like a community. It is good to help the suppliers cut costs because it helps the end product. Schonberger recommends frequent personal contact, as this bonds the business relationship. With the JIT principle, material is stored at the point of manufacture. This avoids handling and damage. And it is best to not manufacture before needed; in this model, the supplier will learn the lesson because overmanufacturing results in higher storage costs; and if (as it does in electronics), the product changes, the supplier "gets stuck" with overproduced goods. You will remember that the world class model likes things close, so worldwide scouring is not a good idea, unless absolutely necessary under this approach. If this is impossible, the worldwide suppliers need to be limited to one or two. [85]

There are discrete differences between Western and Japanese production lines, and those differences offer good lessons for an electronics firm. A different emphasis is presented in Table 2.17. [86]

Schonberger recommends the following tips:

- Try to keep plenty of work ahead of each work center so that the operator can stay there and keep busy.
- Use estimated lead times to control the flow of jobs on the floor. Use some sort of kanban, or material requirements planning. (These are production control techniques outside of the scope of this book.)
- Organize a cell to do manufacturing so that products have to move a shorter distance.
- Have dedicated lines for high-volume product or a long-term contract, or more than one line making the same product (or capable of making it). One line can help another. [87]

TABLE 2.17 Production Lines: Western versus Japanese

Western	Japanese
1. Top priority: line balance	Top priority: flexibility
2. Strategy: stability—long production runs	Strategy: flexibility—rebalance often to match changing demands
3. Assume fixed labor assignments	Flexible labor: move to where needed
4. Use inventory buffers to cushion effects of equipment failure	Employ maximal preventive maintenance to keep equipment from breaking down
5. Need sophisticated analysis to evaluate and cull the options (i.e., computer analysis)	Need human ingenuity to provide flexibility and ways around bottlenecks
6. Planned by staff	Foreman may lead design effort and will adjust plan as needed
7. Plan to run at fixed rate; send quality problems off line	Slow for quality problems; step up when quality is right
8. Linear or L-shaped lines	U-shaped or parallel lines
9. Conveyorized material movement is desirable	Put stations close together and avoid conveyors
10. Buy "supermachines" and keep them busy	Make (or buy) small machines; add more copies as needed
11. Applied in labor-intensive final assembly	Applied even to capital-intensive subassembly and fabrication work
12. Run mixed models where labor content is similar from model to model	Strive for mixed-model production, even in subassembly and fabrication

SOURCE: Reference 86; used with permission.

6. *Design leverage.* Design teams must have a customer focus and must be integrated with the entire organization. Today we call them integrated product teams (IPTs). With IPTs, problems disappear. Concurrent engineering works to reduce part counts and design electronics with modularity and open systems architecture to keep up with the rapidly changing electronics. A marketing-design-manufacturing team, with assistance from purchasing for suppliers, can achieve a fast design and get a winning market to distribution. This has been proven over and over again in the 1990s. The issue of required specifications that Schonberger talks about is being fixed in the late 1990s with the elimination of military specifications and the emphasis on heeding only the customer's requirements. The electronics supplier must look at the company's capabilities and ensure that the capability and capacity is there before accepting a product to

design and/or produce. Design is iterative; check a design aspect with the customer, check with the maker, redesign, check, check, redesign (see Fig. 2.4, Sec. 2.1.1). In 1986, Schonberger suggested that computer-aided design speeds up the process and helps everyone. Forward-looking electronics companies in the late 1990s have computer-aided design and manufacturing tied to a database with proven technologies and parts. [88] This area needs more emphasis in the electronics company that does not have a firm plan to move to more simple systems or equipment that serves world class operations.

7. *Managing the transformation.* Again the key word is simplicity: According to Schonberger there is no need to hire many consultants, no need to wait for elaborate studies. The emphasis should be on quick, visible results: WCM means not only continual improvement but rapid improvement: fast reductions in scrap and rework, elimination of racks full of materials, lowered buying costs. WCM is not about buying expensive equipment/plants. People should be emphasized and offered personal excitement, fulfillment, and rejuvenation. No one is left out. The main risk in many organizations is the resistance to change. This is still going on in the late 1990s, with the change to world class and people empowerment.

WCM prime pursuits should be remembered and kept simple: total quality control, just-in-time, total preventive maintenance, and employee involvement. Customer-oriented measures on cost, lead time, quality, and flexibility should be cultivated. Consulting companies should be hired to train employees in this model. [89] Schonberger's overall strategy may be found in Table 2.18. [90]

8. *Training: the catalyst.* We must remember that the goal is continual and rapid improvement. And we must train for implementation. Decisions must be made on who to train and how to train quickly. Training can be in the departments, or formal, or through professional societies; there is no one way. Western industry must put substantially more resources into training to match the prodigious sums that have been invested in WCM companies in Japan and Germany. Training is the foundation of implementation. Training is everybody's business. [91]

Schonberger sums up the application of simplicity techniques by emphasizing that the goal in WCM is not to employ bottom-up or top-down management; rather, it employs blended management. WCM management is not merely arranging resources in order to produce goods and services. It is marshaling resources for continual and rapid improvement. [92] Do not expect WCM to succeed all at one time. It starts fast, then reaches a plateau, and must be continually addressed (see Fig. 2.14). [93]

TABLE 2.18 Schonberger *World Class Manufacturing* (1986) Strategy Revealed

Manufacturing Excellence According to Schonberger
1. Get to know the customer. 2. Cut work in process. 3. Cut flow times. 4. Cut setup and changeover time. 5. Cut flow distance and space. 6. Increase delivery frequency for each required item. 7. Cut number of suppliers down to a few good ones. 8. Cut number of part numbers 9. Make it easy to manufacture the product without error. 10. Arrange the workplace to eliminate search time. 11. Cross-train for mastery of more than one job. 12. Record and retain production, quality, and problem data at the workplace. 13. Assure that line people get the first crack at problem solving before the staff experts. 14. Maintain and improve existing equipment and human work before thinking about new equipment. 15. Look for simple, cheap, movable equipment. 16. Seek to have plural instead of singular workstations, machines, cells, and lines for each product. 17. Automate incrementally, when process variability cannot otherwise be reduced.

SOURCE: Summarized from Ref. 86, with permission.

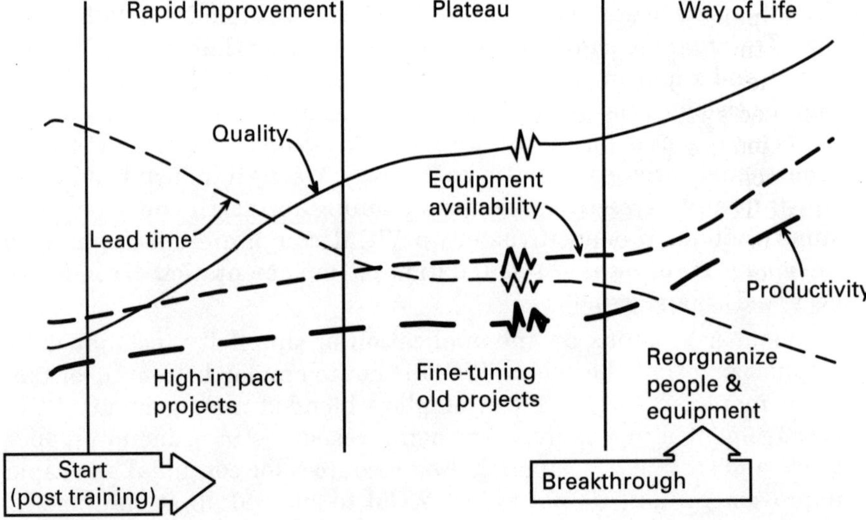

Figure 2.14 Pattern of improvement for world class manufacturing. (*From Ref. 93; used with permission.*)

TABLE 2.19 World Class Management Modes

Mode	Assessment
Management by edict	Inconsistent, wasteful of talent, and out of touch
Management by procedures	More consistent and quicker but wasteful of talent; filled with gaps that force-fit poor solutions, adversely affecting customers
Management by policies	Reflects high-level wisdom but limits broad employee empowerment and organizational learning
Management by principles	Customer-focused, employee-driven, data-based; broadly effective, robust, enduring

SOURCE: Schonberger, 1996, p. 20; used with permission.

Schonberger's most recent book, *World Class Manufacturing: The Next Decade* (The Free Press, New York, 1996) is quite sophisticated. It assumes the reader already understands what he wrote in his previous books, and that is why it is important to detail them here. His new book is summarized as follows:

Schonberger advocates managing by principles: customer-focused, employee driven, data-based, broadly effective, robust, enduring. He sees that twenty-first century management has moved through the phases shown in Table 2.19. He recommends that standard operating procedures make management more systematic.

Companies that follow managing by principles make a long-term commitment to what is commonly known as *total quality management*. This means viewing every part of the business as a process that can be analyzed and systematically improved. The problem is setting up data collection systems and processes so that a company can move ahead.

Principles are outlined in Table 2.20 so that a company can grade itself. This exhibit is quite applicable to electronics companies. Schonberger's matrix is scored tough, so that it can help firms get to the twenty-first century. See the scores in Table 2.21.

Principles are as follows:

General. Team up with customers; organize by families of customers or products (what customers buy/use). Capture and apply customer, competitive, and best practices information. Dedicate your company to continual, rapid improvement in quality, response time, flexibility, and value. Ensure that front-line employees are involved in change and strategic planning—to achieve a unified purpose.

TABLE 2.20a Toward Management by Principles: Five-Step Assessment Tool

Principles of Customer-Focused, Employee-Driven, Data-Based Performance

	General				Design		Operations	Human resources
	1	2	3	4	5	6	7	8
Step	Team up with customers; organize by customer/product family	Capture/use customer, competitive, best-practice information	Continual, rapid improvement in what customers need	Frontliners involved in change and strategic planning	Cut to the few best components, operations, and suppliers	Cut flow time and distance, start-up/change over times	Operate close to customers/rate of use or demand	Continually train everybody for their new roles
5	Customer/client representatives for each focused unit	Broad implementation of better-than-best practices for customer service	Sustained yearly QSFV* improvement rates of 50% or more in all key processes	Frontline teams help develop strategies and set numeric goals, self-monitored	Average reductions of 90% for all products and services	Cross-functional teams achieve 90% average reductions	Entire flow path for key items synchronized to rate of use or demand	80% certified multi-skilled; most also certified trainers
4	Entire enterprise reengineered by customer/product families	All associates involved in customer/competitive/best-practice assessment	95% improvement in QSFV in most key processes	Frontline teams plan/implement cross-functionally with other teams	Average reductions of 80% for all products and services	Experts help achieve 50% average reductions	80% of flow path synchronized to rate of use/demand for key items	50% of associates certified as multi-skilled; most also certified trainers
3	Focused work-flow teams (cells) for key product/customer families	Systematic customer surveys; full-scale benchmarking for key processes	90% improvement in QSFV in most key processes	Frontline teams continuously plan and implement process improvements	Average reductions of 50% for all items	Associates achieve 50% average reductions across all processes	50% of flow path synchronized to rate of use/demand for key items	25% of associates certified as multi-skilled
2	Customer/client representatives on project teams	Gather customer-needs and best-practice data, and non-competitive metrics	80% improvement in QSFV in a key process	Frontline teams assist in planning and implementing changes in own processes	50% fewer parts/operations and suppliers for all key items	In key processes, associates cut get-ready/set-up, flow time and distance 50%	Final process synchronized to rate of use/demand—all key products or services	40 hours of just-in-time (train-do, etc.) training for all associates
1	Cross-functional project teams	Gather customer satisfaction data and competitive samples and metrics	50% improvement in QSFV in a key process	Frontline associates assist in planning changes in own jobs	50% fewer parts/service operations or suppliers for a key product or service	Train associates in readiness, setup/change over, queue limitation	Final process synchronized to rate of use/demand for a key product or service	Key managers and teams receive overview training on process improvements

SOURCE: Reprinted from Schonberger, 1996, with permission.

TABLE 2.20a Toward Management by Principles: Five-Step Assessment Tool (*Continued*)

Principles of Customer-Focused, Employee-Driven, Data-Based Performance

	Human resources	Quality and process improvement	Information for Operations and Control			Capacity		Promotion/marketing
	9	10	11	12	13	14	15	16
Step	Expand variety of rewards, recognition, and pay	Continually reduce variation and mishaps	Frontline teams record and own process data at workplace	Control root causes to cut internal transactions and reporting	Align performance measures with customer wants	Improve present capacity before new equipment and automation	Seek simple, flexible, movable, low-cost equipment in multiples	Promote/market/sell every improvement
5	Profit/gain sharing; stock/stock options	2.0 Cpk; defects below 10 ppm; rework and lateness cut 99%	25+ mostly team/associate suggestions, mostly implemented by associates	Internal transactions cut 99%; 99% of external transactions by fax/EDI[†]	Second-order metrics (e.g., labor productivity, variances) no longer managed	Operators become technicians; downtime cut 80%	90% of equipment owned by focused teams/cells or is highly flexible/movable	Reserve marketing; out of strength, you choose whom to sell to
4	Pay for skills/knowledge; team/unit bonuses (no piecework)	1.33 Cpk; defects below 100 parts per million; rework and lateness cut 95%	10+ mostly team/associate suggestions, mostly implemented by associates	Internal transactions cut 75%; 75% of external transactions by fax/EDI	QSFV are dominant metrics in all processes	Experts teach operators to do repairs; downtime cut 50%	60% of equipment owned by focused teams/cells or is highly flexible/movable	Global/national awards (e.g., Baldrige); over 90% customer retention
3	Investing in employees via training, cross-training, cross-careering	1.0 capability (C_{pk})[‡] for key processes; rework, defects and lateness cut 50%	2 or more suggestions per associate per year	Internal transactions cut 50%; 50% of external transactions by fax/EDI	QSFV are dominant metrics in key support departments	Experts help operators take over their own PM and housekeeping	30% of equipment "owned" by focused teams/cells or is highly flexible/movable	Registrations, certifications, local awards (ISO-9000, Ford Q1, state award)
2	Variety of low-cost/no-cost awards to both teams and individuals	Gather customer needs and best-practice data and noncompetitive metrics	Frontline teams use process analysis, plot trends	Work-flow quality, internal scheduling, and labor transactions cut 25%	QSFV are dominant metrics in key operations	Preautomation (short flow paths, exact placement, housekeeping, etc.)	10% of equipment "owned" by focused teams/cells or is highly flexible/movable	Positive QSFV trends featured in selling, bids, proposals, ads
1	Systematic public recognition/celebration of achievements	Training in and use of "7 basic tools" of statistical process control	Training in measurement, visual management, problem-solving teams	Training in fail-safing, process simplification, root cause control	Training in universal customer wants; QSFV	Training in total preventive maintenance and process simplification	Seek/convert/upgrade marginal equipment to dedicated or high-flex uses	General advertising slogans ("Quality is Job One," "Team Xerox," etc.)

*Quality, speed, flexibility, value.
†Electronic data interchange.
‡Or equivalent.

TABLE 2.20a Toward Management by Principles: Five-Step Assessment Tool (*Continued*)

Scoring: Score one point for each step, for each of the 16 principles. Thus if our organization is at the fifth step for all 16 principles, the total score is 80, the maximum possible.

Assessment:
11–24 points—Eyes open, first steps, early learning
25–38 points—Childhood; trial and error
39–52 points—Adolescence; checklists and guidelines
53–66 points—Adulthood; policies
67–80 points—Maturity; principles

Design. Cut to the few best components, operations, and supplies.

Operations. Cut flow time, distance, and start-up and changeover times all along the chain. Operate close to customers' rates of use or demand.

Human resources. Continually enhance human resources through cross training, job and career-path rotation, and improvements in health, safety, and security. Expand the variety of rewards, recognition, pay, and celebration to match the expanded variety of employee contributions.

Quality assurance process improvement. Continually reduce variation and mishaps. Front-line teams record and own process data at the workplace.

Information for operations and control. Control root causes of cost and performance, thereby reducing internal transactions and reports and simplifying external communications. Align performance measures with universal customer wants: quality, speed, flexibility, and value.

Capacity. Improve present equipment and human work before considering new equipment and automation. Seek simple, flexible, movable, low-cost, readily available equipment and work facilities—in multiples, one for each product/customer family.

Promotion and marketing. Promote, market, and sell your organization's increasing capability and competence—every improvement that results from the application of the 15 principles.

Schonberger picked 127 companies to participate in the initial study of his maturity matrix. A subset of those, electrical and electronics suppliers, is shown in Table 2.21. Electronics companies scored the highest, although the average score was not good. To assist your company, fill out the form in Table 2.22 for today, and again one year from now, and mail it to Schonberger at the address on the table so he can use it for research. If your company falls low, it would be good for you to benchmark with some of the companies in Table 2.21. [94]

TABLE 2.21 ***World Class Manufacturing: The Next Decade*** **Electronics Companies Surveyed**

Sector	Average score	Participants
Electronics (25 participants)	38.1	Alcatel, Richardson, TX—Telecommunication equipment
		Aritech Europe, Dusermond, Netherlands—Security equipment
		ATL, Bothell, WA—Medical ultrasound equipment controls, sensors, recorders
		Digital Systems Intl., Redmond, WA—Call management equipment
		Dover Elevator Systems, Walnut, MS—Elevator controllers
		Eaton Corp., Everett, WA—Optical control devices
		Fluke Corp., Everett, WA—Electronic test instruments
		Ford Electronics, Markham, Ontario—Automotive electronics
		Hewlett-Packard, San Jose, CA—Optocouplers, fiber optics, etc.
		Intel, Fab 6, Chandler, AZ—Semiconductors, microcontrollers
		Linfinity Microelectronics, Garden Grove, CA—Integrated circuits
		Northern Telecom, Calgary, Alberta—Telephones
		Philips Component Group, Juarez, Mexico—Transformers
		Philips Consumer Electronics, El Paso/Juarez—TVs, transformers, etc.
		Philips Modular, Plant 4, Juarez, Mexico—Remote control, PC boards
		Philips Plant 5, TV plant, Juarez, Mexico—TVs
		Physio Control, Redmond, WA—Defibrillators
		Rhomberg Brasler Manufacturing, Capetown—Industrial electronics
		Sentrol Inc., Hickory, NC—Security equipment
		Solectron Washington—Electronic manufacturing services
		Telemecanique-Ireland, Cophridge, Ireland—Contractors
		Tri-Tronics, Inc., Tucson, AZ—Electronic dog-training equipment
		United Electric Controls, Watertown, MA—Temperature & pressure
		United Technologies Automotive, Juarez, Mexico—Wiring harnesses
		Vectron Labs, Norwalk, CT—Crystal oscillators
Electrical (10 participants)	37.3	Baldor Electric, Ft. Smith, AR—Industrial electric motors
		Hill-Rom, Batesville, IN—Medical headwall systems
		Honeywell Home & Building Control, Golden Valley, MN—Heating/ventilation/air-conditioning controls
		Honeywell, Motherwell, Scotland—Heating, cooling, microswitch
		Levical (Leviton), Tecate, Mexico—Wall receptacles, plugs
		Mine Safety Appliance, Safety Products Division, Murraysville, PA—Safety equipment
		Reliance Electric, Flowery Branch, GA—Electrical motors
		Varian, Palo Alto—Nuclear magnetic resonance instruments

SOURCE: Reprinted from Schonberger, 1996, p. 61, with permission.

TABLE 2.22 Toward Management Principles

Five-Step Assessment Tool

Principles of Customer-Focused, Employee-Driven, Data-Based Performance

Steps	1	2	3	4	5	6	7	8	9	10	11	12	13	14	15	16
5																
4																
3																
2																
1																
0																

Assessment (one point for each step, each principle):
11-24 points—Eyes open, first steps, early learning
25-38 points—Childhood: trial and error
39-52 points—Adolescence: checklists and guidelines
53-66 points—Adulthood: policies
66-80 points—Maturity: principles

Total points:__________ Date:__________

Name of company, business unit or plant:__________

Address: __________

Number of employees: __________

Product line: __________

Names and titles of persons doing this assessment: __________

Remarks: __________

Please send to:
Richard J. Schonberger
Schonberger & Associates, Inc.
P.O. Box 66984
Seattle, WA 98166
Phone: 206-433-8066

SOURCE: Reprinted from Schonberger, 1996, with permission.

2.3.3 Competitions in quality: the Baldrige, Deming, and *Industry Week programs*

There are three good programs to get involved in to make your electronics firm a world class electronics company. Even if you do not wish to compete, you can use the entrance data to do internal benchmarks and keep track of improvement. The entry forms are available for a small fee. The Baldrige Award, known for quality improvement, is detailed in Chap. 9. The Deming Prize is detailed in Chap. 5. Details on the *Industry Week* "America's Best Plants" competition follow.

Industry Week. Since 1990, *Industry Week* has been identifying "America's Best Plants" to recognize plants that are on the leading edge in North America to increase competitiveness, enhance customer satisfaction, and create stimulating and rewarding work environments. This could be another definition of a world class company. The publication's goal is to encourage North American firms to adopt world class practices, technologies, and improvement strategies. [95]

Their data are some of the best that can be used for the electronics company to judge itself. For an overall look at criteria, see Table 2.23, a 3-year comparison of companies named to *Industry Week*'s best. Much like the Baldrige Award practice, *Industry Week* prints a statistical profile of the current year and guidelines for the next year in question with a candidate questionnaire. Since 1991, *Industry Week* has printed a statistical profile of the best plants. This serves as industry data for benchmarking excellent, world class companies. Table 2.24*a* and *b* is the input data for *Industry Week,* for your company to use to judge itself. Tables 2.25*a, b,* and *c* shows quality indicators, and Table 2.26*a, b, c,* and *d* show input sheets for quality data. This kind of collection of data is exactly the method for American industry, electronics included, to work with to achieve world class status.

The 1995 finalists were selected from 178 plants. The 25 finalist plants ranged in size from 130 to 2700 employees. In six of the plants (24 percent), workers were represented by unions. Four plants began operations in 1990 or later. Two are primarily devoted to serving defense-related markets. [96]

In 1995, *Industry Week* winners all had emphasized the following:

1. Customer focus
2. Goal setting and metrics
3. Compensation
4. Work teams
5. Quality and productivity

TABLE 2.23 ***Industry Week*** **1995 Statistical Profile of America's Best Plants**

Comprehensiveness of Effort

Three-year comparisons (percent of finalists indicating area is one of major emphasis)

Area	1993	1994	1995
Total Quality Management	96	96	92
Management by policy (breakthroughs)	20	44	52
Cycle-time reduction	92	96	96
JIT/continuous-flow production	96	96	100
Cellular manufacturing	72	80	92
Focused-factory production concepts	84	88	84
Supplier partnerships	96	96	96
Customer-satisfaction programs	96	96	100
EDI links to customers/suppliers	72	80	88
Competitive benchmarking	76	68	96
Continuous improvement	100	100	100
Visibility systems	—	68	—
Employee empowerment	100	100	100
Accelerated worker training	80	68	68
Apprenticeship programs	40	44	56
Use of work teams	100	100	96
Employee cross-training	96	100	100
Use of cross-functional teams	96	100	96
Performance measurement and reward systems	84	100	92
Total Productive Maintenance (TPM)	64	72	68
Advanced process technology	84	84	76
Advanced material technologies	56	72	48
Inventory reduction	96	100	92
Concurrent engineering	92	92	84
Design for assembly/manufacturability	84	96	88
Design for quality	88	88	92
MRP II system enhancement	72	72	52
Delivery dependability	96	100	100
Reducing order-to-shipment leadtime	88	100	96

TABLE 2.23b ***Industry Week*** **1995 Statistical Profile of America's Best Plants (*Continued*)**

Comprehensiveness of Effort

Three-year comparisons (percent of finalists indicating area is one of major emphasis)

Area	1993	1994	1995
Flexible manufacturing methods	88	92	84
Agile manufacturing strategies	—	80	88
Streamlined production flow	88	80	92
Computer-integrated manufacturing (CIM)	76	88	80
Enterprise integration*	64	60	68
Improving union/management cooperation	40	40	20

*Described in 1993 as "cross-functional computer integration."

SOURCE: (Reprinted from *Industry Week,* "Best Plants, 1995," © Penton Publishing Inc., Cleveland, Ohio, with permission.

Based on data from the 25 finalists in *Industry Week's* 1995 search for America's Best Plants.

The 1995 finalists were selected by the *Industry Week* panel of judges from a field of 178 plants nominated for consideration. The 25 plants range in size from 130 employees to 2700 employees. In six of the plants (24%), workers are represented by unions. Sixteen plants are operated by publicly held companies. Four plants began operations in 1990 or later. Two are primarily devoted to serving defense-related markets.

6. Technology (information systems and new technology)
7. Agility—rapid change/adaptation/flexibility
8. Training and education
9. Supply chain—supplier performance (see criteria list in Chap. 7)

Criteria for the 1996 award were:

- A comprehensive effort to achieve world class manufacturing capability
- Strong quality systems and results, including evidence of low product defect rates and good process control capability
- Extensive employee involvement and empowerment programs, especially directed to create high-performance self-directed work teams
- Appropriate use of technology, as required by changing business needs
- Flexible and/or "agile" production systems capable of responding quickly to customer needs and shifts in the marketplace

TABLE 2.24*a* Comprehensiveness of Effort

Please check all activities which have been an area of major emphasis at this facility. (Use NA to indicate not applicable.)

- ___ Total Quality Management
- ___ Management by policy (MBP)
- ___ Cycle-time reduction
- ___ JIT/continuous-flow production
- ___ Cellular manufacturing
- ___ Focused-factory production concepts
- ___ Supplier partnership programs
- ___ Customer-satisfaction programs
- ___ EDI links to customers and/or suppliers
- ___ Competitive benchmarking
- ___ Continuous improvement
- ___ Hoshin/breakthrough strategies
- ___ Employee empowerment
- ___ Accelerated worker training
- ___ Apprenticeship programs
- ___ Use of work teams
- ___ Use of cross-functional teams
- ___ Employee cross-training
- ___ Performance measurement and reward systems
- ___ Visibility systems (visual management)
- ___ Predictive/preventive maintenance
- ___ Total Productive Maintenance (TPM)
- ___ Advanced process technology
- ___ Advanced material technologies
- ___ Inventory reduction
- ___ Concurrent engineering
- ___ Design for assembly or manufacturability
- ___ Design for quality
- ___ Design for procurement
- ___ MRP II system enhancement
- ___ Enterprise resource planning (ERP) systems
- ___ Delivery dependability
- ___ Reducing order-to-shipment lead times
- ___ Flexible manufacturing methods
- ___ Agile-manufacturing strategies
- ___ Streamlined production flow
- ___ Computer-integrated manufacturing (CIM)
- ___ Enterprise integration
- ___ Improving union/management cooperation

SOURCE: Reprinted from *Industry Week*, "Best Plants, 1995," © Penton Publishing, Inc., Cleveland, Ohio.

TABLE 2.24*b* Measurement

In gauging plant performance improvements, what does management measure current performance against?
___ Past plant performance
___ Performance of "best practices" companies
___ The ultimate possible performance
Number of benchmarking studies conducted in last 3 years:[fru5] How did plant obtain benchmarking data? Explain:
If possible, cite benchmarking data comparing recent plant performance with that of other leading operations in the same industry:

SOURCE: Reprinted from *Industry Week*, "Best Plants, 1995," © Penton Publishing, Inc., Cleveland, Ohio.

TABLE 2.25*a* Quality Indicators

Values in Percent unless Otherwise Noted				
Products and components				
	1995 finalists			
	High	Low	Ave.	Median
Finished-product first-pass yield	100.0	90.1	97.2	98.3
Finished-product yield improvement, 5 years†	90.0	0.0	31.6	16.0
Typical-finished-product yield for industry‡	98.9	41.0	86.8	95.0
Finished-product pass-through yield	99.6	75.0	93.6	95.9
Major component first-pass yield	99.9	88.3	97.2	98.0
Component yield improvement, 5 years†	94.0	0.0	31.9	19.7
Typical component yield for industry‡	99.6	65.0	92.7	95.0
Component process capability (C_{pk})§	6.42	1.17	2.47	1.90

*In calculating yield, 23 of the 1995 plants deduct for items requiring rework in the immediate upstream stage of the process. One plant considered itself to be a producer of components, and thus did not report for this category. A second plant did not calculate quality based on this indicator.

†In 1995 "percent" yield improvement was defined to mean percent reduction in rejects.

‡Not all finalists reporting product and component first-pass yields reported industry averages.

§1995 was the first year that entries were asked to report C_{pk} values for manufacturing processes. Twenty plants reported that they calculate C_{pk} to gauge process capabilities. (For definition of C_{pk} see Sec. 3.73.)

SOURCE: Reprinted from *Industry Week*, "Best Plants, 1995," © Penton Publishing, Inc., Cleveland, Ohio.

TABLE 2.25*b* Quality Indicators

Values in Percent unless Otherwise Noted				
Across all processes				
	1995 finalists			
	High	Low	Ave.	Median
Average C_{pk} value across all manufacturing processes	4.09	1.08	1.96	1.60
First-pass yield for all finished products (weighted average)	99.9	76.0	95.7	97.9
Total scrap/rework as a percent of sales	9.0	0.0	1.50	0.8
Scrap/rework reduction in 5 years	91.2	0.0	51.3	50.0
Customer reject rate on finished products, ppm*	16,000	0	3082.5	492
Reject-rate reduction in last 5 years	91.0	35.0	64.6	63.6
Reduction in warranty costs within last 5 years	84.0	6.0	41.9	29.5
Number of benchmarking studies conducted in last 3 years	35	0	10.8	7

*ppm = parts per million.

SOURCE: Reprinted from *Industry Week*, "Best Plants, 1995," © Penton Publishing, Inc., Cleveland, Ohio.

- Management practices geared to achieving breakthroughs in operating performance and customer satisfaction as well as cultivating continuous improvement
- Improvements in manufacturing operations, including shortening of manufacturing cycle times and reduction of work-in-process inventories
- Sound environmental and workplace-safety practices
- Meaningful community involvement and support programs
- A strong customer focus, including formal customer satisfaction programs, customer involvement in product design, employee contact with customers, and efforts to reduce customer lead times
- Effective supplier-partnership programs, including efforts to solicit supplier evaluations of plant practices
- Significant productivity improvement and/or world-class productivity levels
- Reduction of manufacturing costs
- Marketplace competitiveness, such as increasing share of market

TABLE 2.25*c* Quality Indicators

Values in Percent unless Otherwise Noted			
Three-year comparison of medians (in percent unless otherwise noted)			
	1993	1994	1995
Finished-product first pass yield	98.7	98.9	98.3
Yield improvement, last 5 years*	25.0	10.0	16.0
Finished product pass-through yield	—	—	95.9
Major component pass-through yield	98.1	98.1	98.0
Component yield improvement*	20.0	7.5	19.7
Total scrap/rework as percent of sales	0.7	0.9	0.8
Scrap/rework reduction in 5 years	32.0	48.5	50.0
Finished product customer reject rate, ppm	184.0	250.0	492.0
Reject-rate reduction in 5 years	57.0	60.0	63.6
Warranty cost reduction in 5 years	27.2	50.0	29.5
Number of benchmarking studies in last 3 years	7	7	7

*Data report is somewhat inconsistent: Percentage point increase in yield was asked for in past years, but several plants reported percent reduction in defects, since that is how they track yield gains.

SOURCE: Reprinted from *Industry Week*, "Best Plants, 1995," © Penton Publishing, Inc., Cleveland, Ohio.

TABLE 2.26*a* Quality Achievements

Briefly describe quality initiatives and the most significant results:

Has plant sought ISO 9000 or QS 9000 certification? ___ Yes ___ No

Has plant received ISO 9000 or QS 9000 certification? ___ Yes ___ No

Does plant make extensive use of computerized SPC? ___ Yes ___ No

Does SPC provide real-time feedback on process variables? ___ Yes ___ No

Does plant calculate C_{pk} values to gage process capability? ___ Yes ___ No

Are C_{pk} measurements used extensively throughout the plant? ___ Yes ___ No

If no, how does plant gage process capability?

For definition of C_{pk}, see Sec. 3.7.3.

SOURCE: Reprinted from *Industry Week*, "Best Plants, 1995," © Penton Publishing, Inc., Cleveland, Ohio.

TABLE 2.26*b* Quality Indicators for a Typical Finished Product and Major Component

A. *Finished Product* ______________________

Current first-pass yield _______ %

Yield improvement within last 5 years _______ %

Typical yields for your industry _______ %

Pass-through yield (see definition in guidelines) _______ %

B. *Major Component* __________________________

Current first-pass yield _______ %

Yield improvement within last 5 years _______ %

Typical yield for your industry ________ %

C_{pk} value for this component-manufacturing process _______

SOURCE: Reprinted from *Industry Week*, "Best Plants, 1995," © Penton Publishing, Inc., Cleveland, Ohio.

TABLE 2.26*c* Quality Indicators for All Products (Average)

First-pass yield for all finished products (use weighted average)* _______ %

Average C_{pk} value—or comparable measure—across all processes where C_{pk} measurements are applicable (weighted average)* _______ %

If measurement is other than C_{pk} value, explain:

Percent of production (units shipped) represented by new products introduced within the last 12 months _______ %

Scrap/rework as a percent of sales _______ %

Percent reduction in scrap/rework as a percent of sales in last 5 years _______ %

Customer reject rate on finished products (ppm) _______ %

Percent reduction in customer reject rate in last 5 years _______ %

Warranty costs as a percent of sales _______ %

Percent reduction in warranty costs as a percent of sales within last 5 years _______%

*Weighted average should take into account differences in product volumes or value-added.

SOURCE: Reprinted from *Industry Week*, "Best Plants, 1995," © Penton Publishing, Inc., Cleveland, Ohio.

TABLE 2.26*d* Quality Methods

Which of the following quality methods are used extensively?

___ Manual SPC	___ Design of experiments (e.g. Taguchi methods)
___ Computerized SPC	___ Poka-yoke (fail-safing) methods
___ Operators inspect own work	___ Design for quality
___ Quality function deployment	___ Quality circles
___ Total Quality Management	___ Employee problem-solving teams

Other:

Comment:

SOURCE: Reprinted from *Industry Week*, "Best Plants, 1995," © Penton Publishing, Inc., Cleveland, Ohio.

For further information on *Industry Week's* Best Plants program contact *Industry Week,* in Cleveland, OH., 216–931–9344. [97]

All electronic manufacturing companies would benefit from doing a benchmarking exercise with their data or entering the *Industry Week* competition.

2.3.4 Implementing a world class electronics company

World class manufacturing status is achieved not by doing one thing well, but by investment and application of several interrelated breakthrough manufacturing technologies. WCM technologies have been explained in terms of benefits to the entire organization.

The transition to WCM affects every part of the company, but chiefly the culture. The top executive or CEO must make this happen, much in the style recommended by Deming and being popularized now by ISO 9000. The journey to world class is neither smooth nor easy and takes time. If the company wishes to achieve world class status, the CEO must form a "quality council," by whatever name, to be responsible for [98]:

1. Leading the transition
2. Assuming and assigning responsibility for implementation
3. Committing the resources to make it happen

4. Providing the incentives for continued programs
5. Reviewing progress along the way and rewarding accomplishments.

A company strategy must be formed and made known to all the employees. Vision and mission statements must be generated along with a quality policy to keep the entire organization focused. Communication systems must be set up to keep the employees informed on the changes, and training must be planned and completed.

A. Richard Shores, in *Reegineering the Factory; a Primer for World Class Manufacturing,* suggests that good assessments be taken. He tells us that a general assessment of the company's current situation determines a benchmark from which one can measure future improvements. Assessment may cover:

Values. Do company values support a world class environment (customer, teamwork, etc.)?

Investments. How much capital is available for improvements? How much is budgeted for training?

Planning. Is the planning system pervasive across the company? Has the business defined the vision and key success metrics?

Motivation. Are appropriate reward structures in place to motivate WCM performance, i.e., quality, teamwork, continuous improvement?

Customer feedback. Are the company's customer satisfaction measures—i.e., quality, cost, and delivery—clearly defined?

Product design. Is data available and measured on the performance of individual products, that is, material cost, overhead cost, internal and external defects, and on-time delivery?

Organizational responsibilities. Does the team understand what each function's role is in implementing the WCM methods? Are specific skills available when/where needed? Who are internal customers? Are the performance measurements in place to assess results?

Teamwork. Does the organization have an appreciation for team performance? Are teams self-managed and rewarded for success?

Analysis systems. Does the company use a common analysis method like PDCA (Plan, Do, Check, Act)? Are statistical tools commonly used in analysis and decision making in all levels of the company?

It is good to do a survey of all employees for these items to see if management and employees see the business the same way. The rule of thumb, according to Shores, is that companies that have 25 to 30 employees can achieve significant results in less than a year. Larger organizations of 500 or more may take 4 to 5 years to see results. The other variables that affect time are dedication and commitment of resources, and the availability of technical expertise to resolve problems. [99]

2.3.5 Performance measures for a world class company

In order to become a world class manufacturer, the company must have measurements that show performance. In so many companies, old accounting systems continue to operate on mainframes and do not account for the methods that focus on the customer needs and desires, or work to develop low-cost manufacturing centers. It is not the purpose of this book on quality management to define all the methods required to produce performance measurements for a world class company; that has been defined in numerous books, and publications. The reason it is mentioned here is that, without a change in this area, the entire transition of the company to WCM will never truly happen. "In 1989, a full 50% of accountants admitted their cost accounting system was out of date, and 78% say the current cost systems frequently understate profits on high-volume products and overstate profits on specialty items...." [100]

According to Dixon and associates in 1990, there were three main problems with the traditional measurement systems:

- Inconsistency with the new emphasis on quality, just-in-time, and using manufacturing as a competitive weapon
- Outmoded performance measurement systems to compete in today's global marketplaces
- Measurement of current system exceeds the limits of both traditional cost accounting and the "new, improved" cost accounting system. [101]

It is well recognized that conventional accounting systems need to move to what is termed *activity-based costing* (ABC), defined as a method of measuring the causes, performance, and results of work. ABC differs from conventional standard costing in scope and method in that it assigns costs to activities on the basis of their use of resources, and assigns costs to products and customers on the basis of their level of activities. [102]

ABC cost systems allocate overhead in proportion to the activities associated with the product, or a product family. Thus if a product has many parts and operations that cost money, it will cost more overhead than the product with fewer parts and fewer operations. This approach promotes understanding of the whole cost of a value-added process. But to make this work through TQM and JIT processes, companies need to look at cutting parts in a system and the number of transactions to produce a product, along with all the other processes that Schonberger addresses; if that is not done, costs will not be saved. (Data in Ref. 103 was used to support this conclusion.[104])

What should companies do?

In his article entitled, "Activity-Based Costing: Driver for World Class Performance," Peter B. Turney explains an extension of ABC called *workforce activity-based management* (WABM), which applies ABC information to the process of continuous improvement. WABM focuses efforts to adapt business strategies to meet competitive pressure and to improve business operations. The chief benefit of the ABC and WABM systems is that the entire workforce is focused on achieving the company's goals. This focus includes improvements in profitability, in addition to quality and time. The WABM group process and the ABC information combine to enhance employee empowerment and build team skills.

The costs of ABC and WABM depend on the size of the organization, the nature of the implementation plans, and the level of development of the workforce. Internal costs include the time of managerial staff and workers, as well as information systems. External costs include training and software.

Successful WABM implementation requires the following:

1. *Leadership commitment.* Management must articulate how WABM contributes to the goals of the company, and provide resources for developing the system.
2. *Training.* Training is important for both management staff and workers. Increases the likelihood that improvements will be continuous.
3. *Wide availability of ABC information.* The availability of ABC information to the workforce plays an important role in continuous improvement by removing barriers to communication, serving as the sole source of information about cost, time, and quality, and demonstrating how everyone can contribute to meeting the organization's goals.
4. *A group process.* WABM uses a group process to develop and apply ABC information. Workers are divided into teams using storyboards and other visual information. The group process is effective for several reasons. First, the people who do the work have the

most knowledge about what they do. Since WABM is a bottom-up process, the work team develops and maintains the activity-based information, which instills a sense of ownership. Moreover, all employees participate and share in the problem-solving process. Finally, WABM allows everyone to function as his or her own accountant, eliminating the need for large numbers of staff to develop and maintain systems and changing the staff's primary role from that of technician to one of coach. [104]

2.3.6 A case study in world class manufacturing

An excellent example of an electronics product was provided by Alpha Industries and documented by the Supplier Research Group of Shrewsbury, MA.

Alpha Industries manufactures one of the most complex manufacturing products, a microwave semiconductor chip. Each semiconductor chip must meet precise mechanical and electrical requirements. A finished semiconductor chip is formed one layer at a time. Each layer is the result of a unique series of steps such as photolithography, metal, or etches. There may be as many as 300 different process steps to manufacture one semiconductor chip.

Utilizing resources of 29 people, an improvement program produced results that speak for themselves. Over a 12-month period, work in process (WIP) dropped from 1200 wafers to 600 wafers. However, production output remained constant. The average cycle time was reduced from 12 weeks to 6.25 weeks. The yield level increased. The internal customer is pleased. The lead times are predictable within two standard deviations, and the quality levels satisfy the internal customer's requirements.

When a complex product is manufactured, there is a high chance for process variation. A continuous improvement plan to reduce product variation is essential. In addition, a complicated set of process steps creates bottlenecks. Each bottleneck offers storage for WIP, a non-value-added product that hides product defects and variation.

Alpha found that cycle time reduction is an excellent tool for continuous improvement. Hidden quality problems are revealed. Leadership and direction are used to solve the quality issues.

After introduction of improvement teams, the wafer fabrication area of Alpha Industries had a thorough understanding of statistical techniques and problem solving. However, the missing ingredients were activated 2 years after the first improvement team started: leadership, employee involvement, and direction.

The wafer fabrication area at Alpha Industries focuses on two major areas. The first is documentation. The ability to characterize a stable process is a key to any success program. The second is decreasing cycle time and work in process, and forming a closed-loop corrective action system.

Alpha Industries semiconductor wafer fabrication decided to begin an aggressive, focused, continuous improvement program in August 1991. The first step to continuous improvement is to understand the customer needs. The customers were screaming for product that met the quoted delivery date and required specifications. The group's track record for fulfilling these requirements was marginal at best. The semiconductor product was a victim of poor scheduling, process variation, and lack of management direction. Scheduling, at its best, was just a glorified example of expediting. The demanding customer with the "loudest voice" received products in the shortest time frame. Urgent reactions to customer demands cause process errors, and subsequently, lack of time to correct the errors. Internal chaos became the norm. In a chaotic environment, there is seldom time available to properly handle scheduling and eliminate process problems.

In essence, the design of Alpha Industries' semiconductor manufacturing system was to produce late deliveries and poor product. Without a strong intervention from management, the system could never change.

In August 1991, the semiconductor chip manufacturing manager decided to become a leader in continuous improvement. Simply stated, the attitude became:

- Understand customer needs
- Involve all employees in satisfying customer needs
- Select key indices to track progress
- Provide tools for employees to succeed

The group decided to concentrate on world class manufacturing. Schonberger states the goal as "continual and rapid movement." [105] WCM employs the idea of cycle time reduction, to bring actual cycle time of product to ideal cycle time.

The definitions are:

Actual cycle time: The amount of time the product spends in the manufacturing line. This is the time from the start of manufacture of the product until it is finished.

Ideal cycle time: The amount of processing time for the product.

This is the time in which the product would be manufactured if WIP was not in the line.

To quote Schonberger, the goal of WCM is to eliminate any sources of waste. Any source of inventory is adding cost, not value. Examples include counting it, moving it, storing it, expediting it, searching for it (part or tool), taking it out of one container and putting it in another, accumulating it into larger make/move quantities, and inspecting it. [106]

The group looked at existing data. The number of semiconductor chip wafers which started in the manufacturing line was rising. The number of outputs from the line was unpredictable. The production line was producing at a 30 percent reject rate.

Employee involvement with *focused leadership is the only way to accomplish any improvement project.* In August 1991, three teams were formed to accomplish Alpha Industries' plan: Documentation, Cycle Time Reduction, and Yield Improvement.

Documentation team

Overview. The essence of quality system standards, such as ISO 9001, is to baseline and document each operation. Documented systems improve *training,* accommodate process changes, and eliminate differences in interpreting operations.

Charter. The documentation team must document each operation, its work instructions, and product characteristics. The documentation team must document each process instruction sheet for every product.

Data. The documentation team assessed the number of current, usable documents. Only 30 percent of the manufacturing operations were documented. None of the products had proper process instruction sheets. The team met weekly, discussing every document in the manufacturing system.

Successes. The documentation team has successfully baselined each manufacturing operation. Every product currently has a documented instruction sheet. Consolidated manufacturing operations are a product of cycle time reduction. The work instructions and the product instruction sheets may need updating as the operation changes; the documentation team meets weekly to assure that manufacturing operation changes are documented prior to implementation.

Highlights. The semiconductor manufacturing area does not have an active quality assurance department. Instead, the manufacturing operations are responsible for the quality of the products. Visual inspection is a key quality parameter. To assist the operators in making the correct assessment, the documentation team provides visual aids at all the workstations. Each visual aid book is a collection of photographs, representing acceptable product and unacceptable product. The visual aids books are *tools for the employees to succeed.*

Cycle Time Reduction team

Overview. Deliveries to the customer can be improved in one significant way: reduce the time it takes to manufacture the product. Cycle time reduction programs study the actual time it takes to manufacture a product and compare it to the production time. Areas of excess inventory, called *bottlenecks,* are identified and eliminated, Suzaki defines Cycle time as the time between the completion of the last product and completion of the next product, and production lead time as the time between beginning of production and completion of production.

$$\text{Cycle time} = \frac{\text{available production time in a day}}{\text{required production units in a day}}$$

Charter. The Cycle Time Reduction team must reduce the cycle time of every product. The team must be able to quote delivery times to customers. The quoted delivery times must be within 2 standard deviations of the actual manufacturing time.

Data. The team reviewed the existing cycle times. The average cycle time, across all product lines, was 12 weeks. The team detailed two critical categories, work in process and product rejects. To accurately understand the work in process, the team placed logs at every workstation. The operators logged in when the product entered the workstation and when the product left the workstation. The team was interested in identifying the workstations with the highest work in process and the longest cycle time.

The scheduling methods were also analyzed. The team discovered that 25 percent of all the production lots were labeled "hot lots." A hot lot is a production lot which is expedited. The hot lot does not sit at a workstation for extended periods of time. The hot lot can interrupt all production schedules.

Successes. The Cycle Time Reduction team successfully reduced work in process from 1200 wafer units to 600 wafer units. Even with the reduction of active production lots, the final output value increased. At the project's inception, the wafer fabrication process provided 170 wafer units to its customers. Reducing the number of wafer units by half, the final production output nearly doubled. The wafer units provided to the customers increased from 170 units to 300 units.

Since hot lots interrupted the production line, the team decided to eliminate hot lots. Hot lots were eliminated by setting realistic schedules for all product lines. The customers were given realistic and accurate schedule dates. With an accurate schedule, the production line ran smoother. Without hot lots the team was able to collect accurate work-in-process data.

Highlights. The team meets weekly and reviews the work-in-process data. During one meeting, the data indicated a large bottleneck in the glassing area. However, after a couple of attempts, the engineers were unable to reduce the cycle time in the glassing area. The team turned to their most knowledgeable resource, the manufacturing operators. Within a couple of weeks, the operators reduced the number of process steps from seven to two. Product quality indicated a corresponding improvement.

Yield Improvement team

Overview. Improving product performance is a key parameter to customer satisfaction. Understanding production defects is a means to improve product performance. The Yield Improvement team was designed to reduce and/or eliminate product defects.

Charter. Understand current production reject reasons. Initiate a closed-loop corrective action system. Utilize design of experiments to improve product performance. Improve production yields.

Data. To improve product yields, the Yield Improvement team first had to understand the reason product did not meet specification. Therefore, every time a product failed, the appropriate employees filled out a discrepancy report. The discrepancy report detailed the cause of failure and a possible corrective action. At specified times, the data collected from the discrepancy reports are analyzed. A Pareto analysis of the discrepancies is constructed. The team relies on design of experiments to eliminate the discrepancies.

Successes. Product quality and customer satisfaction level has improved. The number of rejected units has decreased.

Design of experiments is an effective tool to eliminate chronic problems. Careful planning of the experiment is critical. The complete execution of the experiment is imperative. The data need to be complete with a thorough analysis. See Sec. 3.7.2.

Highlights. The discrepancy reports highlight a problem at the saw workstation. A design of experiments was constructed. The design variables were selected by using inputs from engineers and operators. Seven variables were selected. A best-case and a worst-case setting was selected and run for each variable. This determined that the proper variables were selected. The worst case had a 65 percent chip yield. The best case had a 100 percent chip yield. A variable search (isolate one variable) was performed, and the critical variable was found. Today, the product has a 100 percent yield through the saw area.

Conclusion. Continuous improvement is an evolutionary process. It requires careful planning, proper implementation, building on the infrastructure for improvement, and training in improvement tools. Continuous improvement also requires leadership. Alpha Industries' semiconductor fabrication area is continuously learning. Every day, an operator, engineer, customer, or supplier provides a new idea, a new angle. However, one thing is constant: the desire to improve, grow, expand, and learn. The operators are now teaching the engineers how to run workstations. The engineers teach operators technical details. In a continuous improvement environment, the spark of enthusiasm is felt immediately. It is a contagious fire which burns within all involved. And it continuously grows. As cycle times reduce, product mix may change. A discrepancy is eliminated, a new one arrives. Predicting the next challenge *is impossible*; predicting the method to solve the challenge is unbroken. The methods of cycle time reduction and design of experiments are proven. They lead to higher customer satisfaction, increased employee involvement, and improved morale.

References

1. Deming, W. Edwards, *Out of the Crisis,* (Reprinted by permission of MIT and The W. Edwards Deming Institute. Published by MIT, Center for Advanced Educational Services, Cambridge, MA 02139. Copyright 1986 by The W. Edwards Deming Institute.) 1995, pp. 133–134.
2. Garvin, David A., *Managing Quality; The Strategic and Competitive Edge,* The Free Press, New York, 1988, p. xii.
3. Poireir, Charles C., and Steven J. Tokarz, *Avoiding the Pitfalls of Total Quality,* AQSC Press, Milwaukee, WI, 1996, p. 66.
4. Poireir and Tokarz, pp. 5–7.
5. Juran, J. M., and Frank M. Gyrna, *Quality Planning and Analysis,* 3d ed., McGraw-Hill, New York, 1993, pp. 6–7.
6. Morgan, Ronald B., and Jack E. Smith, "A New Era in Manufacturing and Service," *Quality Progress,* July 1993, p. 83.
7. "A History of Managing for Quality," *Quality Progress,* August 1995, p. 128.
8. Wilson, P. F., "A Darwinian Future is Looming…," *Quality Progress,* July 1996, p. 45.
9. Sullivan, L. P., "The Seven Stages in Company-Wide Quality Control," *Quality Progress,* May 1986, pp. 77–78.

10. DODI 5000.2, part 6, sec. P, February 23, 1991, p. 6-P-1.
11. Juran, J. M., "The Quality Function," in J. M. Juran, *Juran's Quality Control Handbook,* 4th ed., McGraw-Hill, New York, 1988, p. 2.8.
12. Garvin, pp. 40–41.
13. Garvin, pp. 49–60.
14. DODI 5000.2, part 6, sec. P, February 23, 1991, p. 6-P-1.
15. Gunn, Thomas G., *21st Century Manufacturing,* Omneo, Essex Junction, VT, 1992, p. 77.
16. Godfrey, A. Blanton, and Robert E. Kerwin, "Electronics Components Industries," in J. M. Juran, *Juran's Quality Control Handbook,* 4th ed., McGraw-Hill, New York, 1988, pp. 29.2–29.4.
17. *EIA, JEDEC and TIA Standards and Engineering Publications,* Global Engineering Documents, Englewood, CO, 1996, pp. 1-40, 1-41.
18. Institute of Electrical and Electronics Engineers, Inc. (IEEE), *Standards Project: Standard for Systems Engineering,* IEEE Standards Department, Piscataway, NJ, 1994, p. 47 (Reprinted from IEEE Draft Standard P1220/D1 dated December 1993, Copyright © 1993 by the Institute of Electrical and Electronics Engineers, Inc. The IEEE disclaims any responsibility or liability resulting from the placement and use in this publication. This is an unapproved draft of a proposed IEEE Standard, subject to change. Use of information contained in the unapproved draft is at your own risk. Information is reprinted with the permission of the IEEE.)
19. IEEE, p. 19.
20. ANSI/IPC-RB-276, *Qualification and Performance Specification for Rigid Printed Boards,* p. 13.
21. *IPC Resources,* Fall 1995 catalog, The Institute for Interconnecting and Packaging Electronic Circuits, p. 13.
22. Juran and Gyrna, p. 260
23. Garvin, p. 15.
24. Juran and Gyrna, p. 262.
25. Godfrey and Kerwin in Juran handbook, p. 29.15.
26. Godfrey and Kerwin in Juran handbook, p. 29.17.
27. Fuchs, Jerome H., *The Prentice Hall Illustrated Handbook of Advanced Manufacturing Methods,* Prentice Hall, Englewood Cliffs, NJ, 1988, pp. 336, 337.
28. IEEE, p. 2.
29. DODI 5000.2, part 6, sec. P, February 23, 1991, p. 6-P-1.
30. *Manufacturing Development Guide,* Aeronautical Systems Center, Wright-Patterson Air Force Base, OH, November 30, 1993, pp. 3-43 and 3-44.
31. Juran and Gyrna, p. 101.
32. Juran and Gyrna, p. 102.
33. DODI 5000.2, part 6, sec. P, February 23, 1991, p. 6-P-2.
34. Juran, J. M., "The Quality Function" and Ekings, J. Douglas, "Assembly Industries," in J. M. Juran, *Juran's Quality Control Handbook,* 4th ed., McGraw-Hall, New York, 1988, pp. 2.8 and 30.2.
35. Garvin, p. 98.
36. Garvin, p. 59.
37. ANSI/ASQC Q9001-1994, para. 4.4.8.
38. IEEE, p. 14.
39. Schuler, Charles A., *Electronics, Principles and Applications,* 3d ed., Glencoe, Macmillan/McGraw-Hill, New York, 1993, pp. 6, 7.
40. IEEE, p. 10.
41. Juran and Gyrna, p. 510.
42. Juran and Gyrna, p. 186.
43. Juran and Gyrna, p. 526.
44. Cringley, Robert, "Accidental Empires: How the Boys of Silicon Valley Make Their Millions, Battle Foreign Competition and Still Can't Get a Date," *Harper's Business,* New York, 1992, p. 4.
45. Office of Management and Budget (OMB) (U.S. Government), policy letter 91-2.

46. SD15, *Performance Specification Guide,* Office of the Assistant Secretary of Defense for Economic Security, Washington, DC, June 1995, p. 6.
47. DODI 5000.1, March 15, 1996, part 4, p. 5.
48. Deming, p. 23.
49. Juran and Gyrna, p. 243.
50. *The Best Value Approach to Selecting a Contract Source,* U.S. Army Material Command pamphlet, vol. 5, 16 August 1994, pp. 1, 2, 7, 20.
51. DODI 5000.2-R, March 15, 1996, part 3, p. 5.
52. DODI 5000.2-R, March 15, 1996, part 3, p. 5.
53. SD-2, *Buying NDI,* Office of the Assistant Secretary of Defense for Production and Logistics, 1990, pp. 6, 7.
54. Juran, J. M., *Juran's Quality Control Handbook,* 4th ed., McGraw-Hill, New York, 1988, pp. 16.7, 30.18, and 17.7, plus *Manufacturing Development Guide (MDC),* Aeronautical Systems Center, Air Force Material Command, Wright Patterson Air Force Base, OH, for "key products," p. 3-16.
55. Conway Quality, *The Right Way to Manage: Process Flow Charts and Work Improvement,* Nashua, New Hampshire p. 20.
56. Saylor, James H., *TQM Simplified,* 2d ed., McGraw-Hill, New York, 1996, p. 356.
57. Conway Quality, *The Right Way to Manage: Four Forms of Waste,* revised ed., Conway, 1989, pp. 42, 43.
58. Gunn, p. 65.
59. DODI 5000.1, March 15, 1996, p. 4; DODI 5000.2R, March 15, 1996, part 2, p. 7; part 4, p. 1.
60. ANSI/ASQC Q9001-1994, para. 4.4.
61. Hybert, P., "Five ways to improve the contracting process," *Quality Progress,* February, 1996, pp. 65–70.
62. Clausing, D., *Total Quality Development,* ASME Press, New York, 1994, p. 143.
63. Clausing, pp. 142–144.
64. Purdy, David C., *A Guide to Writing Successful Engineering Specifications,* McGraw-Hill, New York, 1991, pp. 92–94.
65. Conway Quality, *The Right Way to Manage: Four Forms of Waste,* p. 27.
66. Elings, J. Douglas, "Assembly Industries," in J. M. Juran, *Juran's Quality Control Handbook,* 4th ed., McGraw-Hill, New York, 1988, p. 30.11.
67. *Milspec Reform, Results of the First Two Years,* Office of the Under Secretary of Defense for Acquisition and Technology, June 1996, p. 4.
68. MIL-STD 961D, Notice 1, 22 August 1995, *DOD Standard Practices for Defense Specifications,* pp. 5, 7, 96–97.
69. ANSI/ASQC Q9001-1994, para. 4.4.3.
69a. Juran and Gyrna, p. 269.
70. ANSI/ASQC Q9001-1994, para. 4.10.1.
70a. Godfrey and Kerwin in Juran handbook, pp. 29.12 and 29.13.
71. Schmit, Julie, "Chat With the Chief," interview with Gordon Moore, Chairman of Intel, *USA Today,* November 8, 1996, p. 12B.
72. Garvin, D. A., *Managing Quality, the Strategic and Competitive Edge,* The Free Press, New York, 1988, p. 157.
73. Gunn, p. *x.*
74. Wallace, T. F., *World Class Manufacturing,* Omneo, Essex Junction, VT, 1994, p. *ix.*
75. Shores, A. Richard, *Reengineering the Factory,* ASQC Quality Press, Milwaukee, WI, 1994, p. *xii.*
76. Schonberger, R. J., *World Class Manufacturing: The Lessons of Simplicity Applied,* The Free Press, New York, 1986, p. 2.
77. Gunn, p. 11.
78. Schonberger, 1986, p. 2.
79. Schonberger, R. J., *Japanese Manufacturing Techniques: Nine Lessons in Simplicity,* The Free Press, New York, 1982, p. 173.
80. Schonberger, 1986, pp. 14–15.
81. Schonberger, 1982, p. 135.

82. Schonberger, 1986, pp. 39–48.
83. Shores, p. 33.
84. Schonberger, 1982, pp. 158–159.
85. Schonberger, 1986, pp. 155–171.
86. Schonberger, 1982, p. 133.
87. Schonberger, 1986, pp. 101–122.
88. Schonberger, 1986, pp. 144–154.
89. Schonberger, 1986, pp. 190–206.
90. Schonberger, 1986, pp. 217–218.
91. Schonberger, 1986, pp. 207–215.
92. Schonberger, 1986, pp. 216–228.
93. Schonberger, 1986, p. 225.
94. Schonberger, R. J., *World Class Manufacturing: The Next Decade,* The Free Press, New York, 1996, p. 61.
95. *Industry Week,* "America's Best Plants, Guidelines for the 1996 Selection Process," Penton Publishing, Inc., Cleveland, Ohio, p. 1.
96. *Industry Week,* "America's Best Plants: 1995 statistical profile, 1996 guidelines, and candidate questionnaire," Penton Publishing, Inc, Cleveland, Ohio, 1996.
97. *Industry Week,* Guidelines for 1996, pp. 1–2.
98. Shores, p. 80.
99. Shores, pp. 81–82.
100. Gunn, p. 103.
101. Dixon, J. R. et al., *The New Performance Challenge,* Business One Irwin, Homewood, IL, 1990, p. 1.
102. Turney, Peter B., "Activity-Based Costing: Driver for World Class Performance," in T. F. Wallace and S. F. Bennett (eds.), *World Class Manufacturing,* Omneo, Essex Junction, VT, 1994, p. 466.
103. Gunn, pp. 106–107.
104. Turney, pp. 461–466.
105. Schonberger, 1986, p. 2.
106. Schonberger, 1986, p. 48.

Chapter

3

New and Advanced Quality Systems for the New World

Marsha Ludwig-Becker, C. M.

Operations Project Manager
The Boeing Company
Anaheim, California

3.1 Introduction

On April 24, 1995, the U.S. Government and Industry Quality Liaison Panel (G&IQLP) issued the first document addressing overall quality in the United States, *Roadmap for Quality in the 21st Century: World Class Quality* (included in the appendix of this handbook). It outlines a vision to guide both government and industry in efforts to reach world class quality in the twenty-first century. It addresses quality systems, advanced quality concepts, and continuous improvement. In this chapter we will discuss these tenets as they relate to the electronics company.

Quality systems are now all based on a single quality standard, the ISO 9000 series. In addition, the vision is to embrace principles inherent in the Malcolm Baldrige National Quality Award. Companies are encouraged to have a quality system that meets the needs of the company and its customers to improve the quality and value of products and services, along with its international competitiveness. [1]

TABLE 3.1 Items for an Effective New Quality System

ISO 9000
Baldrige Award Criteria or Deming/*Industry Week* World Class
TQM and teams
Management and leadership
Empowering the workforce
Metrics

This, coupled with the management philosophies supporting the big "Q," forms the basis for a very effective quality system (Table 3.1). This is discussed in this chapter so that all companies can take advantage of these breakthroughs.

For electronics companies that wish to move ahead, and are not already practicing the advanced quality methods of quality function deployment (QFD), the Taguchi Method, and statistical process control (SPC), this chapter discusses process capability, variability reduction, and the 6-sigma concept at a management level, with sources for detailed action. These techniques will be vital to the electronics company that wants to become world class and survive in today's markets.

This is extremely important for all electronic system companies so they can gain market share and set up quality management systems to capture a bigger share in four markets: domestic, international, defense, and commercial.

3.2 One Quality System

One quality system was agreed to by the Government and Industry Quality Liaison Panel (G&IQLP), a group of government and industry representatives, formed in 1995 to promote a single quality system for the U.S. government and industry. Twelve federal agencies and multiple industry groups, including the Electronic Industries Association (EIA), have agreed to abolish contract-unique quality requirements in favor of single facility-wide quality management system based on the ISO 9000 series requirements. [2]

This is quite similar to how the Japanese formed a single quality system at a top government level to promote quality in Japan in the 1950s. In the United States, quality has been led by the American Society for Quality (ASQ), primarily a technical group of quality engineers and managers, and not chief executive officers. [3] With top-level government and industry endorsing a single U.S. policy, all the electronics industry benefits by U.S. market growth in the domestic and international world.

To meet their mission, the G&IQLP used the Malcolm Baldrige Guidelines as guiding principles, and ISO 9000 was agreed to as the foundation for the quality management standard. Total Quality Management and Total Quality Metrics had been previously agreed to as master tools. These four items make a focused quality system for the new world that can be based on the company's strategic direction and internal processes. *Roadmap for Quality in the 21st Century* is a milestone for the United States because it defines a standard that American industry can strive for to obtain world class quality.

A key element of this new world quality system is that it is management-based, with the kind of quality leadership that has been promoted for years, primarily by W. Edwards Deming. It also stresses that the customer is always first. These are guidelines for getting to the big "Q" endorsed by the leadership of the United States.

This vision for the twenty-first century allows contractors to decide what kind of management system they need for their products/services and their customers (see Fig. 3.1). This flexible system framework can be used by electronic manufacturers to guide their quality management activities and provide a strategy and unified direction for the future.

The G&IQLP document defines three broad quality management systems (Fig. 3.2) for all businesses to fit into. The systems are basic, advanced, or unique. Depending on a company's products and strategic goals, the company can select a quality system and move toward total implementation. The systems are defined as follows:

Basic quality management system. A defined, documented, and disciplined set of practices that focuses on assuring that the deliverable product (or service) conforms to the performance and config-

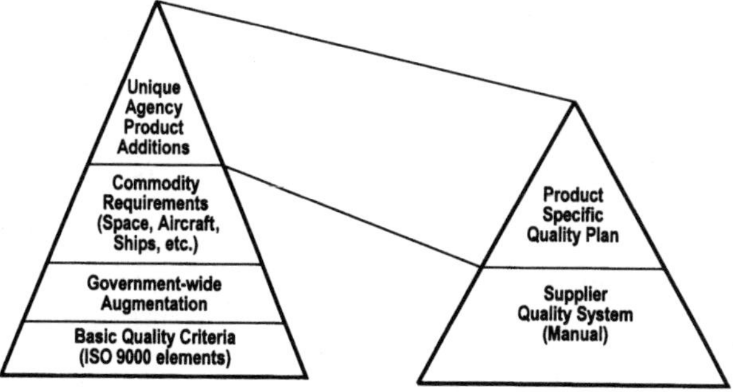

Figure 3.1 One quality system. *From Ref. 1.*

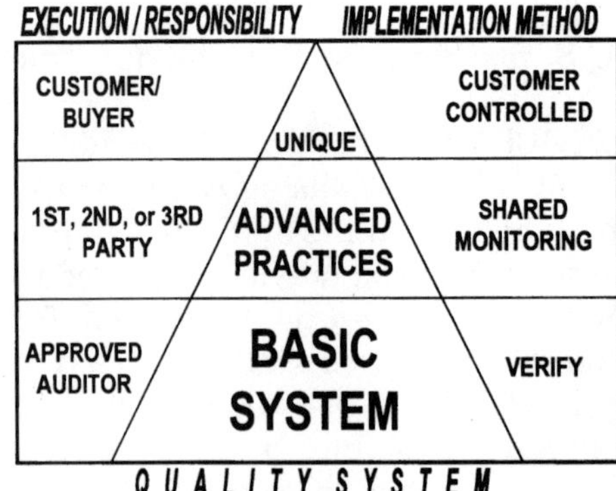

Figure 3.2 Quality system management framework. (*From Ref. 1.*)

uration criteria included in the agreement between buyer and seller. The basic quality management system must provide for the appropriate controls of part and product characteristics and attributes from product design through delivery and include inspection and test criteria/methodology/data used to verify/validate conformance. The key consideration is the delivery of products (or services) that fall within allowable tolerances.

Advanced quality management system. A management commitment to excellence that is manifested in a structured, long-term approach to the continuous improvement of all processes and products. The focus is on the processes that produce a product or service rather than just the product itself. The employees that actually perform the processes are trained to use quality tools to analyze their process, streamline it, stabilize it, measure it, and then, on a continuous cycle, reduce the variability of its characteristics or attributes with the goal of optimizing its output. Advanced quality practices/techniques are complementary to basic quality considerations but do not supplant them.

Unique quality management system. Specific and unique practices that are implemented where special product or service requirements exist. These unique practices could be controlled by the contractor for the unique processes required. This type of management system is applicable to products for space, jet engines, or even things that affect the safety of your automobile.

Since the milestone created in 1996, a contractor can work to one quality management standard, and therefore focus on the management standard that best fits the contractor's business. The electronic company can now look at its existing business and strategic direction, and put plans in place to become a world class quality company. Baldrige criteria, an international specification (ISO 9000), Total Quality Management—all point out that a company needs to look at internal processes and define what it needs to do to succeed. The one quality system had to come from the U.S. government (and industry had to agree) because of the large volume of business that the U.S. government placed with the electronics industry. Until 1994, the government demanded unique requirements. It took 2 more years for a policy to be agreed to and stated. In addition, the Department of Defense (DOD) is pushing to remove military specifications and standards, by its 1996 action.

If you are an electronics company that has multiple government contracts, it is now time to change out all of those systems and move to a single ISO 9000 system. The Single Process Initiative (SPI) was announced in December 1995, and begun in 1996. It is a U.S. government system to deal with the multiplicity of quality systems and military standards and specifications in the United States. It also assists the U.S. government in implementing its quality policy to establish world class companies. It is part of the DOD's acquisition reform thrust toward commercial products and industry-wide practices and technologies.

Common processes are intended to help reduce the contract operation costs and contribute to cost, schedule, and performance benefits for the contractor and its customers, including the government. They also help the contractor focus on facility-wide efforts to standardize process requirements where it makes good business sense. They allow a facility to focus on a quality management approach that can move the company to world company status.

Common processes are all those that affect the quality system and are driven by military standardization specifications. The new DOD 5000 standard gives overall direction on the government's position. Look at your company's processes, especially if several different processes are in place. Block contract changes can be written by your company, accompanied by a cost-benefit analysis that will be evaluated by the government.

If your company needs more information, contact the Defense Contract Management Command (DCMC), which is part of the Defense Logistics Agency. Your DCMC representative will help you understand the process and assist in writing proposals and changing contracts.

Back to the *Roadmap for Quality in the 21st Century:* the 1995 Malcolm Baldrige National Quality Award criteria were used as guidelines to fulfill the mission of the Government/Industry Quality Liaison Panel. These criteria are [4]

1. The customer judges and drives quality
2. Senior leaders are committed and involved
3. Improvement and learning are continuous
4. Employee participation and development are essential
5. Fast response is required
6. Quality is prevention-oriented and design-in-oriented
7. Our view of the future is long-range and global
8. Management is by fact and data
9. Internal and external partnerships are essential
10. Corporate responsibility and citizenship are the basis of action
11. The performance system is results-oriented and value-added-oriented
12. The position of U.S. industry in the internal marketplace will be enhanced
13. Consistency is achieved with commercial and international practices

Complete information on the Baldrige Award is presented in Chap. 9.

The G&IQLP agreed that the foundation for the framework should be a basic quality management system like the ISO 9000. The ISO 9000 series of quality management standards is designed for every industry and for every size company. They are written at a level where they can be applied to almost every situation. They can be tailored to add unique requirements where they are necessary for the customer or unique industries. They are a minimum set of requirements for quality management. [1]

3.2.1 ISO 9000 international quality assurance requirements

The ISO 9000 international standards are founded on the understanding that all work is accomplished by a process (see Fig. 3.3). *Process* is defined in ISO 8402 (the vocabulary standard) as "a set of inter-related resources and activities which transport inputs into outputs." [5] This is the organization of procedures, people, materials, and machines into work activities to produce a specific end result. A process is a sequence of events having measurable input, value-added activities, and measurable output. Processes are communicated to others by a process flowchart, the sequential symbolic description of the organization of people, materials, machines, and procedures into

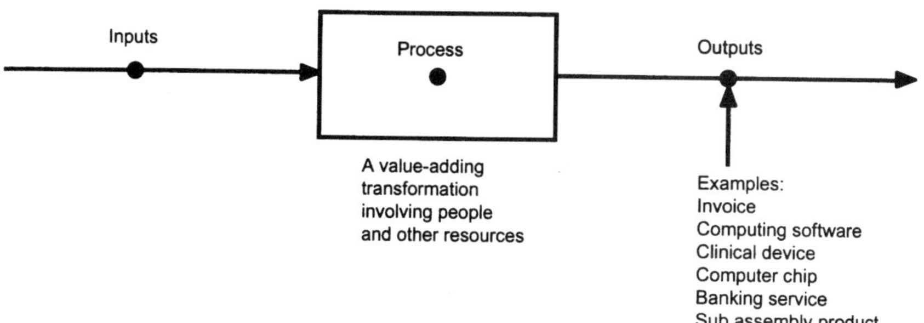

Figure 3.3 Process definition. (*From Ref. 5.*)

work activities required to provide a specific result. It is complete when it includes the element of time—the time required for each event, the time between each event, and the total time elapsed to complete a cycle of the work process. [6] There are opportunities to make measurements on the inputs at various places in the process as well as on the outputs. [7]

The success of a company is based on defining standard processes and then developing metrics that show what is important so those standard processes can be improved. As early as 1983, Kaoru Ishikawa acknowledged that standard processes make the difference in Japan to cut production costs and develop quality products. Steps to standardize are summarized by Ishikawa [8]:

- Company standards should exist in every electronics company for raw materials, parts, and subassemblies, and for meeting customers' requirements by careful analysis (application of QFD).
- Design and engineering standards should be documented and implemented.
- Reliability should be ensured by reducing or eliminating design changes. Ishikawa states that the three measures alone would reduce the number of design hours by one-third to one-half.
- When redesign is necessary, parts and materials should be interchangeable.
- Any changes to increase reliability should be standardized.
- Standard parts should be used and presented in design reviews.
- Industries should be educated about standard parts which will increase cooperation and standardization.

- International standards like those of the International Organization for Standardization (ISO) and the International Electrotechnical Commission (IEC) should be used much as possible.

The ISO quality management standards, in the style of Deming, hold top management responsible and take quality out of the quality department, giving it to the people who contribute most to quality—management, supervision, purchasing, engineers, and production workers. [9] The new quality department is policy makers, specialists such as statisticians, and quality engineers with special skills to train, facilitate, and implement across the electronics company. And, of course, the quality department is often the headquarters for internal quality audits.

The 20 elements of the most complete standard are presented in detail in Chap. 4. This standard is for any supplier that does design, development, manufacturing, and servicing after the product/service is sold. Subsets of these 20 elements can be implemented by a company that does not do all of the above-listed functions. Or if only service is your company's business, all elements of the standard can be applied to that service. It is a business management system that can be used universally. The following list of elements shows how top-level and simple the requirements are. [10] They can be implemented by any kind of a company.

1. Management responsibility
   - Define and document quality policy
   - Establish an organization
   - Assign responsibility and authority
   - Identify verification requirements
   - Provide verification resources
   - Establish a management representative
   - Provide for management review
2. Quality system
   - Document and implement system procedures and instructions
   - Define the quality system in a quality manual
3. Contract review
   - Requirements adequately defined and documented
   - Differences between the order and the contract resolved
   - Verify that the supplier is capable of meeting requirements
4. Design control
   - Establish and maintain procedures to control and verify design
   - Prepare design and development plans

- Identify organizational and technical interfaces
- Identify and document design input requirements
- Identify and document design outputs in terms to validate against design inputs
- Establish design verification procedures
- Establish design validation procedures
- Establish and maintain procedures for design changes

5. Document and data control
   - Establish and maintain procedures to control documents and data
   - Establish document change/modification procedures
6. Purchasing
   - Ensure that purchased items conform to specifications
   - Use only subcontractors that can meet requirements
   - Ensure that purchase documents clearly describe purchase
   - Perform source verification, as required
7. Control of customer-supplied product
   - Establish and maintain procedures to verify, store, and maintain customer-supplied items
8. Product identification and traceability
   - Establish and maintain procedures which allow items to be identified from drawings and other documents and be traced throughout production, delivery, and installation
9. Process control
   - Provide process control plans for those production processes affecting product quality
   - Ensure that procedures for special processes are in place and implemented
10. Inspection and testing
    - Ensure that incoming materials are not used until verified as conforming to requirements
    - Ensure that in-process materials meet requirements
    - Provide plans for final inspection to ensure that product meets requirements
11. Control of inspection, measuring, and test equipment
    - Establish procedures for the control, calibration, and maintenance of equipment
12. Inspection and test status
    - Provide means to identify product conformance or nonconformance to requirements

13. Control of nonconforming product
    - Provide procedures to prevent inadvertent use or installation of nonconforming product
    - Provide procedures for review and authority for disposition of nonconforming product
14. Corrective and preventive action
    - Establish and maintain procedures for all aspects of nonconforming product/service for corrective and preventive action
15. Handling, storage, packaging, preservation, and delivery
    - Establish and maintain procedures for protecting, securing, and preserving product from damage or deterioration
16. Control of quality records
    - Establish and maintain procedures for preserving quality data collected to ensure conformance to requirements
17. Internal quality audits
    - Establish and maintain procedures to ensure that the quality system is being audited to verify that the system is implemented according to procedures and is effective
18. Training
    - Establish and maintain procedures to ensure that all personnel are provided with the tools and understanding needed to carry out the duties of the quality system
19. Servicing
    - Establish and maintain procedures for the provision of contracted service and verification that it meets requirements
20. Statistical techniques
    - Establish and maintain procedures for identifying the need for statistical techniques required for establishing, controlling, and verifying products and processes
    - Establish and maintain procedures to implement and control the application of the required statistical techniques.

3.3 TQM and Teams

For many years, it was acknowledged that the employees who actually do the work should be involved in company decisions. Whether design, development, manufacture, or service, it is the employees who make it happen for the company. Moreover, the Total Quality Management approach starts with establishing processes that are defined and documented, so they can be measured, and then data can be collected. TQM concentrates on customers—both internal and external. TQM has become a proven leadership and management approach since the middle 1980s. It is the basic core of leadership and management to concentrate on processes and continuous improvement with the participation of all the employees, focusing the emphasis on the customer. In 1991, the U.S. General Accounting Office (GAO) reported that TQM had been very successful in making companies successful when it was fully implemented. See Chap. 6 for more information.

TQM takes years to implement, and years to change the culture to a process-centered, employee-empowered, customer-driven organization. TQM is a never-ending process. It is a way of organizational life for survival and growth. TQM goes beyond basic product or service quality. Product and service quality are minimum requirements in today's world. *Total* includes everyone and everything. *Quality* equates to total customer satisfaction. *Management* means the organization's leadership and management approach. It should not be the focus of business; it should be the means to achieve the focus of the business. [11] TQM is built on teams. These teams can be ad hoc problem-solving or process-improvement teams, moderate-term project teams such as integrated product development teams, or long-term product teams such as manufacturing, accounting, and procurement. The type of employee involvement depends on the type, interdependency, and complexity of work. High involvement requires the authority to act. It requires the empowerment of employees (discussed in the next section).

The new phrase, so much in the press, is *integrated product/process development teams.* In some companies this has been the norm for years, without the fancy title. They were known as concurrent engineering teams; basically, they are product teams. Product teams have front-to-back responsibility for their product. A product in this sense is whatever they produce—hardware, software, specifications, services, or concepts. The team is responsible and accountable for a product that meets their customers' (both internal and external) needs and expectations. The team is also accountable for the processes that develop that product or service.

Teams require the latitude to use their creativity and innovation, but need boundaries from management placing limitations on their empowerment. The boundaries, however, must be established so they do not overly constrain the team. Boundaries are like playing areas in athletic events. For example, football would not be a viable competitive sport if the field were 20 yd long and 10 yd wide.

Teams are multidisciplined, since the team is responsible for the product front to back. For example, in the production of an electronic assembly, besides manufacturing people, there would be purchasing, material control, assembly, and testing to name a few. Responsibilities on teams are shared; the team is responsible for meeting their customer needs. Team leadership is a facilitating role rather than a direct and control role.

The ideal evolution of teams in TQM organizations would be self-managing. There is no universally agreed on definition for *self-managing*. In some organizations, self-managing means that a team runs its enterprise like a minibusiness and is empowered accordingly. In a general sense, self-managing means taking on many of the responsibilities that have been traditionally management's. As a minimum, a self-managing team should be involved in planning, goal setting, data gathering and analysis, problem solving, team administration such as meeting administration, overtime, vacations, and budgets. Members of self-managing teams become business owners and stakeholders.

The paradigm has not yet shifted; TQM has started, but in many organizations it has not finished. The big "Q" is still a paradigm shift for many organizations. TQM and the big "Q" are vital to achieve world class quality for electronic systems. TQM is fully discussed in Chap. 6.

3.4 Management and Leadership

The aim of leadership should be to improve the performance of human and machine, improve quality, increase output, and simultaneously bring pride of workmanship to people. Put in the negative, the aim of leadership is not merely to find and record human failures, but to remove the cause of failure, to help people do a better job with less effort. A leader must learn by calculation wherever meaningful figures are at hand, or by judgment; otherwise, who (if any) of his people lie outside the system on one side or the other, and hence either need individual help or deserve recognition. The leader also has the responsibility to improve the system (i.e., to make it possible) on a continuing basis, for everybody to do a better job with greater satisfaction. Yet another responsibility of leadership is to accomplish increasing consistency of performance within the system so that apparent differences between people continually diminish (i.e., reduce variability). [12]

In 1950, a miracle happened in Japan. Before that time, products were poor, but by 1954 Japan started to capture world markets. Why? Top management became convinced that quality was vital for export, and that they could not accomplish the switch without learning about quality. The top management learned to take responsibility for this and then took the lead in world markets. [13]

It is not enough just to improve processes; design of product and services also must be constantly improved. New product, service, and technology must be introduced for a company to survive. All of this is management's responsibility. [14]

Communication not only explains the new initiative but maintains constancy, keeps people focused, and connects them emotionally. Through good communication, leaders overcome a host of pitfalls as they articulate their vision of the organization's future, its values, what is important, the challenges, and their commitments. A good model on communication may be found in Chap. 7.

Lawrence Tabak makes a good observation about how the importance of quality to an organization can be discerned through its communication. "A classic failure occurs when the effort to upgrade quality just dances on the surface, while true corporate culture crawls along unaffected. One of the best, and underutilized barometer of corporate culture is an organization's internal communications—the manner in which a company talks about itself and among itself." [15] Tabak goes on to talk about companies that have television monitors throughout the facility to give "instant" news. Electronic mail (e-mail) through the Internet and recorded hot-line messages are other techniques, in addition to the standard newsletter, which employees seem to like. Many companies are having all-employee or "town hall" meetings. The key here is to develop ways to listen to employees so that they can work with management to strive for excellence. [16]

Leaders need to spend at least 15 to 20 percent of their time on communication. Most management people do, when implementing a new program, but then enthusiasm dies off. Typical management passes down responsibility to middle management and then forgets about it. This will not work. Leaders need to be heard by the employees. This is part of the "walking the talk." A process must be set up so that improvement teams gain status from management; all persons must be encouraged to be part of the improvement process, time must be allowed for meetings to occur. Capital must be provided, people who participate and contribute should be rewarded, and those who do not should be removed. People know the difference. [17]

In a world class quality company there are other leaders besides the CEO; all levels of management must believe in the quality value system. They must communicate through their words and deeds that quality is important to them—through their behavior, the way they do their work, and the way they manage people. [18]

Conway, in *The Right Way to Manage,* defines a leader as a person who has the ability to see and articulate imaginative dreams and visions: the ability to state the mission at hand; the will and determination to see that the course toward those visionary missions is maintained; the courage to take a long-range perspective and to hold on to it despite today's pressures; the drive to remove impediments to success by becoming an enabler; and the capability to display optimism, confidence, and courage. These traits become contagious. The spread of this type of leadership is central to the new world class quality system. After the people at all levels are trained, and strive to produce continuous improvement, then all management acts as leaders. The organization will not bring about this cultural change without strong leadership. [19]

A recent study at the Saturn automobile factory by Thomas Kochran of the Massachusetts Institute of Technology (MIT) and Saul Rubinstein of Rutgers University illustrates the effect of leadership and communication. They found that three significant factors explained 50 percent of variation in quality across the site [20]:

1. Communication/coordination activity
2. Balance of time spent between managing people and managing products
3. Alignment of priorities of work teams (tasks) between partners.

In *Avoiding the Pitfalls of Total Quality,* authors Charles C. Poirier and Steven J. Tokarz emphasize communication and feedback as a way to empower employees: "The deepest principle in human nature is the craving to be appreciated." To apply this principle, they suggest four factors: encourage communication/feedback, catch people doing things right, kill scapegoating, and give appropriate recognition. [21]

Details on management and leadership are found in Chaps. 5 and 6. All chapters speak to how important the management/leadership role is in the new world of the big "Q" and striving to world class quality.

3.5 Empowering the Workforce

Empowerment means to provide employees with greater decision-making, problem-solving, and administrative responsibility and authority. It is required by TQM. It is endorsed by good leaders. Trite phrases like "People are our most important asset" are believed only if the company acts like it means it. In the spirit of ISO 9000, "Walk what you talk." If a company really believes that people are an important part of the organization, then it must prove it on a daily basis.

The workforce is becoming more temporary, but companies that use temporary people use the same people repetitively and make them feel part of the team. Richard J. Schonberger, in his latest book,

World Class Manufacturing: The Next Decade, warns us not to contract, or "out-source" too much, as it can cost the organization even more administrative time, and detract from a unified operation. If you do have to outsource, he suggests you ask your supplier for the same people, again with the emphasis on people working together. Stability and cohesiveness are the key to developing individuals who then become part of a team. A real team shares information and works together to please the customer; it is a culture, and cannot be "managed in." It came first in Japan, and it will come here with the right management and leadership. [22]

All management, especially middle management, must be trained to listen to workers, whether white or blue collar, and function more as "coaches," as described by the philosophy of self-managed teams. This is a change to American industry and is being implemented even in management and union shops. Under study is the strongest empowerment of the workforce at General Motors' Saturn plant, where United Auto Workers Union leaders as well as GM leaders are "not quite sure" that this is a good thing. [conclusion from data in 23] As evidenced in the Saturn study, cooperation between the old style management and workers depends on the philosophy of the leaders. On smaller scales throughout American industry, workers and management are cooperating as they implement the ISO 9001 quality systems and work toward a world class quality system.

It is a change in culture that must come from the executive office, and along with it comes TQM. It is awareness that the workers are part of the quality system, that they participate in the process, recommend improvements, sometimes makes decisions, and get rewards for actions. It is well known that statistical process control works best with a knowledgeable operator who is empowered to regulate his/her process (within boundaries) to produce good product and is the central person who knows what needs to be done to maintain and improve the process with which he or she is empowered.

When the workers are on teams for continuous process improvement, are empowered to control their process as part of a team, and are part of the process of keeping metrics for continuous improvement, the culture does change. It has long been known that people's first desire at their job is to be appreciated.

Employee empowerment has six major characteristics according to Steven Rayner, in Wallace's *World Class Manufacturing* [24]:

1. *Leadership that empowers others.* The leader removes barriers that limit individual and team contributions.

2. *A relentless focus on strategy and results.* There is a clear connection between business strategy, team structure, work design, and individual work roles. The utilization of employee empowerment is recognized as a technique for achieving superior results.
3. *Sharing of relevant information.* Virtually all relevant business information is shared. Teams use the information to help in solving problems, developing innovations, and making decisions.
4. *Borderless sharing of power.* Authority and responsibility are based on the issue being addressed, not on one's position or status. Those most directly affected by an issue, problem, or strategy are given responsibility for it.
5. *Team-based design.* Teams are the primary work units and are responsible for the design of their work, including the determination of the required technology and individual work roles.
6. *Teamwork reinforced by rewards.* Formal and informal rewards reinforce the overall team-based design. There are clear links between improvements made by the team and the rewards received by team members.

Creating a work culture based on these six elements is difficult. Estimates say employee empowerment fails approximately 50 percent of the time because it is implemented only piecemeal. Where there is no recipe for implementing employee empowerment, the strategy in Fig. 3.4 will increase the likelihood of success. Each phase builds momentum, therefore creating a natural pull for the desired change.

Attributes of this strategy are, according to Rayner:

Phase 1—Leadership. The new leader must be someone who can articulate why the change is important and provide a vision for the organization. The champion must understand an empowered organization, exercise the clout to make change happen, and show a clear concern about the people in the organization.

Phase 2—Commitment. In this phase, the effort is expanded to include key opinion leaders within the organization. Initially their efforts focus on clarifying the linkages between employee empowerment and overall business strategy. The opinion leaders then become actively involved in planning the change process and recognizing how roles, including their own, will be affected by the transition.

Phase 3—Communication. The third phase stresses altering the flow of information in order to expand its accessibility to nearly everyone in the organization. Often this involves dismantling the

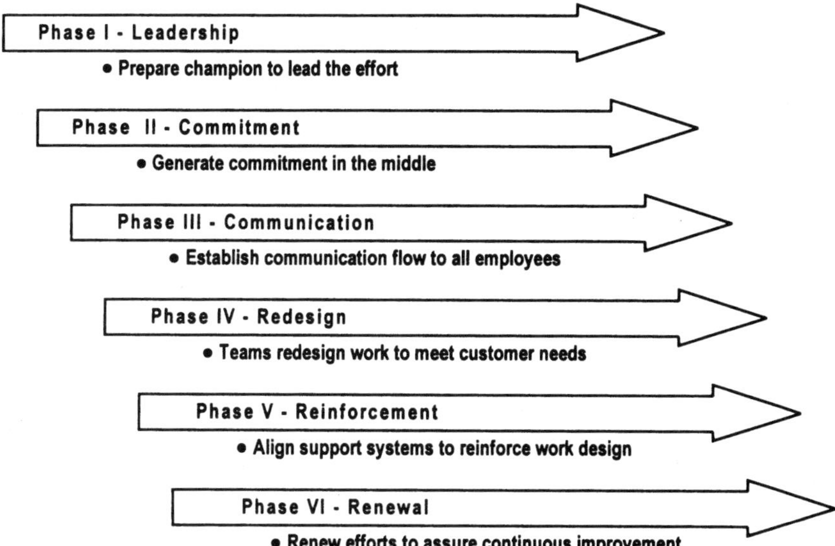
Phase I - Leadership
• Prepare champion to lead the effort
Phase II - Commitment
• Generate commitment in the middle
Phase III - Communication
• Establish communication flow to all employees
Phase IV - Redesign
• Teams redesign work to meet customer needs
Phase V - Reinforcement
• Align support systems to reinforce work design
Phase VI - Renewal
• Renew efforts to assure continuous improvement

Figure 3.4 The transformation pathway. (*From Ref. 24; used with permission.*)

hierarchical information pathway, restructuring reports, and reassessing what information should be considered confidential. As teams begin using the expanded flow of information, problem solving expands.

Phase 4—Redesign. During this phase, the focus shifts to improving the design of the entire organization. Teams take an active role in the redesign effort and address such issues as streamlining processes, improving quality, lessening functional barriers, and expanding roles. At this point in the transition, the organization has begun to alter the way it operates—many procedures, workflow processes, and roles have dramatically changed.

Phase 5—Reinforcement. Next, support systems (such as the compensation system and other internal corporate systems) are the focus of reform. The intent is to change the support systems so they reinforce and strengthen the emerging design.

Phase 6—Renewal. The emphasis in this final phase is on sustaining energy and enthusiasm that ensures continuous performance improvement.

Schonberger, in his latest book, *World Class Manufacturing: The Next Decade,* talks about quality differently than most quality authors, and yet, in this author's opinion, hits the nail on the head: he emphasizes the people aspect of World Class Manufacturing; which is really part of world class quality. He calls his chapter on quality, "Quality. Picture a Miracle." He stresses appreciation of the person—important to empowerment—by using imagineering. Schonberger suggests asking the employee: "Where would you like to be in 10 years? What if a miracle happened here? What would it be?" And then he has the employee visualize, or imagine it.

Imagineering is the most powerful tool of the new management system. It enables employees at all levels to visualize how things would be if all problems were eliminated. Imagineering is central to success in the new management systems. It helps people get to seemingly hidden sources of problems, and begin the process of correcting what might otherwise appear to be an extremely difficult situation. Imagineering opens people's minds to recognize waste, providing the key to improved quality, and productivity. [25]

Imagineering is nearly the same thing that Schonberger calls a "miracle." Based upon counseling techniques and research done by M. Daniel Sloan and Jodi Torpey, *Lowering Healthcare Costs*, ASQC, Milwaukee, 1995, it encourages people to "imagine" a miracle. This approach changes the usual negative effects of working on what is wrong and emphasizes the positive. It is based on the same principles as TQM where the employee is involved. As we well know, appreciation is the most important thing that empowers people. It is used in an interview, in a confidential setting, where the individual is asked, "Where would you like to be in 5 to 10 years? With your expertise, what would you do?" Little miracles provide positive feedback; do not expect big miracles. This is a good way to get good positive feedback and make people feel important and part of the process. [26]

3.6 Metrics

It has long been agreed that what is not measured cannot be managed. Metrics are one of the main tools of Total Quality Management and continuous improvement. They actually form the basis for continuous improvement. Under TQM, processes are documented and references are established for each task. Metrics form an important part of the new quality system. Metrics are no longer just a measure of defects in the manufacturing process, or a yield number. Metrics start in the design process and should measure the program from beginning to end. Metrics should be planned for the overall company and then for each project as it begins. They should center around the following activities:

Sizes. Number of drawings, lines of software code, number of tests, etc.

Quality. Defects, first time yields, audit findings, etc.

Technical Parameters. Technical performance measurements (TPMs), engineering changes

Project Status. Percent of schedule met, budget spent, work completed

Productivity. "Earned value," how much per month for soft tasks, cycle times

Business. Customer satisfaction metrics, award fees

The fact that U.S. manufacturers have repeatedly misread consumers reflects that there is a lack of attention paid to the varied meanings of quality. It also suggests another common quality error: relying on the wrong quality metric. The success of programs to evaluate consumers' reactions to quality depends on the accuracy of the measuring sticks employed. If the metrics used by producers fail to match the interests of the consumers because they are poorly designed, relay technical data that are poor proxies for customer satisfaction, or draw from unrepresentative samples, they will provide little guidance. Internal quality metrics may stay the same as the outside environment changes, and no one notices. [27]

Metrics are measurements made over time that communicate vital information of the quality of a process, activity, or resource. They reflect measures that target continuous improvement. They focus on total customer satisfaction. Metrics must be customer-oriented and communicate a state of health, show where the company is versus where it wants to go, and assess all critical processes and activities leading to success.

Metrics must be designed to be collected as part of the processes. Computerized data systems for collecting data are excellent sources, but they must be designed as part of the system, not as an extra step or extra action.

Metrics is not:

- *Charts.* Just having a chart is not a metric.
- *Schedules.* Schedules can lead to good metrics, but not by themselves.

- *Goals, objectives, strategies, plans, missions, or guiding principles.* Most can be measured but are a means to an end; they are not metrics by themselves.
- *Counts of activity.* Counts can result in metrics, but just counting does not drive the appropriate action.
- *Snapshots or one-time status measures.* These show little trend, and do not provide a real understanding.

A metrics package includes three basic elements: operation and definition, actual measurement, and metric presentation. [28] The following questions need to asked when developing metrics:

What is the process? This is the operation and definition.

What is the objective of the process?

What is being measured in the process that is important to customers?

What is the best way to present this data?

Metrics that are "visual" make the point most effectively and provide the best means to picture the problem and thus create the solution, if the metrics say there is one. [29] An excellent book to help with this is *The Visual Factory: Building Participation Through Shared Information,* by Michel Greif, Productivity Press, Portland, OR, 1991.

Top-level metrics for the company should include quality, performance, cost, and customer satisfaction. Tables 3.2 and 3.3 *a* and *b* show some samples of top-level metrics. For more detailed metrics to ensure continuous improvement, a good example is provided by NASA in its draft manual, *ISO 9000: Implementation and Assessment.* [30] This set includes three categories:

- Costs to assess conformance to requirements
- Costs to correct errors and defects
- Costs to prevent errors and defects

As a company moves from phase to phase in becoming a world class manufacturer, its costs to correct errors should decline while the other two categories stay constant or rise. Costs to assess conformance to requirements might include:

- Design review meetings
- Design verification analyses and testing
- Validation of designs for user applications

TABLE 3.2 Sample Internal Metrics

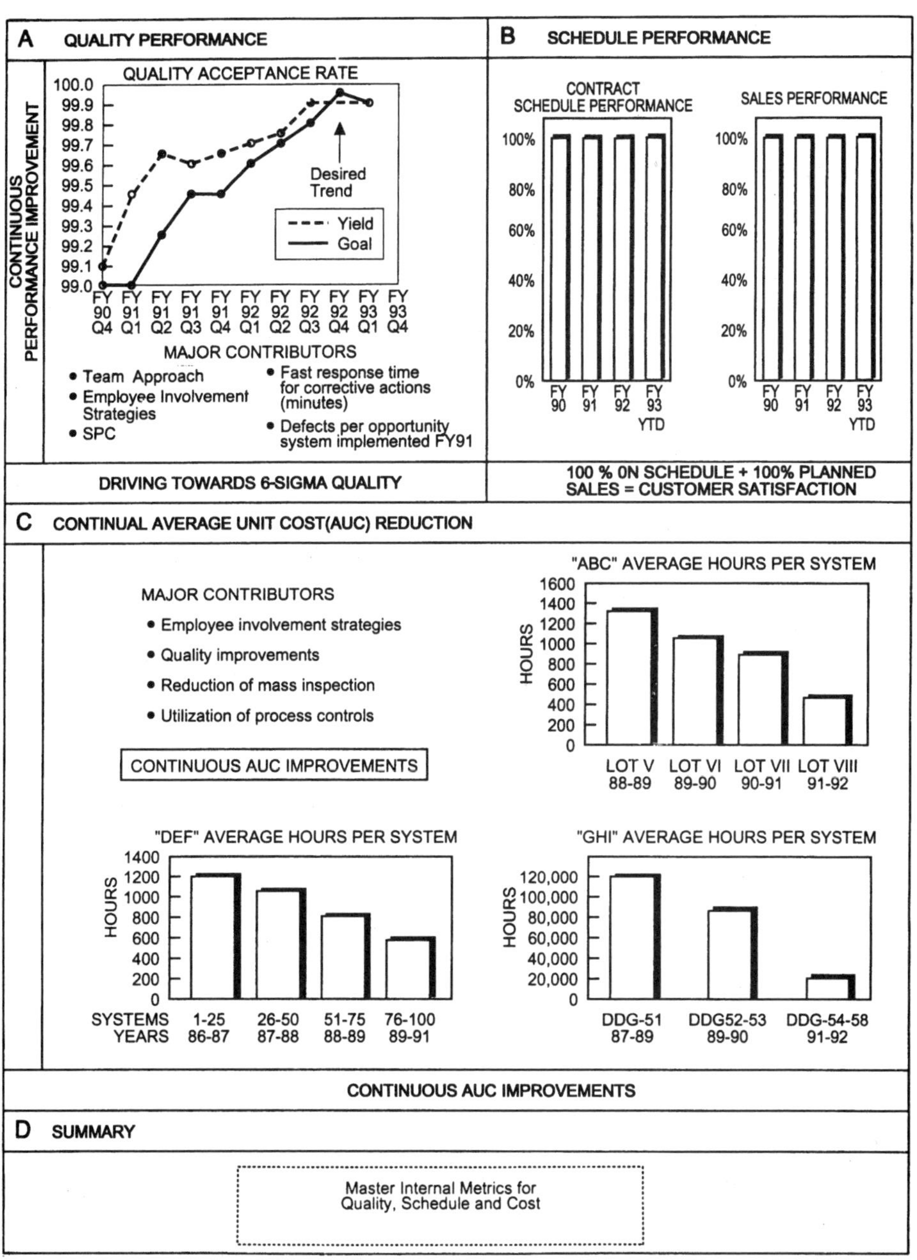

TABLE 3.3*a* Customers' Ratings of Service Category Importance

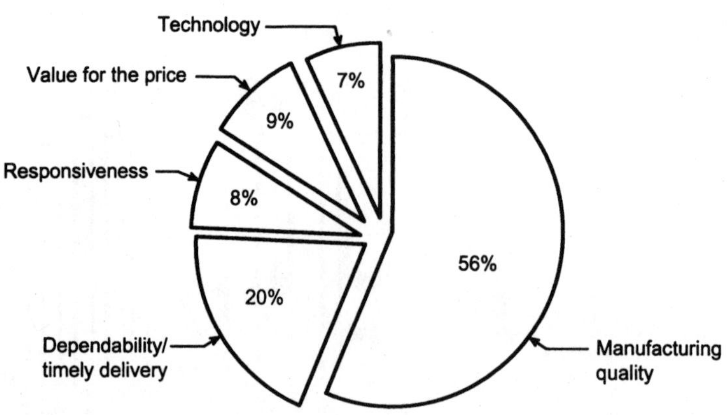

TABLE 3.3*b* Customers' Most Frequently Mentioned Future Requirements

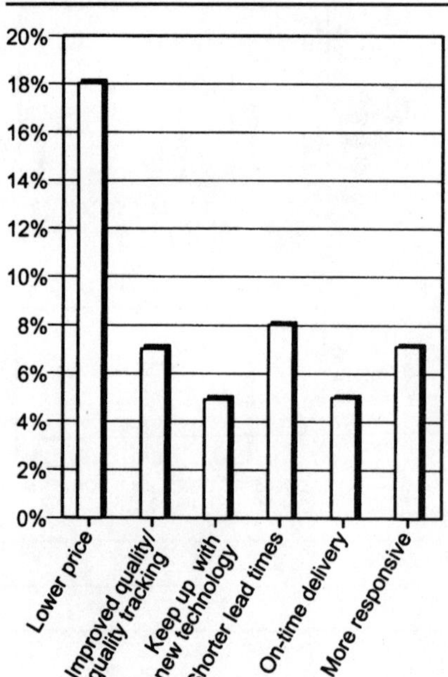

- Gauge or process capability studies
- Management reviews of quality system effectiveness
- Post-drifting check of engineering drawings
- Hardware and software product inspections and tests
- Proofreading of user manuals and other documentation products
- Internal or supplier quality audits
- Support for customer or third-party quality audits

Costs to correct errors and defects might include:

- Hardware, software, and documentation rework and reinspection
- Material review board meetings
- Scrap (material plus labor)
- Expediting fees, unplanned overtime, and premium transportation
- Responding to regulatory agency citations
- Travel and living costs to address field problems
- Documenting, evaluating, approving, and implementing design changes

Costs to prevent errors and defects might include:

- Personnel training and certification
- Creation of design development plans and project quality plans
- Requirements development and review activities
- Design of workstations and fixtures to preclude product damage
- Process improvement team meetings
- Scheduled maintenance
- Customer surveys and focus groups
- Defining and documenting operating procedures

Metrics for tracking the quality of design, quality of conformance, and fitness for use, may be found in Chap. 2 at the end of these sections.

3.6.1 Metrics to get to world class

Advanced quality concepts are the cumulative result of the application of more advanced and sophisticated practices in the areas of design, manufacturing, assessment, corrective action, and continuous improvement. *Industry Week* lists several topics to measure world class quality; see Tables 2.25 and 2.26. These are used as criteria in *Industry Week's* annual selection of America's Best Plants.

If your electronics company is moving toward world class and wants to identify cost drivers to implement an activity-based costing system, the enterprise must first identify cost drivers that accurately reflect the consumption of cost by each activity and the resultant consumption of activities with each product. Examples of such cost drivers are:

- Lead times
- Number of circuit board insertions
- Number of purchase orders
- Number of parts per product
- Number of molds or dies required per product
- Number of material moves
- Number of phone calls to obtain a fully specified customer order

The real value of keeping these metrics is to understand how the product causes costs as it moves through the factory. It is not uncommon for companies to find, when starting to use activity-based costing, that low-volume special products rise in price (some from 200 to 500 percent) and other high-volume standard products decrease 50 percent or more compared to traditional costing techniques. More information on activity-based accounting is available in Sec. 2.3 and in the public marketplace in the publications on world class manufacturing. [31]

3.7 Advanced Quality Concepts/Techniques

Advanced quality concepts are the cumulative result of the application of more advanced and sophisticated practices in the areas of design, manufacturing, assessment, corrective action, and continuous improvement. Advanced quality approaches differ from traditional quality approaches by emphasizing the prevention of defects, rather than the identification and correction of defects after the fact. As such, advanced quality underlies other processes used to develop, produce, deliver, and support products and services that meet customer expectations. Although both approaches share the objective on ensuring that only material that meets customer expectations is delivered to the customer, traditional systems tend to focus on the production phase and rely on inspection and test to sort out defective material. The principal drawback is cost. Inspection, test, segregation, and processing of nonconformances and rework incur costs, yet add no value to the product. In contrast, advanced quality approaches emphasize controlling manufacturing processes, such that only a conforming product is produced.

Advanced quality approaches are intended to prevent defects and are aimed at both development and production. In this context, advanced quality elements are applied during the development phase, normally with integrated product and process development teams, and then focus on defect prevention in manufacturing process control and variability reduction. Examples of tools and techniques utilized to implement these concepts include:

Requirements definition tools—QFD

Designing for robustness—Taguchi Methods

Controlling key product/process characteristics—SPC

Variability reduction—VR.

In order to improve competitiveness in the marketplace, many companies have been applying these advanced tools and techniques. The use of the tools and techniques, and the application of continuous process improvement, has resulted in higher expectations and a raising of the basic quality standard by customers. In addition, this leads to a continuous cycle of improvement, higher expectations, and further improvement. Advanced quality concepts are defined as those practices that are above and beyond the elements defined by the single baseline quality management system. The G&IQLP has a goal to identify, promote, and manage the evolution of advanced quality tools and implement practices which go beyond those in the baseline quality management system. These tools and practices must be integrated effectively into procurement, system engineering, and oversight processes, with consideration to product complexity and cost. This will lead to increasingly higher quality products and services, more commonality between government and commercial industries, and increases in global competitiveness. A system which recognizes and encourages advanced quality may also reduce the need for individual programs that promote quality improvement. Advanced quality concepts affect all aspects of quality—performance, cost, and time. [32] With this kind of leadership in developing quality products and services, the United States will continue to increase its share of the international market and continue to be a world leader with leading world class industries.

3.7.1 Quality function deployment

Quality function deployment is "a method for developing a design aimed at satisfying the consumer. It translates the consumer's demand into design targets, and major quality assurance points to be used throughout the production phase....[QFD] is a way to assure the design quality while the product is still in the design stage....When adequately applied, QFD has demonstrated the reduction of development time by one-half to one-third." [33] Elements of QFD were developed (as early as 1966) in Japan by Yoji Akao. His first work was published in 1978 in a book entitled *Quality Function Deployment,* but was not translated into English until 1994. "QFD has been used by Toyota since 1977, following four years of training and preparation. Results have been impressive....Between January 1977 and April 1984, Toyota Autobody introduced four new van-type vehicles. Using 1977 as a base, Toyota reported a 20 percent reduction in start-up costs on the launch of the new van in October 1979; a 38 percent reduction in November 1982; and a cumulative 61 percent reduction in April 1984. During this period, the product development cycle (time to market) was reduced by one-third with a corresponding improvement in quality because of a reduction in the number of engineering changes." [34] QFD came to the United States in 1985 when it was presented at a GOAL/QPC conference. Many U.S. companies have also begun to use QFD with a fair degree of success. Some notable examples include Ford Motor Company, General Motors, and International Telephone and Telegraph. In 1995, the American Supplier Institute (ASI) estimated that there are literally thousands of QFD applications in progress in the United States. [35] Both GOAL/QPC and ASI are major suppliers of QFD material and software.

The aim of QFD is to create cross-functional interaction and cooperation toward driving the voice of the customer into products and processes. The Japanese concept of QFD entails harvesting the knowledge of production workers, engineers, and managers in a collective decision-making process that leads to continuous improvement of customer satisfaction. This works directly with the cross-functional development process called *concurrent engineering.*

There are different approaches to the performance of QFD; however, the purpose is the same: a cross-functional team focuses on designing to satisfy customer needs to build in rather than inspect for product quality. Approaches begin with identification of internal and external customers with associated clear statements of need, and with a product or service that provides customer satisfaction. This analysis and quantification of the voice of the customer is an important element in QFD's ultimate application. Through a relational matrix, the identified needs of the customer may be matched against product features, and prioritized through analysis. That capability provides a means to obtain direct flow-down of customer needs into product features, manufacturing processes, and process characteristics, and at the same time identification of the most important elements in the production of customer satisfaction. It is the feature that accounts for QFD's ability to design quality into the final product.

Using QFD requires more management time and effort early in the product development and business-planning processes, but reduces total development and planning executive time dramatically because priorities and problems are addressed early, relationships are defined and established, and communication and documentation are improved. QFD identifies key features needed to satisfy needs. QFD identifies initial competing features. QFD provides a window into the issues to be resolved during product or service development, including the processes needed for initial manufacturing procedures.

QFD addresses all the quality of design, quality of conformance, and fitness for use, and therefore supports the systems engineering process for quality assurance used to drive integrated product development.

To do this, the QFD process involves constructing one or more complex matrices. The first matrix is called the *House of Quality* (Fig. 3.5). It displays the customer's needs developed from the "voice of the customer" along the left, and the teams' technical solution to meeting those needs, along the top. The matrix consists of several sections or submatrices jointed together in various ways, each containing information required for the analysis.

There are two main approaches marketed in the United States today: Four Phase by the American Supplier Institute and the Matrix of Matrices by GOAL/QPC.

The American Supplier Institute supports a four-phase approach to QFD patterned after early Japanese approaches. For American industry it was modified by J. R. Hauser and D. Clausing in "The House of Quality," *Harvard Business Review*, May–June 1988, pp. 63–73. Figure 3.6 depicts this process. The first phase, Product Planning, begins with a House of Quality (Fig. 3.5). "Whats" and "whys" are generated from meetings or interviews, or data from the customer; "hows" to accomplish the "whats" are generated by brainstorming. Terminology is defined as

Customer needs	Whats
Customer importance	Whys
Technical requirements	Hows
Target values	How much

To link the House of Quality to the next matrix, the "hows" on the House of Quality become the "whats," and the process begins anew.

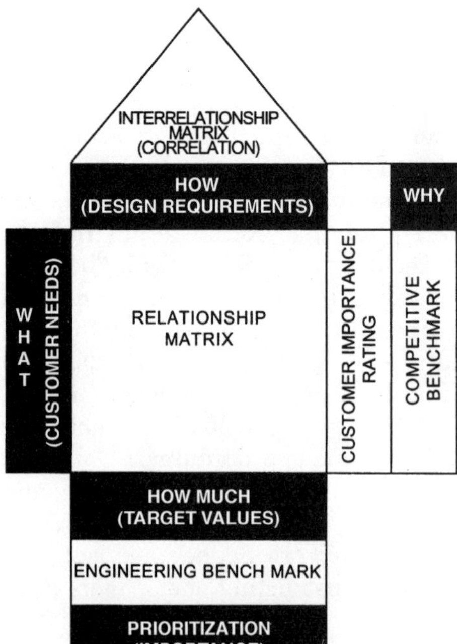

Figure 3.5 Stylized House of Quality.

The process continues in this fashion through the Production Planning matrix. This gets customer requirements all the way to the production floor. [36]

GOAL/QPC's approach, the Matrix of Matrices (Fig. 3.7), is patterned after the original Japanese approach developed by Akao. This set of 30 matrices includes the ones in the four-phase approach. The top-level interrelational matrix of the House of Quality is included as a separate entity. Additional matrices provide options for engineering, technology selection, process planning, risk analysis, and risk mitigation. Not every matrix is a relational matrix and not every element in the system is a matrix. Some elements, such as fault tree analysis and failure mode and effects analysis, utilize hierarchical trees. The matrices also contain matrices supporting the Pugh concept selection methods. [37] These are methods with practical and useful tools for selecting the best of several alternative concepts. The internal methodology for creating and analyzing relational matrices is detailed in a book entitled *Better Designs in Half the Time,* by B. King, GOAL/QPC, Methuen, MA, 1989.

The Matrix of Matrices® includes a set of charts to accommodate performance versus cost, or value engineering. In this fashion, where

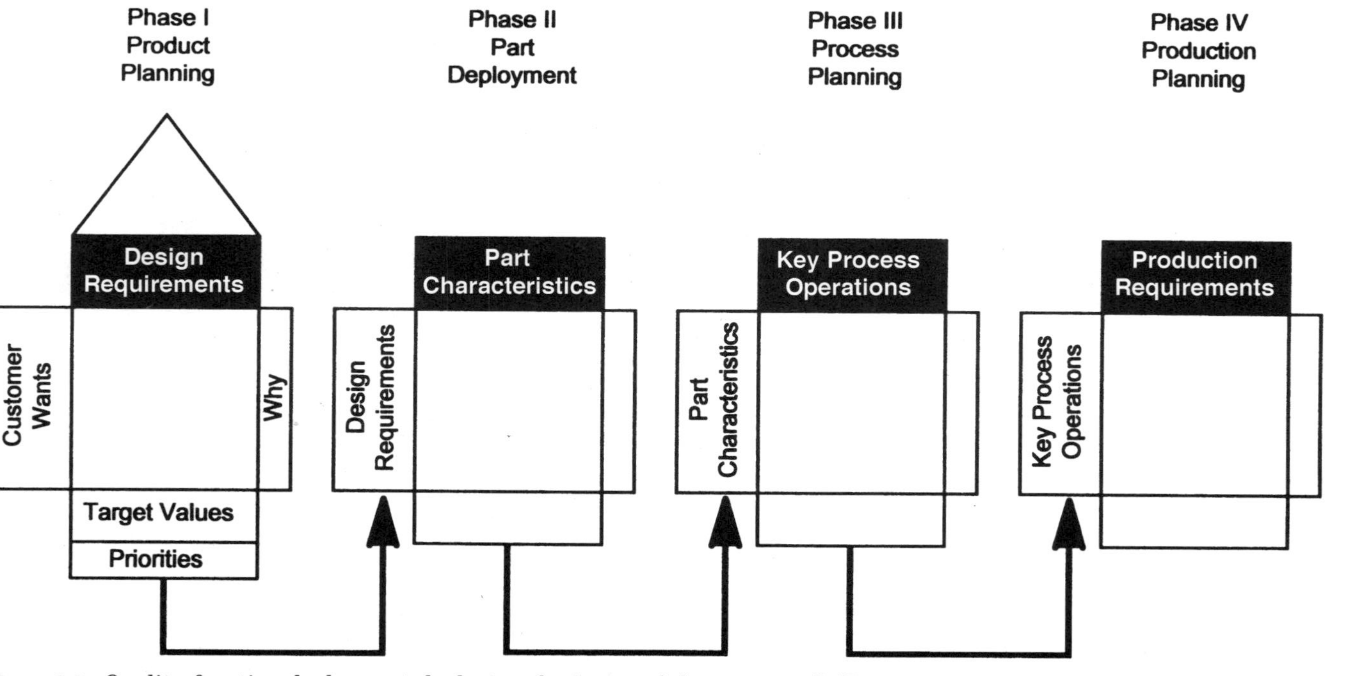

Figure 3.6 Quality function deployment deploying the "voice of the customer." (*From Harvard Business Review, with permission.*)

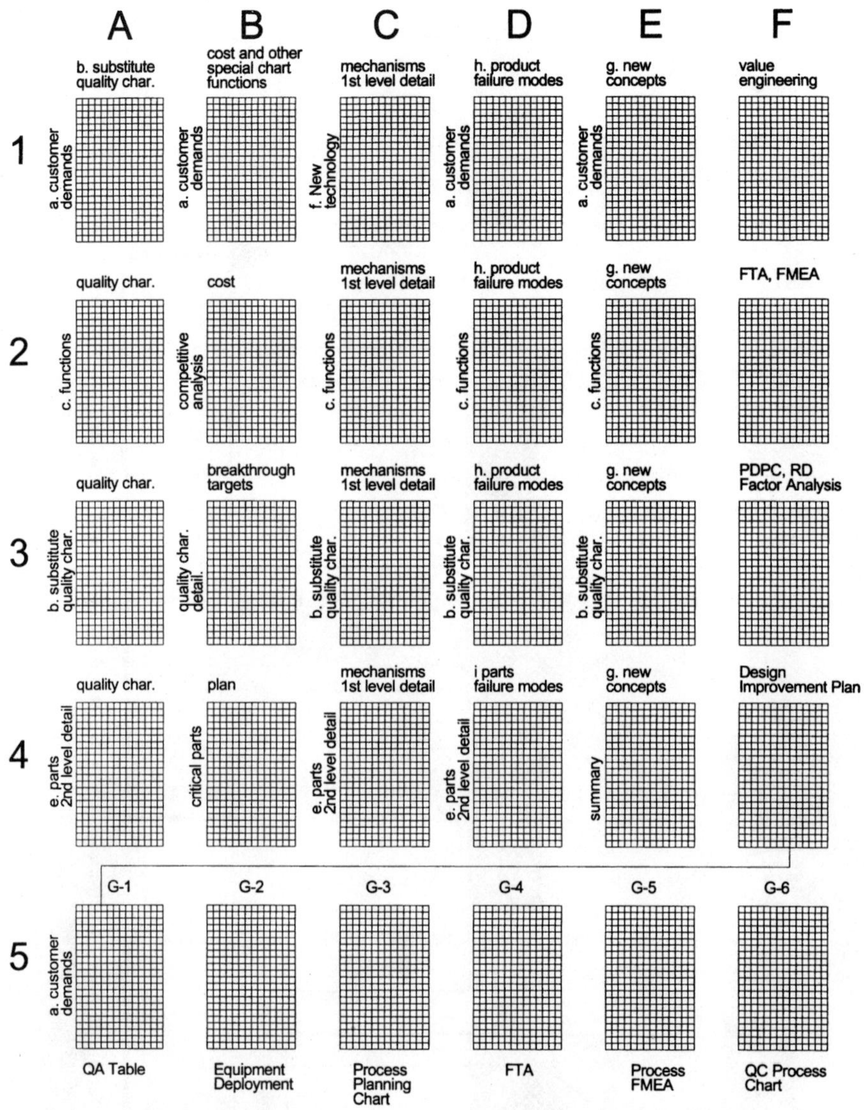

Figure 3.7 The Matrix of Matrices. (*From Technicomp, Quality Function Deployment Application Guide, Cleveland, 1987, p. 4.20; reprinted with permission*)

and how to reduce cost considers not only increased affordability but also increased customer satisfaction. It also supports Taguchi's quality laws:

> We cannot reduce cost without affecting quality.
>
> We can improve quality without increasing cost.
>
> We can reduce cost by improving quality.

GOAL/QPC goes one step further. Their approach includes not only functions, but mechanisms to accomplish those functions.

Many tools are used to work the details of the QFD process. Two of the most important are the affinity diagram (Fig. 3.8) and the tree diagram, which is described in Chap. 6 (Fig. 6.16). The affinity diagram is used to aggregate similar raw data and the tree diagram is used to complete and refine the structure. The affinity diagram is created when there is much disorganization. Both assist in organizing the raw data arising from the voice of the customer and the technical

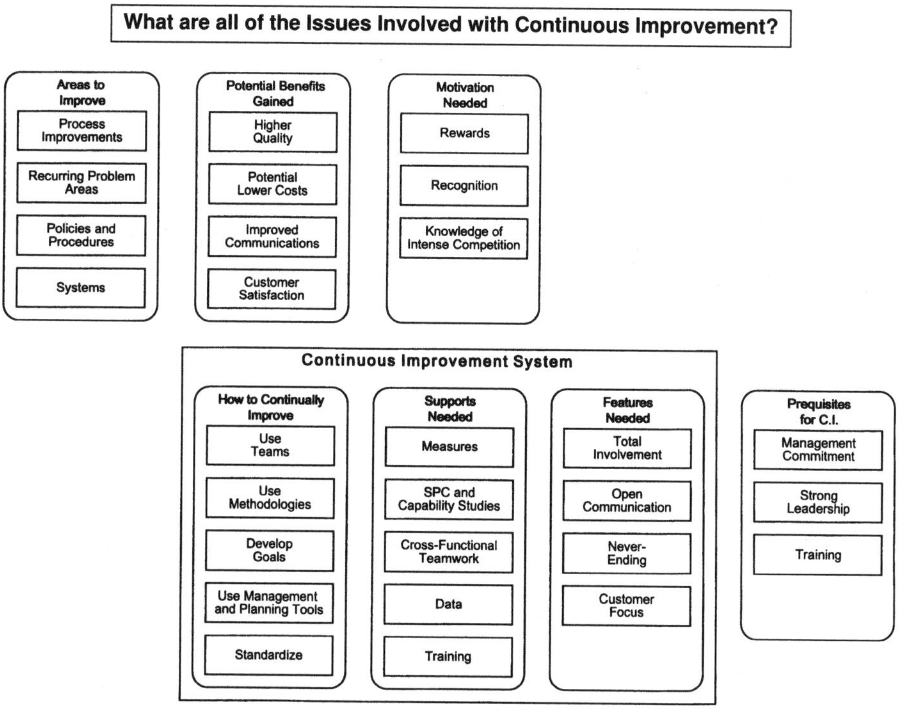

Figure 3.8 Affinity diagram.

barnstorming process. The tree diagram procedure starts at the top and uses logical and analytical thought processes to move down the tree in an orderly manner. This information is used as the input rows and columns for each relational matrix.

The QFD team that develops the QFD matrix will be making key strategic decisions about how a product or service should be. These decisions will affect the company decisions and the customer responses. The persons on the team should be trained so they know their roles and how to listen for the voice of the customer, and thus develop customer satisfaction. They should be experienced in their field and understand the new and advanced concepts of quality, continuous improvement, and customer satisfaction.

The QFD team should include representatives from all functional groups. The ISO 9001 paragraph on design control requires that all required functions are on the design team. Every important decision is made at the beginning. See Chap. 2 for new definitions of quality and Chap. 4 for ISO information on these subjects. A good QFD facilitator should be part of the process and lead the team through the QFD process. The QFD process should be managed and scheduled. It requires time, but there are shortcuts, depending on the time the team has to implement the process. Some software is more applicable to short-cutting the process, and some requires that much detail information be input, so be careful before you buy a package and make sure it serves your company's needs. Major tasks involve:

1. Customer needs and benefits
2. Meeting the needs and the wants
3. Substituting characteristics—product features and characteristics
4. Impacts of those characteristics
5. Relationships of those characteristics
6. Competitive benchmarking

The size of the team and their experience working together will, of course, drive the time. If the customer is well known, this too will shorten the time. If not, a plan will need to be made to collect data, either in person, or by phone, or by another data-gathering approach. Known technology, as compared to state-of-the art, will also drive time; and for all electronic companies today, cost/price is so important that it is another variable. In tasks 2 to 5, if all the customer wants and the ability of the team to characterize and provide them are compatible, the process is quicker. If the wants are high and the price low, many more hours will be spent discussing the "right" approach. The level at which the team wants to drive the QFD process is another factor. Some teams can go to four or five levels, some stay at a high level; again it depends on many variables unique to the company, project, and team. Task 6, competitive benchmarking, is time-consuming, but needs to be done if a competitive procurement is in place; for a commercial company, it must be done prior to spending a lot of money on developing a new product that consumers will not buy, or if there already are sufficient products on the market and your new product adds no value at reduced cost. In addition, the competitive benchmarking should be done by a unique subteam that has *not* defined the product features in tasks 3, 4, and 5; then the competitive benchmarking team can objectively review the original team's recommendations and assist in the right approach to winning the competition, or determining the best product to put in the marketplace.

The QFD process is not difficult, but persons do need to be trained. It is good to introduce the House of Quality at an initial meeting of a team, then work it on a regular basis. The QFD activities should be scheduled as part of the team's plan. It is good to place a large House of Quality on the wall, so the entire team can see it and perhaps mark it up. Figure 3.9 shows a completed House of Quality for large rolls of paper stock used in commercial printing. Table 3.4 shows the actions defined for each block. This example comes from Technicomp, a company in Cleveland that provides training with video tapes and an application guide. [38]

QFD's primary application has been planning and managing product development, but the general ideas of matching "whats" against "hows" can be used for many purposes. Among them:

- Linking TQM and Baldrige strategy to your company's operation
- Strategic product planning
- Doing organizational planning
- Matching the company's work to its objectives
- Cost deployment inside the company, or a project
- Software development—see three cited sources. [39]

In reviewing where a company is in moving to the big "Q" and becoming more world class, QFD is one of the first things that a company should look at to turn itself around. Developing new or revised

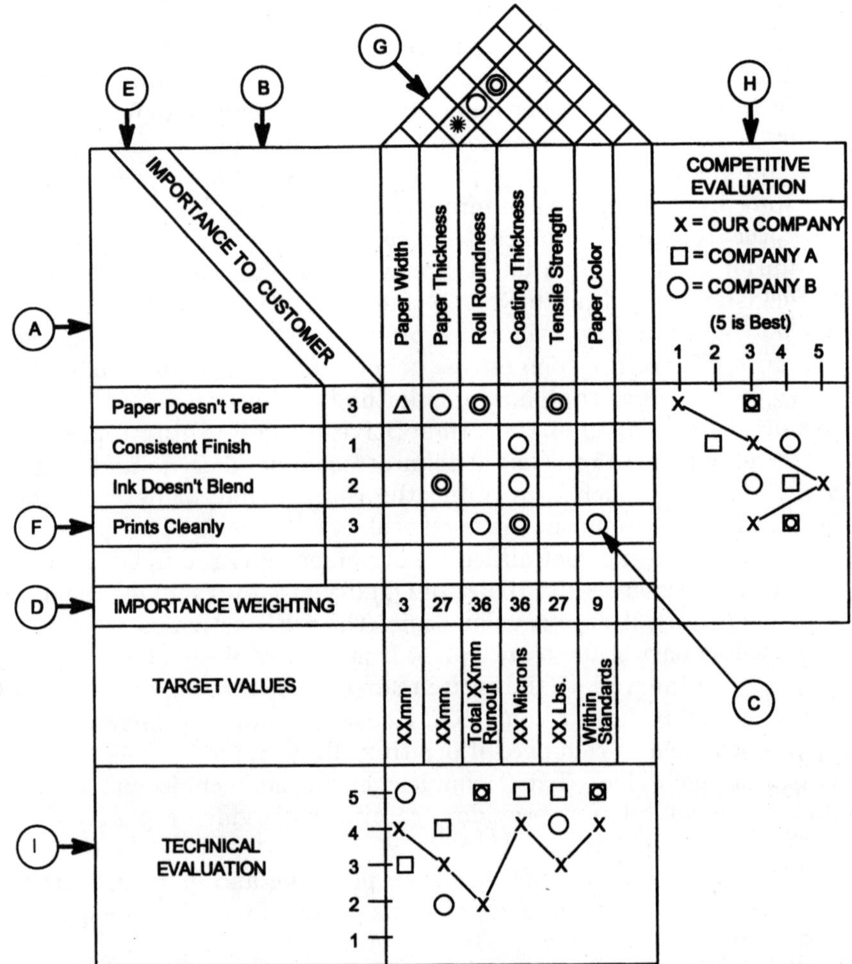

Figure 3.9 House of quality. Circled letters are explained in Table 3.4. (*Reprinted from Ref. 38 with permission.*)

designs that are pleasing to the customer may make the difference in surviving in this competitive environment. Additional information on QFD can be obtained from the QFD Institute, which can be found on the Internet at http://www.nauticom.net/www/qfdi. This institute was founded in 1993 to advance QFD methodology.

QFD is a starting point. Then, the team should move on to doing Taguchi design of experiments (DOE) prior to the product going to the manufacturing floor. It all starts with a good customer understanding

TABLE 3.4 House of Quality Explanations for Figure 3.9

A. *Customer requirements.* Customers' wants and needs, expressed in their own words.
B. *Technical requirements.* Design specifications through which customers' needs may be met, expressed in the company's internal language.
C. *Relationship matrix.* Indicates with symbols where relationships exist between customer and technical requirements, and the strength of those relationships.
D. *Target values.* Show the quantifiable goals for each technical requirement.
E. *Importance to customer.* Indicates which requirements are most important to customers.
F. *Importance weighting.* Identifies which technical requirements are most important to achieve. In this chart, each weighting is calculated by multiplying the "importance to customer" rating by the values assigned to a relationship, then totaling the column.
G. *Correlation matrix.* Indicates with symbols where relationships exist between pairs of technical requirements, and the strength of those relationships.
H. *Competitive evaluation.* Shows how well a company and its competitors meet customer requirements, according to customers.
I. *Technical evaluation.* Shows how well a company and its competitors meet technical requirements.

SOURCE: Reprinted from Ref. 38 with permission.

that can be done under QFD. A maturity matrix for assessing your progress toward quality deployment is found in Table 3.5.

3.7.2 Robust designs with Taguchi's Method

A robust design is one that is strong, healthy, and capable of properly performing its function under a wide range of conditions, including customer misuse. Robust designs are less sensitive to variation in parts, processes, and operating environments. By carefully selecting design parameters, a company can produce products that are more forgiving and tolerant. More robust designs result in higher customer satisfaction, lower production and support costs, and greater production flexibility. Basically, they are easier to manufacture correctly. [40]

Robustness can be thought of as the insensitivity of the product to uncontrollable environmental and process variability, or *noise.* Noise may include such factors as change in temperature or humidity, machine wear, or variability in human performance levels. Robustness minimizes variability of the units produced by making the product so that it is insensitive to these changes. [41]

Dr. Genichi Taguchi has developed a set of techniques to improve quality based on statistical principles and utilization of engineering knowledge. His methodology makes quality decisions based on cost

TABLE 3.5 Quality Function Deployment maturity self-assessment

Management responsibility	System in place	Training in QFD	Techniques and tools	Procedures/processes	Metrics/data
Executive management defines and documents QFD policy. All management participate in process. System is in place for all-new design.	System is in place for all-new design. All new and modification projects use QFD.	All personnel are trained in QFD and the voice of the customer. All personnel know their customer requirements.	Software, hardware in place. Samples for teams to use. Space to post matrices.	Procedures and processes documented and released. Audit of these processes is in place.	Data are collected on every process and used for management review and continuous improvement of the QFD process.
Fully robust system in place: 5	Fully robust system in place: 5	Fully robust system in place: 5	Fully robust system in place: 5	Fully robust system in place: 5	Fully robust system in place: 5
Management systems in place, yet not always practiced: 4	QFD system in place, yet not always practiced: 4	Training for QFD systems in place, yet not always practiced: 4	QFD tools/templates in place, yet not always practiced: 4	Procedures/processes systems in place, yet not always practiced: 4	Metrics/data systems in place, yet not always practiced: 4
Management system in place but rarely practiced: 3	QFD system in place but rarely practiced: 3	Training for QFD in place but rarely practiced: 3	QFD tools/templates in place but rarely practiced: 3	Procedures/processes in place but rarely practiced: 3	Metrics/data in place but rarely practiced: 3
Recognize requirements, planning to implement: 2	Recognize requirements, planning to implement: 2	Recognize requirements, planning to implement: 2	Recognize requirements, planning to implement: 2	Recognize requirements, planning to implement: 2	Recognize requirements, planning to implement: 2
Recognize requirements but no plan to initiate action: 1	Recognize requirements but no plan to initiate action: 1	Recognize requirements but no plan to initiate action: 1	Recognize requirements but no plan to initiate action: 1	Recognize requirements but no plan to initiate action: 1	Recognize requirements but no plan to initiate action: 1

effectiveness. Taguchi's methods are claimed to have provided as much as 80 percent of Japanese quality gains. This is remarkable, when it is remembered that Japanese industry outdid U.S. industry for a time. [42]

Dr. Taguchi's methods were a product of the Japanese post–World War II era. Resources were scarce, financial support minimal, and this period of Japanese history required accelerated learning and giant strides in improvement. From an engineering background, Dr. Taguchi converted his study of statistics and advanced mathematics into a system merging statistical techniques and engineering experiments. Table 3.6 explains the term *design of experiments.*

Dr. Taguchi was born January 1, 1924. His advanced formal training was directed toward textile engineering at Kiryu Technical College, but he devoted extensive personal study to statistics. After World War II, he was hired by the Japanese Ministry of Public Health and Welfare to deploy the country's first national study on health and nutrition. His associations there led him to become involved with Morinaga Pharmaceutical, where efficient experimentation techniques were critical to developing methods for producing penicillin. The realities of deadlines and production helped shape his approach to applying experimental techniques to actual design and production situations. This fostered a growth and appreciation for making assumptions from engineering knowledge to reduce the size of experiments and thereby speeding up the experimentation process. Productivity was reported to have increased tenfold annually during his efforts there.

His continued research took him in the direction of the study of random noise and its effect on variability. He adopted the use of the orthogonal array as an effective experimental design tool. Previously, most experimental studies had been performed by using the classical full fractional approach. His next position was at the Electrical Communications Laboratories of the Nippon Telephone and Telegraph Company. Both Electrical Communication Laboratories and Bell Laboratories from the United States were developing similar cross-bar telephone exchange systems. In addition to facing Bell's

TABLE 3.6 Definitions

An experiment can be defined as any act of observing.
Design of experiments is a series of "experiments" (observations) to minimize variability, improve the product, and increase yield.

greater resources, the Japanese were limited to inferior materials. Despite all the adversity, the new cross-bar system from Japan was rated superior and had a much lower cost to produce. The effect was so dramatic that Western Electric stopped production in the United States and began importing the systems from Japan.

Other Japanese companies learned from these lessons and adopted Dr. Taguchi's techniques, as well. Toyota, Fuji Film, and other Japanese firms followed suit and became the quality and price leaders. [43] DOE has been used by almost all Japanese manufacturing companies since the 1950s and 1960s. Nippon Denso Company, an automotive electrical and electronics subsystem supplier in Japan, reports that its engineers conduct more than 4000 experiments every year. One hundred engineers from Toyota Motor Company take a full 120-hour training course in the design of experiments every year. American companies did not become quality-conscious till the 1980s, but then AT&T, Bell Laboratories, Xerox, and Ford Motor Company adopted the Taguchi methodology. In 1985 Ford Motor Company decided to require suppliers to practice DOE, and in 1988, the Department of Defense highlighted the importance of DOE methods in the report on variability reduction program, which is provided in the appendix of this book. [44]

One of the principal thrusts of Dr. Taguchi was to stress robust design. Robustness is defined in Table 3.7. [45] More-robust designs result in higher customer satisfaction, lower production and support costs, and greater production flexibility.

Traditional approaches to design and production stress compliance with specifications. The ideal "target" value that will translate into the most robust product, and produce the highest customer satisfaction, should be some fixed value within the specification limits. However, as long as products fall within the specification limits, as illustrated by points *a* and *b* in Fig. 3.10, they are considered equally good. If a product falls outside the limits, as in point *c*, it is automati-

TABLE 3.7 Definitions of Robustness

Product robustness: The ability of the product to perform consistently as designed with minimal effect from changes in uncontrollable operating influences.
Process robustness: The ability of the process to produce consistently good product with minimal effect from changes in uncontrollable manufacturing influences.

SOURCE: Reprinted from Ref. 45, with permission.

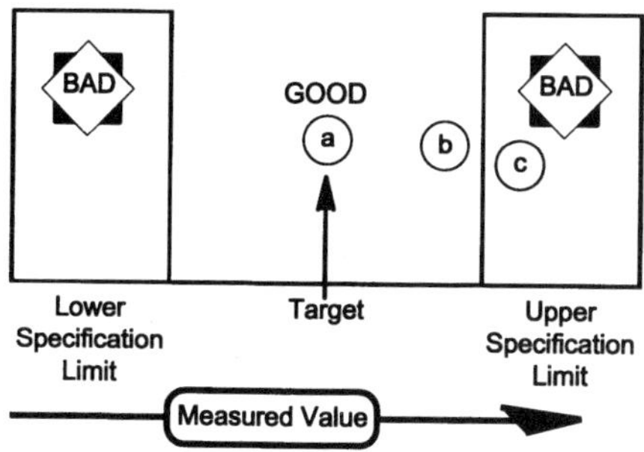

Figure 3.10 Any product within specification is good (old thinking). (*Reprinted from Ref. 46, with permission.*)

cally bad. Observation and common sense reveal that the difference between the "bad" *c* and the "good" *b* can be slight, and the distinction between them is somewhat arbitrary. Likewise the difference between *a* and *b* makes it clear that they are not equal, and *b* the farthest from the target value, must be in some way less "good."

If the target value is indeed an ideal value, product performance will degrade as a function of any deviation from the target. An increasing cost is associated with the product further from the target. For example, the price may be lower reliability, higher warranty costs, less accurate performance, or less tolerance of the environments in which the product is used. Either the customer or the producer may directly bear the increased cost, but ultimately both will be hurt. [46] A dissatisfied customer may choose not to be a repeat customer.

Dr. Taguchi's contribution involved combining engineering concepts and statistical methods to achieve robustness in product design and manufacturing processes.

The Taguchi Method is a different approach from the more conventional statistically designed experiment methodologies because it is highly efficient in its ability to extract relatively large amounts of information from small-scale experiments. This translates to lower material cost, less time for finishing the experiment, less testing time, and less time for analyzing the results. The primary tenet of the Taguchi Method is to improve quality in the most efficient and cost-conscious way.

The Taguchi Method is an engineering tool used prior to putting a product or process into production. By comparison, American quality control tended, in the past, to use tools to control quality during production. The Taguchi Method seeks to minimize the need for production quality control by reducing variability in products and processes, thus increasing the ability to overcome the many uncontrollable, changing conditions experienced during production or use.

The method is applied to develop products and processes with smallest possible variance, while maintaining processes on the target or nominal value of the specification.

The Taguchi Method therefore is a fundamental departure from conventional U.S. quality efforts in its strategy for quality improvement. U.S. manufacturers attempt to improve quality by removing the *cause* of variation. The Taguchi Method reduces variation by making the product insensitive to the causal elements, i.e., *noise*.

The Taguchi Method advocates a three-step design process, often referred to as *quality engineering*. This process may be used to facilitate the systems engineering or concurrent engineering approach to integrated product developments. The steps are

1. *System design.* This step is concerned with the development of any new product or process. The engineer designs a system with a specific function and finds the best technology for the product. In the conventional approach, system design constitutes the total design process, whereas in the Taguchi Method, system design is only the first of three steps. To facilitate system design, QFD may be used as a convenience to guide the development of key product characteristics needed to satisfy customer requirements.

Products or processes can even be optimized during system design. Since the Taguchi Method can be used in simulation experiments with the performance of a system, it can be described fairly well with mathematical formulas. Simulation with the Taguchi Method offers a low-cost way to discover optimum levels for each characteristic in the design.

2. *Parameter design.* This step reduces the *effects* of variability. It attempts to develop the "robustness" of the system through experimentation, to find the parameter values which make the product least sensitive to variation due to noise. The term parameter refers to any aspect of the product or process design that is subject to control by the respective designer. A parameter might be the composition of the materials used in a process, the shape and number of parts in an assembly, the temperature setting for a particular thermostat, or some other factor that can be controlled. Parameter design consists of selecting a set of parameter values for both the product and the process to make them relatively insensitive to noise.

Parameter design, including the use of statistical experiments, has been part of the design process but it has traditionally been practiced when the goal was improvement in performance measure. It has only recently been used in the United States to reduce variability.

As part of parameter design, an important concept in the Taguchi Method is the signal-to-noise (S/N) ratio. It is a noteworthy way to increase productivity, quality, and lower cost. The S/N ratio is one of Dr. Taguchi's concepts that allows analysis to take into account the variability caused by a factor or independent variable. Conventional statistics factors are usually analyzed to determine their effect on the mean (average) performance of some quality characteristic or functional parameter. The Taguchi Method demonstrates, however, that some factors have an effect on changing the mean, while others have an effect on changing *the variability.* The role of factors affecting variability is simply missed by conventional experimental methods. Critical quality characteristics (for example, the input and output characteristics) must be considered as parts of experiments concerning variability.

One major contribution of Taguchi is the adaptation of orthogonal arrays for designing efficient experiments and analyzing experimental data. Orthogonal arrays were originally developed in England by R. A. Fisher for controlling experimental error. Dr. Taguchi has since used the orthogonal array not only to measure the effect of a factor on the average result, but to determine the variation from the mean as well. A primary advantage of orthogonal arrays is that they show the relationship among the factors under investigation in a cost-effective manner. For each level of any one factor, levels for other factors exist, so that all factors occur an equal number of times. This constitutes a balanced experiment and permits the effect of one factor under study to be separated from the effects of other factors. The result is that the findings of the experiment are reproducible.

An enhancement that Dr. Taguchi incorporated, to add flexibility to the orthogonal array, is the linear graph. The linear graph is a graphical representation of the orthogonal array for assigning factors under investigation and corresponding interactions among these factors (see Fig. 3.11). By using these specially designed graphs, the experimenter can effectively study interactions between experimental factors as well as the effects of the individual factors (main effects) themselves. This is possible because the linear graphs provide a logical scheme for assigning interactions to the orthogonal array without confounding the effects of the interactions with the effects of the individual factors being studied. [47]

3. *Tolerance design.* This step reduces the *causes* of variability in the product. The purpose of this area of design is to select the tolerance that must be used to assure minimum loss after the product has been manufactured and is being used by the customer. Only if the product is not acceptable at its optimum level is tolerance design considered.

In this analysis, a function called the *loss function* is used. The loss function equals quality to Taguchi. It establishes a financial measure

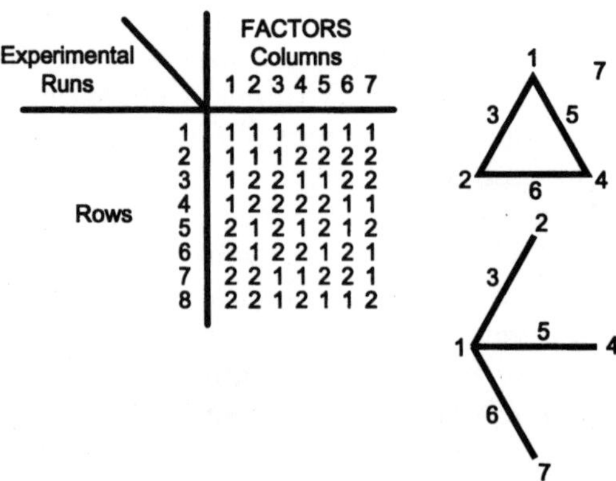

Figure 3.11 Orthogonal array and linear graphics. (*From Ref. 47, with permission.*)

of user dissatisfaction with a product's performance as it deviates from a target value. Statistical techniques are used to compute a loss function, as illustrated in Fig. 3.12, that can be used to determine the cost of producing products that fail to achieve a target value. The tool is also useful for determining the value, or break-even point, of improvement in a process to reduce variability.

The Taguchi loss function is a way to show the economic importance of reducing variation. While QFD is a process to ensure that customer requirements are understood and translated into technical requirements, the loss function is a method to quantify the cost of failing to meet those requirements. Taguchi's loss function is basically a statement that both the product and the customer will inevitably incur increasing costs (losses) as output variation increases around the customer's ideal requirement. [48]

The benefits of robust design have been demonstrated in automobile manufacturing, electronic component production, computer oper-

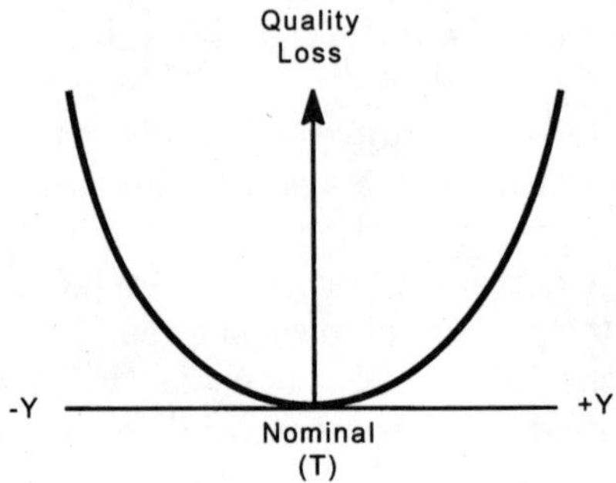

Figure 3.12 Taguchi loss function.

ating systems, engineering design, optimization of integrated circuit (IC) chip bonding processes, ultrasonic weld process optimization, and design of disk brake systems.

Management must endorse the Taguchi Method in the company for it to happen. A great deal of savings can be achieved by examining processes in place for improvements to be made. If there is a process which is more expensive to manage or has too many defects or returned defective parts, that is a good place to start. The continuous process tools of SPC/VR and the Taguchi Method allow a company the opportunity to capitalize on all the savings that are available. On new designs these methods are essential.

Experimental and robust design techniques, coupled with quality function deployment, concurrent engineering, and integrated product/process development (IPPD) are powerful tools to ensure development of robust products and processes. For example, a fundamental requirement of QFD is to determine the strength of relationships between customer and technical requirements, technical requirements and part characteristics, and part characteristics and process characteristics. Simple relationships (between the customer and technical requirements) might already be understood, but complex relationships are not. In these cases, experimental design techniques that use powerful orthogonal arrays can reveal these relationships. As part of QFD, experimental design helps people understand the relationship between customer quality requirements and process parameters. Math models relating input parameters to quality characteristics, along with robust design, increase product reliability. Experimental design used to create robust designs with increased quality and reliability will decrease overall life-cycle cost. [49]

A team should be established to conduct the experiments. The team should establish goals, objectives, a schedule, and methods to quantify savings. Table 3.8 shows the makeup of a typical electronic assem-

TABLE 3.8 Typical Electronic Assembly Experimentation Team

Manufacturing supervisor
Process engineer
Product engineer
Quality engineer
Taguchi advisor/expert
Component engineer
Product technician
Setup technician
Machine maintenance technician
Operator

SOURCE: From Ref. 50, with permission.

bly experimentation team. [50] A team leader should be selected who knows how to keep documentation and remove obstacles from the team members. Management should kick off the activity and continually review the activity end results.

Almost all experimental designs can be done with only the assistance of a calculator and without the use of a computer. However, statistical software packages can simplify the task. It is impossible to write a "current" summary of existing software—because it is changing so often. Each March in the ASQ journal, *Quality Progress,* there is an annual "QA/QC Software Directory" published which lists the latest available. When purchasing software, it is important to remember to consider future data analysis needs. Two sources for software that provide two different types are represented by DESIGN-EASE by STAT-EASE, Minneapolis, which is a PC-based system that takes only 256K of RAM, and the SCA Statistical System, published by Scientific Computing Association, Lisle, IL. This package can be run on anything from a PC to a mainframe, DOS or UNIX, and requires minimum memory of 640K. Most of the software packages that are sold in the United States use the traditional Western ordering of experimental levels as opposed to the Taguchi experimental levels. Lochner and Matar feel that the only difference between the two approaches is the notation and ordering. [51]

To develop robust design, the CEO and the organization must take the following actions:

Develop a clear policy of designing robust products and making production of robust products, services, and processes a high priority

Stress early application of advanced engineering techniques and tools in new product development and business planning

Provide training in advanced design development techniques such as the Taguchi methods

Ensure that all personnel understand and effectively employ the techniques and tools that produce robust designs

Does your company require managers to become actively involved in the design development process, employ statistical methods, facilitate design experiments, and vigorously work to remove design process roadblocks? [52]

Table 3.9 is a maturity matrix on robust design development for your company to use to judge where you are in this process.

There are a vast number of technical books which discuss the design of experiments in detail for the technical people who actually do the DOE studies. The American Supplier Institute, Inc., Dearborn,

TABLE 3.9 Robust Design Maturity Self-Assessment

Management responsibility	System in place	Training in robust design/DOE	Techniques and tools	Procedures/processes	Metrics/data
Executive management defines and documents robust design/DOE policy. All management participate in process. System is in place for all new design.	System is in place for all new design. All new products go through the process. Issues from SPC/VR are referred to DOE.	All personnel get training in DOE. All persons know when to recommend DOE from SPC/VR process issues.	Software, hardware in place. Samples for teams to use. Space to post matrices.	Procedures and processes documented and released. Audit is in place for these processes.	Data are collected on every DOE project and used for management, lessons learned, and continuous improvement in the DOE process.
Fully robust system in place: 5	Fully robust system in place: 5	Fully robust system in place: 5	Fully robust system in place: 5	Fully robust system in place: 5	Fully robust system in place: 5
Management systems in place, yet not always practiced: 4	Taguchi/DOE systems in place, yet not always practiced: 4	Taguchi/DOE systems in place, yet not always practiced: 4	Tools, templates in place, yet not always practiced: 4	Procedures/processes in place, yet not always practiced: 4	Metrics in place, yet not always practiced: 4
Management system in place but rarely practiced: 3	Taguchi/DOE in place but rarely practiced: 3	Taguchi/DOE training in place but rarely practiced: 3	Tools/templates in place but rarely practiced: 3	Procedures/processes in place but rarely practiced: 3	Metrics in place but rarely practiced: 3
Recognize requirements, planning to implement: 2	Recognize requirements, planning to implement: 2	Recognize requirements, planning to implement: 2	Recognize requirements, planning to implement: 2	Recognize requirements, planning to implement: 2	Recognize requirements, planning to implement: 2
Recognize requirements but no plan to initiate action: 1	Recognize requirements but no plan to initiate action: 1	Recognize requirements but no plan to initiate action: 1	Recognize requirements but no plan to initiate action: 1	Recognize requirements but no plan to initiate action: 1	Recognize requirements but no plan to initiate action: 1

MI, phone (313) 336-8877, holds the trademark copyright on the Taguchi Method and has a number of books by Taguchi on the subject. To fully understand Taguchi, it is recommended that one begin with Taguchi's books on quality engineering and system of experimental design.

Motorola University Press, Rolling Meadows, IL, prints two very helpful books especially for Electronics: M. J. Harry, *Electrical Engineering Applications of the Taguchi Design Philosophy,* 1987, and *The Nature of Six Sigma Quality.* Two very current books on the market are available from ASQ. They are G. S. Peace, *Taguchi Methods, a Hands on Approach,* Addison Wesley, Reading, MA, 1993 and R. Lochner and J. Matar, *Designing for Quality; an Introduction to the Best of Taguchi and Western Methods of Statistical Experimental Design,* ASQC Press, Milwaukee, 1990. Numerous books and articles are available from ASQ. Taguchi also is represented on the Internet, http:akao.larc.nasa.gov/dfc/tm.html.

Developing more robust designs is part of an effort to increase quality and become more productive. Efforts to reduce variability must continue across all process areas. It is the combination that yields the most competitive product. Feedback through statistical process control, variability reduction, and continuous improvement may lead the company back to Taguchi's Method to improve the design to be more robust. In the next section we shall address SPC and VR. Continuous improvement is discussed in Chaps. 4 and 6.

3.7.3 Statistical process control and 6 sigma

Statistical process control. In the race to world class manufacturing in a competitive market of electronics, more and more companies are incorporating statistical process control into their operations. SPC uses statistical techniques to create control charts, which aid in the analysis of a process or its outputs for taking appropriate actions to achieve and maintain a state of control, and for assessing where process capability can be improved. [53] An easier definition may be that SPC is a technique for controlling the quality of a process. It allows monitoring a process, detecting changes, and preventing new problems before they occur. It uses data obtained directly from the process. [54] SPC had its beginnings in 1931, when W. A. Shewhart published *Economic Control of Quality of Manufactured Product* (Fig. 3.13). During World War II, the United States used SPC, but then it died out. Since World War II, Japan has used SPC to a significant advantage. Around 1980, U.S. businesses became aware of the need for quality improvement and SPC. SPC is successful because it provides a tool that takes a picture of the

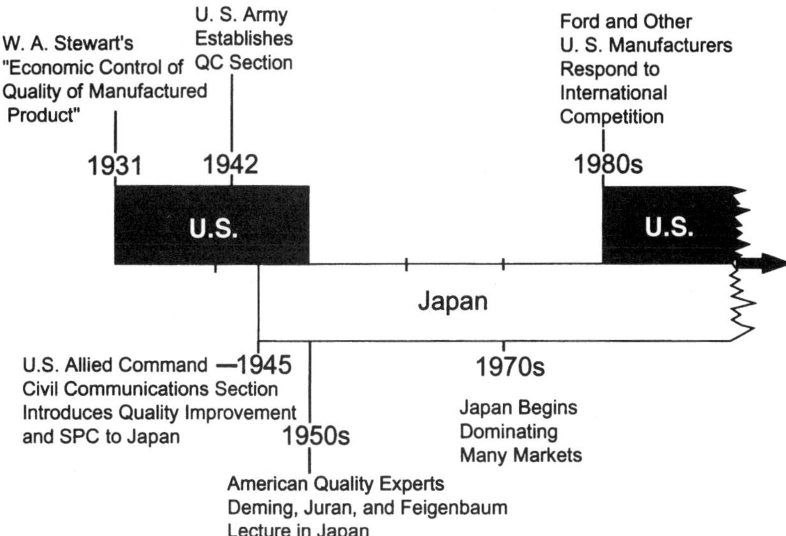

Figure 3.13 Timeline of quality efforts.

process. The data aids in the analysis to understand whether the process is performing as planned. In addition, SPC is controlled by the person who is most knowledgeable about doing the job and managing the process—the employee, not a quality inspector. Data are collected and standards are established. From these standards, one can see if the process is performing correctly.

SPC begins by identifying the variables or parameters that will be measured. Parameters are measured characteristics of a product or a process. Critical and key characteristics (Table 3.10) must be established because they are more important to the process than other variables, or parameters. Critical characteristics are those that have a significant effect (could cause failure) on the process. Key characteristics have a great impact on the process, and must be acknowledged, so they can be focused on rather than other characteristics. [55] Many prime contractors and OEMs will specify critical and key characteristics, or expect their suppliers to do so. General Motors prints a manual entitled *Key Characteristics Designation System,* GM-1805,QN, 1991.

There are basically two forms of data collection for the determination of process quality [56]:

1. *Attribute data,* which is determined by the number of defects present in a sample using typical go/no-go guidelines. Thus we know

TABLE 3.10 Classification of Characteristics

Used for process definition and improvement	
Critical	Critical to safety, operation, or performance in service or manufacturing classification (usually 2–3% of total assembly), e.g., safety equipment for aircraft or automobiles.
Major	Cause of substandard performance; difficult to locate and to correct; costly to repair; requires returning to dealer
Key	Influences performance, supportability, and cost. Determines the production processes.

SOURCE: Adapted from Ref. 54, Chap. 2.

whether we have a good or bad part, but have no indication of the part's actual condition;

2. *Variable data,* which is obtained by making a measurement. By comparing the measured value with a specification or target value, we can determine exactly how far away the process or product is from the specification limit. The power of collecting variable data is that the process mean and the standard deviation (variability) can be calculated and the capability of the processes may be determined. Variability data, therefore, lead to more quantifiable information and should be the preferred data collection system.

SPC is applied to measure and analyze existing process variation. Process variability has to do with the fact all processes are subject to variation. Variation is the reason why specifications have tolerances. In electronics it is important that the processes produce very little variation, and that is why we must use SPC to control them. In manufacturing processes, variability is likely due to the five components that make up a process: people, equipment, materials, methods, and environment.

The goal of SPC is to get the process in control (or repeating itself) by eliminating all the causes of variation. Once a process is in statistical control and is producing parts consistently, it can be compared to the specification limits to see if it is capable of producing quality products.

SPC's main tool is the control chart. A sample control chart is in Fig. 3.14. Charts are used to record, calculate, and plan data and to connect plotted points to find a pattern. They allow you to recognize quickly whether the process is going to produce defects or errors, and

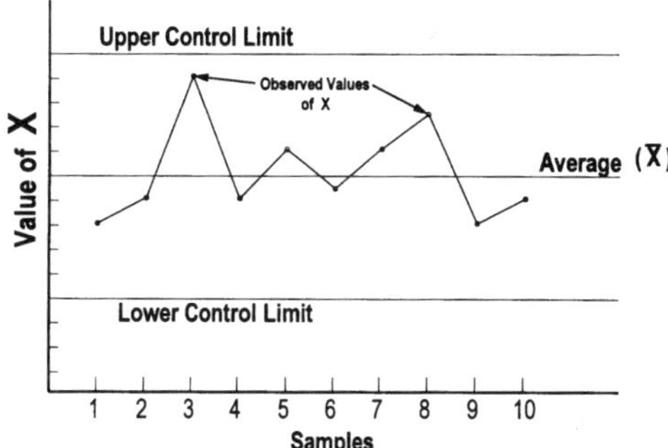

Figure 3.14 Generalized control chart for statistical process control.

as a result, take corrective action. A control chart is a graphic method for evaluating whether a process is, or is not, in a state of "statistical control." The control chart distinguishes between the causes of normal and abnormal behavior in the process. SPC refers to normal behavior as *common,* or *random, causes* and abnormal behavior as *special,* or *assignable, causes.* Table 3.11 shows examples of both kinds of causes.

Dr. W. E. Deming estimated that about 85 percent of all problems in variation stemmed from common or random causes within the processes and were changeable only by management. He said that the remaining 15 percent are due to special, or assignable, causes and are changeable by the employee, mostly right at the workstation. The power of this technique, according to Grant and Leavenworth in 1988, is that the ability to separate out these assignable causes of quality variation brings about substantial improvements in product quality and therefore equates to reduced cost. The improvement in quality relates to an increase in value. Thus, statistical quality control is a double-barreled driver for competitive advantage. [57]

Boeing found that when setting up an SPC system it faced an issue on which type of control charts should be used. (See Table 3.12 for a comparison of control charts and their attributes.) They found that a decision tree process (Fig. 3.15) solved this problem. In using the decision tree, the first thing that must be identified is the type of data being collected: variable or attribute. If variable data are used, two control charts are required; one to monitor the process average and one to monitor process variation. [58]

Increased worldwide competition in the electronics industry has prompted many companies to reevaluate business strategies and begin systems incorporating SPC. These systems use statistical tech-

TABLE 3.11 Examples of Common (Random) and Special Variability

Cause	Common variability	Special variability
People causes	Inadequate management Inadequate recruitment Inadequate training Poor communication	Changes in employees Employee fatigue Employee haste Employee inexperience Inadequate response to training
Equipment causes	Chronically inaccurate machine settings Inadequately maintained machines Machines not suited to requirements Poor quality or poorly calibrated gauges	Change in measuring equipment Not responsible to machine or tool malfunction Not responsive to machine or tool wear
Materials causes	Defective incoming materials Methods unsuited to requirements	Not responsive to unacceptable incoming materials Not responsive to unacceptable output from previous processes of company
Methods causes	Inadequate training Methods unsuited to requirements	Change in measuring method Inappropriate method or method used inappropriately New method used
Environmental causes	Dirt Excessive noise Excessive vibration Extreme heat or cold Inadequate safety precautions Poor lighting Poor ventilation Too high or too low humidity	Not following safety rules Work area not kept clean

SOURCE: Adapted from Ref. 54, with permission.

TABLE 3.12 Comparison of Some Control Charts

Statistical measure plotted	Average $\overline{X}$ and range R	Percentage conforming	Number of nonconformances
Type of data required	Variable data.	Attribute data.	Attribute data.
General field of application	Control of individual characteristics.	Control of overall fraction defective of a process.	Control of overall number of defects per unit.
Significant advantages	Maximum utilization of information. Detailed information on process average and variation for control of individual dimensions.	Data often available from inspection records. Easily understood by all personnel. Provides an overall picture of quality.	Same advantages as p chart but also provides a measure of effectiveness.
Significant disadvantages	Not understood unless training is provided; can cause confusion between control limits and tolerance limits. Cannot be used with go/no-go type data.	Does not provide detailed information for control of individual characteristics. Does not recognize different degrees of defectiveness in units of product.	Does not provide detailed information for control of individual characteristics.
Sample size	Usually 4 or 5.	Use given inspection results or samples of 25, 50, or 100.	Any convenient unit of product such as 100 ft of wire or one TV set.

SOURCE: Reprinted from J. M. Juran and F. M. Gyrna, *Quality Planning and Analysis,* 3d ed., McGraw-Hill, New York, 1993, p. 382, with permission.

niques through product design, process design, manufacturing, and/or the providing of services. The Institute for Interconnecting and Packaging Electronic Circuits (IPC) provides in its ANSI/IPC-PC-90 a systematic path for implementing SPC (Fig. 3.16). As is apparent from the illustration, a company must understand both the product and the process to begin implementation of SPC. Figure 3.17 shows the strategic planning framework from the IPC standard. [59]

One good example of a company needing SPC is Varian's Chromatography Systems in Walnut Creek, CA. As they were moving to a just-in-time (JIT) system they found out they needed SPC to survive. There were problems with printed circuit board failures and contamination problems. Employee morale was low, because customers were acknowledged to want product quality and reliability, and yet the company's manufacturing process was not working. Lines of communication between manager and workers broke down.

SPC became the language that increased communication and that all employees and managers could agree to. Varian decided to insti-

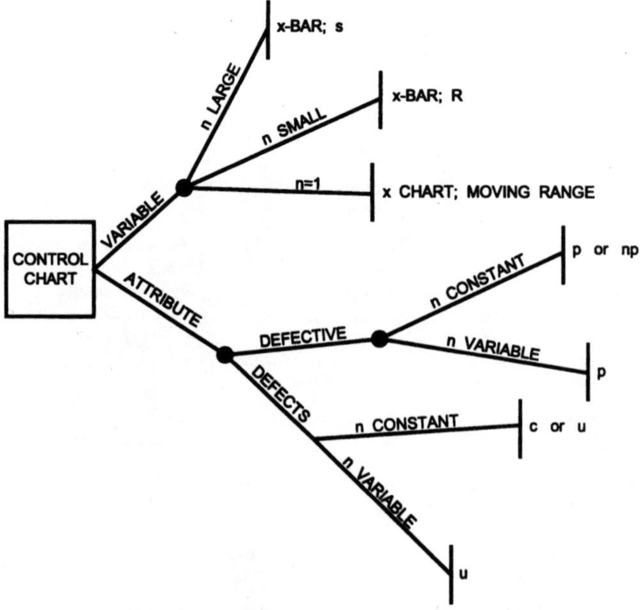

Figure 3.15 Decision tree. (*Reprinted from Ref. 58, with permission.*)

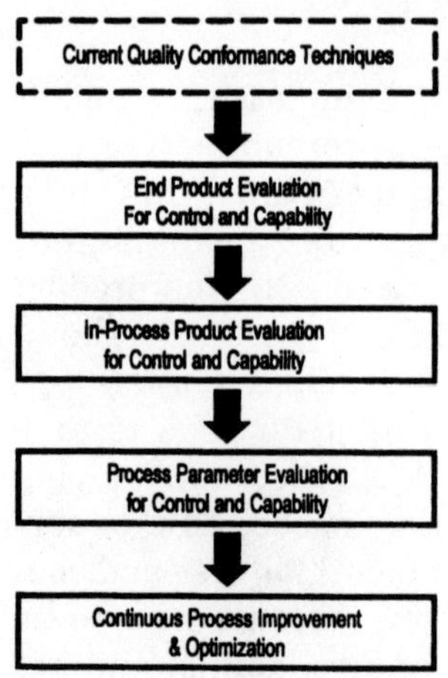

Figure 3.16 Systematic path for implementation of SPC. (*Reprinted from Ref. 53, with permission.*)

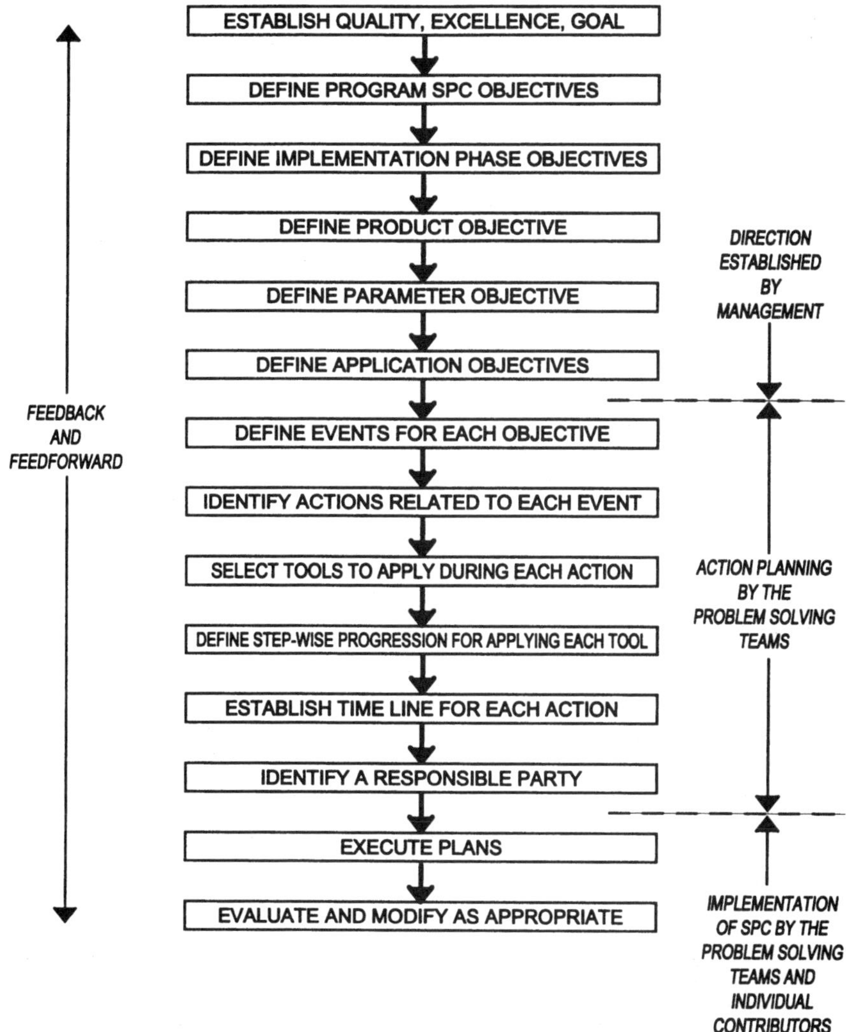

Figure 3.17 SPC strategic planning framework. (*Reprinted from Ref. 53, with permission.*)

tute SPC. Classes were begun. There was some resistance, but the turnaround had begun. Since SPC was implemented, impressive metrics have been collected. Cycle time is down for Varian's instruments from 18 to 64 percent, depending on the instrument; on-time customer deliveries are up 38 percent; manufacturing's cost of quality has dropped 50 percent; productivity (measured by units produced per employee) rose by 28 percent; and factory inventories were reduced 30 percent at a time of sales growth. As good as these concrete measurements are, some of the unmeasurable accomplishments are noteworthy. Employee morale is high, and knowledge of the manufacturing processes and job satisfaction is at an all-time high. Line employees now train the Varian sales force on the benefits of SPC for key customer presentations. One of the biggest problems for Varian to deal with was the fear of change, that SPC would take away jobs, and that the employees could not understand SPC. Training had to occur and employees had to have time to work through their role in SPC and find how the culture changes. [60]

When companies begin instituting SPC, they find that they must go back and work more closely with their suppliers. Varian experienced difficulty at first, until they produced data that showed how much SPC had improved Varian's process, and then developed a process for suppliers that would implement an SPC system to have a longer-term contractual relationship with the company. This is standard in industry.

When moving to autoinsertion equipment in printed circuit board (PCB) assembly, NCR Corporation in West Columbia, SC, found that SPC was the system that could help move from an inspect-and-repair mode to a prevention and design-for-quality mode. [61]

In the Automotive Industry Action Group (AIAG) guide on SPC which is issued as part of QS-9000, good advice is given to companies beginning SPC as a continuous improvement process [62]:

1. Gathering data is not an end in itself. The overall aim is to understand the process, and use that knowledge to improve the process, not just become a technique expert.
2. The basic concept of SPC can be used in any area. It often is associated only with the shop floor, but it can be used for bookkeeping (error rates), gross sales, waste analysis (scrap rates), computer systems (performance characteristics), and material management (transit times). The AIAG guide references a long list of SPC books in the AIAG appendix.
3. Note that SPC stands for statistical *process* control. Many times, SPC is applied only to parts rather than processes. Application of SPC to control output of parts should be only the first step. The processes need to become the focus of the efforts, thereby improving quality, increasing productivity, and reducing cost.
4. Suppliers need to study actual cases from their processes to obtain a deep contact with process control situations, and therefore improve SPC on processes at their locale.

5. Company personnel are encouraged to take formal education in statistical techniques. They need to further their knowledge.

Measurement systems are critical to proper data analysis and they should be well understood before the process data are collected. When such systems lack statistical control, or their variation accounts for a substantial portion of the total variation in process data, inappropriate decisions may be made. More information on this topic is found in the measurement systems analysis (MSA) manual published by AIAG.

Once a process is in statistical control and is producing parts, it can be compared to the specification limits to see if it is capable of producing quality product. The capability of a process is directly related to the ability of the process to produce parts with specification limits. *This is called process capability.* The ability to quantify process capability can be an important tool in controlling and reducing variation. The process variation around the target value can be measured by a process capability index C_p. [63] Process capability provides a quantified prediction of process adequacy. This ability to predict quantitatively has resulted in widespread adoption of the concept as a major element of quality planning. Process capability information serves multiple purposes [64]:

1. Predicting the extent of variability that processes will exhibit. This assists designers in setting realistic specification limits.
2. Choosing from among competing processes that which is most appropriate for the tolerances to be met.
3. Planning the interrelationship of sequential processes. Where one process may distort the precision achieved by a predecessor process, quantifying the respective process capabilities often points a way to a solution.
4. Providing a quantified basis establishes a schedule of periodic process control checks and readjustments.
5. Assigning machines to classes of work for which they are best suited.
6. Testing theories of causes of defects during quality improvement programs.
7. Servicing as a basis for specifying the quality performance requirements for purchased machines.

These seven tasks undertaken as a quality improvement program would yield significant improvement based on the findings of the process capability. This needs to be done for every process in your electronics company.

Process capability should be carried out with the following assumptions [65]:

- The process is statistically stable
- The individual measurements from the process conform to the normal distribution
- The engineering and other specifications accurately represent customer needs
- The design target is in the center of the specification width
- Measurement variation is relatively small

The process capability index C_p portrays the relationship between the upper and lower specification limits and the estimated standard deviation of the process. It does not take into account the location or central tendency of the population. A C_p of 1.0 shows the process is capable. A minimum value of 1.33 is generally used to show that ongoing processes are producing good product. [66] However, C_p does not show how well the process average is centered. The C_{pk} index does this. C_{pk} is commonly used when the process is not centered on the target value. Then the process capability may be calculated by one of the formulas in Table 3.13. [68] Capability indices can be divided into two categories: short-term and long-term. Short-term capability studies are based on measurement collected from one operating run. The data are analyzed with a control chart for evidence that the process is operating in a state of statistical control. If no special causes are found, a short-term capability index can be calculated. If the process is not in control, action regarding the special cause of variation will be required. This other type of study is used to validate the initial parts produced from a process for customer submission. Another use, sometimes called a *machine capability study,* is to validate that a new or modified process actually performs within the engineering parameters.

When a process has been found to be stable and capable of meeting requirements in the short term, a different kind of study is subsequently performed. Long-term capability studies consist of measurements which are collected over a longer period of time. The data should be collected for long enough, and in such a way, that it includes all expected sources of variation. Many of these sources of variation may not have been observed in the short-term study. When sufficient data have been collected, the data are pulled from the con-

TABLE 3.13 Determining Capability Indices

Example steps for determining capability indices for normal distributions.

For normal distributions, the C_p index shows the maximum potential of the process. The C_{pk} index shows the actual process performance. When a C_{pk} number is quoted, it is most important that a C_p number be quoted also. C_{pk} and C_p are related and should be used together. It is important for a process or a product to be on target. Pictures of process data related to the specification limits and target values are useful for reporting the relationship. Sigma and mu are seldom, if ever, known and may be estimated by $\overline{X}$ and σ respectively. Accurate estimates of sigma and mu may come from control charts and product samples.

Calculate the capability indices for the parameter (C_p and C_{pk}).

1. The C_p index may be calculated as

$$C_p = \frac{\text{specification limits or operating limits}}{6\sigma}$$

For bilateral specification limits or operating limits,

$$C_p = \frac{\text{USL}-\text{LSL}}{6\sigma}$$

where C_p = capability index
USL = upper specification or operating limit
LSL = lower specification or operating limit
6σ = capability of the parameter under study (may be approximated as $6s$)

2. The C_{pk} index may be calculated as follows:

For bilateral specification limits or operating limits:

$$C_{pk} = C_p(1-\text{K}) = \min(\text{CPU, CPL})$$

where C_{pk} = capability index (considers centering of the process variability with respect to specification/operating limits)
C_p = capability index
$(1-K)$ = a scaled distance by which the parameter is off center

$$K = \frac{\text{normal}-\mu}{(\text{USL}-\text{LSL})/2}$$

μ = mean of the parameter (may be estimated by $\overline{X}$)

SOURCE: Reprinted from Ref. 68, with permission.

trol chart, and if no special causes are found, long-term capability and performance indices can be calculated. One use for this study is to describe the ability of the process to satisfy customer requirements over a long period of time with many possible sources of variation induced (i.e., to quantify process performance). Automotive companies have set requirements for process capability. [69] The entire electronics industry would be wise to follow this example, and many prime contractors and original equipment manufacturers (OEMs) perhaps have in certain situations. The company needs to communicate with the customer and find out the requirements. It is important to remember that most capability indices include the product specification in the formula. If the specification is inappropriate, or not based on customer requirements, much time and effort may be wasted in trying to force the process to conform. [70] See Chap. 2 for data about specifications.

Many personal computer programs are available for statistical analysis and all types of control charts. Gauges with direct digital outputs can be plugged into hand-held data collection devices. These will store the gauge readings collected on the shop floor and will download them into a personal computer in the office. As an alternative, gauge readings may be entered into the collection device through a keyboard.

The software will calculate the sample parameters, initial control limits, and the control chart. Of course, control limits may be easily recalculated periodically and $\pm 3\sigma$ limits can be calculated as an additional guide. Other software will provide additional summaries and analyses such as listings of the raw data, out-of-specification values, histograms, checks for runs and other patterns within control limits, tests for normality, process capability calculations, Pareto analysis, and trend analysis.

The relatively low cost of personal computers and the availability of this software have contributed substantially to the renewed interest in $\overline{X}$ and R charts. This computer software combination has made it practical to collect large quantities of data and subject it to various and complex forms of analysis. However, process improvement still requires the identification of the vital few, and often unexpected, variables, causing excess variation. Then experiments must be conducted to find the root causes of the variation. Computers are most helpful when the diagnostician has created a template for a spreadsheet or a database program, and least helpful when a prepackaged statistical analysis program is used. Users are rarely familiar with the detailed logic of the prepackaged programs which therefore can lead to erroneous conclusions.

The American Society for Quality annually publishes a directory of software for process control in its March edition of *Quality Progress.* Statistical process control is well covered in the quality press. The documents in the following list are recommended. SPC is clearly a necessity for electronic industry processing. If your company expects to succeed, SPC must be implemented. Benchmarking along with strategic planning has proved the ANSI/IPC standard with industry leaders in electronics. It's essential that a company have a leader who understands SPC and can take the lead in setting up SPC and making sure all the people are trained and the system is operational. Data can be obtained from:

Juran's Quality Control Handbook, 4th ed., J. M. Juran, editor-in-chief, McGraw-Hill, New York, 1988. Sections 23 and 24 deal directly with statistical methods and statistical process control.

Quality Planning and Analysis, 3d ed. J. M. Juran and Frank M. Gyrna, McGraw-Hill, New York, 1993. Chapters cover probability concepts, statistical tools for analyzing data, design for quality statistical tools, statistical process control, and statistical tools for marketing, field performance, and customer service.

Statistical Quality Control, 6th ed., E. L. Grant and R. S. Leavenworth, McGraw-Hill, New York, 1988. An in-depth analysis of statistical control for a statistician.

ANSI/IPC-PC-90, *General Requirements for Implementation of Statistical Process Control,* a standard developed by the Statistical Process Control Subcommittee of the Process Control Management Committee of the Institute for Interconnecting and Packaging Electronic Circuits, Lincolnwood, IL, 1990. It corresponds directly with an SPC system produced by the Electronic Industries Association (ANSI/EIA-557-1989, *Statistical Process Control Systems*). A third document, ANSI/EIA-554-1988, defines the quality report as parts per million in Sec. 3.2.2.3. This is probably the best guide to aid in implementation. It includes information about the SPC system, as well as positions required, and quality checklist and audit information.

Statistical Process Control: SPC, Reference Manual, Chrysler Corporation, Ford Motor Company and General Motors Corporation, 1992, 1995. Copyright by the Automotive Industry Action Group. AIAG may be reached at (311) 358-3570. This is an excellent guide on SPC and process capability.

Measurement Systems Analysis, Reference Manual, Chrysler Corporation, Ford Motor Company and General Motors Corporation, 1992, 1995. Copyright by the Automotive Industry Action Group. AIAG may be reached at (311) 358-3570.

Six-sigma product quality. The 6-sigma concept is a relatively new way to show how "good" a product is. When a product is 6 sigma, it tells us that product quality is excellent. It says the probability of producing defects is extremely low. Motorola, like other world class electronic manufacturers, has established this target for the quality of its products. [71] To better understand, let us define sigma, or σ, the standard deviation. Sigma is the square root of the variance of the distribution of values in the sample being measured. It is a measure of the distribution about the mean value of the sample.

To apply this concept, we must first determine how many opportunities there are for a nonconformity, or "defect," to occur as related to a particular product. Then we must count the number of defects associated with that product during the manufacturing process. With this information, we are now able to determine how many defects there are per million opportunities for a defect. The standard deviation of a process may be obtained through SPC. It also can be converted and expressed as parts per million (ppm). Complete details are illustrated in Harry, *The Nature of Six Sigma Quality.* Using the sigma (or standard deviations) approach, a common vocabulary can be discussed. If the number is small, say 2, the product quality is not very good. Actually, the number of defects produced here is intolerable. But if the number of standard deviations is large, say 6, quality would be excellent (see Fig. 3.18). [72]

Some people think that 6 sigma has become the de facto industry standard for "good" quality in this decade. However, some argue that it is really a quality improvement program to reduce defects. Motorola's concept of 6 sigma, based on the practical uses of the normal distribution, asserts that there are strong relationships between product nonconformities, or defects, and product yield, reliability,

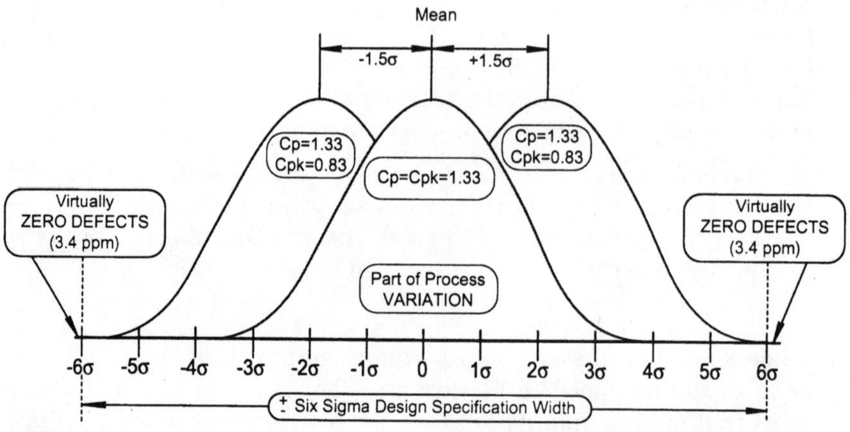

Figure 3.18 Motorola's 6-sigma process. (*From Ref. 71, reprinted with permission.*)

cycle time, inventory, schedule, and so on. More often than not in the industry, statisticians say that 6-sigma quality is associated with 3.4 defects per million parts. Based on these two definitions, it is logical to say that 6-sigma quality is a quality improvement program with a goal to reduce the number of defects to as low as 3.4 ppm. In this calculation, many engineers/statisticians agree that the Motorola model allows the process to be off center in relation to the standard deviation. Engineers/statisticians argue that this does makes sense. The argument is over whether the target value is at the center of the process or not. [73] The real key for an electronics manufacturer would be to set up an SPC system, calculate the sigma, and thus know their objective in improving quality and reducing costs. A company could use this data to predict yields for new product as well. In the Deming philosophy, it is not good to set goals and 6-sigma is the same way as other goals. As in Harry's process, where 6 sigma can be calculated as parts per million, many semiconductor manufacturers are already using parts per million and getting lower than ever thought possible.

Setting up a process to get to 6 sigma, and achieve excellent quality, is not a small task. As Deming says so often, this is management's responsibility. Management must understand that only one facet of setting up SPC is the answer. Many different approaches can be defined. Based on Harry's list, the following recommendations are offered [74]:

- Cycle time manufacturing must be kept, and the shorter cycle time the better.
- Design for producibility and no variability must be in place.
- Statistical process control is a given.
- Statistical process control must be implemented at suppliers.
- Participative management must be practiced (TQM/empowerment of employees).
- Part standardization and supplier qualification must be used.
- Computer simulation must be used.

If these processes were standardized in a company, and practiced as written, the company would have a big "Q" quality system with continuous improvement. Some companies are doing this already: Hewlett Packard, Motorola, and Texas Instruments are examples. It sounds and reads so easy. But the culture at so many companies is not there. The bottom line is that in electronics, the statistical analysis of a company's processes will show where the company fits in the world class scenario, and that costs can be lowered by beginning with SPC and the Taguchi Methods (discussed in the prior section) and further reduced by variability reduction (discussed in the next section).

Variability reduction. "The aim in production should be not just to get to statistical control, but to shrink variation. Costs go down as variation is reduced. It is not enough to meet specifications." [75] Product variation results from three primary sources: insufficient design margin, inadequate process control, and less than optimum parts and materials. [76] Not only Deming, but all of the quality gurus agree that, in general, reduced variability improves the predictability of performance, improves the manufacturing process, and reduces the life cycle of operation. [77]

Systems fail for many reasons. Some components, like tires, wear out. The major cause of eventual failures, however, is variability in the original manufacturing processes. Variability results from changes in the conditions in which items are produced, including differences in raw materials, machines, and their operations. Design methods based on modern quality techniques minimize the significance of variability. Traditionally, the approach has been to tighten the design tolerance and increase inspection. Costs climb and productivity drops as scrap and rework increase. A better method would be to design robust products that are insensitive to the causes of variability in the first place, then eliminate causes of variability through statistical techniques. Reducing variability through statistical techniques and product design is called the *variability reduction process.* [78]

Motorola (as well as other world class companies) believes that in this highly competitive marketplace the ability to effectively and efficiently deal with product variation is crucial. Variation is the enemy to getting to product excellence. [79] Motorola says, "To defeat this enemy, we must all strive to go beyond just 'meeting the specs.'" [80]

It seems that those who really understand quality (the big "Q") and continuous improvement understand that meeting the specifications and improving the existing processes are not enough. Deming sums this up excellently as he says, "It is thus not sufficient to improve processes. There must also be constant improvement of design of product and service, along with introduction of new product and service, and new technology." [81] An example of continuous improvement and design metrics is shown in Figs. 3.19 and 3.20. Sony had stated goals to reduce the size and weight of their portable compact disk player and practiced continuous improvement to increase market share. [82]

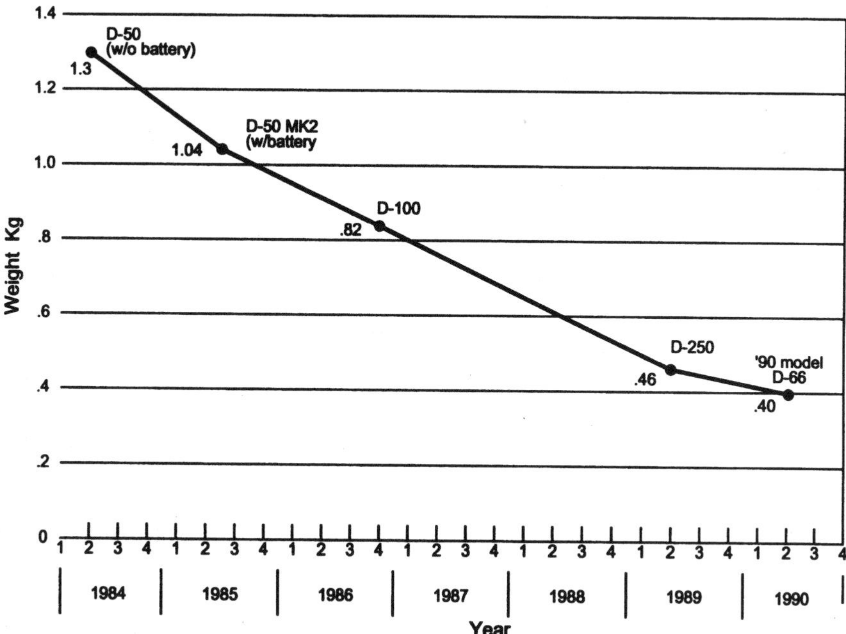

Figure 3.19 Weight of Sony's portable CD models. (*Reprinted from Ref. 82, with permission.*)

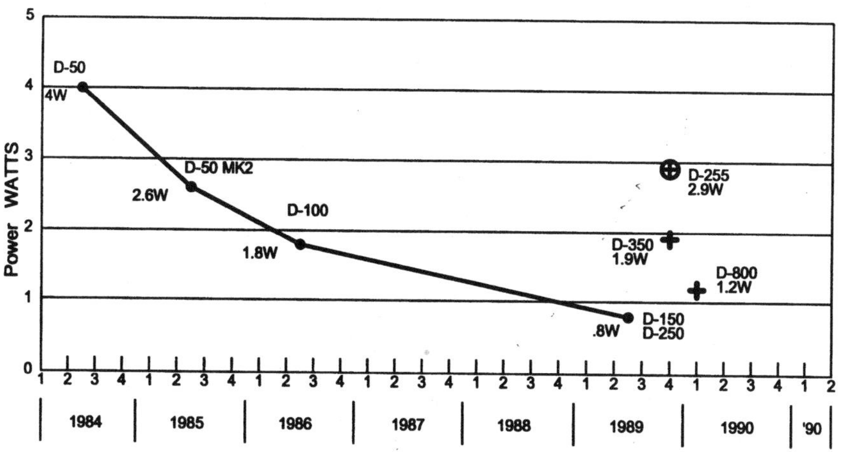

Figure 3.20 Power consumption of Sony's portable CD models. (*Reprinted from Ref. 82, with permission.*)

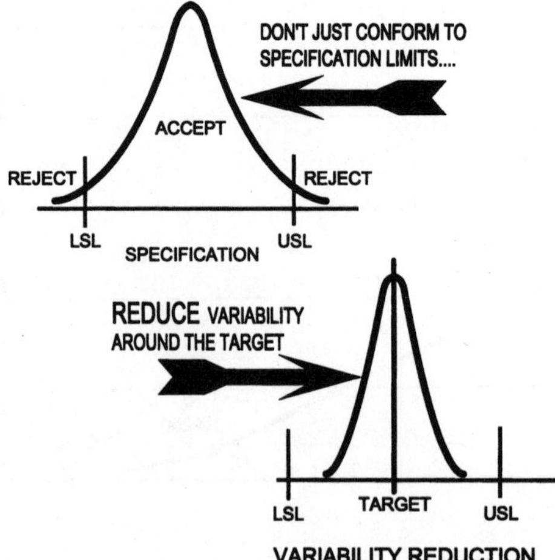

Figure 3.21 Variability reduction approach. (*From USAF R&M 2000.*)

As you embark on a variability reduction program (Fig. 3.21), it will affect the specifications; design support to improve the product will be needed. Companies like Xerox, Sony, and Texas Instruments work to eliminate all variation in the product and process before they manufacture huge quantities of product for consumers. They are continually updating the technology, and provide service 24 hours a day by telephone so that they are in touch with the customers.

To keep a process in control, removable sources of variability must be eliminated and control must be maintained over those that, for some reason, cannot be eliminated.The control chart distinguishes between random and assignable causes of variation through its choice of control limits. These are accumulated from the laws of probability in such a way that highly improbable random variations are presumed to be due not to random causes, but to assignable causes. When the actual variation exceeds the control limits, it is a signal that assignable causes entered the process and the process should be investigated. Variation within the control limits means that only random causes are present. The control chart not only evaluates significant statistics, but also provides an early warning of problems that could have major economic significance. [83] It is important for people to understand what type of variations they are dealing with, otherwise they will waste considerable time and effort chasing problems over which they have no control.

For the supplier, the benefits are many. They include improved yields, reliable scheduling, reduced cycle times, reduced scrap and rework, better cost control, potentially improved profitability, and improved competitiveness, in addition to the intangible benefits of improved customer satisfaction. Emphasis shall be placed on developing manufacturing processes whose variability around target product-critical attributes is minimized rather than on simply being within the product tolerance.

The underlying logic for applying VR is shown in Fig. 3.22. The VR effort should start with, but not narrowly focus on, processes which have been identified as "key processes" (Table 3.10). VR represents a quality philosophy that has broad applicability. Fully implemented, it should push down the decision-making authority to make changes to the lowest level consistent with the requirement to maintain adequate control of the product. It will require a cross-functional support team to adequately evaluate potential changes and implement changes in an efficient manner. The initial effort would be to define a methodology for evaluation of product quality and production process stability and capability, followed by an assessment of possible improvements. During production start-up, it could be the means for implementing the risk abatement plan developed during the prototype phase. Key elements of VR are a culture, an enabling structure, and a well-defined process that encourages and empowers the factory workers and the supporting management infrastructure. The techniques associated with process control and ongoing evaluations of process capability will identify candidates for process variability reduction.

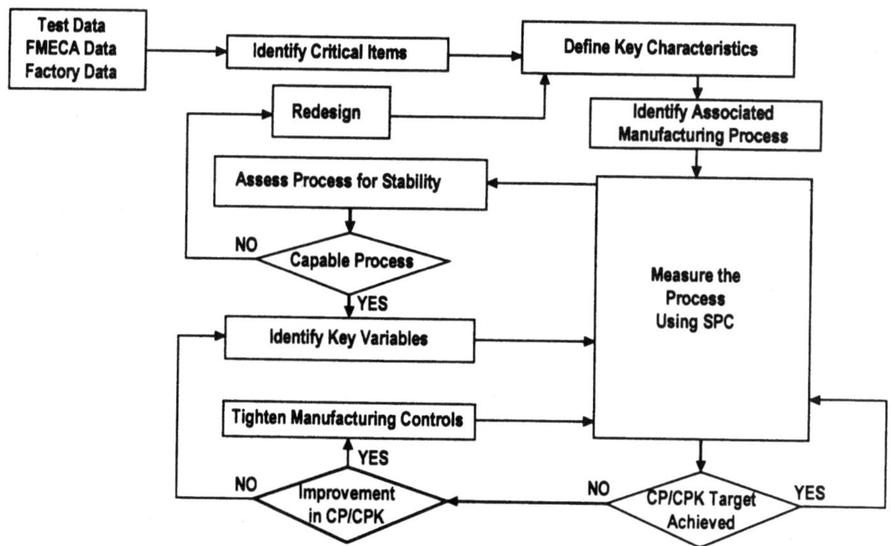

Figure 3.22 Variability reduction process. (*From Ref. 10.*)

It is essential, just as in SPC, that this requirement be passed to the key suppliers. Process variability has a virtually infinite number of causes, only some of which are controllable. Two principal causes of product performance problems are poor product design and poor process design, both of which lead to variations that result in product wear, early failure, unfulfilled expectations, and ultimate customer dissatisfaction. Such problems are costly for the consumer and the producer. Poor quality, high costs, or missed schedules that are caused by process variation produce unnecessary repair costs, lost productivity, frustration, inconvenience, low customer satisfaction, and ultimately lost business. Both the design and the process problems can be minimized with simple, low-cost management methods. Variation in processes can be controlled, and the number of incompatibility problems reduced, by training employees in the proper application of statistical methods.

Variability is a primary source of defects and waste. It exists everywhere in every process and product and it can never be totally eliminated. However, the amount of variability in processes and products can be controlled and reduced by applying several simple straightforward statistical techniques and tools. The concept of reducing variability lies at the heart of classical process control and design methods. Reducing variability is the primary objective of those statistical methods, and it is a core discipline of process improvement activity. Just as we accept that no two fingerprints or two people are exactly alike, we also accept that no two outputs from any process are exactly alike. Measurement detects differences among the outputs, and the amount of variation in a set of measurements can be represented graphically in a chart that shows the distribution of variation around the average input (Fig. 3.23). Since a process produces its output for a customer, the producer must understand as well as possible the customer's real requirements and must translate those requirements into the ideal output (the voice of the customer), as illustrated in Fig. 3.24.

The distribution of output variation represented in Fig. 3.24 reflects the existing inherent inability of the process to achieve the ideal, or target outputs, as defined by the voice of the customer. Then variation in the measured output is, in fact, the voice of the process. This voice of the process can be compared with the customer's requirements (voice of the customer), and inherent process capability to meet customer needs can be determined and quantified. The electronics industry is moving toward tighter and tighter variability reduction by changing from acceptable quality levels to measuring quality as parts per million. [84]

A complete guide to variability reduction, *The USAF R&M 2000 Variability Reduction Process,* may be found in the appendix of this

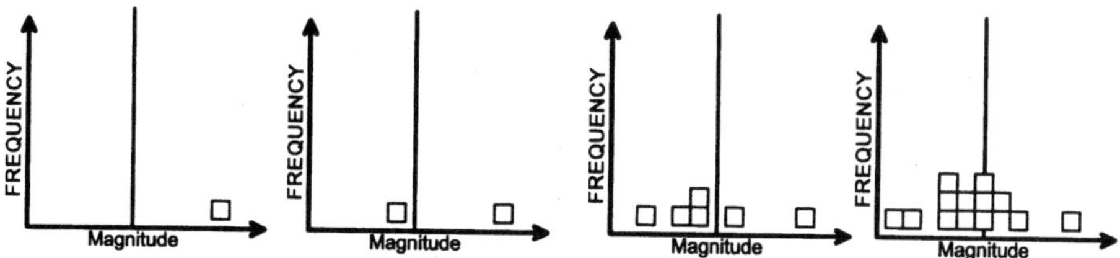

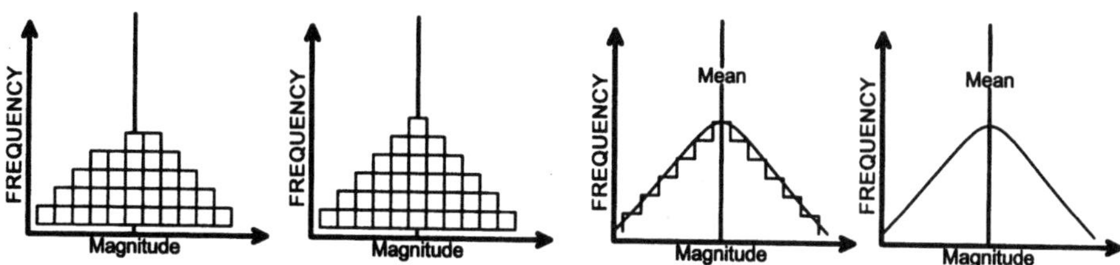

Figure 3.23 Distribution of variation. (*Reprinted from Ref. 84, with permission.*)

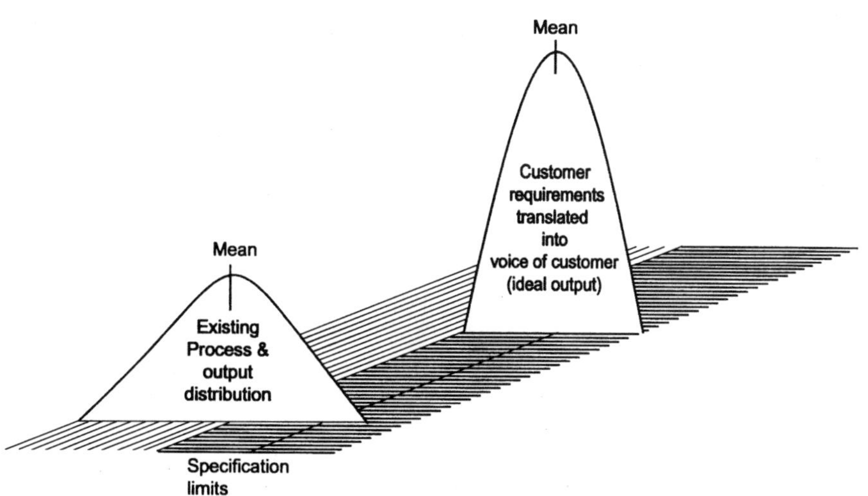

Figure 3.24 Voice of the customer. (*Adapted from Ref. 84, with permission.*)

book. The guide is known as an excellent one for all of industry, although it is no longer available from the U.S. Air Force. It was printed about 1988, and still has excellent data for today.

In *Juran's Quality Control Handbook,* A. Blanton Godfrey and Robert E. Kerwin state [85]:

> The electronics industry is faced with ever-increasing complexity at system, subsystem and component levels, the rapid introduction of new technologies and the shortening of product life cycles.
>
> Increasing complexity at the system and subsystem levels leads to the necessity of the incorporation of self-testing and diagnostics at the design stage, and the requirement for extremely high levels of component quality and reliability. The high quality levels require, in turn, the development of new supplier and user quality programs and measurement and information system to replace traditional Acceptable Quality Level (AQL) and Average Outgoing Quality Limit (AOQL) sampling plans and incoming inspection programs. Subsystem level complexity leads to an increased emphasis on the understanding, measurement and control of the quality and reliability of interconnection processes. At the component level, increased complexity, as represented by VLSI circuits, coupled with relatively smaller production runs, as represented by application-specific integrated circuits, lead to the necessity for built-in self testing at the device level, statistically meaningful fractional test coverage, and dependence on robust process line quality and reliability controls in addition to inspection and testing. Increasing demands on component quality and reliability require increase emphasis on the research and development of quality technology and tools.

The Electronic Industries Association in 1991 released *Zero Acceptance Number Sampling Procedures and Tables for Inspection by Attributes of a Continuous Manufacturing Process* (ANSI/EIA-584-91). EIA states, "Conventional attribute sampling plans based upon nonzero acceptance numbers are no longer desirable. In addition, emphasis is now placed on the quality level that is received by the customer. [86]

In 1995, EIA-554-A, *Assessment of Outgoing Nonconforming Levels in Parts per Million (PPM)* was released. This standard "measures parts per million (PPM) levels in lieu of existing acceptable quality levels (AQL)." [87]

The industry has responded. The update will surely be in the fifth edition of *Juran's Quality Control Handbook.*

3.8 Nothing Left but to Do It

It is now time, if your company has not started, to develop a documented single facility-wide quality management system in accordance with ISO 9000 and your company's strategic direction. Whether you should become compliant or certified is discussed in the next chapter, but the decision is part of your strategic plan. Advanced quality methods make continuous improvement an area where every electronics company can succeed. It will be vital, as the industry moves to the twenty-first century, to embrace these practices, train your employees, and become a world class electronics company. Establish goals, assess needs, make a plan and carry it out. The CEO must lead, but all of his or her staff should have actions so that the entire company is involved.

References

1. *Roadmap for Quality in the 21st Century: World Class Quality,* The Government and Industry Quality Liaison Panel, April 24, 1995, p. *iii.*
2. "Gore Honors ISO 9000-Based Initiative," *Quality Systems Update,* December 1995, pp. 1 and 5.
3. Garvin, David A., *Managing Quality; The Strategic and Competitive Edge,* The Free Press, New York, 1988, pp. 193–194.
4. *Roadmap for Quality in the 21st Century: World Class Quality,* The Government and Industry Quality Liaison Panel, April 24, 1995, p. *vi.*
5. ISO 8402, p. 2.
6. Conway, *Right Way to Manage; Process Flow Charts and Work Improvement,* Conway Quality Inc., Nashua, NH, 1987, p. 18.
7. ANSI/ASQC Q9000-1-1994, p. 4.
8. Ishikawa Kaoru, "Quality and Standardization: Program for Economic Success," *Quality Progress,* January 1984, p. 20.
9. Deming, W. Edwards, *Out of the Crisis,* p. 133. Reprinted from *Out of the Crisis* by W. Edwards Deming by permission of MIT and The W. Edwards Deming Institute. Published by MIT, Center for Advanced Educational Services, Cambridge, MA 02139. Copyright 1986 by The W. Edwards Deming Institute.
10. *Manufacturing Development Guide,* Aeronautical Systems Center, Air Force Material Command, Wright-Patterson AFB, OH, 30 November 1993, pp. 4-1 to 5-7
11. Saylor, James, *TQM Simplified, a Practice Guide,* 2d ed., McGraw-Hill, New York, pp. *xxiv–xxv.*
12. Deming, p. 248.
13. Deming, p. 486.
14. Deming, p. 135.
15. Tabak, L., "Quality Controls," *Hemisphere,* September 1996, p. 33.
16. Tabak, p. 34.
17. Poirier, Charles, and Steven Tokarz, *Avoiding the Pitfalls of Total Quality,* ASQC Press, Milwaukee, WI, 1992, p. 185.
18. Poirier and Tokarz, p. 193.
19. Conway, *The Right Way to Manage, Creating the New Management System,* Conway Quality Inc., Nashua, NH, 1987, pp. 20–21.
20. "Identifying Factors Crucial to Quality," ASQC *Journal of Record,* vol. XI, no. 3, March 1996, p. 5.
21. Poirier and Tokarz, p. 156.
22. Schonberger, Richard J., *World Class Manufacturing: The Next Decade,* The Free Press, New York, 1996, pp. 204–219.
23. Rubinstein, S., and T. Kochran, "Toward a Grounded Stakeholder Theory of the Firm: The Case of the Saturn Partnership," draft, July 15, 1996, pp. 23, 24, 33.
24. Rayner, Steven R., "Making Employee Empowerment Work," in Thomas Wallace, *World Class Manufacturing,* Omneo, Essex Junction, VT, 1994, pp. 154–157.

25. Conway, *The Right Way to Manage, Four Forms of Waste,* Conway Quality Inc., Nashua, NH, 1987, p. 40.
26. Schonberger, p. 204.
27. Garvin, p. 64.
28. Saylor, p. 224.
29. Schonberger, pp. 202–203.
30. NASA, NHB XXXX.X draft, *ISO 9000: Implementation and Assessment,* 1995, pp. 64–65.
31. Gunn, Thomas, *21st Century Manufacturing,* Omneo, Essex Junction, VT, 1992, pp. 104–105.
32. GIQLP, p. 26.
33. Akao, Y. (ed.), *Quality Function Deployment,* Productivity, Cambridge, MA, 1990; quote is from Internet: http://akao.larc.nasa.gov/dfc/qfd.html, Dec. 11, 1996.
34. Sullivan, L. P., "Quality Functional Deployment," *Quality Progress,* June 1986, p. 50.
35. Shores, A. Richard, *Reengineering the Factory: A Primer for World Class Manufacturing,* ASQC Press, Milwaukee, WI, 1994, p. 27.
36. American Supplier Institute, 17333 Federal Dr., Suite 220, Allen Park, MI 48101.
37. Pugh, S., "Concept Selection—A Method That Works," *Proceedings,* International Conference on Engineering Design, Rome, March 9–13, 1981.
38. *Quality Function Deployment, Application Guide,* Technicomp Inc., 2d printing, Cleveland, 1989, pp. 1.8 to 1.11.
39. Cohen, pp. 423–436. Cohen lists these articles for those interested in applying QFD to software: (*a*) "Quality Function Deployment for Software," Richard Zultner, *American Programmer Magazine,* February 1992; (*b*) an English-language description of this approach in "Quality function Deployment Applied to Software at Digital Engineering Corporation," Takami Kihara and Malik Mamdani, working paper at the Thayer School of Engineering, Dartmouth College, Hanover, NH, May 1991; (*c*) Edward Yourdon and Larry L. Constantine, Structure Design: Fundamentals of a Discipline of Computer Program and System Design, Yourdon Press, Englewood Cliffs, NJ, 1970.
40. Mansir, Brian, and Schacht, N. R., *An Introduction to the Continuous Improvement Process: Principles and Practices,* Logistics Management Institute, Bethesda, MD, 1989, pp. 4–75.
41. Institute for Defense Analyses, *The Role of Concurrent Engineering in Weapons System Acquisition,* Institute for Defense Analyses, Washington, DC, 1988, p. 138.
42. Dertouzos, N. L., R. S. Lester, and R. M. Solow, *Made in America: Regaining the Productive Edge,* Harper Perennial, New York, 1989, quoted from the Internet: http://akao.larc.nasa.gov/dfc/tm.html, Nov. 24, 1996.
43. Peace, Glen Stuart, *Taguchi Method: A Hands-on Approach,* Addison-Wesley, Reading, MA, 1993, pp. 1–2.
44. Peace, pp. *xvii, xviii,* 2.
45. Peace, p. 5.
46. Mansir and Schacht, pp. 4-86, 4-87.
47. Peace, p. 4.
48. Mansir and Schacht, pp. 4–74.
49. Blake, S., R. G. Launsby, and D. L. Weese, "Experimental Design Meets the Realities of the 1990s," *Quality Progress,* October 1994, p. 100.
50. Peace, p. 21.
51. Lochner, Robert H., and Joseph E. Matar, *Designing for Quality; An Introduction to the Best of Taguchi and Western Methods of Statistical Experimental Design,* ASQC, 1990, pp. 193–195.
52. Mansir and Schacht, p. A-11.
53. ANSI/IPC-PC-90, *General Requirements for Implementation of Statistical Process Control,* Institute for Interconnecting and Packaging Electronics Circuits, Lincolnwood, IL, 1990, p. 1.
54. Kline, Matthew, *SPC: Decision Support for the Plant Floor,* Kline and Company, Chicago, IL, 1989, p. 1-3.
55. ANSI/IPC-PC-90, pp. 4, 51, 61.

56. Lancaster, Michael C., "Six Sigma in Contract," *Circuits Assembly,* December 1992, p. 36.
57. Grant, E. L., and R. S. Leavenworth, *Statistical Quality Control,* 6th ed., McGraw-Hill, New York, 1988, quoted from the Internet, http://akao.larc.nasa.gov/dfc/sqc.html, Nov. 24, 1996.
58. Munuz, J., and C. Nielsen, "SPC: What Data Should I Collect? What Charts Should I Use?" *Quality Progress,* January 1991, p. 52.
59. ANSI/IPC-PC-90, p. 10.
60. Stone, Edward P., "Employee Support and Interaction Are the Keys to an SPC Program," *Quality Progress,* December 1991, pp. 54–56.
61. Dobbins, J. G., and W. J. Padgett, "SPC in Printed Circuit Board Assembly," *Quality Progress,* July 1993, pp. 65, 67.
62. AIAG, *Statistical Process Control (SPC) Reference Manual,* Chrysler Corp., Ford Motor Co., General Motors Corp. (AIAG), 1992, 1995, pp. 1, 2.
63. *The USAF R&M 2000 Variability Reduction Process,* Special Assistant for Reliability and Maintainability, USAF, 1988, p. 6.
64. Juran, J. M., and Frank M. Gyrna, *Quality Planning and Analysis,* 3d ed., McGraw-Hill, New York, 1993, pp. 393–394.
65. AIAG, p. 57.
66. ANSI/IPC-PC-90, p. 5.
67. Cottman, R. J., *Total Engineering Quality Management,* ASQC Press, Milwaukee, 1993, pp. 126-127.
68. ANSI/IPC-PC-90, p. 45.
69. AIAG, pp. 14–15.
70. AIAG, p. 15.
71. Harry, M. J., *The Nature of Six Sigma Quality,* Motorola University Press, Rolling Meadows, IL, 1988, p. 2.
72. Harry, p. 2.
73. Tadikamalla, Pandu, "The Confusion Over Six-Sigma Quality," *Quality Progress,* November 1994, pp. 83–84.
74. Harry, p. 4.
75. Deming, p. 334.
76. Harry, p. 4.
77. Arnold, Wilbur, *Designing Quality Into Defense Systems: Design, Manufacturing, Support,* Defense Systems Management College, Fort Beauvoir, VA, circa 1988, p. 7.
78. *The USAF R&M 2000 Variability Reduction Process,* p. 1.
79. Harry, p. 1.
80. Harry, p. 4.
81. Deming, p. 135.
82. Susman, Gerald I. (ed.), *Integrating and Manufacturing for Competitive Advantage,* Oxford University Press, New York, 1992, pp. 40–43.
83. Juran and Gyrna, 109–110.
84. Mansir and Schacht, pp. 4-71–4-75.
85. Godfrey, A. B., and Robert E. Kerwin, "Electronic Components Industries," in Juran, *Juran's Quality Control Handbook,* 4th ed., McGraw-Hill, New York, p. 29.17.
86. EIA, JEDEC, and TIA Standards and Engineering Publications, Global Engineering Documents, Englewood, CO, 1996, pp. 1–41 (EIA catalog).
87. EIA catalog, p. 1-40.

Chapter

4

ISO 9000 International Quality Management Standards

Marsha Ludwig-Becker, CM

Operations Project Manager
The Boeing Company
Anaheim, California

4.1 Introduction

In 1992, the ISO 9000 quality management standards became international standards and came to the United States. By some reports, U.S. sites certified to the ISO 9000 quadrupled in the first half of 1992. It was not yet fully accepted in the United States because the U.S. government has a great influence on suppliers in this country and did not accept the standards that year. It was not for 2 years that the entire United States could move ahead to one quality standard.

The year 1994 brought the change to American industry. The ISO standards were accepted by the U.S. government. In February of 1994, then Assistant Secretary of Defense J. M. Deutsch authorized the use of ANSI/ASQC Q9001 for defense contracts. The desire of this U.S. administration was that commercial and defense contractors should work together. This served the electronics community well. The U.S. government needs and buys electronics from the low office end to sophisticated systems. Electronics manufacturers were strung between a myriad of quality standards, and no one addressed one single quality management approach. There was no quality system for commercial electronics to work to.

The worldwide attention being applied to the continual improvement of quality for world trade has created a desire among many countries and companies to set forth minimum standards for assuring quality of products and services. The International Organization for Standardization (ISO) thus created and approved the ISO 9000 quality management standards. These standards (ISO 9000 to 9004) have been adopted by many companies and government agencies that require supplier companies to conform to the ISO 9000 standards as the minimum criterion for supplier selection. Companies that do not conform to the ISO 9000 standards are likely to find that they will not be able to sell into certain markets, and their sales potential will be reduced accordingly. In the *ISO 9000 Survey of Registered Companies, 1996,* the largest group of respondents was from the Electronics and Other Electronic Equipment and Components standard industrial classification (SIC) code. [1] In the world of increasing quality, it is imperative that the electronics supplier aspiring to world class quality become compliant, if not certified, to the ISO standards.

It might be interesting to know a bit of history on the standards: The International Organization for Standardization was founded in 1946. The objectives of ISO are to promote development of standardization so as to (1) facilitate international trade, both goods and services, and (2) develop international cooperation. ISO headquarters is in Geneva, and the organization has over 100 member countries. It prints thousands of standards in all fields except electrical and electronic engineering, which is the responsibility of the International Electrotechnical Commission (IEC). For the United States, the member of ISO is the American National Standards Institute (ANSI), based in New York. Their charter is to adopt/develop, review, and disseminate standards to industry groups. This group works with various industries to produce standards for industry. When, in January 1985, the European Community (EC) announced its plan to write a quality standard, the American Society for Quality Control (ASQC), now renamed the American Society for Quality (ASQ) as of July 1, 1997, informed ANSI that they wanted to be involved. (ASQC was established in 1946 to focus on quality information and standardization.) In 1987, the first ISO quality management standard was released.

The ISO and IEC are the world's largest nongovernmental systems for voluntary industrial and technical collaboration and standardization. In the United States, Underwriters Laboratories (UL) is the representative to the IEC, which is also headquartered in Geneva. The IEC is the authoritative worldwide body responsible for developing global standards in the electrotechnical field. It is dedicated to global harmonization and voluntary adoption in the interests of society, supporting the transfer of electrotechnology, and promoting international trade.

IEC has served the world's electrical industry since 1906, developing international standards to promote quality, safety, performance, reproducibility, and environmental compatibility of materials, products, and systems. For the past 40 years it has also set standards for electronics and telecommunications industries. IEC's present membership of approximately 49 countries includes all major trading nations. Its more than 200 technical committees span virtually all electrotechnical sectors and associated disciplines. The IEC standards are cited in procurements around the world. For example, almost 90 percent of the European Norms (EN) harmonized by CENELEC (the European Committee for Electrotechnical Standardization) and adopted for the European Community and European free trade are either identical or based on IEC international standards. This is important to the electronics supplier in the certification of product; see Sec. 4.2. In the United States, IEC standards are available through ANSI. [2]

In the United States, the Electronic Industries Association (EIA) takes an active part in producing standards and specifications for the electronics industry. Founded in 1924 as the Radio Manufacturers' Association, EIA has more than 70 years of experience in the electronics field. The EIA considers that its specifications and standards facilitates the design and manufacture of electronic components and systems. Its catalog, in association with the Telecommunications Industry Association (TIA), is published by Global Engineering Documents, Englewood, CO. It is a good source for advanced quality assessment standards. [3]

The basic set of ISO 9000 or ANSI/ASQC Q9000 quality management standards involves five documents in the 1994 series. See Table 4.1. ISO Committee 176, which produces the standards, is currently planning an update about the year 2000. At that time the three separate contractual quality standards will be combined. In addition, more requirements for getting results will be in the standards. They are currently criticized for not requiring results in themselves. They stress developing a process, but there is nothing that requires the company to have a good process or create a good product. It must be remembered that the ISO quality management standards are a minimum set of quality management standards. Their power is that they concentrate on management, and baseline set of requirements.

Common and well-documented benefits of an ISO 9000 program are that it:

TABLE 4.1 Cross Reference: ISO ANSI/ASQC

ISO	ANSI/ASQC
ISO 9000 Quality Management and Quality Assurance Standards—Guidelines for Selection and Use	ANSI/ASQC Q9000-1-1994**
ISO 9001 Quality Systems—Model for Quality Assurance in Design/Development, Production, Installation and Servicing	ANSI/ASQC Q9001-1994*
ISO 9002 Quality Systems—Model for Quality Assurance in Production, Installation and Servicing	ANSI/ASQC Q9002-1994*
ISO 9003 Quality Systems—Model for Quality Assurance in Final Inspection and Test	ANSI/ASQC Q9003-1994*
ISO 9004 Quality Management and Quality System Elements—Guidelines	ANSI/ASQC Q9004-1-1994**

* = contractual
** = guidelines

- Accelerates international trade
- Focuses on management and customers
- Controls all processes
- Provides a strong focus on employee morale and teamwork
- Improves overall employee productivity
- Defines authorities and responsibilities for each employee
- Leads to a better understanding of and compliance with customer, partner, and supplier requirements
- Reduces number of customer audits
- Provides continuous identification of areas that need improvement
- Provides public demonstration of commitment to the company quality policy, company procedures, and international quality standards
- Provides compliance with ISO 9000 requirements in requests for vendor proposals and customer contracts
- Emphasizes defect prevention rather than defect control
- Can increase operational efficiency and reduce cost
- Takes a company from the little “q” to the big “Q”

Other benefits are being seen as more and more companies adopt the ISO quality management approach. It is being seen as a communications tool as much as a quality management tool. In the recent past, it has been documented to [4]:

- Build interpersonal communication between managers and employees
- Help resolve political conflicts, work procedure inconsistencies, and conflicts between formal and informal communication
- Train management and employees in communication skills, such as interviewing, writing, and editing
- Create a documentation system and a system for disseminating information company-wide and to all customers
- Provide the basis for a networked communication system
- Lay a foundation for using employees as sophisticated information gatherers and sorters

The general intention of the ISO 9000 standards is to provide the basis of a quality system that attempts to assure the following principles [5]:

> The organization should achieve the quality of the product or service produced so as to continually meet the purchaser's stated or implied needs.
>
> The organization should provide confidence to its own management that the intended quality is being achieved and sustained.

The ISO assumes that there is one system which everyone understands and follows. This includes processes, procedures, and execution thereof. There are three contractual or possibly internally imposed standards from ISO: 9001, 9002, 9003. In the United States, they are published by ANSI/ASQC, and so have those acronyms in the prefix:

ANSI/ASQC Q9000, ANSI/ASQC Q9001-1994, or ISO 9001: *Quality Systems—Model for Quality Assurance in Design, Development, Production, Installation, and Servicing.* The ISO 9001 standard is inclusive of the kind of requirements that would be imposed on a supplier who is designing, producing, and/or installing a product, most probably of custom design, and is controlled by the contractual relationship between the customer and supplier. Most electronics companies will use 9001 because, in order to stay in business, they are updating their design to fast-changing technology. These electronics companies may have subtier suppliers who do not design, but only duplicate the technology. Examples include electronic contract manufacturers like disk manufacturers who do not need the design requirements of ISO 9001; for them ISO 9002 is more applicable.

ANSI/ASQC Q9000, ANSI/ASQC Q9002-1994, or ISO 9002: *Quality Systems—Model for Quality Assurance in Production, Installation, and Servicing.* The ISO 9002 standard is the exact same standard as the 9001 without the paragraphs for design control. This standard is aimed at the quality system for assuring the quality of products that are designed as off-the-shelf commercial and/or defense items. In other words, the manufacturer does not design, but uses another company's specifications, drawings, and/or products definition to meet the requirements of a general market and sells to many different customers with the manufacturer's specifications applied. The change clause is part of the design control paragraphs and could well be tailored into this for an electronics supplier working with a prime, original equipment manufacturer (OEM) or a distributor that is assembling parts. This standard may be for an electronics assembly company, or a duplication type company. The key is that there is *no* design, and whether there is servicing or not depends on contracts—either commercial between primes and subtier suppliers, or primes and government agencies, and the servicing of electronics.

ANSI/ASQC-9000, ANSI/ASQC Q9003-1994, or ISO 9003: *Quality Systems-Model for Quality Assurance in Final Inspection and Test.* This standard is "for use when conformance to specified requirements is to be assured solely at final inspection and test." [6] The text of the standard states that ISO 9003 is to be used when the contract between two parties requires demonstration of a supplier's capability to detect and collect the disposition of any product nonconformity during final inspection and test. This could work in an electronic environment where the details of the internal process were proprietary for a large volume of items on a fast automated line, or where the process is smoothly in place. See Table 4.2 for the differences between the ISO 9001 and ISO 9003.

Table 4.3 explains which standard to use for a specific business. In the next revision, expected about 2000, all three of the above-men-

TABLE 4.2 Exact Differences from ISO 9001 to ISO 9003

Applicability: ISO 9001 is for design, development, production, installation, and services. ISO 9003 is just final inspection and test.

- **Responsibility:** Only related to final inspection and test in ISO 9003.
- **Quality planning:** Only in relation to final inspection and test in ISO 9003.
- **Design control:** None in ISO 9003.
- **Purchasing:** None in ISO 9003.
- **Process control:** None in ISO 9003.
- **Inspection and testing:** Applies only to final inspection and testing in ISO 9003.
- **Preventive action:** None in ISO 9003.

TABLE 4.3 ISO 9001, 9002, or 9003?

What standard should be used? Depends on your business
What do you do? Design, develop, software development, service after delivery ⟶ 9001 Build to print, distribute ⟶ 9002 Final test only, large volumes of product, proprietary ⟶ 9003

tioned standards will be combined, and suppliers will tailor them for their specific use.

There are two additional standards in the basic set:

ANSI/ASQC Q9000-1-1994 or ISO 9000-1-1994: ***Quality Management and Quality Assurance Standards—Guidelines for Selection and Use.*** This is an introduction to the ISO 9000 series and how it should be used. It contains the ISO view of principal concepts, documentation, and quality system situations, and explains the differences of the standards and how they should be used. It also lists supplemental standards for various specialty areas like software, processed materials, and services.

ANSI/ASQC Q9004-1-1994 or ISO 9004-1-1994: ***Quality Management and Quality System Elements—Guidelines.*** ISO 9004 sets forth guidelines for implementing a complete quality management system. The quality management system is, as discussed in this book, the approach and methods used to manage and improve the quality of all the business processes toward higher levels of customer satisfaction.

ISO 9004 approaches its objective by setting forth principles for management, marketing, design, and production. It provides principles for developing quality plans and audits of the quality system in a team environment. It is a top-level overview of a complete quality management system. See Fig. 4.1 for the various activities that have an impact on quality. [7]

An additional good ISO standard, not part of the four but definitively needed for understanding the ISO standards, is ISO 8402:1994: *Quality Management and Quality Assurance—Vocabulary.* This is a standard that lists the formal definitions of terminology used in the ISO 9000 series.

ISO 9001-1994 adequately sets forth standards by which companies can set up a quality management system to ensure assurance of quality of design, manufacture, installation, and improvement. The standards enforce processes, but they do not prescribe a standard of excellence, nor were they intended to do so. Rather, they set acceptable

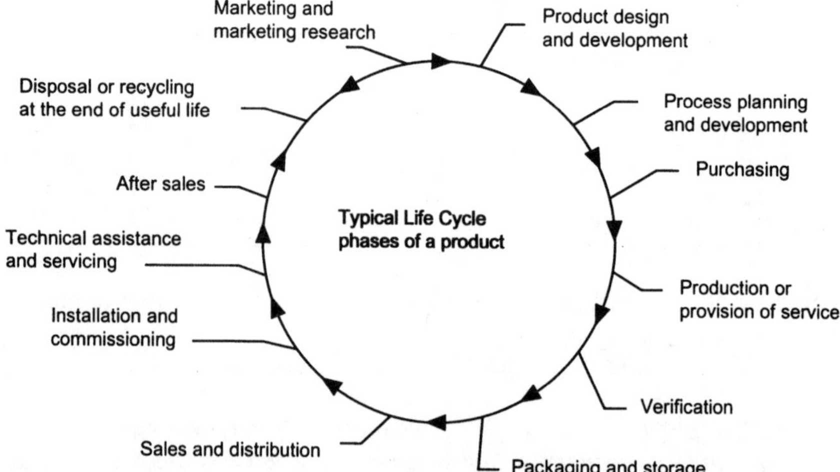

Figure 4.1 Main activities having an impact on quality. (*From Ref. 7; used with permission.*)

standards for doing business in an intentional environment. The ISO 9000 is a starting point to achieve higher standards of excellence.

In the style of the ISO, the committee which prints the standards for the ISO, Committee 176, puts out a series of documents for guidance and added information (see Fig. 4.2). These are good guides for specific quality management and technology guidance, as well as supplements that help develop the quality management system. All are available from ASQ when they are released from ISO. All are currently not in print.

The ISO 9000 standards are continuously reviewed and updated every 5 years. The next revision is in final review and should be due for release around the year 2000. For more information on ISO 9000 standards, contact ASQC in Milwaukee.

4.2 The Elements of ISO 9001

The ISO 9001 standard contains all the requirements for all of the other standards. It contains only 20 elements (see Table 4.4) at a very high level so that it can be used by any industry. Discussions of the 20 elements follow.

4.2.1 Management responsibility

It is the responsibility of top-level management to define the goals of the organization in a stated policy and demonstrate commitment to

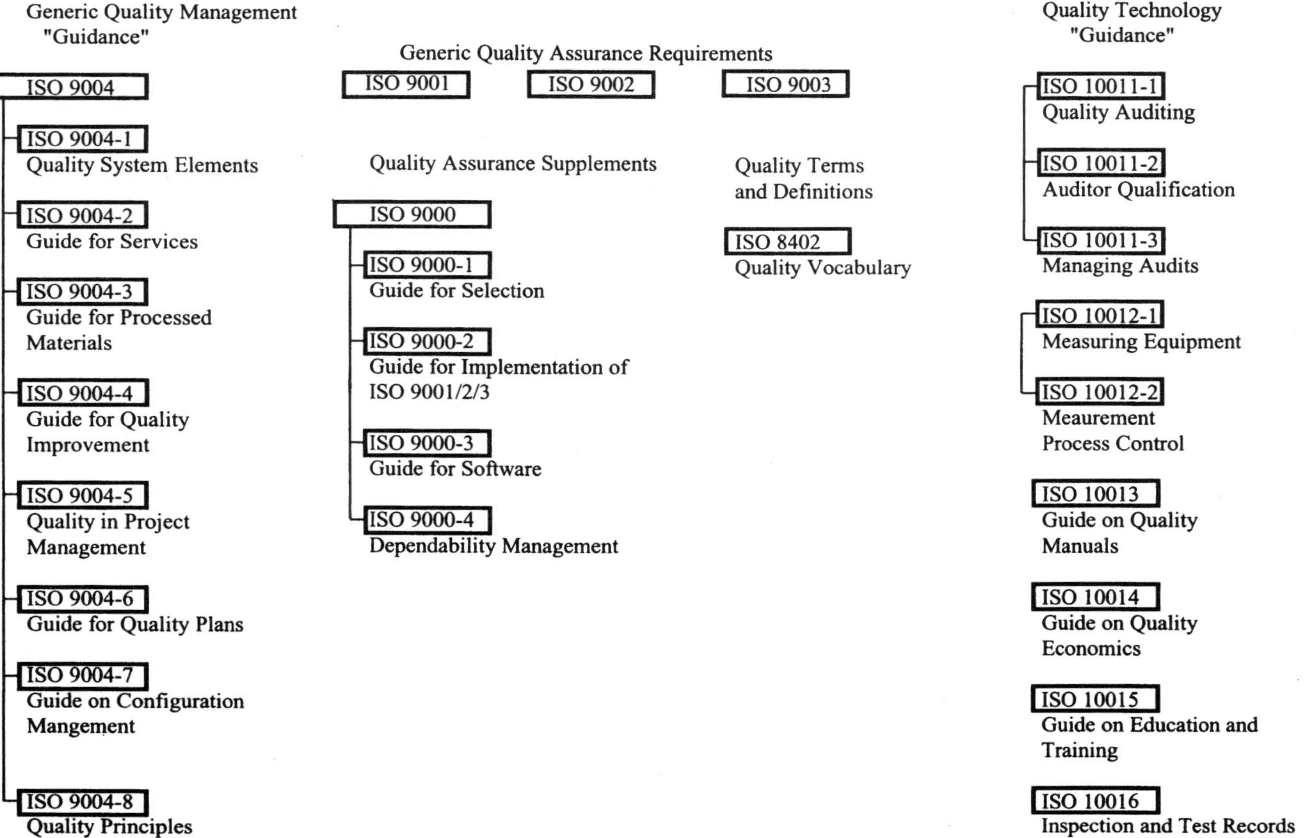

Generic Quality Management
"Guidance"
ISO 9004
ISO 9004-1
Quality System Elements
ISO 9004-2
Guide for Services
ISO 9004-3
Guide for Processed Materials
ISO 9004-4
Guide for Quality Improvement
ISO 9004-5
Quality in Project Management
ISO 9004-6
Guide for Quality Plans
ISO 9004-7
Guide on Configuration Mangement
ISO 9004-8
Quality Principles
Generic Quality Assurance Requirements
ISO 9001
ISO 9002
ISO 9003
Quality Assurance Supplements
ISO 9000
ISO 9000-1
Guide for Selection
ISO 9000-2
Guide for Implementation of ISO 9001/2/3
ISO 9000-3
Guide for Software
ISO 9000-4
Dependability Management
Quality Terms and Definitions
ISO 8402
Quality Vocabulary
Quality Technology
"Guidance"
ISO 10011-1
Quality Auditing
ISO 10011-2
Auditor Qualification
ISO 10011-3
Managing Audits
ISO 10012-1
Measuring Equipment
ISO 10012-2
Meaurement Process Control
ISO 10013
Guide on Quality Manuals
ISO 10014
Guide on Quality Economics
ISO 10015
Guide on Education and Training
ISO 10016
Inspection and Test Records

Figure 4.2 Evolving structure of ISO/TC176 standards.

TABLE 4.4 ISO 9001 Twenty Elements/Paragraphs

1. Management responsibility
2. Quality system
3. Contract review
4. Design control
5. Document and data control
6. Purchasing
7. Control of customer-supplied product
8. Product identification and traceability
9. Process control
10. Inspection and testing
11. Control of inspection, measuring, and test equipment
12. Inspection and test status
13. Control of nonconforming product
14. Corrective and preventive action
15. Handling, storage, packaging, preservation, and delivery
16. Control of quality records
17. Internal quality audits
18. Training
19. Servicing
20. Statistical techniques

these goals. The ISO approach is exactly the approach that Deming wrote about in *Out of the Crisis*. The CEO must be in charge and knowledgeable about a quality management system. He/she must be proactive in improving the quality system for it to happen. Although middle management must strive to make the system happen, it will not do so without the overt direction and commitment of the CEO. The CEO must understand that it is his/her business system for delivering a quality electronic product and/or service. It is well-documented that 85 to 90 percent of problems in a company can be traced directly back to management. Under ISO the emphasis is on management system and procedure stressing customer satisfaction. This is key to the electronic world. The standard makes clear the *top* manager is responsible for quality, not the quality manager. Each element of the standard must be assigned to a person who reports to the top-level manager.

Management's responsibility consists of five main sections:

1. *Quality policy.* Management (actually the CEO) needs to write a company quality policy to operate the business. It is ISO's belief that every company has a quality policy, whether or not it is written down. If it is not written down, the company must develop one. It should be short, as under ISO everyone should know the policy. It is a regular question of auditors.

2. *Responsibility and authority.* The top-level manager is responsible for the quality system, not the quality manager. The CEO is responsible; as Deming states, "He is the one." [8]
3. *Resources.* The CEO must give resources to this quality management system. Remember that quality now is the big "Q"; this is a business management system.
4. *Management representative.* The CEO appoints, from his/her own staff, a member of management to carry out details for implementation and maintenance of the system. The management representative, irrespective of other responsibilities, shall have defined authority for (1) ensuring that a quality system is established, implemented, and maintained in accordance with this international standard and (2) reporting on the performance of the quality system to the supplier's management for review and as a basis for improvement of the quality system. [9]
5. *Management reviews.* The CEO must have reviews of the quality system to ensure compliance and continuous improvement. Fully embraced by a CEO who understands, the internal audit program, which is designed to be very active in an ISO 9000 approach, ensures compliance. Some will argue whether ISO calls out continuous improvement, but it is clearly implied. The collection of data, the internal audits, corrective/preventative are involved under the ISO 9000 umbrella. Data under TQM and other quality initiatives will encompass a quality system that can indeed force continuous improvement, if the emphasis is on data and customer satisfaction.

Records of management reviews shall be maintained so that internal and external auditors can see the management involvement. See Table 4.5 for implementation.

TABLE 4.5 ISO Implementation for Electronics Systems: Management Responsibility

Enlightened management is essential for the electronics company. Enlightened management:

- Understands the big "Q": that quality is the business of the entire company
- Understands that quality is what the customer wants
- Understands that quality is a strategic discriminator
- Understands Total Quality Management

4.2.2 Quality system

A quality management system establishes the effectiveness at which the organization carries out its duties to provide the customer with the desired product. A quality system is measured in various ways: Do personnel follow well-defined procedures and work instructions? Do they use updated control, inspection, and testing techniques? Do workers show a favorable attitude? Documentation is the key to quality. In a quality system, everyone understands what they are supposed to do in order to produce a quality product or service.

The quality system, as defined by the standard, states that *it shall be established and maintained.* In other words, this is a formal system that is written down and traceable through the business. It shall have a quality manual. This is the vehicle that ISO uses to define the quality system.

Documented procedures are required, but don't get hung up on words; flowcharts and pictures provide excellent documentation for some firms. In the electronics world, this is no less true.

Quality planning is a must to ensure that the customers are pleased. Quality plans become the vehicles for detail plans, and in the ISO world, read more like a company or project plan than the traditional "quality assurance" plan, which primarily talked about conformance and inspection. The quality planning here speaks to all things necessary for delivering a product that the customer wants, or that the world would buy.

ISO takes a structured approach to documented procedures and the quality manual. In documenting an ISO 9000 quality management system, the first step is to know about the documentation system that is recommended in the ISO 9000 approach. This approach to documentation makes a company's procedures, policies, and work instructions very clear and easy to read and therefore unites the company in the approach to a workable quality management system. The triangle that represents the documentation structure is shown in Fig. 4.3. [10]

The top level is the quality manual; this is known as Level 1. It is a statement of the company's policy on the quality management system. It defines what is included and excluded in the quality system. The quality manual looks like more of a "company manual" than the traditional quality manual. It is normally from 20 to 30 pages long. Level 2 is the set of procedures that define the "what" of the company's effort. Level 2 documentation is generally three to five pages long. A sample appears in Table 4.6. This procedure set is focused on the top-level processes of the company. Level 3 is the "how to" documentation. It contains work instructions: detailed instructions on how to design,

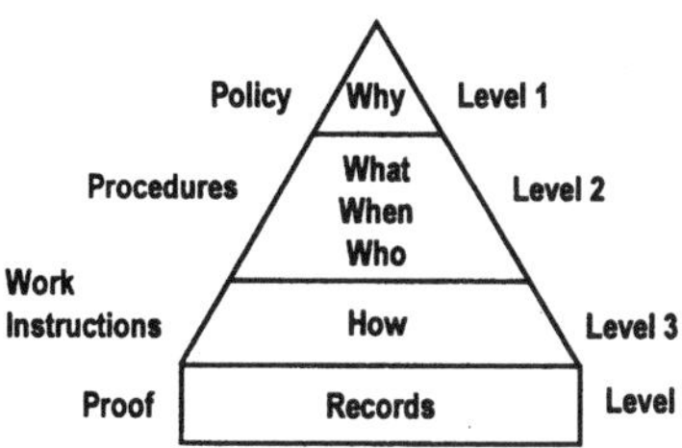

Figure 4.3 Quality system documentation structure. *(from Ref. 11; used with permission)*

develop, manufacture, test, inspect, or service a product. Level 4 collects records or proof.

The quality manual is defined by ISO 8402:1994 as a "document stating the quality policy and describing the quality system of an organization." [11] The pertinent publication is ISO 10013, *Guidelines for Developing Quality Manuals.*

A quality manual should consist of, or refer to, the documented quality system procedures intended for the overall planning and administration of activities which impact on quality within an organization. It should describe, in adequate detail, the same control aspects mentioned in the standard.

Quality manuals have various purposes and may be used by organizations for purposes including but not limited to [12]:

1. Communicating the company's policy, procedures, and requirements
2. Implementing an effective quality system
3. Providing improved control of practices and facilitating quality assurance activities
4. Providing the documented bases for auditing quality systems
5. Providing continuity of the quality system and its requirements during *changing* circumstances
6. Training personnel in the quality system requirements and method of compliance
7. Presenting their quality system for external purposes, such as demonstrating compliance with ISO 9001, 9002, or 9003
8. Demonstrating compliance of their quality systems with required quality standards in contractual situations.

A quality manual should normally contain [13]

1. The title, scope, and the application (What is this manual for?).

TABLE 4.6 An Example Level 2 Procedure

MO015—Electrostatic Discharge (ESD) Control

15.0 Purpose

This procedure establishes electrostatic discharge protection for handling and protecting electronic hardware from damage or degradation due to electrostatic discharge (ESD).

15.1 Scope

This procedure applies to the electronics facility at New York, U.S.A.

15.2 Responsibility

The plant manager is responsible for the implementation and maintenance of the ESD system. Documentation is the responsibility of the manufacturing engineering department, and all persons are responsible for the proper handling of ESD-sensitive material in their area.

15.3 Procedure

15.3.1 All electronic items (components, assemblies, and equipment) procured or fabricated which are ESD sensitive shall be identified, handled, packaged, and stored in accordance with process specification, PSOP 27A, *Electrostatic Discharge Protection.*
15.3.2 Only personnel who are ESD-certified by the training department shall be permitted to handle ESD-sensitive electronic items.
15.3.3 Any item or assembly containing more than one classification of ESD sensitivity will be handled based on the most sensitive item therein.

15.4 Related Procedures

PSOP 27A, Electrostatic Discharge Protection Process Specification MIL-STD-1686, Electrostatic Discharge Control Program for Protection of Electrical and Electronic Parts, Assemblies, and Equipment
TSOP 27, Certification Program for ESD certification for personnel.
CSOP 27, Certification Program for static safe workstations.

15.5 Outputs and Deliverables

A monthly status report will be made by the lead ESD manufacturing engineer and presented to management. A listing of all ESD-certified personnel will be produced quarterly.
ESD symbols to be used to identify ESD certification will be maintained by the training department.

15.6 Records

A monthly surveillance record will be maintained for each manufacturing area. An ESD inventory listing will list all authorized ESD equipment. The training database will identify all ESD certified personnel. The process certification database will identify all ESD-certified workstations, by number.

15.7 Audit Statement

The Quality Audit Department shall audit this system at least twice a year and report to management.

2. The table of contents of the manual. This should follow the table of contents of the ISO standard that the company is going to work to, with paragraphs tailored in. If not applicable, still reference the paragraph title and say "not applicable."
3. The introductory pages about the organization concerned and the manual itself.
4. The quality policy and objectives (or mission) of the organization.
5. The description of the organization's responsibilities and authorities.
6. A description of the elements of the quality system and/or references to quality system procedures.
7. A definitions section, if appropriate.
8. A guide to the quality manual, if appropriate, should be clear enough so that each paragraph can be looked at separately, if the reader desires.
9. An appendix for supportive data, if appropriate.

Examples of two types of quality manuals are in Table 4.7. In one, procedures are listed with the paragraphs; in the other, procedures are cross-referenced at the end. A complete sample manual is provided in the Appendix.

Quality planning shall be consistent with the quality system and documented to suit the method of operation. Quality planning should include:

1. Preparation of quality plans
2. Identification of any controls, processes, inspection equipment, fixtures, total production resources, and skills that may be needed to achieve the required quality
3. Ensuring the compatibility of the design, the production process, installation, servicing, inspection and test procedures, and applicable documentation
4. Updating, as necessary, of quality control, inspection, and testing techniques
5. Identification of any measurement requirement involving capability that exceeds the known state of the art in sufficient time for the capability to be developed
6. Identification of suitable verification at appropriate times
7. Clarification of standards of acceptability for all features and requirements

TABLE 4.7 Two Formats of Quality Manuals

A. Paragraphs listing procedures

4.10 INSPECTION AND TESTING

4.10.1 PURPOSE
Incoming, in-process, and final inspections perform process monitors, audits, and physical inspection on products/processes to ensure that they meet all applicable workmanship standards and processes.

4.10.2 PROCEDURE/METHOD/COMPLIANCE
Company ABC ensures that all incoming raw material received into the facility meets all physical and/or electrical characteristics prior to releasing such material to stock. Documentation must be received and in place with new material. In-process inspections and tests are performed during manufacturing. Final inspection and test are completed before the products are shipped to ensure all acceptance tests/inspections are passed. All inspections and tests are documented throughout the manufacturing flow. Reports of inspection and test records are scrutinized to detect trends and incorporate improvements, as required.

Quality Procedure and Operating Manual Procedures

Receiving Inspection and Testing:

PP-001-003-1	Receiving Inspection Process

In-Process/Final Inspection/Test and Records

SOP 4-10-1	First Article Inspection
SOP 4-10-2	In-Process Inspection
SOP 4-10-3	Product Test Processes
SOP 4-10-4	In-Circuit Test Processes
SOP 4-10-5	Functional Test Process
SOP 4-10-6	Burn-in Process
SOP 4-10-7	Systems Test Process

B. Procedures cross-referenced at end of Manual

Cross-reference from quality system paragraph number to standard operating procedures

System paragraph	Procedure
4.10	PP-001-003-1
	SOP 4-10-1
	SOP 4-10-2
	SOP 4-10-3
	SOP 4-10-4
	SOP 4-10-5
	SOP 4-10-6
	SOP 4-10-7
4.11	PP-001-004-1
	SOP 4-11-1
	SOP 4-11-2
	SOP 4-11-3
	SOP 4-11-4
4.12	SOP 4-12-1
	SOP 4-12-2
4.13	Manual NCP 13-1
	SOP 4-13-1
	SOP 4-13-2
	SOP 4-13-3

8. Identification and preparation of quality records
9. Development of metrics for product quality, service responsiveness, and customer satisfaction.

Quality plans are developed to ensure that specified requirements for a product, project, or contract are met. A quality plan may be part of a larger overall plan. A plan is necessary for new products/processes, or significant changes.

The quality plan should define:

1. The quality objectives
2. The steps in the process
3. The specific allocation of responsibilities, authorities, and resources
4. The specific documented procedures and instructions
5. Suitable testing, inspection, examination, and audit programs at appropriate stages
6. A documented procedure for changes and modifications of the quality plan, as the project/process/product progresses
7. A method for measuring the achievement of the quality objectives
8. Other actions necessary to meet the quality objectives

Quality plans may be included or referenced in the quality manual and/or documented in procedures. [14] See Table 4.8 for implementation.

TABLE 4.8 ISO Implementation for Electronics Systems: The Quality System

A documented quality system (the big "Q") is essential for all electronic suppliers, large or small. For a small company, the quality system can even be documented with pictures, automated. Key elements include: ▪ Ensuring customer requirements are met ▪ Quality products and services delivered ▪ Total quality management and continuous improvement ▪ Audits and corrective action. ▪ All elements of ISO 9001/9002/9003 that apply to your electronics company. Essential to an electronics quality system is quality planning. What is your system for creating plans to ensure quality for each product, each service, and each customer?

4.2.3 Contract review

Contract review is understanding what the customer wants and determining whether or not your company is able to meet those requirements. The review, as well as any amendments, are documented and then explained to all responsible personnel. Requirements need to be adequately defined and documented, and consciously reviewed for capability, to meet the contract. A system to review any contract changes also needs to be in place. Any issues must be resolved, contractually. This is linked in the standard to design reviews. A note in the standard reminds that channels for communication and interfaces with the customer's organization in these contract matters should be established.

Specification review is a very important aspect of contract review for electronic systems. In the software ISO documentation, much language is used for the review of both customer specifications to the prime contractor and vendor specifications to the subcontractors to ensure all persons understand. See Table 4.9 for implementation.

4.2.4 Design control

Design control employs techniques to verify that every product design meets customer requirements and expectations. From input to output, everything is well planned, design activities are called out properly, and documents planning the design are clear and easily understood.

The standard defines *design control:* "The supplier shall establish and maintain documented procedures to control and verify the design of the product in order to ensure that the specified requirements are met." [15] Table 4.10 shows a method of tracking which design control documents are used and what they are based on.

Design control consists of eight subelements:

1. Design and developmental planning—how the company plans the design.

TABLE 4.9 ISO Implementation for Electronics Systems: Contract Review

Contract review and/or purchase order review is essential for the electronics company. Includes specifications, drawings, all engineering documentation
Contract requirements must be visible to entire production or design team.
Also necessary to include system to review changes to contract with team.

TABLE 4.10 Design/Documentation Table

Design task	Requirement	Company process	Company procedure
Design documentation	IEEE P1220/ D1, December 1993	Process 4.4	SOP 1-4.4
Drawings/ specifications/ schematics	ASME Y14.24M, ASME Y14.1, ASME Y14.34M	Process 4.4.1	SOP 1.4.4.1
Configuration management	MIL-STD-973	Process 4.8	SOP 1.4.8
Quality system for design	ISO 9001-1994	Process 4.1	SOP 1.4.1

2. Organizational and technical interfaces—the people and the skill types required to design, and make sure customer requirements are met. Describes concurrent engineering at a high level.
3. Design input: all the things required for input, statutory as well as customer manufacturability/testability—everything.
4. Design output: how will the design be documented; what will the customer see; schematics, block diagrams, user manuals.
5. Design review—a vehicle for reviewing design and various phases.
6. Design verification—checking on the performance requirements.
7. Design validation—checking on what the customer ordered.
8. Design changes—there must be a system for handling the changes that occur in a design; assumes a baseline.

Design control requirements are new quality requirements to defense industries, which previously worked to government statements of work and military standards and specifications. They must now define processes and procedures for design. Commercial industry followed processes, whether documented or not, but to become world class, processes need to be standardized and documented with metrics collected on them.

Under design control, quality aspects of design should be discussed up front in the design process. So often only technical evaluations are done up front. In the electronics world, the items that show up on the quality function deployment (QFD) matrix belong here. Key, critical, and major (see Table 4.11) characteristics should be identified. [16] If there are any new or significantly modified manufacturing or tests to do, it should be planned at the same time. The standard requires the establishment and maintenance of documented procedures to control and verify the design in order to ensure that the specified requirements are met. The chief

TABLE 4.11 Classification of Characteristics

Used for process definition and improvement	
Critical	Critical safety or operating failure or major performance, service or manufacturing classification (usually 2–3% of total assembly); e.g., safety equipment for aircraft or automobile
Major	Cause of substandard performance; difficult to locate and to correct; costly to repair; excessive; requires returning to dealer
Key	Influences performance, supportability, and cost; determines the production processes

SOURCE: Reprinted from Ref. 16.

goal of design process is to translate customer requirements into the technical specification and/or drawings for output.

The subparagraph "Design and developmental planning" requires design plans for each design/development, with specific responsibilities assigned to qualified personnel equipped with adequate resources. [17] It is good to have a design checklist as shown in Table 4.12. The plan needs to be updated as the design evolves. The Taguchi Method should be part of the plan. Here the detailed customer focus comes into being. What are the costs to produce, how many, and what are the operational requirements? All are required to be known to complete the design plan.

Organizational and technical interfaces define different groups which input into the design process. This is really the concurrent engineering paragraph in the ISO standard. The standard states that the necessary information shall be documented, transmitted, and reviewed regularly. For electronics, electrostatic discharge (ESD) and electromagnetic interference (EMI) design parameters come into the plan.

Based on process design, the first element after the design plan is *design input.* What documents need to be identified and reviewed? What process documents need to be reviewed? What are the statutory and regulatory requirements? The standard states the need to resolve incomplete/ambiguous/conflicting requirements, and, of course, coordinate with contract review activities. [18]

Again, based on process design, *design output* is next to ensure that the "specified design" meets the design input requirements. They should contain reference acceptance criteria and identify product characteristics that affect safety and "proper" functioning. The standard requires that the company reviews these documents prior to release. These outputs should be standard in a company: drawings, specifications, block diagrams, and specific kinds of analysis. [19]

TABLE 4.12 Design Plan Checklist

1. Design requirements—Does it meet:
   - Customer implied and specified requirements
   - Design specification
   - Applicable industry standards/regulations—federal, state, local
2. Product/system description—Does it contain:
   - Electrical description and configuration
   - Block diagrams with top-level partitioning
   - Sublevel interface descriptions
   - Mechanical description and configuration
   - Specification/drawing tree
   - Software descriptions/specifications/documentation
   - Functional descriptions—top level and sublevel
   - Software/hardware integration approach
3. Functional requirements—Does it meet:
   - Overall functional and physical requirement
   - Mechanical strength
   - Shock loading
   - Motion requirements
   - Memory requirements
   - Speed/operational requirements
   - Capacity requirements
   - Operating time requirements
   - Size/weight requirements
   - Temperature, thermal requirements
   - Critical electrical requirements
4. Environmental requirements—Does it meet:
   - Temperature (operating, storage)
   - Humidity extremes
   - Vibration (operating, shipping)
   - Corrosion (saltwater, alkaline, or acid)
   - Submersion (brackish water, soil, chemicals)
   - Seals (water, soil, air, gases, vacuum)
   - Production (cleaning solvents)
5. Design analysis—Does it meet:
   - Design analysis plan
   - Mechanical—thermal/stress/vibration
   - Electrical-circuit parameter/component
6. Production requirements—Does it meet:
   - Use of standard parts or assemblies
   - Contain COTS/NDI*
   - Production tolerances
   - Well-defined materials
   - Proven processes
   - Value-added considerations
   - Ease of assembly
   - Inspection/testing standards
   - Finishes, protective
   - Component interchangeability
   - Special part considerations
   - Quantities planned/scheduled
   - Producibility
7. Testing requirements—Does it meet:
   - Engineering development tests
   - Verification testing
   - Reliability/growth/demonstration/test
   - In-circuit tests
   - Line proofing tests
   - Acceptance test
   - Test equipment design/built/certified
8. Operational requirements—Does it meet:
   - All fitness-for-use requirements
   - Field installation assessment
   - Instruction sheets
   - User maintenance manuals
   - Field service assessment
   - Special tools for servicing
   - Prototype device experience
9. Reliability/maintainability requirements—Does it meet:
   - Reliability requirements
   - MTBF requirements†
   - Availability requirements
   - Failure modes and effects analysis
   - Maintenance philosophy
   - MTTF requirements†
   - Built-in-Test (BIT) requirements
   - Testability guidelines
   - Human factors guidelines
   - Personnel safety requirements
   - Product liability considerations
10. Cost requirements—Does it meet:
    - Cost avoidance guidance
    - Nonrecurring cost guidelines
    - Product cost guidelines/goals
    - Capital equipment costs
    - Design/cost trade studies
    - Life-cycle cost guidelines
11. Design reviews—Are they planned?
    - Customer and internal scheduled reviews
    - Special consideration peer reviews
    - Subcontractor design reviews
12. Risks—Are they mitigated?
    - Technical risks
    - Schedule risks
    - Cost risks

*COTS/NDI = commercial off-the-shelf/nondevelopmental item.
†MTBF = mean time between failures; MTTF = mean time to failure.
SOURCE: Adapted with permission from D. Leech and B. Turner, *Engineering Design for Profit*, Wiley, New York, 1985, p. 287.

Design reviews are formal and documented, and include representatives from all functions, as well as specialists. The review philosophy is the same as the systems engineering philosophy. Sometimes the reviews are internal, sometimes external, but they need to be done to verify that customer requirements are being met. Table 4.13 can be used to do an internal assessment of your company's reviews.

Design verification is held "at appropriate stages of design; design verifications shall be performed to ensure that the design-stage output meets the design-stage input requirements." [20] Suggestions are given in addition to design reviews. They include alternative calculations, similarity analysis with other equipment, tests and demonstrations, and performance testing. [21] A good example of design verification may be found in Table 4.14.

"Design validation shall be performed to ensure product conforms to user needs and requirements." [22]

Design validation follows design verification and is performed under defined operating conditions. It is normally performed on the first production unit. Multiple validations are performed if there are multiple uses. [23] In the American vernacular, the functional configuration audit (FCA) and physical configuration audit (PCA), with first article inspection, and factory acceptance test are considered validation.

The last paragraph under "Design Control" recognizes that as part of design there are design changes. This paragraph requires design change procedures. This is one part of configuration management called out in ISO 9001. Changes may be applied to build print contracts, and this paragraph may be tailored into ISO 9002 for those contracts so that a good change program is set up at the beginning. See Table 4.15 for implementation.

4.2.5 Document and data control

The documents that define the quality management system's policies, procedures, and work instructions will be rigorously controlled so everyone is working to the same guidance. Any documents important to the quality system and the customer, including drawings, master documents, and procedures, are maintained and changed under controlled conditions.

Document control involves issue, approval, change, and modification. Automated systems are good to use but must have definition. Documents must be numbered, dated, and removed upon obsolescence. Documents must be easily obtainable and usable and changes to documents must use the same loop as the original approval. In other words, if six departments signed the original document, then six departments

TABLE 4.13 **Design, Testability, and Manufacturability Reviews:**

Rate your company by this matrix:

Attribute	Rating 1 (20%)	Rating 2 (40%)	Rating 3 (60%)	Rating 4 (80%)	Rating 5 (100%)
Thoroughness	No formal meetings held.	Only internal peer review held, no minutes taken.	Formal meeting held in one functional group, minutes published.	Only one formal meeting held with more than one functional group attending, minutes of meeting published.	Formal meetings held as needed and all the functional support groups attended. Formal feedback including corrective actions published.
Timeliness	Not in time for any corrective action.	Meeting held too late to prevent any problems in first production.	Meeting held prior to beta deliveries, customers were informed of pending changes.	Meeting held after design was completed and prototypes built.	Meeting held throughout the product development cycle, several preventive actions were identified and implemented.

TABLE 4.14 VLSI Design Process and Design Verification

Design phase	Model	Tool	Quality assurance
Architectural	Behavioral simulator	C language	Tests, reviews, analysis
Functional	Functional simulator	C, AIDE, PLA tool, checker	Tests, reviews, regression
Logic	Logic model	LAMP, DRAW, checker	Tests, synthesis, regression
Circuit and timing	Circuit model	MOTIS, ADVICE, checker	Tests, regression, analysis, synthesis
Layout	Layout model	Circuit extractor, LTX, CTL-LCC	Analysis
Wafer and chip integration and system certification	Wafer Chip	Test machine Test facilities	Tests, fault analysis Tests, regression
Product engineering	Chip	Test machine	Tests, analysis, parametric characterization

SOURCE: Goksel, Sekino, and Troutman, 1986, copied with permission from *Juran's Quality Control Handbook,* p. 29.9.

TABLE 4.15 ISO Implementation for Electronics Systems: Design Control

Essential for knowing customer requirements, both specified and implied, in new or modified design.

The electronics company should know the marketplace and what the customers want.

The design plan, inputs/outputs, must be written and meet public laws.

Standard processes only increase productivity and do not inhibit the creativity of designers.

Over time, every design changes; especially in electronics, the design control plan needs to address changing design from the first set of baselined documents.

In 1993, Analog Devices and DY 4, both electronic suppliers, reported that by using the ISO design control requirements, they standardized a consistent set of design processes. This benefited designers because the design tools are so complex (Analog), that by tailoring the supplier selection and control, and the way they specified design requirements with suppliers, the process was simplified. By doing this, DY 4 reduced defects for component suppliers at receiving inspection by "orders of magnitude." (*Military & Aerospace Electronics, March 15, 1993, pp. 30–31, with permission.*)

must sign the change. This is one good reason for streamlining a company's document change procedures. Under ISO, document changes need to be handled swiftly, with changes to the processes. A master list, or equivalent document control procedure identifying the current revision status of documents, shall be established and be readily available to preclude the use of invalid and/or obsolete documents.

This control shall ensure that [24]:

1. The pertinent issues of appropriate documents are available at all locations where operations essential to the effective functioning of the quality system are performed.
2. Invalid and/or obsolete documents are promptly removed from all points of issue or use, or otherwise assured against unintended use.
3. Any obsolete documents retained for legal and/or knowledge-preservation purposes are suitably identified.

One issue that arises during ISO 9000 implementation is that of differentiating documents and records. Table 4.16 provides some guidance that is not found in the ISO documentation. [25] See Table 4.17 for implementation.

4.2.6 Purchasing

This element applies to items and products that are purchased by a supplier or subcontractor. All purchasing data is kept, including veri-

TABLE 4.16 Differentiating Documents and Records

Documents	Records
Specifications/blueprints	Engineering change requests
Inspection instructions	Inspection data
Test procedures	Test reports
Work instructions	Production data
Quality manual	Management review meeting notes
Calibration procedures	Calibration data
Supplier quality requirements	Purchase orders
Qualified suppliers list	Completed supplier survey forms
Product development procedure	Development team meeting notes
Process instructions	SPC control charts

SOURCE: Craig, p. 133; printed with permission.

TABLE 4.17 ISO Implementation for Electronics Systems: Document Control

Create a master list of documents that your company works to:

- What documents do you work to? Make a list.
- What specifications/standards do you use? Make a list.

Documents need to be controlled?

- How are revisions released to both design and production/coding personnel?
- How do your employees know they have the "latest revision?"

A good system is to keep all documents electronically and in an automated system. When a document prints, it states "uncontrolled."

If all documents are not electronic, the masters can be kept under configuration control, and the titles and revisions appear in the master list.

A central location needs to maintain document status and distribute documents.

fication of the purchased product. This should include evaluations of contractors and vendors. The supplier shall establish and maintain documented procedures to ensure that purchased product conforms to specified requirements.

The supplier shall:

1. Evaluate and select subcontractors on the basis of their ability to meet subcontract requirements, including the quality system and any specific quality assurance requirements.
2. Define the type and extent of control exercised by the supplier over subcontractors. This shall be dependent on the type of product, the impact of subcontracted product on the quality of final product, and, where applicable, on the quality audit reports and/or quality records of the previously demonstrated capability and performance of subcontractors.
3. Establish and maintain quality records of acceptable subcontractors.

Purchasing documents shall contain data clearly describing the product ordered, including, where applicable:

- The type, class, grade, or other precise identification
- The title or other positive identification, and applicable issues of specifications, drawings, process requirements, inspection instructions, and other relevant technical data, including requirements for

approval or qualification of product, procedures, process equipment, and personnel

- The title, number, and issue of the quality system standard to be applied

The supplier shall review and approve purchasing documents for adequacy of the specified requirements prior to release. Where the supplier proposes to verify purchased product at the subcontractor's premises, the supplier shall specify verification arrangements and the method of product release in the purchasing documents.

The supplier's customer or the customer's representative shall be afforded the right to verify at the subcontractor's premises if the contract so states. Such verification shall not be used by the supplier as evidence of effective control of quality by the subcontractor. Verification by the customer shall not absolve the supplier of the responsibility to provide acceptable product, nor shall it preclude subsequent rejection by the customer. [26] See Table 4.18 for implementation.

TABLE 4.18 ISO Implementation for Electronics Systems: Purchasing

Having systems to track suppliers and their product is the key.
Define the system for selecting subcontractors and suppliers in your company. Document it. This includes setting up types, grades, and classifications of things that you buy (paper to high-reliability components).
Define the database to keep track of suppliers, deliveries (by part number).
Tie this information to your supplier rating system.
Define the control system that your company has over subcontractors/suppliers (inspection/SPC?).
Purchase orders must contain the product you will buy, with the type, grade, or classification, the engineering documentation, and the quality system required at the supplier.
Engineering documentation (drawings, specifications) that accompanies purchase order needs to be under configuration control with purchasing documentation.
What about electronic communication? Should be in planning or already in place.
Suppliers should be closely tied with their customers and part of the team.
Supplier partnerships should be encouraged to limit the number of suppliers.
See Chap. 7 for more data.

TABLE 4.19 ISO Implementation for Electronics Systems: Control of Customer-Supplied Product

The first question here for implementation is, "Does your company handle customer-supplied product?" If it does, then this paragraph must be addressed in your quality manual and your procedures/processes; but if not, then mark this paragraph "not applicable" in your quality manual. If you do get customer-supplied product you need to: ▪ Track it—so you know what you have ▪ Record the condition that it arrives in ▪ Keep it in a secure location ▪ Mark it so that all employees know the difference between customer and company product ▪ Record actions on that product on your standard documentation ▪ Record when it leaves

4.2.7 Control of customer-supplied product

Customers can sometimes provide items to be incorporated into, or used in the development of, a final product. These items must be controlled and any nonconformances reported back to the customer. Documented procedures shall be established and maintained for verification, storage, and maintenance of purchaser-supplied product provided for incorporation into the supplies or for related activities. If any such product is lost, damaged, or is otherwise unsuitable for use, it shall be recorded and reported to the customer. Verification by the supplier does not absolve the customer of the responsibility to provide acceptable product. See Table 4.19 for implementation.

4.2.8 Product identification and traceability

This is another part of configuration management: product identification. Whether it be a purchased product, a customer-supplied product, or a final company product, proper identification is used in the electronics business to provide an easy means of identification. This can include such things as documentation, tags, or any other markings which are understood by all and serve to identify each product as unique and distinct. Again, documented procedures shall be established and maintained for identifying the product from applicable drawings, specifications, or other documents, during all stages of production, delivery, and installation. See Table 4.20 for implementation.

TABLE 4.20 ISO Implementation for Electronics Systems: Product Identification and Traceability

The electronics industry does use part numbers; so part numbers are applicable for electronic components and systems.
This same identification is attached to the part and used throughout its life cycle.
If your company does design, your engineering drawing practices should specify a way to establish part numbers and keep track of parts from component suppliers by using electronic component part numbers as well.
Good manufacturing and servicing practices necessitate part identification on their work documents.

4.2.9 Process control

This is the heart of the ISO production system. A good process control system serves to prevent problems from occurring before the final product is delivered. Processes shall be used in the production of products or service to ensure a consistent way of doing things and to help each employee understand the job. This paragraph does address production or servicing only, as processes affecting design and development are under the "Design Control" paragraph. "Process yields, especially in the competitive electronics markets, are major determinants in the manufacturing costs." [27]

For process control, it is required that there be "...documented work instructions defining the manner of production and installation, where the absence of such instructions would adversely affect quality." [28] This includes "soft," or support, processes, and equipment as well. What is documented depends on people and processes. ISO feels that it is a company's decision as to which level of document fits "the process." Generic flows are good for standard paper-type tasks that support production under this paragraph. This is the place to define the advanced quality system for continuous improvement as well. Criteria for workmanship, compliance with standards, and maintenance are covered here as well.

The production process, in broad terms is: (1) receipt of material or components, (2) in-process material flow, (3) assembly of minor and major subassemblies, (4) final assembly, (5) tests and/or screens, and (6) assessments of product quality. [29] Processes/procedures and training must be in place for all of these activities under this section. Generally, Level 2 procedures can be written to determine policy for manufacturing, or production, and a manufacturing plan can be written for each product or product line. Standard processes in the production facility can be documented in Level 3 processes/procedures

TABLE 4.21 Production Assembly Criteria

- Engineering model build, included in initial production build
- Configuration control in place
- Process availability in place
- Plant facilities in place
- Material handling in place
- Tools, gauges, test equipment in place
- Labor plans in place
- Training plans in place
- Tryout schedule complete
- Quality-sensitive parts plans in place
- Material replenishment and control in place
- Safety requirements reviewed
- Central corrective action in place
- Product packaging in place
- Quality information system in place
- Audit procedure issued
- Process analysis studies planned/completed
- Controlled build planned/scheduled

SOURCE: Adapted from Ref. 30 with permission.

which the manufacturing plan can tailor, if necessary. Table 4.21 illustrates the criteria for actions to be taken before beginning to assemble a new product, or a new product line. This can also apply to later stages of assembly in the pilot plan. [30]

For electronics, screens or "burn-ins" may be part of the process. Various run-in and burn-in techniques (or screens) are described in *Juran's Quality Control Handbook*, 4th ed., McGraw-Hill, New York, 1988. [31]

One hundred percent burn-in of electronic modules was used at Xerox to protect the customer from early failures. The original standard for all complex electronics modules was to burn-in under a profile that required 96 h of operation and an 8-h thermal excursion, ranging from 25 to 70°C with power applied to the module. Capital equipment was necessary to do this, and so the second key objective was to reduce or eliminate this costly equipment. The key ingredient was to collect data, define the problems, and implement corrective action so that (1) incremental capital equipment would not be required, (2) the modules' inherent reliability characteristics could be upgraded, and (3) production throughput would be improved. This was not easily done. On one program, which had a mix of 12 to 14 very complex assemblies, experiments were conducted over 18 months to assess the shape of the "bathtub curve" (the period). Corrective actions were then implemented, which resulted in the reduction of burn-in from 96 h to 48 h and ultimately to zero hours (no burn-in) for some modules. The same approach should be applied where there is need for an initial assembly level burn-in to simulate some period of operation at the customer's site. [32] For small quantities for specialized high-reliability purposes, and for initial production units, a burn-in should be used to ensure that failures in the field do

TABLE 4.22 Examples of Special Processes for Electronic Systems

- Plating, protective coating
- Electrostatic discharge sensitivity
- Automatic test equipment
- Magnetic particle inspection
- Eddy current inspection
- Ultrasonic inspection
- Radiographic (x-ray) inspection
- Wiring and crimping
- Soldering
- High-reliability soldering
- Corrosion control processes

not occur. In one case, with high-reliability units, the customer needs the electronic units not to fail, contractually. In initial production to send a new product to market, early failures on the new product would inhibit sales on the follow-on product. Therefore, Xerox's model shows how their method leads to increase sales, and a reduction of burn-in over time, with proven product.

A special process is a process that cannot be fully confirmed through testing or inspection so that a defect surfaces only when the product is used by the customer. Table 4.22 shows examples of special processes for electronic systems. Continuous monitoring and/or compliance with documented procedures is required to ensure that the specified requirements are met. Special processes are defined as needing to be "qualified."

The requirements for any qualification of process operations, including associated equipment and personnel, need to be documented. The standard suggests that such processes require prequalification. This is extremely important for electronics. A good electronics company must have qualification of certain processes. Certain processes may be "special processes" and not be confirmable through inspection and testing. Other "special type" processes may be controlled because the manufacturer feels they are critical or important to the operation.

Complicated processes begin with a process specification, which defines the process and verification techniques. This specification is "released" into the controlled documents of the electronics company under specific numbers (with revisions). Process certification begins with a checklist of various quality peripherals (see Table 4.23) that must be evaluated in association with the requirements of the process specification. The checklist needs to be drawn up by a process and a

TABLE 4.23 A Quality Checklist for Process Certification

Quality system	Environment	Supervision
Effective CM	Water/air purity	Clear quality goals
Engineering release	Dust/chemicals control	Clear instructions
Equipment calibration	Temperature/humidity control	Combining tasks
Preventive maintenance	Human/product safety	Natural work units
Built-in equipment diagnostics	Lighting/cleanliness	Client relationships
Visible, audible alarms for poor quality	Electrostatic discharge control	"Ownership" through vertical job enrichment
"Poka-yoke," fool-proof inspection	Storage/inventory control	Feedback of results
Neighbor and self-inspection		Encourage suggestions
No partial build policy		Coach, not boss
Worker authority to shut down poor-quality line		

SOURCE: Reference 33; reprinted with permission.

quality engineer. Then a team, including the design engineer "certifies" the process before production starts. Key to this certification is the training and certification of the operators. All of this needs to be documented. Once the design of experiments has identified and reduced the important variables, the SPC can be applied to continuously monitor the product going through the process. [33]

Electronic companies need a list of critical skills/tasks requiring certification of employees. Typical examples are soldering, electronic assembly and test, and crimping, wiring, and connector assembly. Specialized processes are defined according to the companies' needs. The certification of the employees, for the processes that they are qualified to work, is part of the training records of the company.

Processes, as well as the operators, need to be recertified at least annually, depending on the process.

One of the most standard of all processes in the electronics industry is found in the electronic components industry. The Electronics Industries Association, along with representatives from the U.S. government, came up with a standard process that is widely used for Department of Defense (DOD) components. For many years, the military subsidized the testing of semiconductors and integrated circuits that fell under its qualified product list (QPL). Suppliers, in turn, were bound to the self-declaration of conformity once their product passed the initial testing process. Testing was specialized and had to be approved by the DOD.

There was no standardization of testing. Moreover, with budget constraints mounting, the military could not afford to subsidize testing of supplier products. In response to these concerns, the semiconductor/integrated circuit industry developed a new testing system called quality manufacturers list (QML). See Table 4.24 for the benefits of QML.

While responding to military budget constraints, QML's aim is to allow for the introduction of new technologies and design, offer flexible response to problems, better utilization of customer feedback, stronger management support, development of common processes for contracts, and support of U.S. government acquisition reform. In supporting standard products, QML qualification simplifies the procurement process, reduces lead times, and encourages off-the-shelf availability. The Defense Electronics Supply Center (DESC) certifies the QML and ISO 9000 system at the supplier. Most certifications are for ISO 9002, except where they need to be ISO 9001 because the supplier does the design. The qualified manufacturer's list exists for various kinds of components. For example, a specification hybrid microcircuits meeting MIL-H-38534 is published as QML-38534-20. QML specifications are often cross-referenced to supplier part numbers for formal standard military drawings (SMDs) and to MIL-STD-883, aimed to specify and qualify military components. Specifications are validated annually. Qualification is based on:

- Validation of the manufacturer's design
- Fabrication

TABLE 4.24 QML Characteristics

- Performance/output-driven
- Flexible response to problems
- Processing commonality
- Qualification by manufacturing line
- Utilizes standard methods
- Short introduction cycle
- Manufacturer-prepared specifications
- QML and TQM control quality
- Supports U.S. acquisition reform
- Allows plastic packages
- Manufacturer's Review Technology Board approves changes
- If a product does not always meet specialized U.S. government requirements, then the government agency must send for special testing, screening.

TABLE 4.25 ISO Implementation for Electronics Systems: Process Control

Process control for both hard and soft processes needs documentation.
Process control can lead to huge savings in the paper processes that support the manufacturing process when standardized and tracked.
Process yields must be calculated.
Processes should all be flowcharted and under statistical process control (see Chap. 3).
Taguchi method should be used to optimize processes prior to statistical process control (SPC) to achieve zero variation (Chap. 3).
Burn-in should be used for high-reliability parts and to verify production, and then eliminated.
Special processes must be documented and certified, including processes, personnel, and equipment.
See Sec. 3.6.

- Assembly/test operations
- Quality assurance processes

The Defense Electronic Supply Center audits this process to the ISO 9001 standard. The QML was the first process where the government allowed the manufacturers to use "best commercial practices (ISO/TQM/SPC/DOE)" as the first of a standard process at a supplier-creating product. The EIA is in process of issuing a new quality standard for electronic process certification to be published as EIA-599-1. It is currently in draft version. In addition to the components of the ISO 9000 quality systems, it contains processing certifications for JESD-46, "Requirements for User Notification of Product/Process Changes by Semiconductor Suppliers" (another EIA standard), plus requirements for C_{pk} reporting and Positrol applications (see Chap. 3 for discussion of C_{pk}). It requires control plans, but cites only minimum requirements, and provides two control plan examples in the draft standard: one for plastic and one for hermetic integrated circuit packages. [34] See Table 4.25 for implementation.

4.2.10 Inspection and testing

When a product is received at the company, inspection verifies that it has not been used or damaged. This section discusses requirements for in-process inspection, as well as testing. This serves to better identify the product and to document that requirements have been met.

This section defines receiving inspection as the place to "inspect" or "otherwise verify as conforming to specified requirements" in accordance with the quality assurance plan or documented procedures. Verification can take many forms and does *not* have to be always done in "receiving inspection." The supplier has the responsibility to ensure that incoming product is not used or processed until it has been inspected or otherwise verified as conforming to specified requirements. Verification shall be in accordance with documented procedures.

Inspection and testing, in-process and final, must be documented. The product is specifically "held" until it passes all tests. Nonconforming product must be identified and segregated and records of inspection and test must be established and maintained. The required inspection, testing, and records shall be detailed in the quality plan, or documented procedures.

Some suppliers think that cutting out receiving inspection in the just-in-time (JIT) supply chain is good business, and it might be if there is a good supplier partnership and a good record of quality parts deliveries from subtier suppliers. In determining the amount and nature of receiving inspection, consideration shall be given to the amount of control exercised at the subcontractor's premises and the recorded evidence of conformance provided. A period of demonstrated quality receivals should be established before receiving inspection is eliminated. What many suppliers do not do is track the upturn in cost when there is JIT delivery and defects are found later in the production cycle. Then many products must be reworked, and "hoards" of people must run "wild" to replace the defective components or assemblies. The cost escalates dramatically as opposed to checking at receival or certifying suppliers.

Where incoming product is released for urgent production purposes prior to verification, it shall be positively identified and recorded in order to permit immediate recall and replacement in the event of nonconformity to specified requirements. In-process products shall be inspected, tested, and identified as required by the quality plan and/or documented procedures. Products shall be held until the required inspections and tests have been completed or necessary reports have been received and verified. The quality plan and/or documented procedures for final inspection and testing shall require that all specified inspection and tests, including those specified either on receipt of product or in process, have been carried out and that the data meets specified requirements. No product shall be dispatched

TABLE 4.26 ISO Implementation for Electronics Systems: Inspection and Testing

Inspection and testing is the key to delivering a quality electronic product.
Starting with the electronic components, sampling is no longer the way that components should be inspected, as defective electronic components may create an entire system that "does not work," which is guaranteed to produce customer dissatisfaction.
At this time, some good component suppliers are using 100 percent inspection with automated inspection systems.
Some electronics, like printed wiring boards, still require source inspection for high-reliability components. These processes need to be put under statistical process control so that the process can better be controlled and inspection costs can be saved.
For electronic assembly supplies, a set of inspections and tests during the product process ensures that a quality product gets to final inspection, and then to the field for distribution.
A comprehensive inspection and test program is a must for electronic suppliers.

until all the activities specified in the quality plan and/or documented procedures have been satisfactorily completed and the associated data and documentation is available and authorized. The supplier shall establish and maintain records which give evidence that the product has passed inspection and/or test with defined acceptance criteria. Where the product fails to pass any inspection and/or test, the procedures for control of nonconforming product shall apply. [35] See Table 4.26 for implementation.

4.2.11 Control of inspection, measuring, and test equipment

Testing will be effective only if the tools used are consistently accurate and precise. Procedures will ensure that products have been tested by equipment which is controlled and calibrated to the accuracy deemed necessary. All measurements and results are documented and each piece of equipment used in testing is properly identified.

This is the longest clause in the Q9001 specification. Implementation will depend on your product. Many small electronic suppliers may subcontract calibration to the manufacturer of the test equipment or an independent laboratory. Be sure the laboratory works to the ISO standards (Guide 25); the supplier is still responsible, even if the work is subcontracted out. Compliance to former MIL-STD-45662 makes this paragraph easier to comply with. Those who

TABLE 4.27 ISO Implementation for Electronics Systems: Calibration

Take the ISO requirements directly from the standard, and match to your documented procedures. In other words, do a GAP analysis. (See Table 4.44.) Get a copy of either the ANSI or ISO standard, and do a GAP analysis of your procedures/processes. Many systems must be set up if you are already not in compliance: ▪ Training ▪ Proof of calibration ▪ Tests and software used ▪ Specifications/standards used ▪ How product is tagged on the floor If you are doing this anew, this is a good benchmarking project, after a GAP analysis in your company. If the supplier subcontracts this effort out, then the quality manual says "not applicable." However, include information on how your company verifies that equipment used for producing product is in calibration. The supplier is still responsible to ensure equipment used is calibrated.

are not familiar with this standard will find it more difficult, as will those who have not been audited to this standard. The replacement for MIL-STD-45662 is ANSI/NCSL Z540-1-1994, *American National Standard for Calibration—Calibration Laboratories and Measuring and Test Equipment—General Requirements,* or ISO 10012-1, *Quality assurance requirements for measuring equipment—Part 1: Metrological confirmation system for measuring equipment.* These standards define a calibration system, including records, tests, and trained personnel. This is an area that should be discussed with the auditor, prior to the audit, if you are unsure that your company is ready for the audit. See Table 4.27 for implementation.

4.2.12 Inspection and test status

Test status indicates whether or not a product has been inspected, accepted, or rejected. Regular reporting methods will help your company, and customers, determine the exact test status of any developmental product. The product will be marked with proper identification to indicate status. Procedures need to cover engineering integration and software, as well as manufacturing. See Table 4.28 for implementation.

TABLE 4.28 ISO Implementation for Electronics Systems: Inspection and Test Status

During the manufacturing process, it is necessary to have documented on the traveler or production paper which tests/inspections are required, and then which have been passed or failed. If a test/inspection has been failed, the system should show the reason for failure and corrective action. In electronic assembly, during the integration and proofing of systems phase, testing may go on for long periods of time. Knowledge of test status, of the required integration tests plus what is passed and what has been failed, again for corrective action, is vital to making sure that the customer requirements have been met.

4.2.13 Control of nonconforming product

A nonconforming product, whether it is the beginning, intermediate, or final product, is one that does not meet specifications. Every employee is asked to ensure that no product is delivered that does not meet customer expectations. A method must be available to identify and *separate* the nonconforming versions of a product as it is developed. Control of nonconforming products applies to all industries. Repair and/or rework depends on contract. Repaired and/or reworked product shall be reinspected in accordance with the quality plan and/or documented procedures. Nonconformities can be found internally or externally. The customer (especially government) must declare whether it wants to be involved. Nonconforming product shall be reviewed in accordance with documented procedures. It may be

- Reworked to meet the specified requirements
- Accepted with or without repair by concession
- Regarded for alternative applications
- Rejected or scrapped

Where required by the contract, the proposed use or repair of product which does not conform to specified requirements, shall be reported for concession to the customer or customer's representative. The description of the nonconformity that has been accepted, and of repairs, shall be recorded to denote the actual condition. [36]

In the world of electronic components, it is perfectly legal for components that do not pass certain tests to be sold as other products, but they must be so identified, and the tests they have passed should be documented. A set of defined specifications with increased levels of

TABLE 4.29 ISO Implementation for Electronics Systems: Control of Nonconforming Product

Again, the system here must serve the customer.
Does the customer accept nonconforming product? (With waivers, material review actions?)
Does the customer want to be involved when nonconforming product is produced? If so, this must be defined in the contract or purchase order.
Nonconforming must be segregated.
When a product is allowed to be sold as another product, it must be clearly shown what requirements have been passed.
A separate area needs to be established to store nonconforming product.
Where the product is re-marked and sold under other part numbers, clear areas of demarcation should show that the product is segregated, so than no confusion can occur.

testing would allow a component supplier to sell, for various purposes, a product that fails the next higher test. See Table 4.29 for implementation.

4.2.14 Corrective and preventive action

Corrective action is what is actually done to correct a nonconformance. Procedures will define who is responsible for correcting the problems and what measures will be taken to prevent recurrence. The quality management system provides the mechanisms for continuous improvement through corrective and preventive actions on nonconforming product and corrective action to prevent recurrence. The goal is to eliminate potential causes (find root causes) by analysis of processes and work operations and audits of customer complaints. Then, management must apply controls to ensure corrective/preventive action is taken. Good record keeping of changes to procedures resulting from corrective action and their implementation provides for continuous improvement.

The procedure for corrective action shall include [37]:

1. The effective handling of customer complaints and reports of product nonconformities, investigation of the cause of nonconformities relating to product, process, and quality system

TABLE 4.30 ISO Implementation for Electronics Systems: Corrective and Preventive Action

Corrective and preventive action is the key to continuous improvement.
Beginning with internal quality audits and continuing with metrics on inspections and tests passed and failed, receiving inspection metrics, customer service metrics, good metrics and audits can (along with a TQM system) provide for continuous improvement.
Applies to all processes—business, support, and product hardware, software, and services.
Corrective and preventive actions should be visible to all employees.

2. Recording the results of the investigation, including copies of the quality records
3. Determination of the corrective action needed to eliminate the cause of nonconformities
4. Application of controls to ensure that corrective action is taken and that it is effective

The procedure for preventive action shall include:

1. The use of appropriate sources of information such as processes and work operations which affect product quality, concessions, audit results, quality records, service reports, and customer complaints to detect, analyze, and eliminate potential causes of nonconformity
2. Determination of the steps needed to deal with any problems requiring preventive action
3. Initiation of preventive action and application of controls to ensure that it is effective
4. Ensuring that relevant information on actions taken is submitted for management review.

See Table 4.30 for implementation.

4.2.15 Handling, storage, packaging, preservation, and delivery

This section has requirements to ensure that products are controlled through all phases of handling, storage, packaging, preservation, and delivery. These measures will ensure that damage or deterioration of the product is prevented.

TABLE 4.31 ISO Implementation for Electronics Systems: Handling, Storage, Packaging, Preservation, and Delivery

The supplier is responsible for getting the electronic product to the consumer (whether commercial or defense) in the condition where it will "work" to "specified requirements."
Again, these are documented, auditable, and established processes with specifications/standards.
Applies to all processes—business, support, and product hardware, software, and services.
Internal procedures, including flowcharts, should be generated for this area.
Each company should have standard practices that all employees use.
Consumer products are a good example. Items like a television or home appliance have lots of packaging and plastic over areas that could scratch and styrofoam to prevent crushing.

Documented procedures ensure the product remains undamaged and will not suffer deterioration. Raw material must be secured and segregated in storage from finished materials. There needs to be documented procedures for transfer of materials. Formal and documented packaging procedures, marking requirements, and instructions must be under controlled systems. And of course, personnel must be trained and the delivery of the product must be in accordance with the contract. See Table 4.31 for implementation.

4.2.16 Control of quality records

This element contains procedures that cover record identification, indexing, storage, access, and disposal. Good records demonstrate that procedures were followed and products or services are of the desired quality.

> The supplier shall establish and maintain documented procedures for identification, collection, indexing, access, filing, storage, maintenance, and disposition of quality records.
>
> Quality records shall be maintained to demonstrate conformance to specified requirements and the effective operation of the quality system. Pertinent subcontractor quality records shall be an element of these data.
>
> All quality records shall be legible and shall be stored and retained in such a way that they are readily retrievable in facilities that provide a suitable environment to prevent damage or deterioration and prevent

TABLE 4.32 Quality Records Required by ISO 9001

- Management reviews
- Contract reviews
- Design reviews
- Design verification records
- Control of customer-supplied product
- Product identification
- Assessment of subcontractors
- Processes: qualification status, equipment, personnel
- Receiving Inspection and test records
- Inspection, measuring, and test equipment (metrology, calibration records)
- Inspection and test status
- Quality records of build/as built
- Nonconformities of products accepted
- Corrective/preventive actions investigated/solved
- Internal audits
- Training needs/records/certifications

> loss. Retention times of quality records shall be established and recorded. Where agreed contractually, quality records shall be made available for evaluation by the purchaser, or the purchaser's representative for an agreed period. [38]

Records are mentioned in 13 of the 20 Q9001 paragraphs (see Table 4.32). It is good to automate, in a database for all empowered employees to see, and make paperless. It is also a good idea to have a matrix of quality records for your quality manual, showing the process owners and the documents involved; Table 4.33 shows an example. See Table 4.34 for implementation.

4.2.17 Internal quality audits

The ISO audit program is a structured, comprehensive audit program to verify that the quality system is working. This program judges the effectiveness and health of the quality system. It is scheduled on the basis of importance of the activity and performed in accordance with documented procedures. The results are briefed to responsible management personnel and to the CEO. Corrective action must be timely on any deficiencies. It is tracked, briefed, and closed.

Internal quality audits employ methods prescribed to perform self-assessments and to assess the effectiveness of the quality system. They are performed on a scheduled basis, used for self-improvement, and used by management to evaluate the overall quality system. An audit function improvement process is shown in Fig. 4.4.

TABLE 4.33 Quality Records (Sample Table)

Records	Purpose	Process owner	Location/retention
Product specification	Defines requirements	R&D manager	Product file/5 yr
Process specification processes	Defines manufacturing after last order	Manufacturing manager	On-line database/10 yr
Validation	Identify problems; show results	R&D manager	On-line database/5 yr
Engineering change notices (ECNs)	Document product changes	Document control	On-line document control database/10 yr
Manufacturing work instructions	Define manufacturing processes	Document control	On-line database/10 yr after last order

TABLE 4.34 ISO Implementation for Electronics Systems: Quality Records

"You can't manage what you can't measure." Make a list of your records.

Remember the listing is a "minimum list" and the records must show things important to your product or service.

Under each type of record, your company needs to define the records that show the quality of your "outputs," both internally and externally.

This is a significant difference from former quality systems, like MIL-Q-9858A or MIL-I-45208. The intent is that the supplier is responsible for itself, takes the responsibility and authority, and doesn't wait for an outside agency to find that the quality system is not correct and suggest continuous improvement. *Now* the supplier takes responsibility for system and improvements by *internal* audit.

When implementing the system, audit new subjects first; audit subjects that have most findings more often; and audit all 20 paragraphs at least once a year. In the current system, registrars who have registered/certified companies reaudit every 6 months.

Good internal audits depend on the auditor's familiarity with the standard and the company's processes. The human relations skills of the auditor are important because auditing should be a learning activity under the ISO plan. A complete checklist for standards is found in Table

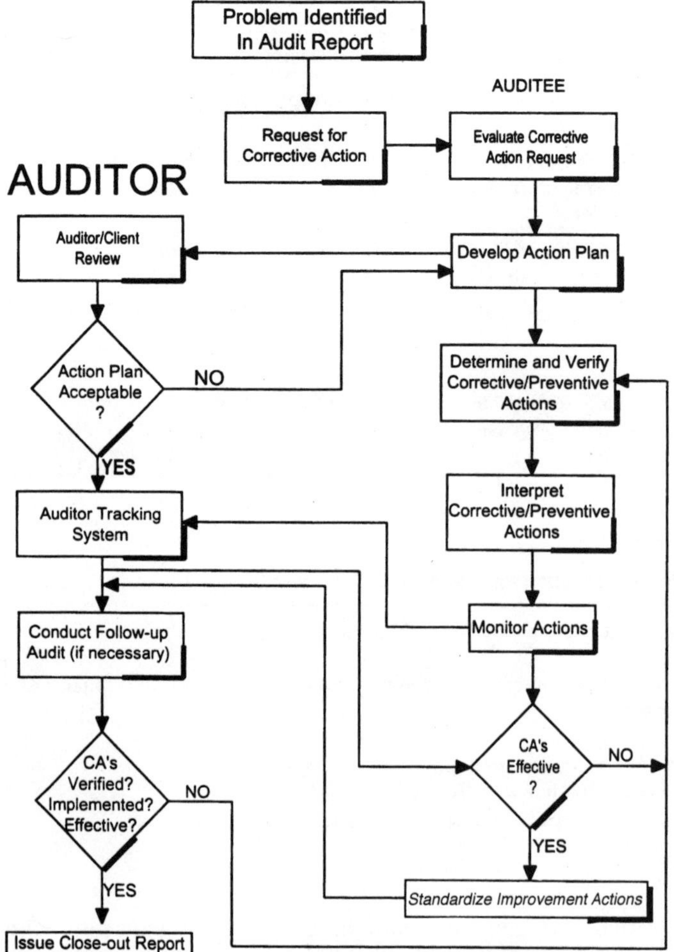

Figure 4.4 Audit function improvement process flowchart.

4.51. The right-hand side of the page is blank so that the auditor can make notes. A sample system audit report is presented as Fig. 4.5.

Prior to the audits, the scope of each audit should be planned, including resources. Audits are two-way communications.

Notes in the standard recognize that follow-up activities shall verify and record implementation and effectiveness of the corrective action taken. The results of internal quality audits form an integral part of the input to management review activities. For the ISO standards that give guidelines on ISO audits, see Table 4.35. See Table 4.36 for implementation of internal audits.

SYSTEM AUDIT REPORT Number______

TO: ______________________________ To be answered by: Date__________
Department Name ______________________ Department Phone Number__________
Team / Auditor:______________________________

FINDING(S):

Procedure Number/Title ______________________________
ISO Clause Number______________________________
Discrepancy Agreement Signature______________________ Date__________

CORRECTIVE ACTION PLAN:

Date of Implementation__________ Signature______________________ Date__________

Response Acceptable ☐ ______________________ Date__________
Check for yes and sign (Auditor)

Follow-up Audit Date______________

☐ System Audit Report (SAR) Closed (check if yes)

Comments:______________________________

______________________ Date__________
(Follow up auditor)

Figure 4.5 Sample system audit report.

4.2.18 Training

Proper training of all personnel ensures confidence, workmanship, and the quality of the final product or service. Records will be maintained to prove that training is adequate and current. This requirement is relevant to safety, environmental, and technical training.

> The supplier shall establish and maintain documented procedures for identifying training needs and provide for the training of all personnel performing activities affecting quality. Personnel performing specific

TABLE 4.35 ISO Auditing Standards

ISO 10011-1:1994, Guidelines for Auditing Quality Systems, Part 1: Auditing
ISO 10011-2:1994, Guidelines for Auditing Quality Systems, Part 2: Qualification Criteria for Quality System Auditors
ISO 10011-3:1994, Guidelines for Auditing Quality Systems, Part 3: Management of Audit Programs

TABLE 4.36 ISO Implementation for Electronics Systems: Internal Quality Audits

Internal quality audits judge the health of your quality system.
The first order of internal audits is independent internal auditors who oversee all these audits and report to the CEO, through the ISO management representative.
In addition, internal quality audits should be made within departments, by independent personnel (e.g., an engineer or technician on one product or program audits another), and functions should cross audit (e.g., finance should audit engineering tasks, production persons should audit quality tasks).
Train several people in each department to be auditors.
Set up an audit findings database.
By cross-training employees as auditors, with lots of audits and a findings database, all employees get involved and take ownership. Audits become the focus of improvement and not a negative experience.
Audits are the checking of the standard processes for improvement of those processes.

> assigned tasks shall be qualified on the basis of appropriate education, training and/or experience, as required. Appropriate records of training shall be maintained. [39]

Lack of training is a huge industry issue as we move into the twenty-first century. The Software Engineering Institute (SEI) at Carnegie Mellon University, with its emphasis on training in the software industry, serves as a good example for the electronics world. On-the-job training should be documented and records kept. All departments should document requirements for positions. Records need to be maintained to show validation periods, certifications, and completion.

L. P. Sullivan believes that the single most important contributing factor to Japanese success in world markets is their system for educating and training all employees on a continuing basis. [40]

Extensive training in quality improvement has occurred since 1980. One excellent example is the Xerox Corporation. Xerox's Leadership Through Quality strategy has, as part of its initial phase, a 40-h train-

TABLE 4.37 ISO Implementation for Electronics Systems: Training

Set up a system for assessing what training each employee needs.
Plan for that training annually.
Document training courses and course material.
Keep track of completed training and certifications.
Keep track of certification expiration to plan new certification.

ing module that all company personnel must attend, from the CEO on down. This quality improvement process has been used effectively in many ways, from redesigning staff meetings to addressing technical problems with limited photoreceptor life in the field. [41] See Table 4.37 for implementation of a training program to ISO requirements.

4.2.19 Servicing

This section requires that the quality system apply procedures to postcontract service and maintenance of a product for its entire life cycle. The development of policies and procedures will ensure quality services to our customers after fielding or distribution.

Servicing, see Table 4.38, may be your company's only business. Then it needs procedures, metrics, and audits. The servicing company needs to go back through the prior 18 standard paragraphs and apply it to servicing. Also, good servicing advice is available in the guide from ISO 9004-2-1991, *Quality Management and Quality System Elements—Part 2: Guidelines for Services.* The emphasis here should be on customer satisfaction. What kind of servicing would make your customer want to keep you? Servicing can be almost anything: providing a "service," consulting, or technical support. If this was your business, the quality manual would be a servicing manual. See Table 4.39 for implementation.

TABLE 4.38 Servicing

Servicing	Servicing needs
▪ Can be any activity	▪ Procedures
▪ Customer service	▪ Metrics
▪ Field support	▪ Audits
▪ Consulting	

TABLE 4.39 ISO Implementation for Electronics Systems: Servicing

First, define what is your servicing or customer service policy? ▪ To whom? ▪ By what means?
Are procedures in place that explain what the policy and actions are?
Do you keep data on your customer servicing? You should.
Define what you do in your quality manual.
Metrics should be kept and reviewed by management.

4.2.20 Statistical techniques

This section requires us to identify the need for statistical techniques. These techniques can include such things as graphics methods, data analysis, and process control and capability studies. The control of techniques will be documented in the system documentation.

The first step is to determine when to use statistical techniques. Then, if the supplier chooses to use them, and they are appropriate, there will be procedures to establish and maintain. Statistical techniques verify acceptability, and in recent times, sometimes are contractually stated. These statistical techniques include software metrics as well as statistical process control.

Statistical techniques are an absolute requirement for electronic systems to produce quality products. As part of the SPC approach, it is necessary to have a good variability reduction (VR) program as well. Variability reduction is a systematic approach to reducing product and process variability to improve cost, schedule, and performance. Deming states in his famous *Out of the Crisis,* "Costs go down as variation is reduced. It is not enough to meet specifications." [42] It represents a cultural change toward the quality of the product to introduce the idea that just falling within specification limits (goal posting, pass/fail, attribute testing) is not the best measure of quality. Rather, the variability of a key process (from variables data) and its relationship to design limits (process capability) become a measure of merit. See Chap. 3 and, in the Appendix, *The USAF R&M 2000, Variability Reduction Process.* See Table 4.40 for implementation.

4.2.21 One Quality System for the Automotive Industry: QS-9000

QS-9000 is a new quality system model that was developed by Chrysler, Ford, and General Motors. The standard applies to all suppliers, internal and external, who provide production and service

TABLE 4.40 ISO Implementation for Electronics Systems: Statistical Techniques

This is absolutely essential for electronics, but more importantly, variability reduction.
This is the only way than an electronics supplier can achieve a world-class environment, or in a contract manufacturing environment, with ISO 9002, the only way to succeed.
Start with SPC on the system that needs most improvement. Hire a consultant, buy a book, go to training. Just get started. It will be worth it.
The company must hire SPC/VR specialists and then empower personnel producing product to understand control, and train in the SPC/VR process. This is an area where, if you do not have a suitable person in your organization, you must get and train, a consultant with enough in-house knowledge to maintain the system.
In a small company, find a person who has had some statistics training, or likes math, and train him/her to become your SPC/VR leader.

parts and materials to the "Big 3." It is the most widespread standardization effort in the history of the automotive industry. In 1995 its use became international. Its goals are to pursue continuous improvement, prevent defects, and reduce variation and waste in automotive products. Beginning with ISO 9000 (the international model explained above), the Big 3 added supplemental requirements from their former quality systems: Chrysler's "Supplier Quality Assurance Manual," Ford's "Q-101 Quality System Standard," and General Motor's "North American Operations Targets for Excellence." The Big 3 formed a group called Automotive Industry Action Group (AIAG) to coordinate the QS-9000 requirements and to gain inputs from other automotive companies, like Volvo. If embarking on the QS-9000 requirements, it is necessary to have the following four standards, which AIAG has printed as well. They are

1. *Production Part Approval Process* (*PPAP*) was created to determine if all customer engineering design record and specification requirements are properly understood by the supplier and that the process has the potential to produce product meeting these requirements during an actual production run at the quoted production rate. [43]
2. *Potential Failure Mode and Effects Analysis* (*FMEA*) was created to issue a common FMEA and give general guidance on the application of the technique. [44]
3. *Advanced Product Quality Planning and Control Plan* (*APQP*) was created to communicate common product quality planning and control developed jointly so that such plans please the Big 3. They

expect the following two benefits of writing this manual: (*a*) a reduction in the complexity of product quality planning for customers and suppliers and (*b*) a means for suppliers to easily communicate product quality planning requirements to subcontractors. [45]

4. *Measurement System Analysis Reference Manual* was created to have a common set of measurement system analysis tools at a high level. It presents guidelines for selecting procedures to assess the quality of a measurement system. The primary focus is measurement systems where the reading can be repeated on each part. [46] This guide seems to be mostly the way to show the statistical process control (SPC) output that is described in the *Statistical Process Control (SPC) Reference Manual,* which is also printed by the Big 3.

These guides are excellent and would provide good guidance for these processes for any company in the electronic world, even if the company does not deliver product to the automotive industry. All are available from the Automotive Industry Action Group (AIAG), Troy, MI, phone (810) 358-3570.

Suppliers get certified first, to the ISO standard of their choice: ISO 9001, 9002, or 9003. Then, additional quality system requirements must be complied with, labeled either *sector-specific* or *customer-specific* requirements. The sector-specific requirement involves standard processes such as part approvals, engineering changes, productivity, continuous improvement activities, process planning, tool design and fabrication, and tool management. The customer-specific requirement in QS-9000 refers to the individual requests that Chrysler, Ford, and General Motors have stipulated for their suppliers. For example, General Motors may have a specific prototype process for material that Ford and Chrysler do not have.

Registrars must be accredited for QS-9000 to certify companies for the Big 3. [47]

4.3 Product Certification

When the ISO system was set up in Europe, one part of the program not directly related to the quality system, but conducted in concert with it, was product certification. There is much interest now in product certification in the United States as international trade is increasing. The United States is now signing mutual recognition agreements so that U.S. labs can perform testing of U.S. products for sale internationally. European laws have been changed to allow more international involvement, and the ISO/IEC standards allow for international trade. This is the beginning of a new system in international trade and all the issues are not 100 percent solved, as you will see.

4.3.1 In Europe

In Europe, products are either regulated or nonregulated (see Table 4.41). [48] Only 10 to 15 percent are regulated, generally for reasons of product safety and liability. Industry groups are recognized, and in charge of determining for each product:

- European Community directive details
- Essential requirements
- Conformity requirements
- Testing/assessment periods

The industry group, through its lab or inspection house, is registered to provide the "CE" mark. This mark, French for *Conformité Européenne,* is required by the directives. The CE mark replaces all national marks previously used in concert with the European community to show compliance with legislative requirements for regulated materials. If you have not yet seen the CE mark, it is prevalent on computer parts (integrated circuits) and toys—look inside your computer and look at toy boxes for the mark.

The industry may or may not require ISO 9001 quality system registration. This is product certification, and not a review of the quality system. The emphasis here is on the *product.* U.S. manufacturers selling or planning to sell products in the European Union (EU) must understand the EU requirement to make sure their plans include compliance with the EU specifications. If the product is on the list of products which must have the CE mark, it cannot be sold in Europe without it, since it certifies that a product conforms to European health, safety, and environmental requirements. According to the International Trade Association, it will eventually be required on two-thirds of all U.S. exports to Europe. Noncompliance is far worse than that for ISO 9000. Products not in compliance may be restricted, even prohibited from sale, or even withdrawn from the market. CE marking is the law in Europe.

A good current source book on international standards has just been published. It is *International Standards Desk Reference: Your Passport to World Markets,* by Amy Zuckerman, American Management Association, New York, 1997.

4.3.2 In the United States

The move to increased international trade is good for electronics. The issue now becomes to what technical standards electronics is built to. The European Economic Community (EEC) speaks of "harmonized

TABLE 4.41 Products Affected by the European Union and Associated Directives—Require CE Mark!

Simple pressure vessels	87/404/EEC
Safety of toys	88/378/EEC
Construction products	89/106/EEC
Electromagnetic compatibility	89/336/EEC
Safety of machinery	89/392/EEC
Personal protective equipment	89/686/EEC
Gas appliances	90/396/EEC
Nonautomatic weighing instruments	90/384/EEC
Active implantable medical devices	90/385/EEC
Hot water boilers	92/42/EEC
Motor vehicles and trailers	70/156/EEC
Low voltage	73/23/EEC
Mobile machinery	91/368/EEC
Recreational craft	OJL 100,1994
Equipment for explosive atmospheres	94/9/EEC
Lift machinery	93/44/EEC
Satellite earth station equipment	93/97/EEC
Elevators	95/16/EEC
Industrial trucks	74/150/EEC
Pending	
Pressure equipment	
In vitro diagnostic medical devices	
Furniture flammability	
Measuring instruments	
Community telecom equipment	
Portable fire extinguishers	
Appliances using fuel other than gas	
Roading requirement for construction machinery	
Marine equipment	
Fasteners	
Child safety articles	
Cableways	
Lifts	

SOURCE: Reprinted from Ref. 48, with permission.

standards," but then it has become clear that the definitions used by the EEC do not necessarily match the definitions by the ISO and IEC. [49] Therefore, if the electronic supplier wishes to do business in Europe, it is essential to become familiar with the so-called new approach, and become familiar with CENELEC, the European Committee for Electrotechnical Standardization. CENELEC was set up in 1973 as an international nonprofit organization, under Belgian law, to help create an electrotechnical marketplace in Europe that is fully open to home-based manufacturers and producers, and also to non-European suppliers. CENELEC publishes European standards for the electrotechnology domain (which they estimate as 22 percent of all products). Under the EEC, these are the standards for the European Community. CENELEC publishes a catalog of all the electrotechnical standards needed for the establishment of the open market in Europe, available in the United States from ANSI. [50] There are a good deal of IEC standards in the CENELEC catalog; however, before you attempt to increase market share in the EEC, it is recommended that you closely scrutinize the CENELEC standards for your particular product.

Lessons learned come from two cases that involve Dormant (a small manufacturer of rubber hoses), and General Electric (GE). Dormant is finding that it must continue to meet national product requirements in several countries, including France and Britain, and GE is involved in a dispute with German authorities. All are being resolved in the courts. [51]

If you wish more information, it is available from the U.S. Commerce Department. Request information on Directive 93/68/EEC of 22 July 1993 from the U.S. Department of Commerce, Office of European Trade, or from the International Trade Association, Office of European Union and Regional Affairs at (202) 482-5276.

Today the United States Underwriters Laboratories (UL) is moving to increase U.S. product certification. Product certification began in the United States in 1991 with the International Electrotechnical Commission Quality (IECQ) System, after an agreement between the IEC and UL. Originally established in 1982, IECQ was initially marketed to large defense contractors rather than commercial original equipment manufacturers (OEMs).

Electronic components from the following areas or families can be certified by IECQ:

- Passive components
- Active Components
- Hybrid integrated circuits

- Printed wiring boards
- Electromechanical components
- Electromagnetic components
- Electrooptical components
- Wire and cable
- Piece parts and materials

When it is properly utilized, IECQ can provide stronger assurance that an electronic component meets the quality requirements of a buyer-seller specification. It can effect important and continuing cost savings for customers and suppliers. It can become an effective tool for improving market penetration and enhanced customer satisfaction. And it can help build bottom line profitability with minimal investment.

For firms that buy electronics, IECQ prints a register of firms and products certified to the IECQ system. The guide lists approved manufacturers, distributors, and test laboratories in one part, and in another part lists qualified products, with the approved specification number, capability approvals, and index to product codes.

This system is good for electronic manufacturers who want to buy quality products. It gives the electronics integrator a wider choice in suppliers of components of consistent quality, with reduced costs in acquiring sourcing information. It is internationally recognized, but on individual components for European Union products, it is good to check the CENELEC catalog. For further information on the U.S. product certification system, contact Underwriters Laboratories in Melville, NY.

4.4 Becoming Compliant: Getting Organized

Becoming compliant to the ISO 9000 series that you select is the right thing to do. It will help you have a quality system that is recognized domestically and internationally, in all industries, but especially in electronics. Management must clearly understand that a quality management system needs, in the words of Deming, "not a substitute, but the CEO." [52] If only an underling starts the compliancy effort, the commitment is not there and the ISO standards will not help you establish a quality system that can lead to a world class company. The CEO must be the one defining what is needed by the company, and make time available to let the employees know that the CEO is involved and in charge. [53]

We define some common terms in auditing: *first-party auditors* are a company's internal auditors; *second-party audits* are customer audits, audits of companies paid for by customers [including for the U.S. government, the Defense Contract Management Command (DCMC) audits for the Defense Logistic Agency]; and *third party audits* are audits performed by a company in the business of doing audits for pay, that is registered by an accreditation agency.

The message that a new quality management system is to be put into place must reach every person in the company, and the reason must be clear. In the current environment, where in many companies the Quality Department is still the police officer of the company, the ISO philosophy is a significant change. Again, to quote Deming, "The people doing the work, will end up in the ISO system being the ones that are responsible for quality." [54] Often a company does not have the right infrastructure for an ISO system. It takes time for any company that does not have a quality management system to set up the ISO infrastructure; the current quality systems that are in place are, for the most part, not management systems, so most contractors will not have the management infrastructure or the totality of the ISO. Nor will they have all the procedures and processes documented. They will have gaps. Even the quality organizations have to change, with self-inspection in manufacturing and internal audits set up in departments under the quality department. This is a real change in culture. ISO drives a system of management with procedures that all employees work to. If employees do not currently work to procedures, which is not uncommon in industry in the mid 1990s, then it will be more difficult to transition. It is another part of the learning process.

The question becomes: Do we get certified or registered? In Europe, the implementation of the ISO system was defined by being "registered" or "certified" by an independent audit company. These companies have become known as "third party auditors" or registrars. Mainly, companies become certified because their customers demand it, or their marketplace demands it. If your company is a player in the international world, it will surely want to get certified, as certification is recognized internationally as indicating a good quality management system. In some cases, it is required for getting contracts with certain foreign governments. If your company is a supplier of electronics or telecommunications in the United States, and feels a good quality system is in place, then it might be the right thing to become aligned with an alternative movement to change the certification system. You might want to hold off on registration/certification and find out about the system level assessment (SLA) that Motorola and Hewlett-Packard are trying to get adopted at the ISO TC 176 committee level (see the next section). Table 4.42 shows advantages/benefits of certification.

TABLE 4.42 Benefits of ISO 9000 Registration

- In some countries, registration is a mandated requirement to market internationally
- Customers are more receptive to partnerships
- Many companies are requiring certification from their suppliers
- ISO 9000 allows smaller companies to complete with large ones, as it can ensure "approved vendor" status
- Customers have more confidence in the supplier's quality management system
- Preventive techniques are in place to foster continuous improvement
- Forces disciplined to well-documented procedures are in place
- Increased operational awareness by all employees of the quality system
- Personnel are adequately trained
- May help gain a market edge

The *ISO 9000 Survey: Comprehensive Data and Analysis of US Registered Companies, 1996* reported that:

> An overwhelming majority of registered companies in the US said they have experienced both external and internal benefits as a result of registration. In a composite ranking of the most significant external benefits, 98.3% cited higher perceived quality; about 70% said competitive advantage; about 56% cited reduced customer quality audits; and, about 29% cited improved customer demand. About 18% of companies cited increased market share....In a composite ranking of internal benefits, nearly 88% of the respondents listed better documentation as an internal benefit, about 83% cited greater quality awareness by employees, about 53% stated enhanced internal communications, about 40% pointed to increased operational efficiency/productivity, and about 19% cited reduced scrap/rework expenses. About 5% pointed to documented sales gains.
>
> An overwhelming 77% ranked quality benefits among their most important reason for pursuing registration; about 73% cited market advantage, and 68% cited customer demands or exceptions. Approximately 27% cited a corporate mandate or large strategy as among the most important reasons. [55]

A survey done by Veritas Labs and ISO Environmental Consultancy Inc., in cooperation with Dowling College in Oakdale, NY, of 48 ISO-certified companies in the New York area found out the number one reason the companies chose registration is their customers required it; the number two reason was that the ISO implementation has been shown to increase operational efficiency and reduce costs. More than half of those surveyed said employees had increased motivation, ability for job performance, and increased productivity, with strong involvement of employees with the company processes. The companies found themselves to be more proactive, and found employees were more likely to "buy in" to the company mission and continued mission growth. [56]

Under the certification registration contract, the third party auditors do continuous assessments usually at 6-month intervals, which keeps the company's ISO system "on line" with continuous independent auditors. And, over a 3-year period, a complete assessment of the quality systems is done by the registrar.

Commercial companies, especially small companies, or those that have grown up fast, may not have their system documented. A large company has more issues than a small one, simply because there is more going on and more people to reach, and generally more bureaucracy. A small company can easily have an all-employee meeting, set up a communication system, and transition to the ISO standard, while large, established organizations seem to take more time to change.

If multiple sites of the same organization are becoming compliant at the same time, it is better to stagger them and develop the easiest first. A manufacturing plant, not in the same location as the main plant, that is already working to the MIL-Q, would have a rather easy change to the ISO 9002, so that is a good place to start. Again, coming from the old style, the other sites could work on engineering processes/procedures, or for electronics, another big area, special processes. In a DuPont study (Fig. 4.6) [57] it took about 18 months to become compliant. [58] In that same study, the benchmarking shows that the management representative who sets up the system works on the system 80 percent of the time during implementation; and once the system is set up, the maintenance drops to approximately 20 percent. This author believes that it takes 100 percent of a manager's time to implement the system in a large organization.

Once the CEO has given the "go" for the company to change to the ISO 9000 system, he/she needs to appoint a management representative and to form a team. Representatives from each department, with the authority and responsibility to make decisions for the department, need to assembled. These representatives become the steering committee. This team needs to be trained on the 20 paragraphs in the standard and then discuss implementation at the company. Table 4.43 is a good look at the prime and support decisions of a sample company team. One of the first decisions is whether your company needs a consultant. That depends on a lot of things—mostly, how much experience already exists in your company, and will your company's managers listen better to someone from the outside? It is recommended that in a large company, your entire steering committee goes to lead assessor training. In a small company, send a representative number. Again, as part of a large organization, you may get assistance from your sister divisions. Small companies can share among themselves or suggest industry groups to share.

TABLE 4.43 Team Responsibilities by Department, Example

Q9000 Paragraph	Company Procedure	Management	Quality Assurance	Engineering	Contracts	Purchasing	Operations/ Manufacturing	Logistics/ Field Support
4.1 Management Responsibility		P						
4.2 Quality System		S	P	S	S	S	S	S
4.3 Contract Review		S	S	S	P	S	S	S
4.4 Design Control		S	S	P	S	S	S	S
4.5 Document Control	T	P	S/SOME	S/SOME	S/SOME	S/SOME	S/SOME	S/SOME
4.6 Purchasing Data		S	S	S		P		S/SOME
4.7 Purchaser Supplied Product		S	S	S	S	S	P	S
4.8 Product Identification and Traceability		S	S	P	S	S	S	S
4.9 Process Control		S	S	S		S	P	S
4.10 Inspection and Testing	B	S	P=INSPECT/ S=TEST	P=TEST			S	S
4.11 Inspection, Measuring, and Test Equipment		S	P	S			S	
4.12 Inspection and Test Status		S	P=INSPECT	P=TEST				
4.13 Control of Non-Conforming Product		S	P					
4.14 Corrective Action	D	S	P	S	S	S	S	S
4.15 Handling, Storage Packaging, & Delivery		S	S	S	S		P	S
4.16 Quality Records		S	P	S	S	S	S	S
4.17 Internal Quality Audits		S	P	S	S	S	S	S
4.18 Training		P	S	S	S	S	S	P/SOME
4.19 Servicing		S	S	P	S	S	S	P
4.20 Statistical Control		S	S	P/SW			P	

P = Prime, S = Support, SW = Software

The 18 to 20 months that DuPont quoted is a good measure (see Fig. 4.6). [59] The time it takes really depends on several factors:

1. How good are the procedures that define what your company does?
2. How disciplined is your company? When the CEO says "We will change," how fast does it happen?
3. How good is your official communication approach?
4. How good is your training? If procedures must be changed, can employees be trained in them?
5. How much competition is there with other initiatives in your company? If your company is downsizing or trying to move to a just-in-time philosophy, commercialization, TQM, or the Baldrige award, it may be good to carefully plan how 9000 compliance fits in.

Irwin and Dunn & Bradstreet conducted a survey, printed as *ISO survey, Comprehensive Data and Analysis of US Registered Companies, 1996.* The group received replies of their survey from 1880 companies. From those companies, three different measurements were obtained [60]:

- Number of months from learning of ISO 9000 until senior management commitment—averaged about 12 months
- Number of months from senior management commitment until passing final audit—averaged about 15 months
- Number of months from learning of 9000 until passing final audit—total time averaged 26 to 27 months.

One of the first steps is to do a needs, or GAP, analysis to determine the schedule. A GAP analysis compares your procedures with those required by the 9000 standard that you have selected. See Table 4.44 for a GAP analysis form for the first few paragraphs of the standard. The audit checklist in Table 4.51 and the ISO standard may be used to build an entire GAP analysis. This will tell you how many procedures/processes are in place in your company to be made compliant to the ISO quality management system. For every procedure/process that is missing, a schedule will be made to produce those procedures/processes. If your company is not already working to procedures, then another type of learning must occur.

If your company was MIL-Q–compliant, the main areas that are missing will be

- Management procedures to be part of the quality management system

Road Map to ISO 9000 Registration
Months 1 3 6 9 12 15 18 20
1
Management Awareness and Commitment
2
ISO 9000 steering committe chartered and trained
3
IIT chartered and trained
Lead auditor training
4
Manager training
Communicate to entire organization
5
Personnel training
6
Existing system evaluation
Gap analysis
Adequacy and compliance determination
Define areas for improvement
7
Write quality system manual
Upgrade procedures
Circulate and approve procedures
Provide training
8
Write second-tier documents
Procedures
9
Write third-tier documents
Work instructions
10
Internal auditing
Corrective actions
Compile objective evidence
Upgrade quality manuals
First management review
11
Preliminary registration assessment
Corrective deficiencies
Compile objective evidence
Second management review
12
Clear non-conformances
Correct deficiencies
Compile objective evidence
13
Registration
Continue improvement
Continue internal audits, corrective actions, management review, and surveillance audits

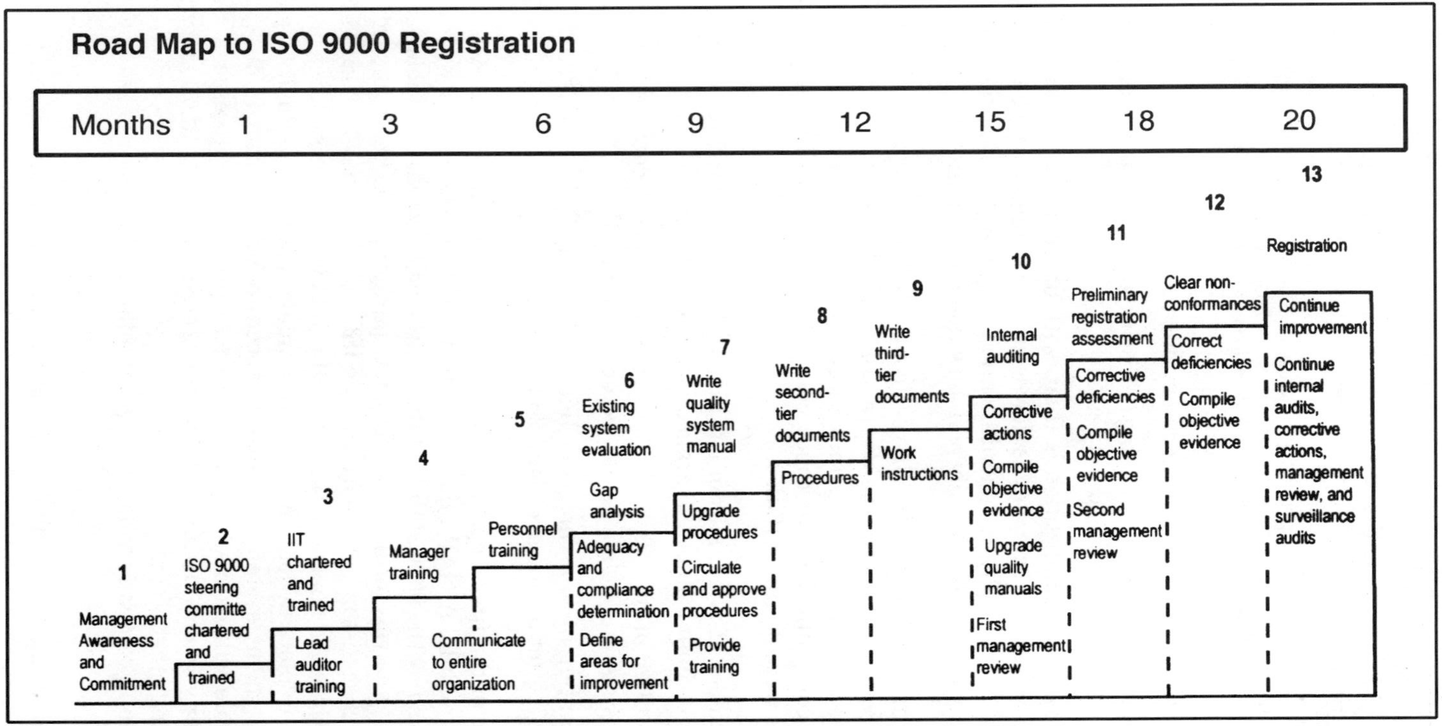

Figure 4.6 Road map to registration. (*Reprinted from Ref. 59 with permission.*)

TABLE 4.44 GAP Analysis

ISO 9001 paragraph/title	Requirement	Company procedure numbers
4.1 MANAGEMENT RESPONSIBILITY		
4.1.1 Quality Policy	*a.* Have a documented quality policy, relevant to goals/expectations/customers *b.* Management with executive responsibility shall define commitment to quality *c.* Management shall ensure the policy is understood, implemented, and maintained	TBD
4.1.2 Organization	Responsibility, authority, and the interrelations of all personnel who manage, perform, and verify work affecting quality shall be defined and documented, particularly for personnel who need the freedom and authority to: *a.* Initiate action to prevent the occurrence of any nonconformities relating to the product, process, and quality system *b.* Identify and record any problems relating to the product, process, and quality systems *c.* Initiate, recommend, or provide solutions through designated changes *d.* Verify the implementation of solutions *e.* Control further processing, delivery, or installation of nonconforming product until the deficiency or unsatisfactory condition has been corrected	TBD
4.1.2.2 Resources	Shall identify resource requirements and provide adequate resources, including the assignment of trained personnel, for management, performance of work, and verification activities including quality audits	TBD
4.1.2.3 Management representative	The company's management with executive responsibility shall appoint a member of the company's own management who, irrespective or other responsibilities, shall have the defined authority for: *a.* Ensuring that a quality system is established, implemented, and maintained in accordance with this international standard *b.* Reporting on the performance of the quality system to the supplier's management for review and as a basis for improvement of the quality system	TBD
4.1.3 Management Review	*a.* The CEO shall review the quality system at defined intervals sufficient to ensure its continuing suitability and effectiveness in satisfying the requirements of this standard and the suppliers stated quality program *b.* Records shall be maintained	TBD

- Engineering procedures
- Document control
- Quality records

Where there is no infrastructure, a simple process can be set up: (1) establish area for improvement, (2) plan the improvement, (3) document, (4) train, and (5) audit. Then move on to the next process. If resources permit, several teams could be established. Many of the paragraphs are overlapping and a priority should be set so that the work can get done to the established schedule.

A good tool to work is the implementation plan. See Table 4.45.

The management representative will lead the ISO 9000 implementation team. It is good to kick off the effort in an all-employee meeting. Large companies may need several sessions, but it is a good tool to get a unified message to the company on strategy, goals, and a quality management system. The management representative should have experience in setting up new systems or have a project manager to assist. A strategic plan needs to be developed to determine what standard, how to handle multiple sites, and how and when to involve all management and perhaps customers. The implementation plan needs a schedule, and contains the entire plan for completing the effort. Detail plans in each area may need to be made, depending on your company, and the current quality system.

There are risks that need to be addressed: First and foremost is the resistance to change. This is a natural phenomenon in life, especially with middle management. It requires a cultural change. The boss is now the coach. The ISO system is to involve all the employees, so the more people the ISO implementation project can get involved, the better. Make sure that all middle management agrees and does not just give a handshake and maintain a latent tension. Change must occur in

TABLE 4.45 Implementation Plan
(Suggested contents)

1. Background
2. Objectives
3. Scope
4. Overview of the system
5. The quality team
6. The initial steps
7. The tasks to be carried out
8. Support from departments/functions
9. Training plan and schedule
10. Schedule and resource plan

attitudes, values, and self-images. It is not easy to unlearn established patterns, and is not uncommon to hear, "We've always done it this way." If the company does not currently follow the procedures that are established, it is a risk that must be dealt with. Are internal audits in place, and paid attention to? In most companies, under 9000, the internal audit system needs to be beefed up, with improvement both in resources and infrastructure. It is a good approach to have internal auditors in each department. You may need to start autocratically and move to more democratic ways, once you get the company on your side. The teams you establish need to be empowered to make decisions and make changes. If your company is not into a structured total quality management (TQM) or continuous project improvement (CPI) program, then it's more difficult. The standards look easy at the beginning, but cause much discussion within the company. They are written at a very high level, to address any industry and across traditional functional disciplines. Interpretation is not so easy if the company has not worked with a cross-functional team and knows how it operates.

4.4.1 Tasks to schedule to become compliant or get certified

The following actions need to be addressed as part of the implementation plan.

Management decision and commitment. This may seem to be an easy task, but you cannot go ahead until the CEO is on board and wants to devote time and energy. Actual data shows it takes about 15 months to get a management decision, once the manager knows of the new management system. [61]

Develop strategic plan. One of the chief strategic moves is to decide whether to get registered/certified or simply compliant, and join with a group for SLA. For electronics, this will depend on several things. If your company wishes to do international business, then assuredly, you will become registered. If you have a strong quality/total quality management program and believe you can demonstrate management commitment and internal audits, you may want to become compliant; declare your compliance and wait to see what the ISO will do about the SLA. Even if you only deal domestically, the competitive market is driving companies toward good quality systems, and the perceived quality of a registered ISO 9000 house is viable for new business. [62]

Develop implementation plan. A detailed implementation plan is necessary to carry out the compliance/certification on schedule. Table

4.46 gives sample implementation tasks. If your company is small (5 to 50 employees) then it might be good to combine procedures/work instructions, and/or elements of the standard. A small company needs to look carefully at what it does and what is subcontracted out. It is a good idea to involve all the workers. When their horizons expand, they become responsible and therefore anxious to carry out the implementation. Do not necessarily hurry for third party registration. Be firm on your strategy position. Moreover, for all, make the system simple, and perhaps paperless. Local-area network (LAN) systems make compliance with document control much easier.

Plan and schedule training. Training is so important because this is a new concept. Everyone is responsible for quality. There is no longer a quality "cop." There are increased documentation requirements and many quality records. All employees need some training, perhaps through normal staff meetings. The steering committee and functional representatives who are really responsible for setting up the system need a more thorough training. It is good to send them to lead assessor training as well. Be sure to remember to train a set of auditors while planning the internal audit program. All management, executive and middle, need to be trained to understand how they fit into the new quality system.

Establish a steering committee. This committee is usually made up of departmental representatives who actually put together the plan, procedures, communication, etc., that implement the ISO 9000 system. These people should be good communicators and problem solvers and work closely together, as the 20 clauses of the standard are quite functionally mixed by traditional standards.

Address new processes. If there are any new processes, new infrastructure, or systems set up because of ISO implementation, employees have to be trained, especially in the new documentation requirements, the quality records, and the internal audits. For management, the management review of the quality system and implementation of Q9000, is necessary to see how the quality manual fits in.

Write a quality manual. A quality manual is the artifact that shows you have an ISO 9000 system. See Sec. 4.2 for instructions. A sample quality manual is included in the Appendix of this handbook.

Perform a GAP analysis. This is the area that shows you how much work must be done. Every procedure/process that is required, and not written, takes time that must be scheduled. It takes time to write the procedure, release it, and implement and audit it. This is the key to developing the schedule. (See Table 4.44.)

Procedure upgrades. There will be procedures that address part of the standard requirements, but not all of them. Collecting records is a good example. Before the ISO structure, there was no industry standard on how a company's procedures needed to be organized. Many companies may just have a "batch" of procedures that are not clearly defined in this manner. The Level 2 and 3 procedures will need to be defined (see Sec. 4.2).

Audit implementation of procedures. The audit plan needs to be started at the beginning. As procedures are written, upgraded, released, and implemented, they need to be audited. Priorities for audit need to be set. New processes/procedures are always high on the list. There may be complicated processes that are not documented. There are some processes that are more critical to the company's quality system. This establishes the priority of the auditors. Make the auditors part of the process from the beginning. Have them do an implementation audit to judge how close your company is to implementation. Then you are ready to do the next item.

Corrective action for implementation audit. This lets you see lessons learned and validate that the schedule will be met. Reaudit those areas for implementation of corrective action on a periodic basis.

Establish management review and add the quality audit findings. Ensure that an ongoing plan, procedures, and records are written and the processes are in place as required. The formal quality system audits should be at the executive level at this point.

Correct deficiencies. As part of this process, the GAP analysis needs to be reviewed and completed. Audits that have been held need to have their corrective action closed and re-audited. By this time, process control and customer surveys should be set up and integrated with the quality management system. This item lets the team correct any deficiencies from the original plan, either procedurally or with process implementation. Now the ISO implementation project can be integrated with the company's continuous improvement process, and the ISO implementation team is disbanded.

Approve documentation. The quality manual, procedures, and process documentation should be in an approval process. Be sure the approval process is documented. Check that the process matches procedures. If a process is automated, procedures are written and process changes are documented.

Complete training. There should be a training plan as part of the implementation plan, and a method for all employees to get training. This is a check of that training. Training can take many forms: outside/in-house seminars, staff meetings with an item of the quality management system included, or status briefings once a month. Communication systems (i.e., newsletters, videos, e-mail, and bulletin boards) are all part of training.

Conduct preassessment audit. At some point in time, once the system is forecasted to be completely implemented, an implementation audit is a good idea. If possible, it is good to get someone not within your organization to do this. Perhaps a sister division, a partner in business, or a prime or subtier supplier. Third party auditors will do it for a fee. This allows your company to have an independent survey, and from that, corrective action can be written. At this point, your company may declare itself "compliant." This is called *self-declaration.* From that point, the rest is part of the strategic decision.

Complete the plan. Depending on the strategic plan, have a registrar, DOD, or customer audit. Your strategic company plan has defined what path your company is going to take. If you have planned registration, many more tasks will be involved in the implementation plan. Those include selecting a registrar and the related activities.

Table 4.46 shows a summary of things that need to be done. Figure 4.7 is a flowchart for a standard improvement process that can be used.

TABLE 4.46 Summary: Implementation Tasks

- Make schedule
- Make detailed plans
- Establish a project manager—point of contact with experience in getting things done
- Kick off an all-employee meeting
- Gain commitment of all department managers
- Establish briefings, training sessions

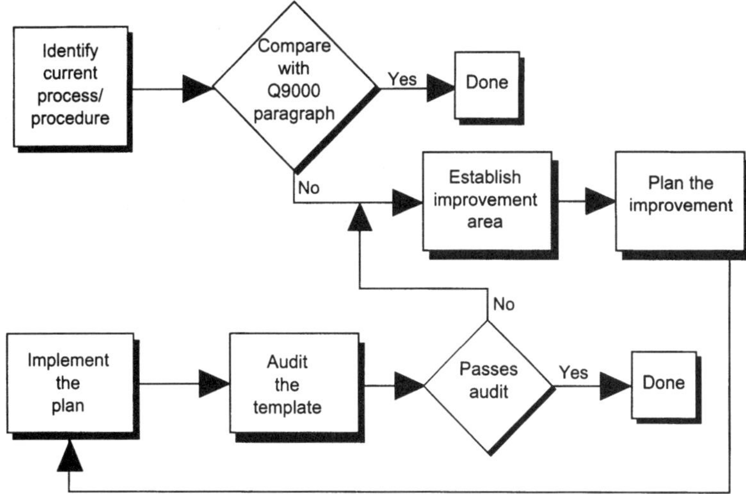

Figure 4.7 Q9000 improvement/implementation flow.

In summary, establishing a quality management system, the big "Q," is not difficult, but it is a change, and must be addressed as such. It must be planned, people must understand the plan and be trained in the new style, and audits must verify the implementation of the system.

4.4.2 Self-evaluation/assessment

It is important to understand where your company is in the process of moving to an ISO management system. Doing a self-evaluation/assessment is useful. Table 4.47 is a good top-level tool. [63] It is based on the Software Engineering Institute's matrix to judge the maturity of the system. A more detailed maturity matrix, keyed to each ISO 9000 paragraph, is presented in Table 4.48*a, b,* and *c.*

4.5 Preparing for Certification/Registration

Certification and registration are synonymous words in the United States. Both refer to a system where a third party audit company comes into a supplier's facility and assesses the company's readiness to operate in accordance with the 9001, 9002, or 9003 quality system. Certification/registration is branching out to include ISO 14000, on the environment and safety, and, of course QS-9000. About 60 registrars operate in the United States. Most have been accredited by an

TABLE 4.47 Quality Management System Maturity Model

Level	Name	Characteristics	QMS status
1	Initial	Ad hoc, unpredictable, and even chaotic processes. Little formalization or control. Success invariably depends on individual effort.	No QMS. Little process documentation or process management. No formal continuous improvement controls. Quality immature.
2	Repeatable	Basic management processes. Sufficient process consistency to ensure repeatability.	QMS established and independently registered. Quality conformance.
3	Defined	Company processes qualitatively formalized, standardized, and integrated. Processes fairly well understood.	Period of continuing assessments by registrar to support maturation phrase. Quality conscious.
4	Managed	Company processes quantitatively controlled and integrated. Effective use of performance measures.	Mature QMS eligible for self-declaration of conformity provided certain criteria are met. Extensive use of qualitative metrics. Quality competitive.
5	Optimizing	Focus on continuous process improvement and process change management.	QMS eligible for quality awards such as Malcolm Baldrige Award. Total quality culture.

SOURCE: From Ref. 63, used with permission.

accreditation body in accordance with EN 45012, which delineates the system for registrars to certify companies. [64] In the United States, the registration authority is the Registrar Accreditation Board (RAB) which can be reached at the same number as the ASQ. There were roughly 10,000 certified registered contractors in the United States as of July 1996. [65] The largest part of these companies are in the electronics business, so all electronics companies that want to succeed need to get involved.

It is typically a three-step process:

1. Assessment of the procedural compliance
2. Examination of the disciplines implementation
3. Postassessment or follow up and issuing a certification, or actions to complete certification

4.5.1 Selecting a registration company

Because a company will have a long-term relationship with the registrar, the registrar should be picked with care. A company needs to pick a registrar known in its industry and in the part of the world in which the company desires to sell its product or service. If more than one certificate type is needed, the company may need to check that the registrar can perform multiple certifications, (e.g., QS-9000 and environmental).

Criteria should be established to select a registrar. Some general criteria may include [66]:

- The registrar's qualifications and experience
  - The number of companies registered
  - The structure of customer bases
  - The registrar's experience in specific industry sectors
  - The registrar's qualification requirements for its auditors
  - Continued training of its auditors
- Recognition of the certificate
  - Cooperation agreements with other registrars
  - The certificate's value and how it is perceived in the market
  - Recognition by and potential customers
  - References
- The registration process
  - Does the registrar have a structured approach tailored to your needs?
  - The contract's design
  - Service and support
  - The quality of informational material
  - The registrar's objectivity, neutrality, and independence
  - The consistency of the audit team
  - The combination of quality and environmental audits available
- The time spent for registration
  - Lead time
  - Time required for the initial on-site audit
  - Time expenditure for surveillance
- Registration costs
  - Initial registration costs
  - Surveillance costs
  - Annual fees
  - Traveling costs
  - Other fees (e.g., application, registration, administration, etc.)

TABLE 4.48a Baseline Quality System Maturity—Self-Assessment

(Organization) (Date)

Management responsibility	Quality system	Contract review (less 9003)	Design control (less 9002 and 9003)	Document and data control	Purchasing (less 9003)	Control of customer-supplied product (less 9003)
Executive management defines and documents quality policy. Responsibility and authority of personnel who manage, perform and verify quality operations defined and documented. Management reviews the quality system at defined intervals to ensure its effectiveness. Associated documents and records are systematically maintained.	A quality system is established and documented to ensure product conforms to specified requirements. Quality manual covers requirements of this international standard. Quality system procedures, consistent with ISO requirements, effectively implement the quality system. Quality planning procedures define how the requirements for quality will be met. Associated documents and records are systematically maintained.	Procedures for review of contract or order to ensure that requirements are established and documented. Processes to resolve differences between contract requirements and company tender are in place. Capability to meet the contract requirements assessed. Amendments to a contract are similarly reviewed. Associated documents and records are systematically maintained.	Procedures to control and verify product design are established and documented to assure that specified requirements are met. Design and development planning is assigned to qualified personnel with adequate resources. Requirements of organizational interface, design input, design output, design review, design verification, design validation, and design changes are proceduralized. Associated documents and records are systematically maintained.	Procedures to control all documents and data related to the requirements of the ISO standard are established and documented. Procedures identify current revision status of documents and a method to preclude the use of invalid/obsolete documents. Changes to documents and data reviewed and approved by same function that performed original review and approval, unless designated otherwise. Nature of change identified in document or attachment. Associated documents and records are systematically maintained.	Procedures to ensure that purchased product conforms to requirements are established and documented. Evaluation, selection and control of subcontractors based on their abilities and quality system, along with the impact on the quality of the final product. Quality records of acceptable subcontractors are maintained. Purchasing documents clearly describe documents/systems needed. Procedure for verifying product at the subcontractors' premises are established and documented. Associated documents and records are systematically maintained.	Procedures for the control, verification, storage, and maintenance of customer-supplied product for the incorporation into the final product or the related activities are established and documented. Associated documents and records are systematically maintained.

Fully robust management system in place: 5	Fully robust quality system in place: 5	Fully robust contract review system in place: 5	Fully robust design control system in place: 5	Fully robust document and data control system in place: 5	Fully robust purchasing system in place: 5	Fully robust system for customer products in place: 5
Management systems documented and in place, yet with weakness in one or more elements: 4	Quality systems documented and in place, yet with weaknesses in one or more elements: 4	Contract review system documented and in place with weaknesses in one or more elements: 4	Design control system documented and in place with weaknesses in one or more elements: 4	Document and data control system documented and in place with weaknesses in one or more elements: 4	Purchasing system documented and in place with weaknesses in one or more elements: 4	Customer-supplied product systems documented and in place with weaknesses in one or more elements: 4
Management in place, yet ineffective: 3	Quality system in place, yet ineffective: 3	Contract review system in place, yet ineffective: 3	Design control system in place, yet ineffective: 3	Document and data control system in place, yet ineffective: 3	Purchasing system in place, yet ineffective: 3	Customer-supplied product system in place, yet ineffective: 3
Fully recognize requirements, on track toward planned implementation: 2	Fully recognize requirements, on track toward planned implementation: 2	Fully recognize requirements, on track toward planned implementation: 2	Fully recognize requirements, on track toward planned implementation: 2	Fully recognize requirements, on track toward planned implementation: 2	Fully recognize requirements, on track toward planning implementation: 2	Fully recognize requirements, on track toward planned implementation: 2
Recognize the need for management systems, but no process initiated: 1	Recognize the need for quality systems, but no process initiated: 1	Recognize the need for contract review system, but no process initiated: 1	Recognize the need for a design control system, but no process initiated: 1	Recognize the need for document and data control, but no process initiated: 1	Recognize the need for a purchasing system, but no process initiated: 1	Recognize the need for a system, but no process initiated: 1

TABLE 4.48*b* Baseline Quality System Maturity—Self-Assessment

(Organization) (Date)

Product identification and traceability	Process control (less 9003)	Inspection and testing	Control of inspection, measuring and test equipment (IM&TE)	Inspection and test status	Control of nonconforming product	Corrective and preventive action (less 9003)
Procedures are established and documented to establish unique identification of individual products or batches for traceability-required applications. Associated documents and records are systematically maintained.	Procedures are established and documented to identify processes that directly affect quality and ensure these processes are carried out under controlled conditions. Processes are carried out by qualified operators. Continuous monitoring and control of process parameters are in place. Requirements specify any qualification of process operations, e.g., personnel/associated equipment. Associated documents and records are systematically maintained.	Quality plan/procedures are established and documented for inspection and testing activities to verify specified requirements for product are met. Receiving inspection verifies incoming product conforms to specified requirements prior to inspection and test completion. Final inspection to complete establishment of conformance of the finished product to specified requirements. Inspection and test records show the product pass-fail results of inspection/testing to acceptance criteria. Associated documents and records are systematically maintained.	Procedures established and documented to control, calibrate, and maintain inspection, measuring, and test equipment used to demonstrate conformance of product to specified requirements. Measuring and test equipment identified that can affect product quality—calibrate at specified intervals. Process defined for calibration of measuring and test equipment—certified equipment used is traceable to recognized national standards. Environment for measuring and test equipment is such that accuracy and fitness for use are maintained. Associated documents and records are systematically maintained.	A means to indicate conformance or nonconformance of product with regard to inspection and tests performed is established and documented. Status is maintained throughout production, installation, and servicing of product. Only product passing required inspections and tests can be used. Associated documents and records are systematically maintained.	Procedures to ensure that product which does not conform to specified requirements is prevented from unintended use or installation. Controls provide for identification, documentation, evaluation, segregation, and disposition of nonconforming product. Associated documents and records are systematically maintained.	Procedures for corrective/preventive action to eliminate the causes of nonconformities identified are established and documented. Process to effectively handle customer complaints, to ensure a corrective action is taken and is effective. Preventive action includes use of appropriate sources of information to detect, analyze, and eliminate causes of nonconformities. Associated documents and records are systematically maintained.

Fully robust ID and traceability system in place: 5	Fully robust process control system in place: 5	Fully robust inspection and test system in place: 5	Fully robust IM&TE system in place: 5	Fully robust inspection and test status system in place: 5	Fully robust nonconforming product system in place: 5	Fully robust corrective/preventive system in place: 5
ID and traceability system documented and in place, yet with weakness in one or more elements: 4	Process control system documented and in place, yet with weaknesses in one or more elements: 4	Inspection and test system documented and in place with weaknesses in one or more elements: 4	IM&TE system documented and in place with weaknesses in one or more elements: 4	Inspection and test status system documented and in place with weaknesses in one or more elements: 4	Nonconforming product system documented and in place with weaknesses in one or more elements: 4	Corrective/preventive system documented and in place with weaknesses in one or more elements: 4
ID and traceability system in place, yet ineffective: 3	Process control system in place, yet ineffective: 3	Inspection and test system in place, yet ineffective: 3	IM&TE system in place, yet ineffective: 3	Inspection and test status system in place, yet ineffective: 3	Nonconforming product system in place, yet ineffective: 3	Corrective/preventive system in place, yet ineffective: 3
Fully recognize requirements, on track toward planned implementation: 2	Fully recognize requirements, on track toward planned implementation: 2	Fully recognize requirements, on track toward planned implementation: 2	Fully recognize requirements, on track toward planned implementation: 2	Fully recognize requirements, on track toward planned implementation: 2	Fully recognize requirements, on track toward planned implementation: 2	Fully recognize requirements, on track toward planned implementation: 2
Recognize the need for an ID and traceability system, but no process initiated: 1	Recognize the need for a process control system, but no process initiated: 1	Recognize the need for an inspection and test system, but no process initiated: 1	Recognize the need for an IM&TE system, but no process initiated: 1	Recognize the need for an inspection and test status system, but no process initiated: 1	Recognize the need for a nonconformance system, but no process initiated: 1	Recognize the need for a corrective/preventive system, but no process initiated: 1

TABLE 4.48c Baseline Quality System Maturity—Self-Assessment

(Organization) (Date)

Handling, storage, packaging, preservation and delivery (HSPP&D)	Control of quality records	Internal quality audits (less 9003)	Training	Servicing (less 9003)	Statistical techniques
Procedures are established and documented for handling, storage, packaging, preservation, and delivery of product. Handling methods prevent damage/deterioration. Designated storage area or stock room controls prevent damage or deterioration of product pending delivery. Packing, packaging, and marking processes ensure conformance to specified requirements. Appropriate methods of preservation and segregation of product being used. Product after final inspection and test awaiting delivery is protected. Associated documents and records are systematically maintained.	Procedures for the identification, collection, indexing, access, filing, safe storage, maintenance, and disposition of quality records are established and documented. Records conform to specified requirements. Associated documents and records are systematically maintained.	Procedures to implement a program of planned/scheduled audits that determine the effectiveness of the quality system are established and documented. Audits are performed by personnel independent of activity audited. Audit results are recorded and responsible area managers take timely corrective action on deficiencies. Closed-loop follow-up verifies implementation and effectiveness of corrective actions. Associated documents and records are systematically maintained.	Procedures for identifying training needs and the means to provide the training to all personnel performing tasks that affect quality are established and documented. Associated documents and records are systematically maintained.	Procedures are established and documented for performing, verifying, and reporting that servicing meets the specified requirements. Associated documents and records are systematically maintained.	Procedures are established and documented for the statistical techniques required for establishing, controlling, and verifying process capability and product characteristics. Associated documents and records are systematically maintained.

Fully robust HSPP&D system in place: 5	Fully robust quality record system in place: 5	Fully robust audit system in place: 5	Fully robust training system in place: 5	Fully robust servicing system in place: 5	Fully robust statistical system in place: 5
HSPP&D system documented and in place, yet with weaknesses in one or more elements: 4	Quality record system documented and in place, yet with weaknesses in one or more elements: 4	Audit system documented and in place with weaknesses in one or more elements: 4	Training system documented and in place with weaknesses in one or more elements: 4	Servicing system documented and in place with weakness in one or more elements: 4	Statistical system documented and in place with weaknesses in one or more elements: 4
HSPP&D system in place, yet ineffective: 3	Quality record system in place, yet ineffective: 3	Audit system in place, yet ineffective: 3	Training system in place, yet ineffective: 3	Servicing system in place, yet ineffective: 3	Statistical system in place, yet ineffective: 3
Fully recognize requirements, on track toward planned implementation: 2	Fully recognize requirements, on track toward planned implementation: 2	Fully recognize requirements, on track toward planned implementation: 2	Fully recognize requirements, on track toward planned implementation: 2	Fully recognize requirements, on track toward planned implementation: 2	Fully recognize requirements, on track toward planned implementation: 2
Recognize the need for an HSPP&D system, but no process initiated: 1	Recognize the need for a quality record system, but no process initiated: 1	Recognize the need for an audit system, but no process initiated: 1	Recognize the need for a training system, but no process initiated: 1	Recognize the need for a servicing system, but no process initiated: 1	Recognize the need for a statistical system, but no process initiated: 1

There are established registrars that are familiar with the electronics industry. Table 4.49 is a listing of registrars for electronic firms. [67] Information on the registrar, and on the registration process, is readily available from the registrar, simply by requesting it. To understand how the registrar conducts an audit, it is good to attend a lead assessor course. Lots of questions need to be asked; see Table 4.50. Audit guidelines and representative questions are listed in Table 4.51 on page 284.

Once a registrar is picked and contracted by the contractor, an opening meeting will be held with the subject of the audit. The accredited registrar will explain the exact methods by which the audit will be held. A sample flow is shown in Fig. 4.8. Methods of operation will vary slightly with company, industry, and the registrar's policy. Then procedures will be reviewed, and a schedule for implementation audit set. Once all this is carried out, the contractor will be notified of pass or nonconformance.

On the original contract, or purchase order, the normal practice is for the registrar to do one complete audit at the beginning to certify a company, then come back and perform surveillance audits every 6 months for 3 years. The surveillance assures the registrar that the companies quality system is "on track."

Costs for registration are coming down. According to the 1996 survey, companies are reporting average total costs associated with registration of $187,000, or $58,200 less than 3 years ago. While overall costs are down, the survey found that U.S. companies reported a decline in the average savings. In 1996, annual savings were reported to be $117,000, down from $179,000 in 1993. They also reported an average one-time savings of $77,000 in addition to the annual savings, something which was not reported in 1993. [68]

The main thing to do in this area is to start early and not rush. Because of the high demand, there may be some wait time, so planning needs to begin early in the process of registration. If registration is in the future, it is good to get the registrar in early in the implementation process so the team will be more successful.

4.5.2 Alternatives to the certification/registration process: 1996

The subject of certification/registration is an open one. Internationally, multiple organizations are debating what is the best method to have international cooperation, including product and quality system registration and the methods thereof. There is criticism that the ISO certification has become a multitude of audit and

TABLE 4.49 Registrars for Electronics

Name, address, phone	Notes
ABS Quality Evaluations, Inc. 16855 Northcase Drive Houston, TX 77060 (713) 873-9400	Internationally known, based in Houston; does automotive QS-9000. Has offices off site in Mexico, London, Singapore, and Brazil
British Standards Institute 9000 Towers Crescent Drive, No. 1350 Vienna, VA 22182 (703) 760-7828	World's largest, most experienced 9000 registrar. Handles ISO 9000, QS-9000, ISO 14000, EN 46000, and TickIT. Also designated under most European CE marking directives.
Bureau Veritas Quality International North American Central Offices James, NY 14701 (716) 484-9002	International registrar with eight offices in the United States, and more than 50 worldwide. Direct, multiple accreditation, and emphasis on electrical and electronics
DLS Quality Technology Associates 108 Hallmore Drive Camillus, NY 13031 (315) 468-5811	Accredited only in the United States, small company that focuses on electronics and instruments, and is customer-oriented and quite responsive to customers.
DNV Certification 16340 Park Ten Place Houston, TX 77084 (713) 579-9003	Holds 13 accreditations for registration. U.S. certification in five offices from Atlanta to Long Beach, CA.
Electronic Industries Quality Registry Inc. 2500 Wilson Blvd. Arlington, VA 22201-3834 (703) 907-7563	Established by the Electronics Industries Association in January 1993. Primarily serves electronic and telecommunications industries and customers
GBJD Registrars Ltd. 9251-8 Yonge St., Suite 310 Richmond Hill, Ontario L4C9T3, Canada (416) 218-5594	Holds accreditation from United Kingdom, Europe, and United States. Notified body of Europe for medical devices/explosive atmospheres products. Wide scope including electronics.
Instituto Mexicano de Normalización y Certificación A.C. Manuel Maria Conteras No. 133, ler. Piso, Col. Cuauhtémoc Ciudad de Mexico, DF 06470, Mexico 525-566-4750	Mexican registrar founded in 1993 to offer standardization and third party certification services in quality management, metrology, and others. Used to fulfill the requirements of the Mexican government in official certifications. Emphasis on electronics.
Kema-Registered Quality Inc. 4379 Country Line Road Chalfont, PA 18914 (215) 822-4258	Registered by RvA (Dutch) and RAB (U.S.). Notified body for medical devices, telecommunications, machinery and EMC.
Lloyds Quality Registrar 33-41 Newark St. Hoboken, NJ 07030 (201) 963-1111	LRQA is one of the leading registrars in the United States. It is a full service supplier with multiple accreditations, including environmental, QA-9000, EC directives, and CE marking.
PECS QA 533 Main St. Acton, MA 01720 (508) 263-4811	Customized service for professional, automotive, health care, and electronics. PECS' Small Business Division caters to companies employing 10 persons or fewer, ensuring the most cost-effective certification for small businesses.

TABLE 4.49 Registrars for Electronics (*Continued*)

Name, address, phone	Notes
TUV America, Inc. 5 Cherry Hill Dr. Danvers, MA 01923 (508) 739-7075	More than 120 years of experience in testing, inspection, product safety, and electromagnetic compatibility in addition to ISO 9000 registration and CE conformity assessment services.
Underwriters Laboratories Inc. 333 Pfingsten Road Northbrook, IL 60062 (847) 272-8800	Leading registrar headquartered in the United States. Certifies products to International Electrotechnical Commission Quality Assessment for Electronic Components
Warnock Hersey Professional Services Ltd. 8810 Elmslie St. La Salle, Quebec, H8R 1V8, Canada (514) 366-3100	Canadian registrar; does electrical products to nationally recognized standards or specifically designated requirements of jurisdictional authorities.

SOURCE: Data printed from Ref. 67, with permission. Globus Information Services Inc. maintains a database of consultants and registrars. Globus may be reached at (800) 710-9038 or via e-mail at globus@io.org.

TABLE 4.50 Questions to Ask the Prospective Registrar

- How much experience in our industry?
- What is your pass/fail rate?
- How many findings before we fail?
- If it looks like we're failing, do you stop the audit?
- Do we get charged for all?
- If a major finding is found, do we fail?
- What is the appeal process if we disagree with a finding?
- What costs are included in the quote that you have given?
- What about travel, certificate, other costs?

certification schemes, instead of concentrating on quality systems and improving the quality products/processes of industry. [69]

The first known alternative is one called *self-declaration.* In this alternative, the supplier conforms internally to the ISO standard of choice. After the supplier does a self-assessment and feels that it has a good ISO quality management system, it simply puts out data that says it self-declares, that it is an ISO 9001/9002/9003 quality supplier. The customers may audit the supplier for this compliance, but this is outside the third party registration system. Since there were not a lot of quality management systems in place prior to the ISO 9000 standards, it is doubtful that many companies could achieve this declaration alone. Many proponents of third party registration feel that

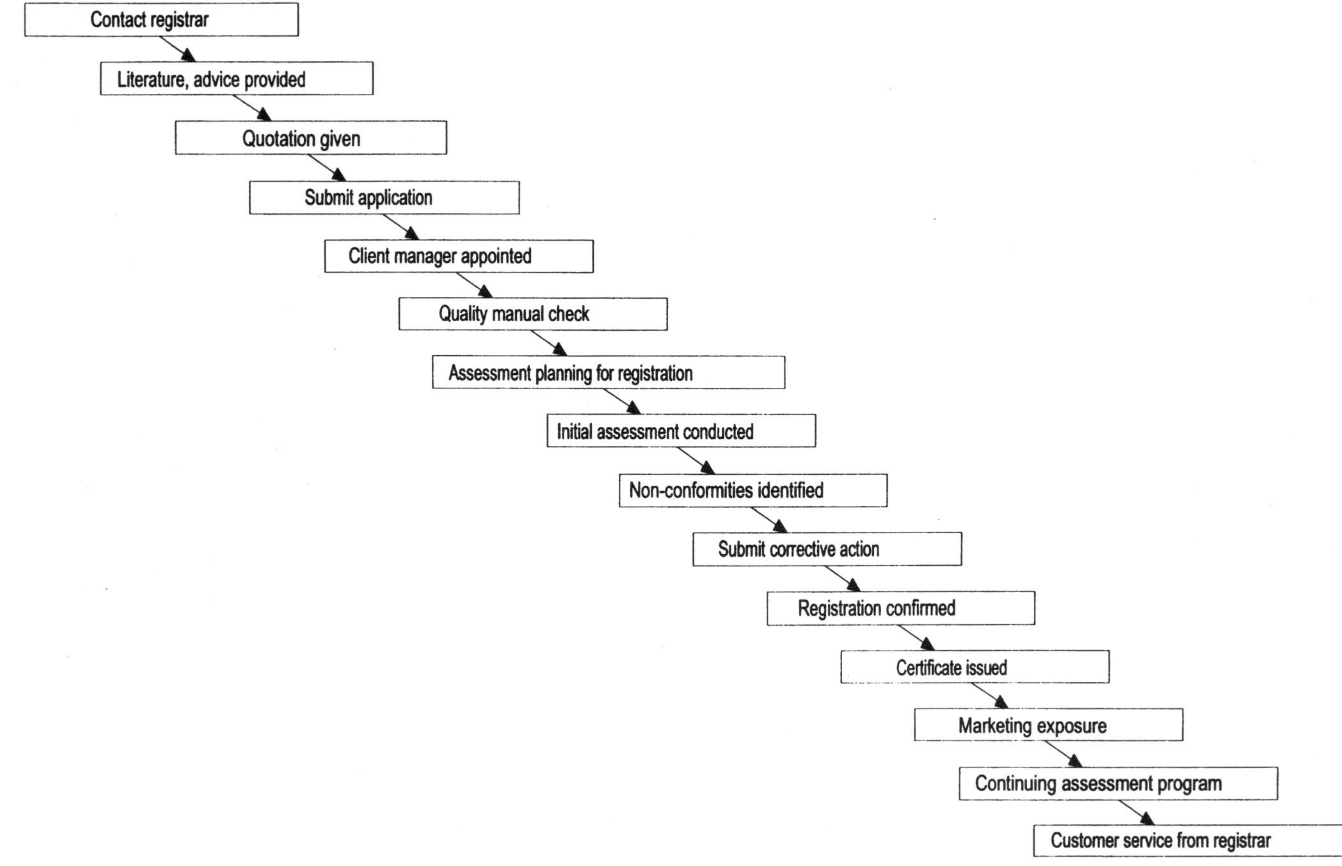

Figure 4.8 The registration process, with a third-party registrar.

it provides the necessary focus for companies to implement a quality management system. This may be another option for the RAB to incorporate in the next option. [70]

In late 1994, Hewlett-Packard and Motorola proposed an alternative approach [71] based on a belief that certification costs money and does not make any allowance for the previous quality system and/or Total Quality Management (TQM). They maintain that a long history of quality programs in the United States has led to sophisticated quality programs and these are not recognized by the current auditing practice of going through a complete implementation audit. They consider the complete ISO 9000 implementation audit and certification costs as "non-value-added" to the quality system. Thirty-nine electronics and telecommunications companies joined with them to propose an alternative and started a pilot program to gain recognition. By November 1996, they had gained signatures of 47 companies, mostly electronics and information companies, including Microsoft, Xerox, Digital Equipment, Whirlpool, and Intel. [72] They briefed both nationally and internationally—before getting the U.S. RAB to take the proposal to the ISO committee in Geneva. They based their argument on the fact that Europe has more than one system for product certification, so why should there not be more than one system for quality system certification? [73]

Their system is called system-level assessment. [74] Originally it was called supplier audit certification (SAC). The name was changed to ensure that all understood that this approach was not trying to eliminate third party auditors. [75] Their approach follows:

> The company declares compliance. Third party audit (registrar) verifies the company's internal auditors and process, issues a certification and then checks periodically. The responsibility for the maintenance of the quality system is with the company, which must perform extensive internal audits to show the registrar. The fact that the supplier is compliant to the standards must be apparent. This means that there is visible internal audits, a corrective/preventative action process, and a management review to keep compliance to the system. A working paper to keep up with this movement may be found on the Internet at http://www.corp.hp.com/Publish/iso9000-sac/rab.htm. [76]

The SLA system is being studied and debated, along with the whole issue of should there be multiple registration schemes, or just one? [77]

It is not known at this time if the ISO and European Community will approve of the new system. However, the ISO, in the implementation of the ISO 9000 quality management systems, did say that each country should have its own registrar accreditation board. The U.S. accreditation board, the Registrar Accreditation Board, which is associated with ASQC, has chosen to endorse the current European method, and is also considering SLA criteria. [78] It would seem to this author that the United States could have its own alternatives within the United States. The issues are the specifications and standards that would be recognized for world trade. There is already, in the automotive industry, evidence that U.S. registrars say that the United States is not bound to EN 45102, which governs registrars. The fact is, there is no U.S. law at this time on this subject. In addition, the U.S. Department of Defense (DOD) has stated and documented in its procurement regulations (the DODI 5000 series) that it will not require registration and that an agency of the Defense Logistics Agency, the Defense Contract Management Command (DCMC), will audit to the proper Q9000 specification for the supplier. All of this is yet to be defined, as the regulations affecting the ISO 9000 quality standards across the United States have not yet been written into U.S. law. It is the responsibility of the U.S. Commerce Department.

Again, this, for a company, must be a strategic decision. If the company is well-positioned in the marketplace, and wishes to join an alternative movement, there are options to do that. However, most of companies involved are large, multinational corporations. The supplier that is trying to establish a place in the marketplace needs to have a firm strategy of the quality system that will be implemented and for what reason. This strategy can be reviewed yearly, and depending on what the international and U.S. organizations determine, can change.

In conclusion, there are many decisions for an electronics company to make. However, one that knows its strategic direction and has a vision will have an easier time of planning actions to become compliant with the new ISO quality management standards and the new world of voluntary standards and specifications. What the electronics company cannot do is choose to do nothing. The competitive market in electronics is growing so fast, and businesses every day are seeing so many benefits of these new approaches, that to stay competitive a company has to be ready to change. It is important though, that because of all the changes, your company does need to subscribe to electronic and quality trade journals, so that your company can keep abreast of the changes. Recommended are the publications of the Electronics Industry Association, the American Society of Quality Control, and the ISO 9000 updates. All are available through the Internet.

TABLE 4.51 ISO 9000 Audit Checklist

General Audit Guidelines

A. Contact department manager and request a representative, a time, and a place to conduct the initial interview.

B. Interview the department management to determine compliance with ISO paragraph 4.1 (management responsibility).

   1. Determine which ISO paragraphs are applicable to this department.

      Question: What do you do?

C. Interview auditee to determine compliance with all applicable ISO paragraphs, per checklist, attached.

   1. Ask questions to department how the department complies with the applicable ISO paragraphs.

      Question: How do you do it?

   2. Determine if methods and procedures are properly documented.

      Question: Where and how is it documented?

   3. Collect documented evidence of compliance or noncompliance.

      Question: Do you have an example that you can show me?

D. Review any deficiencies with auditee and departmental representative and come to an agreement as to what was observed.

E. Complete system audit report.

   If you have questions, request assistance from the lead assessor.

F. Have the auditee sign the system audit report, have the departmental representative initial or sign, and provide them a copy before leaving the area.

   Note: Inform them that this is preliminary.

G. Turn in system audit report, checklist, work sheets, and notes to audit coordinator.

ISO PARAGRAPHS

4.1 MANAGEMENT RESPONSIBILITY

4.1.1 Quality policy

- Q: What is your company's quality policy?
- Q: Where is your quality policy documented?
- Q: Does the policy address organizational goals? Customer expectations and needs?
- Q: How do you ensure that the quality policy is understood, implemented and maintained within your department?

4.1.2 Organization

4.1.2.1 Responsibility and authority

- Q: How is your responsibility and authority to perform your assigned duties defined?
- Q: Does the organizational structure define authority and responsibility to deal with process problems? Quality system problems?

TABLE 4.51 ISO 9000 Audit Checklist (*Continued*)

4.1.2.2 Resources

Q: Describe your verification activities relating to inspection, tests, and monitoring of the design.

Q: How do you assign and select qualified personnel?

- Are resources identified and provided for management?
- For performance of work?
- For verification activities?

4.1.2.3 Management representative

Q: Who assigned the quality management representative? (Appointment by senior management?)

Q: Who is the person representing management? (Member of company management?)

Q: Does the management representative have the defined authority to:

1. Establish the quality system?
2. Report on the performance of the quality system for review and improvement?

4.1.3 Management review

Q: What are the measurements you use to determine if your functions/processes are in control?

Q: How frequently do you review the results and with whom?

Q: Can you show an example of corrective action taken to correct a problem or negative trend?

4.2 QUALITY SYSTEM

4.2.1 General

Q: What documents the quality system?

Q: Show me where it is documented and how it is maintained.

Q: What procedures or instructions are applicable to your work?

Q: Demonstrate/provide documentation on how you manage your program to meet intent of the quality manual.

4.2.2 Quality system procedures

Q: How/where are the procedures for the 18 remaining clauses of the standard documentation?

- Look for documentation trail.

4.2.3 Quality planning

Q: Does the company do quality planning?

Q: Where is the process of quality planning defined and documented?

Q: Where are the records of the results of quality planning maintained?

Does the quality planning activity ensure the identification of all necessary resources for products, projects, or contracts?

4.3 CONTRACT REVIEW

4.3.1 General

Q: How do you ensure that all contract requirements are resolved, defined, and documented?

Q: How do you ensure the proper resources are identified to meet the contractor requirements?

Q: Demonstrate that the contract review records are maintained.

TABLE 4.51 ISO 9000 Audit Checklist (*Continued*)

4.3.2 Review
- Q: Are contract review activities applied to tenders or contracts and orders?
- Q: Where and how is the procedure documented?
- Q: Does the procedure for contract review deal with verbal orders?

4.3.3 Amendment to a contract
- Q: Does the procedure for contract review deal with amendments and communication of the changes?

4.3.4 Records of contract reviews shall be maintained (see 4.16)
- Q: Do you have evidence of contract reviews, including functional groups reviewed?

4.4 DESIGN CONTROL

4.4.1 General
- Q: Explain your procedures that control and verify the design of your product.

4.4.2 Design and development planning
- Q: How is design and development planning responsibility identified and updated?
- Q: How are design and verification activities planned and assigned to qualified personnel with adequate resources?

4.4.3 Organizational and technical interfaces
- Q: How are intergroup activities with input to the design defined and documented?

4.4.4 Design input
- Q: How are your product design input requirements identified and documented?
- Q: Do the defined design inputs include:
  - Statutory and regulatory requirements?
  - Results of contract review activities?

4.4.5 Design output
- Q: How are your product design output requirements documented?
- Q: How do you verify and validate the design output against the design input?
- Q: Are design output documents reviewed before release?
- Q: What groups are represented at the review? (All involved with design stage?)
- Q: Show me a record of a review.

4.4.6 Design review.
- Q: What type of documented, formal reviews of your designs were conducted?
- Q: Show some evidence. (Records of review maintained?)

4.4.7 Design verification
- Q: How do you ensure and document that your product's design is verified?

4.4.8 Design validation
- Q: Does the company do design validation?
- Q: How do you ensure that the product resulting from a design conforms to the defined user needs and requirements?

4.4.9 Design changes
- Q: How do you manage change during the design and production of your product?
- Q: Are all design changes reviewed and approved before their implementation?
- Q: Do you use a configuration management plan?

TABLE 4.51 ISO 9000 Audit Checklist (*Continued*)

4.5 DOCUMENT CONTROL

4.5.1 General

Q: Does the document control procedure apply to documents generated outside of the company?

4.5.2 Document approval and issue

Q: How do you control your document?

Q: Who is authorized to approve your documents?

Q: Where and how is the approval and issue procedure documented?

Q: Show me the master list or equivalent document control procedure which identifies the current revision status of working document. (Is master list readily retrievable?)

Q: Do you retain obsolete documents? Show them to me. (Check for identification.)

4.5.3 Document and data changes

Q: How can you determine the current revision of a document?

4.6 PURCHASING

4.6.1 General

Q: How are specified requirements for purchased products communicated to the vendor?

Q: Who maintains the documented procedures?

4.6.2 Evaluation of subcontractors

Q: How are vendors selected?

Q: Are records of acceptable vendors maintained?

Q: Can you show me where and how the procedure is documented?

4.6.3 Purchasing data

Q: Do you have evidence that purchasing documents have been approved and reviewed prior to release?

Q: How and where is the review and approved procedure documented?

Q: Do the company's purchasing documents identify all quality system standards to be applied?

4.6.4 Verification of purchased product

4.6.4.1 Supplier verification at subcontractor's premises

Q: Is purchased product verified at the subcontractor's premises?

Q: Where is the verification arrangements and the method of product release documented?

4.6.4.2 Customer verification of subcontracted product

Q: Is this included in your contract?

Q: If so, how do you verify the quality requirements prior to the customer?

Q: Can you show me the procedure for this?

4.7 PURCHASER-SUPPLIED PRODUCT

Q: Do you have any purchaser-supplied products and if so, how are they controlled?

Q: Can you show me where the controlling procedure is documented?

4.8 PRODUCT IDENTIFICATION AND TRACEABILITY

Q: How are products identified during design and where is this documented?

Q: How are products identified during each stage of production, delivery and/or installation?

Q: Where is the procedure which documents the process maintained?

TABLE 4.51 ISO 9000 Audit Checklist (*Continued*)

4.9 PROCESS CONTROL

Q: Do you have a production plan?
Q: How are processes that affect quality controlled?
Q: How and where is workmanship criteria identified?
Q: What documented procedures define process control?
Q: How is the equipment chosen and maintained?
Q: What environmental controls are employed?
Q: What methods of monitoring the process and product are used?
Q: Do you have any special processes as defined in, and if so, how are they controlled?

4.10 INSPECTION AND TESTING

4.10.1 General
Q: What document details the planned procedures for inspection and testing requirements?
Q: Established records?

4.10.2 Receiving inspection and testing

4.10.2.1 Q: How do you ensure that incoming products are not used or processed until they have been verified as conforming to specified requirements?
Q: Can you show me where and how the procedure is documented?

4.10.2.2 Q: What procedure do you use to determine the amount and nature of receiving inspection that is needed on the product?
Q: Recorded evidence?

4.10.2.3 Q: Does incoming product ever get released for urgent production purposes prior to verification?
Q: What procedure is used?
Q: How do you identify the product?
Q: How do you keep track of the product?
Q: When does it receive verification?

4.10.3 In-process inspection and testing
Q: How can it be determined that all required inspection and testing of the finished product has been completed?
Q: How and where is the process documented?

4.10.4 Final inspection and testing
Q: How can it be determined that all required inspection and testing of the finished product has been completed?
Q: How and where is the process documented?
Q: What controls are in place to ensure that a finished product has satisfactorily completed all required activities and that associated data and documentation is available?

4.10.5 Inspection and test records
Q: Do you have documented evidence that the product has passed inspection and/or test with defined acceptance criteria?
Q: Could you show me an example?
Q: How and where is this documented?
Q: Does the failure of product to pass inspection and testing initiate the procedures for the control of nonconforming product?
Q: Do inspection and test records identify the authority for the release of all product?

TABLE 4.51 ISO 9000 Audit Checklist (*Continued*)

4.11 CONTROL OF INSPECTION, MEASURING, AND TEST EQUIPMENT

4.11.1 General

Q: How is inspection, measuring, and test equipment (IM&TE) that is used to demonstrate the conformance of product to specified requirements controlled?

Q: How and where is the procedure documented?

Q: Are the measurement uncertainties (or tolerances) of calibration of IM&TE and inspection tooling documented?

Q: Is this documented uncertainty considered in the selection and control of IM&TE and inspection tooling for demonstration of product to specified requirements?

Q: Are the allowable environmental and specific conditions of use documented for IM&TE and inspection tooling?

Q: Is inspection, measuring and test equipment that can affect product quality calibrated against certified equipment having a known valid relationship to nationally recognized standards?

Q: Show me documented evidence.

Q: Do you use test software or test aids to perform inspection?

Q: How do you ensure that they are capable of verifying the acceptability of the product?

Q: How is the software or test aids checked and how often?

Q: What records are kept to show evidence of control? (Show me.)

Q: Is there a requirement to make data pertaining to the inspection, measuring, and test equipment available to the customer?

Q: What mechanism do you have in place to accommodate?

4.11.2 Control Procedures

Q: What determines the measurements to be made and the required accuracy?

Q: Who selects the appropriate inspection, measuring, and test equipment?

Q: How do you determine capability of equipment?

Q: Is IM&TE and tooling periodically recalled for calibration?

Q: Are calibrations traceable to national measurement standards?

Q: Where is this policy documented and can you show me evidence of traceability?

Q: Do you have calibration procedures for each piece of equipment that comply with all of the requirements stated in ISO 9002, Para. 4.10.c?

Q: Can you show me some examples?

Q: How can you determine the calibration status of a piece of inspection, measuring, and test equipment?

Q: How are inspection, measuring, and test equipment calibration records maintained?

Q: Is the system documented?

Q: How are suitable environmental conditions for the calibrating, measuring, and testing of equipment ensured?

Q: What procedures/precautions are in place to ensure that the handling, preservation, and storage of equipment is such that the accuracy and fitness for use is maintained?

Q: How are adjustments which would invalidate the calibration setting safeguarded?

TABLE 4.51 ISO 9000 Audit Checklist (*Continued*)

4.12 INSPECTION AND TEST STATUS

Q: What method or procedure is used to identify the inspection and test status of a product as related to its conformance or nonconformance?

Q: How and where is this procedure documented?

4.13 CONTROL OF NONCONFORMING PRODUCT

4.13.1 General

Q: How are products that do not conform to specified requirements prevented from inadvertent use or installation?

Q: How/where is this procedure documented?

Q: How are nonconforming products identified and controlled?

4.13.2 Review and disposition of nonconforming product

Q: How is a nonconforming product disposed?

Q: Who has disposition authority and where is it documented?

Q: What documented procedure states the process for review of nonconforming product?

Q: Does the customer require to be notified of the proposed use or repair of product which does not conform to specified requirements?

Q: What procedure do you have in place to accomplish the reporting?

Q: What procedure documents the reinspection of repaired or reworked product?

4.14 CORRECTIVE AND PREVENTIVE ACTION

4.14.1 General

Q: What procedure documents the process for implementing corrective and preventive action?

Q: How do you determine appropriate corrective or preventive action has been implemented?

Q: Do you keep records of corrective and preventive action? Show me.

4.14.2 Corrective action

Q: What kind or type of corrective action is used within your area or responsibility?

Q: Can you show me where and how it is documented?

Q: Who determines corrective action effectiveness?

Q: What follow-up activities do you use to ensure corrective action is taken and effective?

4.14.3 Preventive action

Q: What kind or type of preventive action is used within your area or responsibility?

Q: Can you show me where and how it is documented?

Q: Who determines preventive action effectiveness?

Q: What follow-up activities do you use to ensure preventive action is taken and effective?

Q: Are implemented preventive actions submitted for management review?

TABLE 4.51 ISO 9000 Audit Checklist (*Continued*)

4.15 HANDLING, STORAGE, PACKAGING, PRESERVATION, AND DELIVERY

4.15.1 General

4.15.2 Handling

Q: What methods are used to prevent damage or deterioration of product?
Q: Is this method documented?

4.15.3 Storage

Q: Do you have designated storage areas?
Q: What method do you use for receipt and dispatch from these areas?
Q: Are conditions of product in stock checked periodically?
Q: Are these methods/procedures documented? Where and how?

4.15.4 Packaging

Q: How do you control the packing process, including materials used, of product to ensure conformance to specified requirements?
Q: How is the product identified?
Q: Where are the packaging procedures documented?

4.15.5 Preservation

Q: How do you ensure preservation and segregation of the product?

4.15.6 Delivery

Q: How is the product protected from damage after final inspection and test?
Q: Does the contract specify that product be protected from damage in delivery?
Q: What do you do to ensure product is not damaged?

4.16 CONTROL OF QUALITY RECORDS

Q: Can you show me how records that demonstrate achievement of quality requirements and the effective operation of the quality system are maintained?
Q: How and where is the quality records retention system documented?
Q: How is the access to quality records controlled?
Q: How are quality records stored? (Suitable environment?)
Q: Can you retrieve one or two records for me?
Q: Where are the retention times of quality records documented?

4.17 INTERNAL AUDITS

Q: What procedure documents the planning and implementation of your internal quality audits?
Q: How often and in which areas do you perform internal quality audits?
Q: What determines the audit schedule?
Q: Show me a schedule.
Q: Can you show me some examples of documented internal quality audits?
Q: Can you show me some examples of documented corrective action taken on deficiencies found during an audit?
Q: What follow-up activities are required?
Q: How do you follow up on implementation and effectiveness of corrective action?
Q: Show me an example.

TABLE 4.51 ISO 9000 Audit Checklist (*Continued*)

4.18 TRAINING

Q: How is the training needs for personnel performing activities that affect quality determined?

Q: What kind or type of training is provided for your people?

Q: Where and how is the training policy documented?

Q: Can you show me some examples of training records?

4.19 SERVICING

Q: Is servicing a specified requirement for your customer?

Q: If so, what procedure documents the process?

4.20 STATISTICAL TECHNIQUES

4.20.1 Identification of need

4.20.2 Procedures

Q: What statistical data do you collect and maintain?

Q: Can you show me some examples?

Q: How and where is the collection of this data documented?

References

1. *ISO 9000 Survey; Comprehensive Data and Analysis of U.S. Registered Companies, 1996,* Irwin and Dun & Bradstreet, New York, 1996, p. 23.
2. International Electrotechnical Commission, *Catalogue of IEC Publications, 1994,* pp. 4–5.
3. *EIA, JEDEC, and TIA Standards and Engineering Publications,* published by Global Engineering Documents, Englewood, CO, 1996, Annex C, p. 1, and Introduction section, letter from P. F. McCloskey, president EIA.
4. Zuckerman, A., and A. Hurwitz, "How Companies Miss the Boat on ISO 9000," *Quality Progress,* July 1996, pp. 23–25.
5. ANSI/ASQC Q9000-1-1994, *Quality Management and Quality Assurance Standards—Guidelines for Selection and Use,* ASQC, 1994, p. vii, summarized, paraphrased.
6. ANSI/ASQC Q9003-1994, *Quality Systems—Model for Quality Assurance in Final Inspection and Test,* p. vii.
7. ANSI/ASQC Q9004-1, 1994, *Quality Management and Quality System Elements—Guidelines,* p. 4.
8. Deming, W. Edwards, *Out of Crisis,* Massachusetts Institute of Technology, Center for Advanced Engineering Study, Cambridge, MA, 1982, p. 21, paraphrased.
9. ANSI/ASQC Q9001-1994, *Quality Systems—Model for Quality Assurance in Design, Development, Production, Installation, and Servicing,* para. 4.1.2.3.
10. Benson, Roger S., and Richard W. Sherman, "ISO 9000: A Practical Step-by-Step Approach," *Quality Progress,* October 1995, p. 77.
11. ISO 8402, *Quality management and quality assurance—vocabulary,* ISO, 1994, p. 19.
12. ISO/IDS 10013, *Guidelines for developing quality manuals,* para. 4.2.1.
13. ISO/DIS 10013, para. 7.1, 7.9.
14. ANSI/ASQC Q9004-1-1994, pp. 5, 6.
15. ANSI/ASQC Q9001-1994, para. 4.4.1.
16. Juran, J. M, *Juran's Quality Control Handbook,* 4th ed., McGraw-Hill, New York, 1988, pp. 16.7, 30.18, and 17.7 plus *Manufacturing Development Guide (MDC),* Aeronautical Systems Center, Air Force Material Command, Wright Patterson Air Force Base, OH, for "key products," pp. 3–16.
17. ANSI/ASQC Q9001-1994, para. 4.4.2.
18. ANSI/ASQC Q9001-1994, para. 4.4.4.

19. ANSI/ASQC Q9001-1994, para. 4.4.5.
20. ANSI/ASQC Q9001-1994, para. 4.4.7.
21. ANSI/ASQC Q9001-1994, para. 4.4.7.
22. ANSI/ASQC Q9001-1994, para. 4.4.8.
23. ANSI/ASQC Q9001-1994, para. 4.4.8.
24. ANSI/ASQC Q9001-1994, para. 4.5.2.
25. Craig, Robert J., *The no-nonsense guide to achieving ISO 9000 registration,* American Society of Mechanical Engineers (ASME), 1994, p. 135.
26. ANSI/ASQC Q9001-1994, para. 4.6.
27. Godfrey, A. B., and R. E. Kerwin, "Electronics Components Industries," in Juran, J. M. and Frank M. Gyrna, *Juran's Quality Control Handbook,* 4th ed., McGraw-Hill, New York, 1988, p. 29.5.
28. ANSI/ASQC Q9001-1994, para. 4.9a.
29. Ekings, J. Douglas, "Assembly Industries," in J. M. Juran and Frank M. Gyrna, *Juran's Quality Control Handbook,* 4th ed., McGraw-Hill, New York, 1988, p. 30.21.
30. Ekings, pp. 30.16–30.17.
31. Ekings, pp. 30.24 and 30.49.
32. Ekings, pp. 30.24–30.25.
33. Bhote, Keki R., *World Class Quality: Design of Experiments Made Easier; more cost effective than SPC,* AMA Management briefing, American Management Association, New York, 1988, pp. 149, 150.
34. Electronic Industries Association, Standards Proposal No. 3501, Proposed New Standard "Process Certification Standard for Semiconductor Devise Assemblers" (if approved, to be published as EIA-599-1), pp. 1, 2, 7.
35. ANSI/ASQC Q9001-1994, para. 4.10.
36. ANSI/ASQC Q9001-1994, para. 4.13.
37. ANSI/ASQC Q9001-1994, para. 4.14.
38. ANSI/ASQC Q9001-1994, para. 4.16.
39. ANSI/ASQC Q9001-1994, para. 4.18.
40. Sullivan, L. P., "The Seven Stages in Company Wide Quality Control," *Quality Progress,* May 1986, p. 81.
41. Ekings, p. 30.47.
42. Deming, p. 334.
43. Reprinted from *Production Part Approval Process,* 2d ed., 1995, Chrysler Corporation, Ford Motor Company, General Motors Corporation, p. 1., with permmission
44. Reprinted from *Potential Failure Mode and Effects Analysis Reference Manual,* 2d ed., 1995, Chrysler Corporation, Ford Motor Company, General Motors Corporation, p. 1., with permission.
45. Reprinted from *Advanced Product Quality Planning and Control Plan (APQP),* 2d printing, 1995, Chrysler Corporation, Ford Motor Company, General Motors Corporation, p. 1., with permission.
46. Reprinted from *Measurement Systems Analysis (MSA),* 2d ed., 1995, Chrysler Corporation, Ford Motor Company, General Motors Corporation, p. 4., with permission.
47. Garrison, A., *QS-9000: Chrysler, Ford and General Motors' New Quality Standard, An Executive Overview,* Perry Johnson, Inc., El Segundo, CA, 1995, pp. 2, 6, 11.
48. *Continuous Improvement,* June 1996, p. 2.
49. CEN, *The New Approach, Legislation and Standards on the Free Movement of Goods in Europe,* CEN, 1991, p. 9.
50. CENELEC, *Catalogue of European Standards,* 1996, pp. *v, xi.*
51. Zuckerman, A., "European Standards Officials Push Reform of ISO 9000 and QA-9000 Registration," *Quality Progress,* p. 134.
52. Deming, p. 21.
53. Deming, p. 148.
54. Deming, pp. 133–134.
55. *ISO 9000 Survey,* p. 9.
56. "Survey Finds ISO 9000 Registration Market Driven," *Quality Progress,* March 1996, p. 23.

57. Hockman, K. K., R. Grenville, and S. Jackson, "Road Map to ISO 9000 Registration," *Quality Progress,* May 1994, p. 41.
58. Hockman, Grenville, and Jackson, pp. 39–40.
59. Benson and Sherman, p. 76.
60. *ISO 9000 Survey,* p. 57.
61. *ISO 9000 Survey,* p. 57.
62. *ISO 9000 Survey,* p. 66.
63. *Quality System Update,* December 1994, p. 21.
64. Heinloth, Stephan, "Selecting a Registrar," *Quality Digest,* September 1996, p. 33.
65. *Quality System Update,* July 1996, "From the editor," p. 2.
66. Heinloth, pp. 35, 36.
67. "Survey of ISO Registrars," *Quality Digest,* May 1996, pp. 46–58.
68. *ISO 9000 Survey,* p. 7.
69. Zuckerman, A., "European Standards Officials Push Reform of ISO 9000 and QA-9000 Registration," *Quality Progress,* September, 1996, p. 132.
70. Middleton, David N., "Supplier's Declaration Not Appropriate for All Companies," *Quality System Update,* December 1994, p. 23.
71. Scicchitano, P., "Alternative Registration," *Quality Digest,* June 1996, p. 21.
72. *Continuous Improvement,* November 1996, p. 5.
73. Ling, David, Supplier Audit Confirmation Route Package, February 27, 1995, unpublished.
74. *Continuous Improvement,* November 1996, p. 5.
75. *Continuous Improvement,* November 1996, p. 5.
76. *Continuous Improvement,* November 1996, p. 5.
77. Zuckerman, p. 133.
78. Zuckerman, p. 133.

Chapter

5

Managing the Deming Way

Karl Haushalter

President of Optimization Works, providing education, training, and consulting in the philosophy of W. Edwards Deming

5.1 Introduction

"There is no substitute for knowledge...."

W. Edwards Deming [1]

What is the best or optimal way to manage? This is an important question for (1) anyone engaged in the profession of management and (2) companies challenged by today's competitive world. Competition in the electronics industry is about as tough as it gets. We are in the midst of a period of significant change in the way we manage. Many describe the change as being from "command and control" to "coach and counsel." But that is only part of the change, part of the total transformation in the way we manage that is required to take us to the next level of productivity.

Electronics managers need to look at (1) increasing attention given to the system, (2) using data more effectively, (3) increasing the speed and extent of continuous improvement, and (4) tapping into the presently untapped potential of all of the people within their organization. The question now is not whether an organization is improving, but is it improving faster than its competition? This chapter explains how to make the needed changes, the transformation.

Electronics companies and managers who want to be the best, to successfully compete, and sometimes simply to survive, need to employ a whole new paradigm, an entirely new way, of managing. W. Edwards Deming's theory of management is the new paradigm. It provides the foundation for management's new way of thinking and a basis for selecting optimal management actions. Deming's theory is based on a holistic approach—looking at an organization's entire system over time. Deming's theory is a very comprehensive approach to the question of "What is the best or optimal way to manage?"

The aim of this chapter is to introduce Deming's theory of management and provide guidance on how to put it to work for electronics managers and their companies. This chapter is written for electronics managers at all levels, from the very top of a company to the first line supervisor. It is very applicable to electronics companies striving for world class quality and productivity. It is necessary for survival in today's global electronics industry.

5.1.1 Who was Deming?

W. Edwards Deming is best known as the American statistician and management theorist who went to Japan in 1950 to help rebuild its economy. He gave a series of lectures on quality management which top management, representing 80 percent of the capital of Japan, attended. He told the Japanese that, if they followed his theory, within 5 years their exports would start to flow. And, as they say, the rest is history.

These lectures were part of an effort that started in the mid- to late 1940s when U.S. engineers were helping the Japanese to improve quality in electronics. While Deming was not the only one teaching quality in postwar Japan, he was clearly the most famous. For additional background information on Dr. Deming, see Chap. 6.

Deming, and his theory which major Japanese industries adopted, changed the economic history of the world. The Japanese went from a war-ravished country to a world economic power. *U.S. News & World Report* in its April 22, 1991 issue named him one of history's nine hidden turning points.

Dr. Deming started consulting in 1950 with a number of Japanese companies including the Toyota Motor Company. He was its consultant until his death in December 1993. Toyota is generally acknowledged as being the most productive manufacturing company in the world. An indication of Toyota's appreciation for Dr. Deming is in the lobby of the corporate headquarters, where three pictures hang: the founder's, the present chairman's, and Dr. Deming's. And Dr. Deming's picture is the largest and the only one in color.

The Deming Prize is Japan's most coveted award. It is given every year to a company or division of a company that has achieved distinctive performance improvements through the application of company-wide quality control which Dr. Deming introduced to Japan. The Deming Prize was created in 1951 by the Union of Japanese Scientists and Engineers (JUSE) to commemorate the friendship and contribution of Dr. Deming to Japanese industry. Additional description of the Deming Prize and its criteria is found in Sec. 5.8.

Dr. Deming often said that profound knowledge must come from the outside and by invitation. Dr. Deming was a management outsider. He received his Ph.D. in mathematical physics from Yale in 1928. He worked as a statistician for various companies, the Bureau of the Census, and the United States Department of Agriculture. In the early years of World War II, Dr. Deming worked with engineers on improving productivity. He taught a way of managing that involved statistical process control and constantly improving the production process. The engineers then applied this way of managing in the American war industries. After World War II, the United States was the only major country in the world with its manufacturing capability still intact. American manufacturers could sell anything and everything they made. Dr. Deming tried to get them to adopt the management theory so successfully used in American war industries but since they were doing so well without it they were not interested. They had no time to listen to Dr. Deming.

Between 1945 and 1949 engineers from the United States, from Western Electric and Bell Labs, pursued such varied activities as upgrading Japanese working environments, establishing the Electrical Testing Laboratory to certify that quality standards were being met, and advising Japanese business leaders on questions of production management. [2] In 1947 Dr. Deming worked on the census in Japan. The Japanese knew about the superiority of American airplanes in World War II, Deming's role in their manufacture, and that he had already been in Japan working on the census. In 1950 they asked for his help in rebuilding their economy, especially their export economy.

It is interesting to note that Deming's theory was followed in Japan's major export industries, but has not had as significant an impact on some other parts of Japan. Many American managers are quick to dismiss the success of Deming's theory in Japan as being dependent on the Japanese culture. That is just not true. Deming felt very strongly that his theory worked in Japan *in spite of* their culture, not because of it.

In 1980 an NBC White Paper, "If Japan Can, Why Can't We?" brought Deming's theory of management to the attention of American business. American managers were now much more willing to listen. American business was becoming noncompetitive in a global market that had become increasingly competitive. In the early 1980s Dr. Deming started working with Ford, General Motors, and other large American companies. He also taught his theory at New York University's Stern School of Business and to the public in his 4-day seminars until his death in December 1993.

One of the best examples of the successful application of Deming's theory of management in the United States is NUMMI. NUMMI stands for New United Motor Manufacturing Inc., a joint venture between General Motors and Toyota. General Motors closed its Fremont, CA, plant in 1982. Two years later, the plant reopened under this joint venture. Using the Toyota (Deming) management approach and essentially the same workforce, the plant went from one of General Motor's worst to one of its best in productivity and quality.

Deming's impact on American management thought is starting to be felt at an accelerating rate. More and more of his thinking is creeping into American management thinking. For example, the Ford Taurus is known inside Ford Motor Company as the "Demingmobile." His theory is an excellent starting place for electronics managers striving to be the best.

While many have attributed to Deming the role of inspiring Total Quality Management (TQM), in his later years he was not comfortable with that association. Deming's concerns with TQM were that TQM was not clearly defined and had become many different things to many people. Also he felt some of what was done in the name of TQM was harmful, doing more damage than good. And, finally, TQM was not getting at the real problem—the way management manages. Management needed to be guided by his theory.

Deming was focused on something larger than just quality. To him quality was just a by-product of good management. He felt there must be a transformation of management thought and practice to have any real lasting impact (improvement). Deming focused on management and the organization's entire system in order to improve an organization's productivity, quality, competitive position, and profitability.

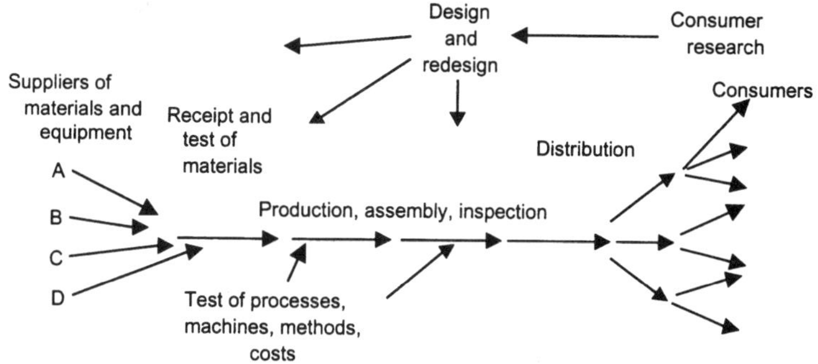

Figure 5.1 Deming's flow diagram. (*Reprinted from Ref. 3, with permission.*)

5.1.2 What was the spark that ignited Japan?

According to Dr. Deming, the spark that ignited Japan and fueled their economic turnaround after World War II was his famous flow diagram. This chart (see Figure 5.1) [3] was first used in August 1950 at a conference with top management at Hotel de Yama on Mount Hakone in Japan. Deming used this flow diagram to explain his concept that "management's job is to optimize the entire system over time." [4]

Deming saw that success of an organization is not determined by how well each component does by itself, but rather how well the components interact. The components are interdependent. Results are not determined by the sum of the parts, but rather from the interaction of the parts. Thus, management must manage the interaction of the parts.

Optimize means to make something the best it can be—the best possible—to get the most out at the least cost. Deming's own definition of optimization was "a process of orchestrating the efforts of all components toward achievement of the stated aim." [5] Optimizing entails:

- Everyone winning, no losers
- Delighting customers—building loyal customers
- Building proud employees—being an employer of choice
- Taking the waste out of the system
- The entire system working better and better and better
- Having long-term enhanced financial results

5.1.3 Deming's aim

In his 1993 book, *The New Economics for Industry, Government, Education,* Dr. Deming said:

> The aim...is to provide guidance for people in management to successfully respond to the myriad changes that shake the world. Transformation into a new style of management is required. The route to take is what I call profound knowledge—knowledge for leadership of transformation. Transformation is not automatic. It must be learned; it must be led.
>
> The transformation will lead to adoption of what we have learned to call a system and optimization of performance relative to the aim of the system. Individual components—teams, departments, divisions, plants—will not compete. Instead, each area will make choices directed at maximum benefit for the whole organization. An organization that seeks profound knowledge is already poised for the transformation. [6]

5.1.4 What is management?

Management includes two major components—leadership and supervision. Or to phrase it another way, management involves deciding where to go and making sure you get there. At the top of an organization management is mostly leadership and, at the front line, managers spend most of their time supervising. But both do some of each.

From another perspective, management is responsibility—a responsibility to utilize, as best as possible, the organization's resources to achieve the aim or purpose of the organization. Managers have a responsibility to optimize their organization's entire system over time.

Much has been written about management, some of it more useful than others. Bookstores have entire sections stocked with books on management ranging from the cartoon character Dilbert to serious scientific studies. Almost all current literature describes various management practices without mention or explanation of their underlying theory. It is Deming's emphasis on theory and then using the theory to make distinctions regarding management practices as optimal or not optimal that distinguishes Deming's approach. A deep understanding of Deming's theory of management is required to make the distinctions necessary to decide what management practices are consistent with optimization of the entire system over time.

Deming's theory has the greatest impact when it is understood and implemented organizationwide. However, even managers in an organization that has not adopted Deming's theory of management can still use the theory and principles in their own areas of responsibility.

5.1.5 Overview of rest of this chapter

This chapter will first describe Deming's theory of management. Flowing from the theory are 16 specific operating principles for management. The theory and the principles will be translated into the seven action steps for high-performance electronics managers. Common management practices will be reviewed in the light of Deming's theory with suggestions for alternative actions. Leadership and other key concepts for success will be outlined as they relate to Deming's theory. For reference, information and checklists for the Deming Prize will be listed before the chapter closes with the call to action: There is nothing left but to do it.

5.2 What Was Deming's Theory?

Dr. Deming's second theorem was "We are getting ruined by best efforts. We would be better off if most managers came in late and read the newspaper. At least they would not be tampering." [7] He felt that many, if not most, managers were working very hard, with the best of intentions, at doing the wrong things. Best efforts are necessary but they are just not sufficient. Instead, Deming felt, "Management needs to be guided by the correct theory, by profound knowledge:

Part # 1—Appreciation of a System

Part # 2—Theory of Variation

Part # 3—Theory of Knowledge

Part # 4—Psychology [8]

All four parts are a system, each part necessary, interrelated, and inseparable. The four parts are not a menu from which to pick and choose. Rather, the real power of Deming's theory comes from the interaction of the four parts. Deming called his theory of management "profound knowledge" because of the profound results achievable when his theory is followed.

The aim of profound knowledge is transformation of management. Embracing his theory involves a completely new way of managing. Deming saw the system of profound knowledge providing a lens through which the entire system of an organization can first be understood and then optimized. Table 5.1 presents some ideas for implementing these ideas in an electronics company. Each part of Deming's theory is explained in greater detail below:

TABLE 5.1 Ideas for Starting to Put Deming to Work in Your Electronics Company

1. Sketch a flow diagram of your organization's entire system.
2. Read on, start to understand Deming's theory. See if it makes sense for you and your organization.
3. Start to focus on optimizing—not just improving, but optimizing.

5.2.1 Part 1—Appreciation of a system

Everything done is done as part of a system. A system comprises a series of interdependent processes. Doing something always involves a process. The creation and improvement of processes is where managers need to devote most of their attention. In almost all situations, the manager's biggest leverage point is the system and its related processes.

You will get what the system will deliver. You will only get what the system will deliver—that is, what it is capable of producing. This is a simple yet important and profound concept. If you stop to think about it, it does make sense, doesn't it? If you want a different result, you need to change the system. In fact, insanity has been defined as doing the same thing over and over and over again, expecting a different result.

At least 94 percent belong to the system. On the basis of his experience, Dr. Deming estimated "At least 94% of troubles and possibilities for improvement belong to the system (the responsibility of management) and 6% are attributable to special causes." [9] Special causes are explained below. Others have felt that 94 percent was too conservative an estimate, that it is more like at least 99 percent. The exact percentage does not matter as long as the electronics manager remembers that the vast majority, almost all, of the troubles and possibilities for improvement belong to the system.

So when something goes wrong where should a manager look? The system is the logical answer. Yet in real life it seems like most traditional American managers look first to the person. We simply need to move the "w" to change the question from "Who?" to "How?"

What is a system? According to Deming, "A system is a network of *interdependent* components that work together to try to accomplish the aim of the system." [10] It is the interdependencies that Deming's theory suggests managers need to focus on. That is not an easy task. As Peter Senge points out in *The Fifth Discipline,* interdependencies are not always obvious. Actions and consequences (cause and effect) may be widely separated in time and space." [11]

A company's or organization's system is made up of all the processes involved in its operations, including, but not limited to:

1. The hiring process
2. The orientation process
3. The training process, especially the on-the-job training process
4. The job assignment process
5. The resource allocation process
6. The marketing process
7. The sales process
8. The purchasing process
9. The production (of goods, services, or information) process
10. The supervision process
11. The leadership process
12. The customer service process

Each of these processes can be further broken down into subprocesses and steps.

85 percent of impact is in first 15 percent of the process. To improve a process, where should an electronics manager focus attention? The intuitive answer for most people is at the beginning. If something is started right, it usually turns out pretty good. Conversely, if something starts poorly, it is usually very hard to correct midstream.

Figure 5.2 portrays Deming's assertion that 85 percent of the impact is in the first 15 percent of the process. [12] This front end of the process is the time for accumulating all the bright ideas, carefully planning, testing alternative ideas, finding out what the customer really wants, setting it up to be successful, etc. Focus on the design and improvement of the process and the results you want will follow.

By what method? As generally practiced, where do approaches such as establishing goals, management by objectives, and management by results focus? They tend to emphasize the end result of the process—the wrong end to have real impact. More on this later in this chapter. Of course it is helpful to know the desired results or output of a system. But the real question is "By what method?" As Deming pointed

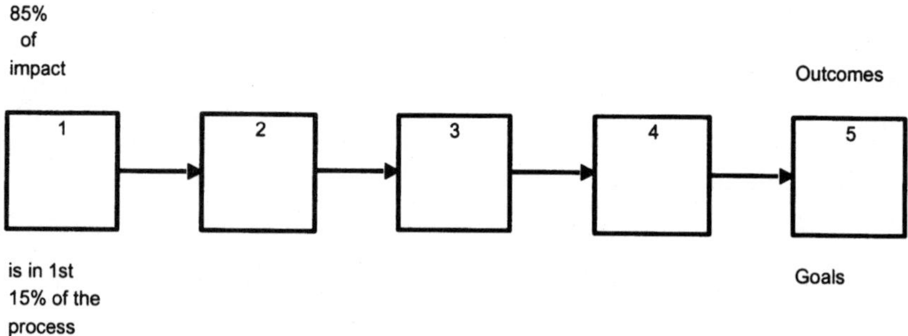

Figure 5.2 Where to focus attention.

out, "A goal without a method accomplishes nothing." [13] It is focusing on the optimal (best results at least cost) method or process that is of lasting benefit to a company. Focusing on the goal alone can lead to short-term gains but long-term disaster.

Whenever managers say they are going to do significantly better than before, the question to be asked is "By what method are you going to do that?" "By trying harder" is the most common, but not very good answer. Were they not trying hard before? Without improving the process there is little chance for lasting improvement in results.

Dissolving problems. When a problem occurs and management blames the persons involved, it loses sight of the need to fix the system of related processes. As long as it is "Joe's fault" there is no need to fix the system of related processes, just to find another "Joe." The problem is that unless the system is improved, the problem will most likely return. Table 5.2 shows ways of handling a problem. [14] Which is the most powerful? Most agree the most powerful approach is to dissolve a problem. This takes enormous discipline.

TABLE 5.2 Four Ways to Handle a Problem

1. **Absolve** yourself of it—"It's not my problem."
2. **Resolve** it—remember what worked before with a similar problem, modify as appears necessary, and do just enough to satisfy the present situation (do the minimum so the symptoms will go away for now).
3. **Solve** it—do study and research to find the best solution to the problem.
4. **Dissolve** it—both solve it and prevent its recurrence. Change the system or process so the system or process no longer has the problem. This is a key part of the Deming approach.

SOURCE: Paraphrased from Ref. 14 with permission.

Optimization requires cooperation. What does it take for management to accomplish its job of optimizing the entire system over time? Deming defined optimization as the "process of orchestrating the efforts of all components toward achievement of the stated aim." [15] Thus, optimization of the entire system requires cooperation of all the interdependent components. The components need to work together in the spirit of "win/win." Management must lead and foster cooperation between all components from design to sales to production to customer service. Deming referred to this as managing the interaction between the components of the system.

A key problem with ranking and rewarding people and departments (or functions) according to their own performance is that such actions are not focused on the whole system, but rather on the individual pieces. If management wants its people and departments to cooperate, why pay them to compete? A key problem with bonuses based on management by objectives (MBO), as practiced, is that it is very hard to get the objectives of the various components coordinated so that in total they will optimize the entire system. In addition, components have a tendency to focus more on being sure they make their own objectives and less on being sure the entire system is optimized.

Optimizing the pieces won't optimize the whole. Performance of the whole results from the interaction of the parts, *not* just from the sum of the parts. Things need to come together in harmony, not by each part doing its best. An example is an orchestra which performs best when the various sections of instruments aren't all trying to play the best solos they can. Rather it is how well they together make great music.

A system must have an aim. Deming pointed out that a system must have an aim and management sets the aim with input. Everyone must understand the aim of the system. Without an aim for the entire system, the components will have a tendency to focus on themselves, often to the detriment of the organization as a whole.

Optimization over time requires both continuous improvement and innovation. Continuous improvement alone is not enough. Having the world's best system for production of vinyl records did not guarantee success of a record company when cassette tapes and compact discs came in. Management must have the foresight to design products and services that will delight customers and entice them to buy. Innovation in an electronics company's products and services is one form of innovation essential to success. Another form is process innovation—doing things in a new way.

Loyal customers are the focus. Loyal customers are customers who will come back again and again, bring friends, wait in line, and be a little forgiving. The essence of every business is the production of loyal customers. Building loyal customers is a key focus of optimization.

The system is management's responsibility. The people who work in the process can provide a lot of help improving the system, but management is ultimately responsible. Management controls the resources, provides the direction, and says, "OK, produce with this system and related processes," or "No, we need a stronger system and related processes." Management needs to work on and ahead of a company's systems (not in them). Organizations need to lead with their system and related processes.

Sometimes in the electronics industry managers are asked to do the impossible, particularly at the end of a month or quarter. The solution is to build sufficiently robust systems to be able to handle whatever one might be asked to produce.

Strong processes will control work and the workers if it is really clear what they are supposed to do and why they are doing it. By letting the process generally control the workers in a process, managers can free themselves to do more-value-added work. These more-value-added activities include but are not limited to (1) coaching and developing workers, (2) planning, (3) continually improving processes, and (4) working with other components of the system, including working with customers. Electronics managers need to help their people focus on the process, not just the results.

Ways to start focusing on the system. One of the best ways of starting to focus on the system is to draw a flow diagram of your organization's entire system. Use Deming's flow diagram (Fig. 5.1) as a guide. Figure 5.3 is an example of a flow diagram from a high-tech electronics company.

Another way to get started is to develop flowcharts of the processes. A flowchart is a simple picture of a process showing the procedures or steps involved. It is a great tool for communication. It is very difficult, if not impossible, for people to discuss a process without the aid of a flowchart.

More on appreciation. Another way to start appreciating your system is to start observing it more closely. By tracking all items that inter-

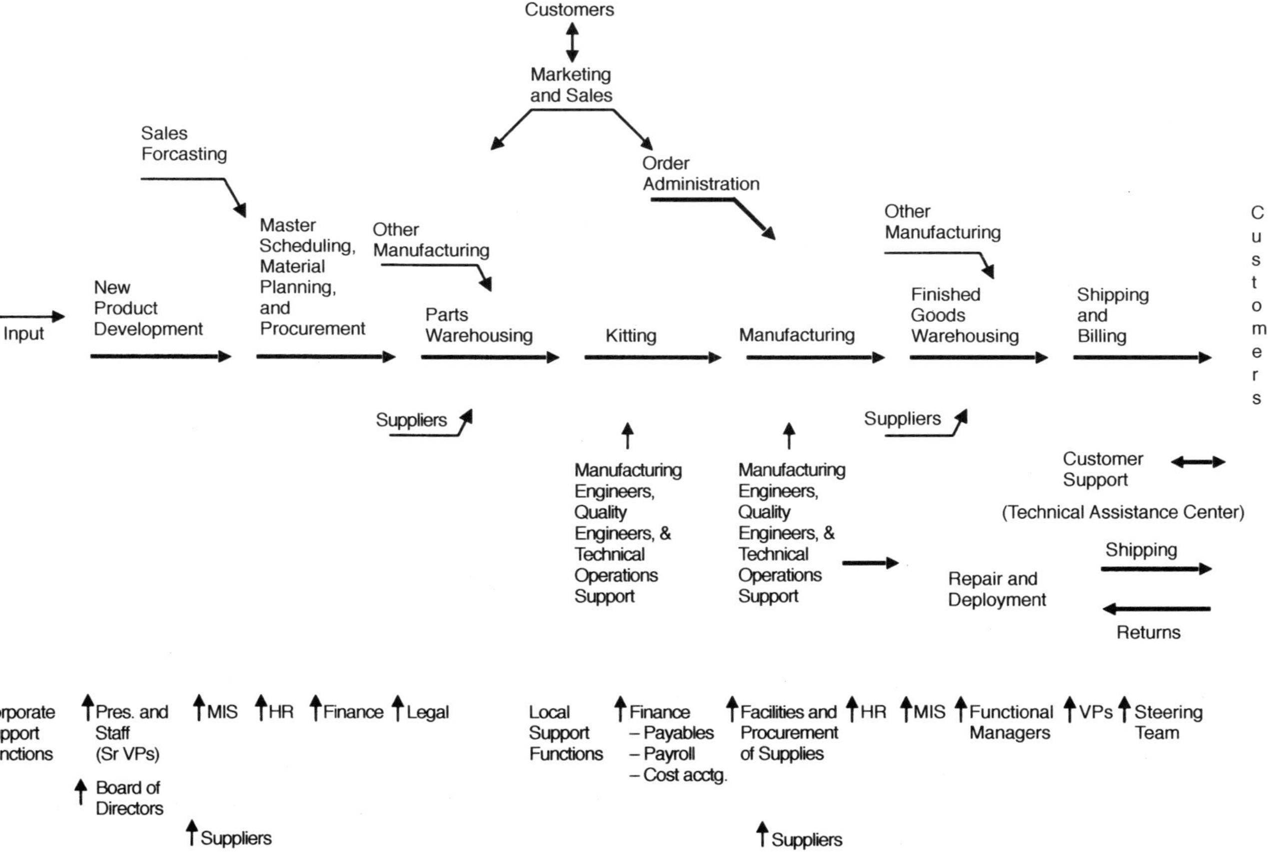

Figure 5.3 An electronics company as a system.

TABLE 5.3 Ideas for Starting to Put Deming to Work in Your Electronics Company

1. When something goes wrong, look first to the system and related processes.
2. Draw a flowchart of the processes for which you are responsible.
3. Identify all of the major processes making up your organization's system.
4. Start tracking all items that interrupt the smooth flow of your processes.
5. With your people, start working on those processes, making them the best they can be.

rupt the smooth flow of a process, it will become clear how well the process is working and where to start improving it. Table 5.3 lists some ideas for putting Dr. Deming's "appreciation for a system" to work in your company.

5.2.2 Theory of variation

No two things are exactly alike. Variation happens. In the output of any process, there is always variation. Sometimes it is obvious, and other times the variation is so small you cannot see it without some aid, like a micrometer or laser.

Variation. Variation is the voice of the process. Knowing how to listen to it can significantly help in managing processes. Variation is also the villain of quality. Customers do not like variation; they want everything the same (or as close to the same as possible) and on target. Understanding Deming's theory of variation will change the way managers act. Using statistics, one can make what is otherwise invisible visible. By charting data, one can see the variation including the patterns and types of variation.

The two causes of variation. Basically the causes of variation can be from either of two sources:

1. From the process—part of the normal or existing process
2. Outside the process—not part of the normal or existing process.

Common causes. Causes of variation from the process are called *common causes*. Common causes are causes of random variation in a process that are inherent in the process. They have no specific, identifiable separate cause. Since common causes are part of a process (and thus part of the system), and management is responsible for the entire system, common causes are the responsibility of management.

Special causes. Causes of variation not from the process are called *special causes.* Walter Shewhart called them *assignable causes.* Special causes are causes of nonrandom variation of a process due to *fleeting* events. [16] Changing a process could be a special cause. Identification and correction of specific special causes are generally best done by the person directly responsible for a process, since that person knows more about the process than anyone else. Management needs to then consider the need for a change to a higher-level process to prevent the special cause from recurring.

Special causes can be either beneficial or detrimental. The objective is to (1) eliminate and prevent recurrence of any detrimental special causes and (2) incorporate into the process by standardizing [including in standard operating procedures (SOPs)] any beneficial special causes.

The simple example shown in Fig. 5.4 is one that an 11-year-old boy used to plot the arrival times of his school bus. [17] Dr. Deming used this example in his public seminars besides citing it in his book. Common causes of the variation in the arrival times of the school bus might include such factors as traffic flow, traffic lights, weather, mood

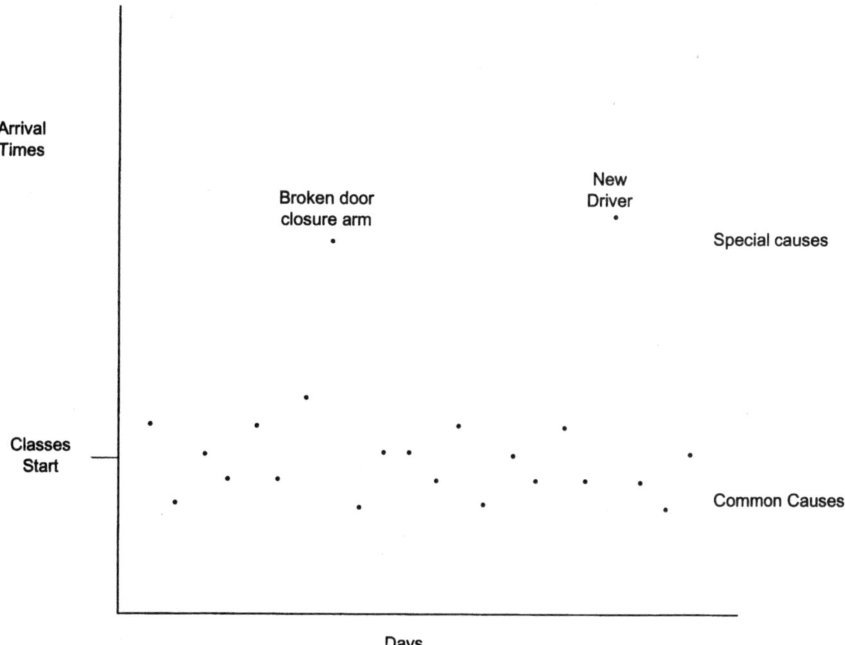

Figure 5.4 School bus arrival times. (*Reprinted from Ref. 17 with permission.*)

of the driver, and arrival times of the children. Special causes are readily identifiable as the broken door closure arm and the new driver. A word of caution here: when taking corrective action on the special cause of the variation, management should be sure to get to the root cause. Ask why at least five times "Why was the door closure arm broken?" The root cause might go all the way back to the method used to buy new buses (lowest bid), not just the specific incident that broke the arm.

The basic strategy. The basic strategy, based on Deming's theory of variation, is to first remove any special causes and then (1) shrink, shrink, shrink the variation from common causes and/or (2) move the process average toward target (ideal or optimal value).

Optimal managerial action varies with the type of cause. The optimal managerial action depends on the type of cause, common or special. Dealing with common causes differs from dealing with special causes. A manager must know when to act and when not to act. When something happens a manager does not like (a negative variation in the output of a process), there is a tremendous temptation to fix it, to do something immediately. In fact, there is an old adage, "Don't just stand there, do something." Unfortunately, such reactive behavior is not optimal. A more sophisticated approach is needed, an approach that first identifies whether the undesirable variation is from common causes or a special cause and then uses the appropriate strategy—the common cause strategy if the variation is from common causes and the special cause strategy if the variation is from a special cause.

The common cause strategy. Since common causes are inherently part of the process, the common cause strategy is to improve the process. The common cause strategy is *not* to act immediately, but rather to use a more studied approach to improve the process. The manager should initiate and lead the improvement effort. The basic tool of plan–do–study–act should be used along with the other quality tools such as fishbone, flowchart, and Pareto chart (see Chap. 6) to work on improving the process. In working to improve a process, all data points are relevant, not just the ones you do not like. The following story illustrates this point:

> A manufacturer was having a quality problem with its product. Engineering studies were conducted on the defective products and they were found to have a high level of contaminants. The only way to reduce this level of contaminants was to produce the product in a "clean room."

> The decision was made to construct a clean room at a substantial cost. Prior to actual construction, someone suggested that nondefective products also be given the same engineering studies. Guess what. The level of contaminants was about the same for nondefective and defective products. The root cause of the defects was not the contaminants and would not have been resolved by moving the manufacturing process into a clean room.

Common cause variation can hardly ever be reduced by attempts to explain differences between high and low data points when the process is stable. A stable process is one that operates over time within its calculated control limits and does not display nonrandom patterns. Stable does not mean good, desirable, or ideal; it merely means the absence of special causes.

Working on only a piece or portion of a stable process instead of working on the entire process can make the process worse. Deming called that tampering.

The special cause strategy. The special cause strategy is to act, to take immediate action to identify what was different and then act appropriately. If the special cause was beneficial, you want to make it part of the process so it is no longer a special cause, but rather a common cause. If the special cause was detrimental, the special cause needs to be eliminated. Management needs to then consider the need for a change to a higher-level process to prevent the special cause from recurring.

Table 3.11 shows typical common and special causes and may be found in Sec. 3.7.3 on statistical process control.

Only two mistakes. How does a manager distinguish common causes from special causes? In 1925 Walter Shewhart, Deming's mentor, solved that problem. Table 5.4 lists the only two mistakes a manager can make according to Shewhart.

Minimize net economic loss. One could eliminate mistake 1—acting when you should not—by simply never acting. That would mean treating all variation as common causes. But if one never acted, it

TABLE 5.4 The Only Two Mistakes a Manager Can Make

Mistake 1: Acting when you should not (that is, treating a common cause as if it were a special cause).
Mistake 2: Not acting when you should (that is, treating a special cause as if it were a common cause).

would mean the maximum loss from mistake 2—not acting when you should. One could eliminate mistake 2 by simply always acting. That would mean treating everything as a special cause. But if one always acted, it would mean the maximum loss from mistake 1.

Shewhart came up with a way to make mistake 1 sometimes but not too often and make mistake 2 sometimes but not too often. The result is a way to minimize the net economic loss from both mistakes.

To minimize the net economic loss, Shewhart created what he called control charts and the tool known as statistical process control (SPC). His control charts are based on statistical theory. Statistical theory says that variation from a stable process will appear random and almost always within a certain or defined range. In other words, if the variation is from the process, from common causes inherent in the process, data points on the process will almost always be within the control limits (that certain or defined range) and random. This makes the detection of a special cause fairly easy:

1. A data point is outside the range designated by control limits. Shewhart called these *action limits.*
2. A series of data points reveals a pattern that is nonrandom.

Control charts. Control charts are run charts to which have been added the average or centerline and control limits (that certain range). Table 5.5 identifies the most common types of control charts. For additional information on control charts and SPC, see Table 3.12 in Sec. 3.7.3.

Application to management. While Shewhart created control charts, it was Deming who saw their application beyond the production floor to management. In fact, Dr. Deming felt that using control charts on the shop floor "constitutes only a small fraction of the needs of industry, education, and government. The most important application of Shewhart's contribution is in the management of people." [18]

TABLE 5.5 Common Types of Control Charts

1.	XmR chart: individual and moving range control chart
2.	$\overline{x}$R chart: average and range control chart
3.	np chart: number nonconforming control chart
4.	p chart: proportion nonconforming control chart
5.	c chart: number of nonconformities control chart
6.	u chart: number of nonconformities per unit area of opportunity

What to measure. A frequent question is what to measure. What data should we plot on the control charts? The answer is (1) key quality characteristics data (data that will tell you how well the process is working) and (2) key problem identification data (data on what is going wrong, i.e., customer complaints). To decide what to measure, an electronics manager should ask these questions:

1. What data would tell me how well the process is working?
2. What data would tell me where the process is not working?

For both questions, managers should consider what *customers* care about.

Managers need to consider both process data (especially measurements on the first 15 percent of the process) and outcome data measurements as to how close to target is the output of the process. Target is a clearly stated and specific ideal outcome (the quality attributes) of a system or process in written and/or visual and/or numeric form. Examples of data to collect and analyze include:

- First pass yield percent
- Productivity
- Physical characteristics such as length, diameter, weight
- Cycle time—the interval of time it takes to complete one cycle of a process (time to produce one output of goods, services, and/or information)
- Number and type of defects
- Amount and type of rework
- Customer (both internal and external) complaints

Measurement, data collection, and analysis are a key part of process improvement and managing the Deming way. They are an essential part of the plan–do–study–act cycle (see next section of this chapter), of continuous improvement and TQM (Chap. 6), and of statistical process control (Sec. 3.7.3). There are many companies that have gone just about as far as they can go without statistics, without Deming's theory of variation, without making the invisible visible. Table 5.6 highlights some action items that electronics industry managers can use to start putting Deming's theory of variation to work for them and their departments or companies.

TABLE 5.6 Ideas for Starting to Put Deming to Work in Your Electronics Company

1. Expand the use of data to guide your decisions.
2. Learn how to construct and use a control chart (see Chap. 3).
3. Make a control chart for some key process for which you are responsible.
4. Teach others how to construct and use a control chart.
5. Expect/require those working for you to present process and outcome data in control chart format.

5.2.3 Theory of knowledge

How do we know what we know? What is our theory? While these may sound like philosophical questions, they can provide major guidance to the electronics manager. A lot of people in business state their opinions as if they were facts. Very powerful questions electronics managers should ask are:

- How do you know that?
- Where is your data?

Deming felt that, because managers were not sufficiently vigorous in their thinking, a lot of what managers think they know is not or may not be so.

Importance of theory. Theory is the way of thinking about something. Your theory or theories determine how you think about a given subject. There can be, and frequently are, multiple theories for a given subject. Some theories are more accurate and more useful than others.

Without theory, experience has no meaning. You need a way of thinking to interpret experience. Without theory, one has no questions to ask. You need a way of thinking to ask meaningful questions. Hence without theory, there is no way to grow in understanding. Thus, theory is the window to the world. Theory leads to an ability to make predictions. Without theory and prediction, experience and examples teach nothing.

To copy an example of success, without understanding it with the aid of theory, frequently leads to disaster. A good example of this is quality circles (QCs) in America. At one point substantially all of the Fortune 500 had quality circles, but today very few do. American companies copied the Japanese companies' practice of QCs but did not understand the theory. Without understanding the theory, American companies did not get QCs set up right and they fell way short of their potential.

Thinking for ourselves. As Peter Drucker said, "Unlearning is the most difficult learning of all." [19] Many times we glibly follow traditional management behaviors without questioning them because we know that "they work." It is not what you do not know that is hurting you, it is what you know that is not so. The problem here is that when you know something you never even stop to question it. And if you never question it and it is less than optimal, you will manage less than optimally. Instead we need to be more rigorous in our thinking, asking "How do we know what we know?" looking for ourselves at what is our theory, and studying results of various ways of doing things to determine what is the best method. This leads us to the next part of Deming's theory of knowledge.

How does a manager build knowledge? Knowledge means the capacity for effective action. Building knowledge, or learning, means enhancing the capacity for effective action. The way a manager builds knowledge is with the plan–do–study–act (PDSA) cycle of learning as shown in Fig. 5.5. This is the essence of learning and is based on the scientific method. Table 5.7 shows the steps included in the basic learning cycle of PDSA.

Knowledge (the capacity for effective action) is based on a theory that is tested over time using PDSA. Interpretation of data from a test or experiment can increase knowledge and increase the degree of

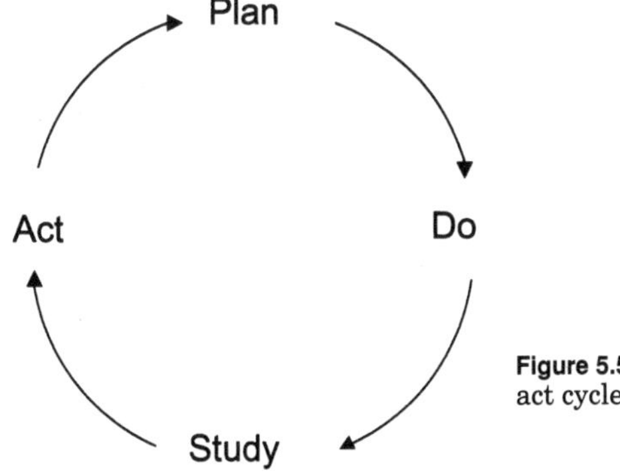

Figure 5.5 The plan–do–study–act cycle.

TABLE 5.7 Basic Steps in Plan–Do–Study–Act

Plan 1. Choose what to improve 2. Determine present condition (the problem, relevant data, etc.). 3. Develop solution. 4. Make a prediction. 5. Identify the theory underlying the prediction. 6. Plan the actions to test the prediction. **Do** 1. Carry out the plan, preferably first on a small scale. 2. Collect data.	**Study** 1. Analyze the data. 2. Compare results with prediction. 3. Consider what was learned and adjust theory if necessary. 4. Conclude whether the solution worked. **Act** 1. Decide on next plan: ▪ If it did not work, **plan** and **do** something else. ▪ If it worked, plan next **do** to (1) "test" on larger scale or (2) standardize (create a standard operating procedure) and roll out.

belief in the theory and in the method of doing something. Rational prediction requires theory and builds knowledge through systematic revision and extension of theory based on comparison of predictions with observations of tests (the basic learning cycle of PDSA).

The learning cycle has been described a lot of ways:

- Dewey's—discover, invent, produce, reflect (leading to new discovery)
- Shewhart's—specification, production, inspection
- Deming's—plan, do, study, act
- Deming Prize criteria's—plan, do, check, act (PDCA)

While the Deming Prize Criteria use plan–do–check–act (see Sec. 5.8), Deming preferred plan–do–study–act. *Study* is more descriptive of what is involved than *check*. Study involves analyzing the data, comparing the results to the prediction, considering what was learned, and concluding whether the improvement worked. *Study* emphasizes what was learned.

Operational definitions (used whenever data is collected). A key to using PDSA is having consistent data. To ensure data is consistent, Deming suggested the use of operational definitions. An operational definition is a statement used to define the measurement process to reduce variation in data collected. Table 5.8 shows the six necessary parts of every operational definition. Operational definitions are always and only used when the definition involves measurement. Operational

TABLE 5.8 Six Necessary Parts of Operational Definitions

1. What is being measured (the process, product, or service being measured)
2. Why this is being measured
3. The target [a clearly stated and specific ideal outcome (its quality attributes) of a system or process in written and/or visual and/or numeric form]
4. The method of measurement to determine how close to target the system or process is
5. A specific procedure to tabulate and evaluate the results (the data)
6. Agreement on the above between both the customer of the data and the supplier of the data

definitions are needed because information is relative. There is no true value of any characteristic, state, or condition that is defined in terms of measurement. Changing the procedure for measurement (changing the operational definition) produces a new number.

An operational definition is a procedure agreed on for translation of a concept into a measurement of some kind. Operational definitions help to link what I am seeing and thinking with what you are seeing and thinking. Operational definitions are required for the communication and cooperation that is necessary for optimization. Consistent data collected according to agreed-upon procedures (operational definitions) is a major cooperative tool.

Working definitions (same as dictionary definitions). Also needed for clear communication (and thus cooperation) are working definitions. A working definition is a statement that clarifies a specific term or terms in order to reduce variation in communication. Working definitions should be established any time a term is used that can have multiple meanings or there is a possibility for misunderstanding. A working definition is used in every instance except at data feedback points, where operational definitions are required. The key thought underlying both operational and working definitions is that clarity is power.

Management is prediction. Now, for some clarity on management. What is the essence of management in one word? The most frequent answers given are leadership, control, planning, direction, and vision. However, Deming's answer was prediction. What did Deming mean by management is prediction? Some answers are

- Managers deal with the future; inherent in every decision is a prediction about the future.

- Management is being able to predict what a company's processes will produce.
- Hiring is prediction—a prediction that the person hired will be better than the other candidates available.
- Promotion is prediction—a prediction that the person promoted will be better than the other candidates.

Deming suggested managers needed to focus more on prediction. Managers should work on their ability to make predictions, continually improve their ability to make predictions. Some guidance follows:

- Stable processes are predictable. Eliminate special causes to get to stable processes.
- Use plan–do–study–act to build knowledge.
- Once a process is stable, then reduce the variation to make predictions within a smaller range.

Ideas for starting to put Deming's theory of knowledge to work in an electronics company are given in Table 5.9.

5.2.4 Psychology

Everything a manager does is done through people. A manager has to be "good with people." Part 4 of Deming's theory of management, psychology, provides guidance in this area.

People want to do a good job. The key point of psychology, according to Deming, is that people want to do a good job. People want to contribute, to make a difference, and to have pride and joy in their work. This is true for the vast majority of people. Deming said, "Give the work force a chance to work with pride, and the 3 percent that apparently don't care will erode itself by peer pressure." [20]

To manage people, managers need to know something about:

- People's needs

TABLE 5.9 Ideas for Starting to Put Deming to Work in Your Electronics Company: Put Knowledge to Work

1. Start using the plan–do–study–act cycle of learning in everything you do.
2. Get your people to start using the plan–do–study–act cycle of learning in everything they do.
3. Inquire how data was collected—ask for the operational definitions.
4. Start examining more vigorously how you know what you know.

- People's skills, strengths, weaknesses, and development needs
- The way people are motivated
- The way people interact
- What brings out the best in people
- What robs people of their joy in work
- What will challenge people

People need to understand the "big why." In order to do a job, a person needs to know why they are doing it (the "big why"). This is true whether they are working on optimizing the entire system over time or just accomplishing some simple task. The example Deming liked to use was washing a table. His point was you could not wash a table unless you knew the purpose. Obviously to do the job right varies on why you are washing it—to play cards on it, or to eat dinner, or to perform surgery.

It is the manager's job to ensure their people know the big why of what they are doing. Understanding the big why is helpful in these ways:

- It gives the workers a context of what they are doing and allows them to adjust as things come up.
- It ensures the workers understand the importance of their job—why their contribution matters.

People are different. People are different from one another. They learn in different ways and at different speeds. One size does not fit all, nor does one way of training fit all. People also contribute in different ways. Electronics managers must recognize the differences in their people and use such differences to optimize their organizations.

Extrinsic and intrinsic motivation. There are basically two kinds of motivation: extrinsic and intrinsic. *Extrinsic* motivation is a force, stimulus, or influence that causes a person to act that comes from outside that individual's inner resources. Examples of extrinsic motivation include incentive compensation, bonuses, awards, and fear. *Intrinsic* motivation is an inner drive, impulse, or intention that causes a person to do something or act in a certain way. Intrinsic motivation is not dependent on external circumstances. Deming suggested we need to tap more into intrinsic motivation (people's self-esteem and dignity). Deming saw that a principal problem with our prevail-

TABLE 5.10 Ideas for Starting to Put Deming to Work in Your Electronics Company

1. Be aware of fear and consciously drive out fear.
2. Give your people a chance for pride and joy in their work.
3. Involve everyone in optimizing the system.

ing traditional management approach is overreliance on extrinsic motivation, which can smother intrinsic motivation.

Drive out fear. Managers must drive out fear and build trust. Why drive out fear? Fear causes people to shut down to some extent. In the worst situations it can actually be paralyzing. As Deming said, "No one can put in his best performance unless he feels secure." [21]

How does a manager drive out fear? First and foremost is to stop doing those things that lead to fear. Examples of fear-causing behaviors include (1) being secretive, (2) being reactive, and (3) being judgmental. People need to feel that they understand what is going on and why. Deming did not say just drive out the fear a manager may have caused; he suggests a manager must drive out the fear arising from other sources too.

Building trust. How does a manager build trust? First, by being trustworthy. Sharing information is another way to build trust because it demonstrates trust. People share information with people they trust. Table 5.10 lists ideas for putting the psychology part of Deming's theory to work in an electronics company.

5.2.5 Deming's theory as a system

Deming saw his theory as providing a lens "by which to understand and optimize the organizations that we work in, and thus to make a contribution to the whole country." [22] Deming's theory of management is also known as *profound knowledge* because of the profound results achievable by following his theory. Profound knowledge must be viewed as a system (way to look at, think about business). The theory is holistic. Deming thought that each part (appreciation for a system, theory of variation, theory of knowledge, psychology) is necessary, interrelated, and inseparable. If there is any magic to Deming's theory of management, it is in the interaction of the parts. It is like having four lenses or filters—one for each part of Deming's theory—and being able to view problems, issues, and challenges from each of the four vantage points simultaneously.

TABLE 5.11 Questions an Electronics Manager Might Ask about Unexpected Results

Based on appreciation for a system:
1. Why didn't the process deliver as expected?
2. Were the standards (standard operating procedures) followed?

Based on theory of variation:
1. What did the data look like?
2. Was the variation from a special cause (which means the manager should act immediately) or just from common causes (which means the manager should lead improvement of the process)?

Based on theory of knowledge:
1. Were our assumptions wrong?
2. How could the process have worked?
3. What did we learn?

Based on psychology:
1. What coaching is needed?
2. How can my team help dissolve this problem (both solve the problem and prevent its recurrence)?

To illustrate the use of Deming's theory, consider this scenario. A manager's function fails to reach expectations on a key output of their process. The manager following Deming's theory of management would ask questions like those in Table 5.11.

5.3 The Operating Principles for a Manager for Managing the Deming Way

There are certain operating principles based on Deming's theory of management for managing the Deming way to guide a manager. In the past Deming was known by many people for his 14 points for management (see Table 5.12). [23] This listing of dos and don'ts for managers was taken by many as the absolute answer and they followed it without thinking—unfortunately they missed Deming's point that managers need to think for themselves. Others have seen it as a menu from which to pick and choose those they liked and ignore the others; unfortunately they missed the point that the 14 points represented a system, not a menu.

Deming strongly recommended that managers think for themselves and develop their own list of principles by which they would manage their organization. The following list is presented in the same spirit—for readers to use as a guide and model from which to build their own

TABLE 5.12 Dr. Deming's 14 Points for Management

1. Create constancy of purpose toward improvement of product and service, with the aim to become competitive and to stay in business, and to provide jobs.
2. Adopt the new philosophy. We are in a new economic age. Western management must awaken to the challenge, must learn their responsibilities, and take on leadership for a change.
3. Cease dependence on inspection to achieve quality. Eliminate the need for inspection on a mass basis by building quality into the product in the first place.
4. End the practice of awarding business on the basis of price tag. Instead, minimize total cost. Move toward a single supplier for any one item, on a long-term relationship of loyalty and trust.
5. Improve constantly and forever the system of production and service, to improve quality and productivity, and thus constantly decrease costs.
6. Institute training on the job.
7. Institute leadership (see point 12). The aim of supervision should be to help people and machines and gadgets to do a better job. Supervision of management is in need of overhaul, as well as supervision of production workers.
8. Drive out fear, so that everyone may work effectively for the company.
9. Break down barriers between departments. People in research, design, sales, and production must work as a team, to foresee problems of production and in use that may be encountered with the product or service.
10. Eliminate slogans, exhortations, and targets for the work force asking for zero defects and new levels of productivity. Such exhortations only create adversarial relationships, as the bulk of the causes of low quality and low productivity belong to the system and thus lie beyond the power of the work force.

11*a*. Eliminate the work standards (quotas) on the factory floor. Substitute leadership.

11*b*. Eliminate management by objective. Eliminate management by numbers, numerical goals. Substitute leadership.

12*a*. Remove barriers that rob the hourly worker of the right to pride of workmanship. The responsibility of supervisors must be changed from sheer numbers to quality.

12*b*. Remove barriers that rob people in management and in engineering of their right to pride of workmanship. This means, *inter alia,* abolishment of the annual or merit rating and of management by objectives.

13. Institute a vigorous program of education and self-improvement.
14. Put everybody in the company to work to accomplish the transformation. The transformation is everybody's job.

SOURCE: Reprinted from Ref. 23 with permission.

operating principles for an electronics company. The 16 operating principles in this list are a representative compilation of what a few electronics companies came up with as their operating principles after studying and committing to Deming's theory.

1. *Adopt Deming's theory and these operating principles.* We are in global economic competition. We must embrace the challenge, learn our responsibilities, and lead the change to optimize our company.

2. *Create constancy of purpose.* Our constancy of purpose is directed toward optimizing our organization's entire system over time thus improving our products and services, with the aim to become more competitive, to stay in business, to provide secure jobs, and to prosper.
3. *Constantly optimize system.* We must improve constantly and forever our system of production and service to improve our quality and productivity and thus constantly decrease costs. Our objective is to become the highest-quality, lowest-cost producer.
4. *Build quality in.* We will eliminate the need for inspection on a mass basis by building quality into our products and processes in the first place, from the beginning of the process.
5. *Minimize total cost.* Instead of awarding business on the sole basis of price tag, we will work to minimize *total* cost. We also will move toward few suppliers and even a single supplier for items based on a long-term relationship of trust, loyalty, and dependability.
6. *Manage the interaction between departments and break down any departmental barriers.* We must work together. For example, people in research, design, sales, and production must work as a team to foresee problems of production and in use that may be encountered with our products or services.
7. *Provide everyone with opportunity for pride in his / her work.* We will focus on having processes sufficiently robust so everyone can be proud of the output. We will not sacrifice quality for quantity.
8. *Drive out fear.* We must drive out fear, so that everyone may work effectively for our company. Our people must not be afraid of their manager, of asking questions, or of making suggestions.
9. *Provide supervision.* The aim of supervision is to help people and machines do a better job. We will vary the level of supervision on the basis of need.
10. *Provide leadership.* Leadership provides the direction for our company. Leadership creates and communicates our aim, vision, and mission. Leadership makes sure we are doing the right things. Leadership makes sure we are focused on optimizing our entire system over time. Our leaders lead by example.
11. *Provide training on the job.* We will ensure all our people are adequately trained to do their jobs. Our training is facilitated by having well-documented processes to train to.
12. *Institute education.* We will provide and encourage a vigorous program of education and self-improvement for all employees. The theme will be lifelong learning.

13. *Management by system.* We manage by system, which means we build our system and related processes to be able to achieve the results we want to achieve. We are leading with our system. The strength of our system gives us our strategic advantage.
14. *Build loyal customers.* We are focused on building loyal customers because it is the best way for our company to have consistently higher profits and grow faster. Research in other industries has shown this to be true. [24]
15. *Focus on target.* We are focused on target, which is the ideal or optimal output of our processes.
16. *Involve everybody.* We will put everybody in our company to work to optimize our entire system over time. Optimization is everybody's job.

5.4 Translating the Theory into Action

So far this chapter has covered Deming's theory of management and the principles of management flowing from his theory. If management's job is to optimize the entire system over time, the question is then how do managers do that? How can managers add the most value to their organization? Their department? Their employees? Their own performance? A manager can make the greatest contribution by studying and understanding Deming's theory of management, translating that theory into a personal action plan, and then following the action plan in all interactions that affect the organization.

High-performance managers utilize seven basic actions, based on Deming's theory of management, in this new paradigm of management. The seven behaviors which define the basic role of a manager in the new paradigm are described in Table 5.13. Each of these behaviors/roles is explained in greater detail below.

TABLE 5.13 Role of an Electronics Manager in the New Paradigm

1. Thinking optimization
2. Creating a team focused on being the best
3. Educating, training, coaching, and counseling
4. Involving the workers in decision making and optimizing
5. Interacting with the other components of the system including suppliers and customers
6. Getting the job done
7. Constantly improving the system

5.4.1 Optimization thinking

High performing managers utilize *optimization thinking* as their primary way of thinking. Optimization thinking involves systems thinking, using data to guide actions, building knowledge, and tapping into the potential of their people. Examples of each component of optimization thinking are given below:

Systems thinking (appreciation for a system) includes:

- Focusing on management's job of optimizing the organization's entire system over time.
- When something goes wrong, looking first to the system, rather than blaming individuals. The vast majority of troubles and possibilities for improvement belong to the system, at least 94 percent according to W. Edwards Deming. [25]
- Since the system and related processes are management's responsibility, working on the organization's system and related processes. This involves building clearly defined processes and related standard operating procedures for repetitive tasks. You will only get what the system will deliver.
- Spending the time up front on new projects or processes (85 percent of the impact is in the first 15 percent of the process). [26]
- Focusing on the aim or purpose of both (1) the organization and (2) the activities/functions/departments for which the manager is responsible.
- Realizing that optimization requires both continuous improvement and innovation.
- Communicating and cooperating with the other components of the system towards achieving the aim of the entire system.
- Building loyal customers—customers that appreciate what the organization does for them.
- In short, managing by system (system dependent management, not people dependent management).

Using data to guide action includes:

- Properly collecting data on processes—both process and outcome data.
- Using statistics to make the invisible visible—show patterns and types of variation by charting the data to see the variation. Such variation is the voice of the process.

- Using data to tell whether changes made are improvements or just changes.
- Distinguishing between special causes (which are not part of the process and require immediate action) and common causes (which are part of the process and require a more systematic, studied approach to improving the process).
- Acting according to type of cause and, thus, avoiding tampering.
- First removing any special causes of variation to create a stable, predictable system and then shrinking, shrinking, shrinking variation (reducing variation reduces costs) and/or moving the process average to target (ideal or optimal value).
- Deming's theory of variation applies whether there is data or not—even without data being guided by the theory.

Building knowledge using the plan–do–study–act cycle of learning includes:

- Learning, which is building the organization's knowledge, and knowledge is the capacity for effective action.
- Using the plan–do–study–act approach in everything the manager does, especially building the manager's own knowledge (capacity for effective action).
- Using operational definitions wherever measurements are involved to ensure data is consistent and reliable.
- Being aware of one's own assumptions and asking "How do I know what I know?" Being open-minded to new ways of looking at things and to testing new ways of doing things.

Tapping into potential of their people includes:

- Using psychology to tap into the full potential of all their people.
- Realizing that people want to do a good job, make a difference, contribute, and have pride and joy in their work.
- Respecting their people and treating their people with respect.
- Tapping into intrinsic motivation (self-esteem, dignity, and desire to learn) of their people.
- Recognizing the differences in their people and using such differences to optimize everyone's contribution to the system.
- Driving out fear by building trust and providing leadership. Having no one afraid to make suggestions or ask questions.

- Having fun.

In these ways, high-performing managers use Deming's theory as a way to look at and think about business—as a lens through which they first understand and then optimize their organizations.

5.4.2 Creating a work team focused on being the best

By creating the work team, managers can involve everyone and tap into more of the potential of their people. Managers are responsible for creating the working environment and culture. How the manager acts sets the tone and expectations. The work team approach allows all members to support each other, build on each other's thinking, and learn from each other. Ideal interaction between team members will produce synergy, where the results are greater than the contributions of individual team members. *Warning:* It is also possible to have negative interactions, where the results of the team are less than the contributions of individual team members.

The suggested aim for the work team is to accomplish their mission, to continually improve how they accomplish their mission, and to be "the best" at what they do. Each member of the work team needs to understand the work team's mission, how that mission fits into their organization's larger system, and how they are an important part of the larger system. Individual members of the team also need to see their job as both getting their job done right and continually improving how the job is done. Getting the job done right means the way the customer wanted it, with minimum variation, in minimum time, and at minimum cost. Having the aim of being the best gives the team a focus and a source of pride and will help bring out the best in each team member.

5.4.3 Educating, training, coaching, and counseling

One of the key roles of an electronics industry manager is to educate and train people. Education has to do with teaching people how to think while training involves teaching people the skills to do their job. In many situations it is appropriate to supplement what a manager provides with additional training, both internally and externally. The manager also needs to look for employees who are outside the system and in need of special help. The manager must figure out what special help is needed and then provide the help or see that it is provided by others.

Educating, training, coaching, and counseling are interrelated and real-time activities consisting of:

- Being sure the workers understand the aim, mission, and vision of the organization and of their work team.
- Being sure the workers understand their role in the organization's larger system.
- Providing the workers the necessary training, including on-the-job training, so they have the required skills to do the job.
- Setting it up so the workers collect and chart data from the processes they are involved with, thus ensuring feedback to them from the processes.
- Observing the workers' performance.
- Supporting the workers, including reinforcing desired behaviors and acknowledging their efforts and accomplishments.
- Challenging the workers by varying their responsibilities and providing cross-training.
- Teaching the workers how to think.

With a strong system and related processes including standards (standard operating procedures), the need for the manager to be a cop or boss (as in telling people what to do) goes way down or is even eliminated.

The judging of people is an activity that adds little or no value and may actually cause harm. For many people, just knowing you are going to be judged causes fear.

5.4.4 Involving the workers in decision making and optimizing

This means asking workers for their input on how the process can be improved and how they would do things, listening to what the workers suggest, and exploring their thinking by saying: "Tell me more" or asking "What's your reasoning?" Next, whenever possible the manager (and/or the work team) should run tests using the plan–do–study–act cycle of learning. Rather than intellectually trying to figure something out, run some tests, preferably first on a small scale. Be sure to collect data. Managers ultimately have to decide what to do on the basis of their own experience and wisdom, the results of the tests run, and on the input received from the workers.

Most important, managers should explain why they made the decision that was made. People want to know why their input was not fol-

TABLE 5.14 Why Should a Manager Involve the Workers in Decision Making?

- It leads to better decisions.
- The workers' involvement leads to their commitment.
- It is a way of showing respect.
- It creates a motivating environment.

lowed and what the manager sees that they did not see. Explaining decisions goes a long way toward reducing resistance and increasing buy-in and commitment.

Listening to and involving the workers in decision making does not mean the manager gives up any authority and responsibility. It is possible to listen and still be in charge; it is called "leading by listening." Table 5.14 gives some of the key reasons for involving workers in the decision-making process.

5.4.5 Interacting with the other components of the system including suppliers and customers

One of the key actions of an electronics industry manager is to interact with the other components of the system with the aim of continually enhancing the way the entire system performs. The first step is managing the interaction between the components of a process or of the entire system. It is amazing what just getting the components to talk to each other will do. Many problems are quickly identified and easily eliminated. After a while a more formal agenda may be helpful, including items such as:

- Coordination issues—how can we make things go even smoother?
- Anticipating, identifying, and eliminating problems.
- Enhancing what the components in the supplier-customer chain do for each other. The discussions should focus on (1) here is what I can do for you and (2) here is what you could do for me.

5.4.6 Getting the job done

A manager is expected to get the job done. If a manager fails this expectation, it can be detrimental to one's managerial career. How does an electronics industry manager in the new paradigm ensure the job gets done? First, by focusing on the first 15 percent of the activity, whatever it is. This includes the planning phase. It also includes ensuring that the process is capable of producing the desired results. Second, by supervising—by observing actual production, looking at the data from the process, and intervening if and when appropriate.

Occasionally it may not be possible to produce desired results with the present system and related processes. There are three choices:

- Do a quick process improvement. This may not be possible in all cases.
- Override the system. Work around or outside the system if necessary, then go back and fix the system. Why was the override necessary?
- Say "No, that cannot be done with the resources presently available."

A manager must manage the old (get the job done) and create the new (work toward optimizing the entire system over time). Part of getting the job done is also handling the administrative or paperwork requirements.

5.4.7 Constantly improving the system

The electronics industry manager needs to be sure the work force is not only getting the job done but also constantly improving the way the job gets done. This needs to be an integral part of managing. This needs to be an expectation every person on the team understands. On the portion of the system that the manager controls, the processes he or she is responsible for, the manager can make this happen. The basic tool is plan–do–study–act, and the other quality tools such as fishbone diagram, flowchart, Pareto chart, and control chart will be helpful.

The basic approach for improving a process is:

Step 1. Clearly define the process including standards (standard operating procedures).

Step 2. Collect and analyze data from the process.

Step 3. Implement continuous improvement and innovation, including working with suppliers (internal and/or external) and customers (internal and/or external).

There will also be processes that cross functional lines. The manager should take the lead in forming cross-functional teams and/or participating on such teams. This is one way the manager can influence those processes he or she does not have direct control over. Remember, management's job is to optimize the entire system over time. The seven actions are the way to be a high-performance manag-

TABLE 5.15 The Four Parts of Deming's Theory

Part 1	Appreciation for a system
Part 2	Theory of variation
Part 3	Theory of knowledge
Part 4	Psychology

er in the new paradigm and in today's competitive world. While they are not that hard, for most people they do represent, to varying degrees, a very different way of managing. What is being suggested is that electronics industry managers learn and live the theory by:

1. Understanding Deming's theory—all four parts and how they interact as a system. See Table 5.15.
2. Creating their own operating principles for management, in their words. See Table 5.12 for Dr. Deming's suggested operating principles.
3. Creating their own list of key management actions, in their words. See Table 5.13.
4. Applying plan–do–study–act to their own process of managing so they can continually improve how they manage.

Coupling this foundation, this fundamental guide, with their knowledge of their organization and their subject matter expertise will enable electronics managers to become high performance managers.

5.5 A Review of Common Management Practices

This section examines some common traditional management practices in light of Deming's theory of management. Deming felt "The present style of management is the biggest producer of waste, causing huge losses whose magnitude cannot be evaluated, cannot be measured. It is these losses that must be managed." [27] The question is not does a particular management practice work, but rather is it optimal? Will it best move the organization toward optimization? The specific traditional management practices this section examines are:

- Proliferation of unnecessary paperwork
- Reactive management
- Reliance on numerical goals
- Performance appraisal and ranking

- Pay for performance and other forms of incentive pay
- Management by objectives (as practiced)—managing the pieces, not the whole.

For each, the possible problems with the traditional management practices are identified and then suggestions as to what to do instead are given. These suggested practices will move the organization further toward optimization based on Deming's theory.

It should be noted that these traditional management practices were designed by and used by managers doing their best. Again, as Deming said, "Putting forth best efforts, best efforts with hard work. Our ruination." [28] What are the causes of these suboptimal traditional management practices?

- Lack of a theory of management.
- Not distinguishing between special and common causes of variation.
- Not thinking for ourselves.

5.5.1 Proliferation of unnecessary paperwork

Almost every manager I have ever spoken to about waste has commented on unnecessary paperwork. Never has a manager ever complained to me that their organization did not have enough paperwork. Unnecessary paperwork is a major source of waste in most organizations. Automating unnecessary paperwork may reduce the time spent on it, but if it is unnecessary, it is still waste. See Table 5.16 for what is wrong with proliferation of unnecessary paperwork. To reverse and prevent the proliferation of unnecessary paperwork, electronics industry managers should consider the steps outlined in Table 5.17.

5.5.2 Reactive management

When something happens, "Don't just stand there. Do something," is a good description of reactive management. Reactive managers spend

TABLE 5.16 Some of the Problems with Proliferation of Unnecessary Paperwork

- It wastes time, time that could be used to work on optimizing the system.
- It causes confusion, especially confusion as to what is important.
- It is discouraging and demoralizing—no one likes their time wasted.
- It kills trees.

TABLE 5.17 What an Electronics Manager Can Do to Reduce Proliferation of Unnecessary Paperwork

- Review all reports and study how each report is used. Most organizations that conduct such a review significantly reduce the number of reports produced.
- Use control charts and other graphic representations instead of reams of computer reports.
- Insist that data be presented in context and graphically.
- Stop asking for explanations of common cause variation.
- Take action yourself (or form a team) to identify and eliminate any unnecessary paperwork.
- Stop trying to control with paperwork.

TABLE 5.18 Some of the Problems with Reactive Management

- It has a high probability of making matters worse.
- It wastes valuable time that could have been spent working on optimizing the entire system over time.
- It sends mixed messages as to what is really important—one day it is this and the next day it is that.
- It causes fear.
- It encourages short-term thinking.

TABLE 5.19 What an Electronics Manager Can Do to Reduce Reactive Management

- Distinguish between common causes and special causes of variation, and act when you should and do not act when you should not.
- Manage toward optimization of the entire system over time.
- Be a strategic manager—be in control as opposed to being reactionary, pause between stimulus and response.
- Do more long-term planning—focus on the long term.

most of their time first creating (or allowing others to create) and then fighting fires. Table 5.18 lists some of the problems with reactive management. Suggested alternatives to reactive management are given in Table 5.19.

5.5.3 Reliance on numerical goals

A lot of managers rely on numerical goals. Figure out what you want, compute the numbers, and tell those who report to you what they have to produce. As Dr. Deming was fond of saying, "I wish management were as simple as that." [29]

When you give someone or some group or department a goal they will most often accomplish it, especially if you do not count the costs. If the goal represents output greater than the system presently delivers, they have three alternatives:

1. Improve the system and related processes.
2. Distort the system.
3. Distort the data.

An example of distorting the system is the well-publicized case of the Sears automobile service representatives selling and performing unnecessary repairs to customers' vehicles in order to meet their quotas (goals), keep their jobs, and receive their bonuses. Other problems with overly aggressive goals:

- They cause frustration.
- They create fear.

If goals are too easy, they can lead to complacency. People have a tendency to stop when they reach the goal. And, if goals are just right, they probably have little or no effect. What else is wrong with reliance on numerical goals? They encourage short-term thinking; "Just meet the goal" becomes the message. "A numerical goal accomplishes nothing. Only the method is important, not the goal. By what method?" as W. Edwards Deming often said. [30]

What to do instead of relying on numerical goals? Instead of relying on numerical goals, managers should manage toward optimization of the entire system. Manage the components for optimization of the aim of the system. This would include having all significant process in control (except to the extent they are continuously improving). A numerical goal is not needed to improve a process. All that is needed is actual data from the process and a clear definition of target (ideal or optimal value) of the process.

If a manager is in an organization that is managed by goals, the manager should treat the goals as minimums and be sure the processes are sufficiently capable to produce desired outcomes even on a "bad day." This is the essence of having a method to accomplish each goal. And then, get on with the real work of shrinking variation and moving the process average toward target. Again, target is a clearly stated and specific ideal outcome (its quality attributes) of a system or process in written and/or visual and/or numeric form.

TABLE 5.20 Some of the Problems with Performance Appraisal and Ranking

- It is inherently biased and defies clear definition.
- It assumes that most people would not do their best without it.
- It creates fear.
- It creates boss pleasing.
- It gives the manager a lever (a tool) to control workers and lets managers substitute such control for leadership.
- It creates unseen costs.
- It creates winners and losers. Who wants to be a loser? Who wants to work with a loser?
- It is not possible to separate the impact of the system from the impact of the person.
- Results are substantially determined by the system, not the individual(s).
- It creates competition, not cooperation.

5.5.4 Performance appraisal and ranking

Periodic performance appraisals and then ranking of employees is a very common traditional management practice. Usually there is some form of labeling as part of the ranking process such as outstanding, above average, average, needs improvement, and unacceptable. See Table 5.20 for what is wrong with performance appraisal and ranking.

What to do instead of performance appraisal and ranking? Instead of performance appraisal and ranking, Deming's theory of management would have the manager educate, train, coach, and counsel his or her people on a real-time basis. The objective is to develop the people. Developing individuals includes cross-training, assigning more complex tasks, and challenging them. Providing coaching includes ensuring people get feedback from the processes they are involved in on a real-time basis. Imagine what would happen if managers took the time previously spent on performance appraisal and ranking and spent it instead on process appraisal and ranking the priorities for process improvement.

5.5.5 Pay for performance and other forms of incentive pay

Pay for performance and other forms of incentive pay are fairly common ways of compensating individuals in an organization.

What's wrong with pay for performance and other forms of incentive pay? See Table 5.21 for what is wrong with pay for performance and other forms of incentive pay.

TABLE 5.21 Some of the Problems with Pay for Performance and Other Forms of Incentive Pay

- It puts quantity ahead of quality.
- It assumes that most people would not do their best without it.
- It creates fear.
- It creates boss pleasing.
- It gives the manager a lever (tool) to control workers—lets managers substitute control for leadership.
- It creates unseen costs.
- It creates winners and losers. Who wants to be a loser? Who wants to work with a loser?
- It is not possible to separate the impact of the system from the impact of the person.
- Results are substantially determined by the system, not the individuals.
- It can result in overpaying.

What to do instead of pay for performance and other forms of incentive pay? The following three-part system of pay is suggested as an alternative to pay for performance and other forms of incentive pay:

1. Pay fair market value. Pay what the market pays for the various positions in an organization.
2. Organizationwide gain sharing (or profit sharing).
3. A small premium for seniority.

The guiding principle of a system of pay should be fairness.

People performing within the same process should be basically paid the same. However an outlier (someone who consistently outperforms the rest) represents some interesting questions. If outliers have a better way of doing something, you want them to train the other workers. Making them team leaders with the responsibility to train the others, possibly with an increase in pay, may be the answer. Just paying them more may discourage sharing their better way with the others because then they may not clearly be the best.

5.5.6 Management by objectives (as practiced)—managing the pieces, not the whole

Management by objectives, as practiced, generally sets some objectives for the various components of an organization and manages to those objectives. Frequently some part of the compensation process is based on how well a manager does compared to the designated objectives.

What's wrong with management by objectives (as practiced)—managing the pieces, not the whole? Some of the problems with management by objectives (as practiced)—managing the pieces, not the whole—are

- Suboptimization
- Creates and encourages competition rather than cooperation
- Can cause fear
- Encourages short-term thinking

What to do instead of management by objectives (as practiced)— managing the pieces, not the whole? Instead of managing by objectives (as generally practiced), Deming suggests adopting his theory and managing the entire system toward optimization.

5.6 More about Leadership

Leadership is a key ingredient in optimization. As Deming said, "The transformation is not automatic—it must be learned—it must be led." [31] This section explores the leadership required.

5.6.1 Deming on leadership

Deming felt that understanding of his theory of management "will lead to transformation of management. The transformation will lead to adoption of what we have learned to call a system, with a stated aim. The individual components of the system, instead of being competitive, will for optimization of the system, reinforce and support each other." [32]

Deming also felt that the transformation in any organization will not be spontaneous. Deming saw "the job of a leader is to accomplish transformation of his organization. The leader uses knowledge, personality, and persuasive power to lead the transformation." [33] How may the leader accomplish transformation? According to Deming:

> First, he has theory. He understands why the transformation would bring gains to his organization and to all the people that his organization deals with. Second, he feels compelled to accomplish the transformation as an obligation to himself and to his organization. Third, he is a practical man. He has a plan, step by step.
>
> But what is in his own head is not enough. He must convince and change enough people in power to make it happen. He possesses persuasive power. He understands people. [34]

TABLE 5.22 Other Things That Good Managers Do

- Act with integrity
- Respect others
- Manage yourself
- Embrace change
- Embrace the paradox
- Do what is important but not urgent
- Take personal responsibility
- Have a passion for learning

5.6.2 Brightness, darkness, and frequency

One strategy for leaders is to paint two pictures of the future. First, brightness of the future. This is frequently called a vision, a compelling picture of a preferred future state. The other is the flip side of brightness—darkness of the future. What is likely to happen if the people do not follow the leader? This includes the facts of life, such as "If we do not eliminate our operating losses, we will go out of business." Both messages need to be delivered frequently—over and over and over again.

5.7 Other Key Concepts for Success—Other Things Successful Managers Do

There are other little things that good managers do. These recommendations are not earth-shattering but they will make an electronics industry manager more effective (see Table 5.22).

5.7.1 Always act with integrity

Integrity is first and foremost. Truly successful people have integrity.

5.7.2 Respect others

Having respect for others is second only to integrity. When you have respect for others you treat them differently than you would if you do not respect them. Respect is something everyone craves; no one likes to be treated disrespectfully.

5.7.3 Manage yourself

The manager must be strategic, thoughtful, and in control. It is not a reactionary job. An example of managing oneself is the first law of holes—if you are in one, stop digging. We all manage ourselves, at least unconsciously. To be more effective, a high-performing manager does it consciously. And then such a manager, using data, continuously improves personal performance. Managing yourself includes having a process for time management that works for you.

5.7.4 Embrace change

With respect to change in today's competitive world, is it speeding up or slowing down? Just about everyone says it is speeding up, going faster and faster. A good question managers need to ask: "Is my company changing as fast?" And a good question managers need to ask themselves is "Am I changing as fast?" So what is the optimal personal strategy with respect to change? Well, it is probably not "resist change." Even "accept change" lacks the enthusiasm to be optimal. The optimal personal strategy is to "embrace change." *Embrace* means to welcome and get your arms around it.

How to take the risk out of change? Or, at least minimize the risk? The answer is to use plan–do–study–act, especially acting on a small scale. Another way is to be sure to involve your people in the change. People will support what they have had a part in. They can also help identify the potential problems and frequently have good ideas on how things can be done better.

5.7.5 Embrace the paradox

Managers need to manage the old and create the new. Which is more important? The answer to that question is both. While at times it may be possible to manage the old and create the new simultaneously, most of the time these are mutually exclusive activities. Therefore the answer "both" is a paradox. Managers need to embrace this paradox, be sure that both get done.

5.7.6 Do what is important but not urgent

The problem with working on the system (and the reason most managers devote little or no time to it), is that while it is extremely important, it is seldom urgent. There will almost always be something more urgent, more pressing. This problem leads to a very frequently asked question: "Where do I (or we) get the time to work on the system?" The first response should be "Do you understand that working on the system is an integral part of your job?" A side note: it would be helpful if more job descriptions contained words to that effect. Once there is a clear understanding and agreement on the importance of working on the system, the next response might be "find it," or "make it," or

TABLE 5.23 Stages of Awareness of Problems

Stage 1: Unaware. People are so unobservant that they are clueless, totally unaware of the problem.
Stage 2: Denial. Denying that what has occurred is any sort of problem or concern.
Stage 3: Lay blame. Lay blame on someone else, anyone else.
Stage 4: Justify. Find an excuse or reason for the problem.
Stage 5: Shame. Once they are aware of the problem, cannot deny it, cannot blame someone else for it, nor justify it—then shame is left.

"make it a priority," or "by eliminating something that wastes time," or "just do it." Steven R. Covey does a nice job of describing this problem and suggesting how to deal with it in his book, *The Seven Habits of Highly Effective People.*

Failure to devote time to the important but not urgent most often leads to the matter eventually becoming urgent and then getting handled in a crisis or fire-fighting mode. It is in this way that some of an organization's best fire fighters are probably also its arsonists. This mode of ignoring the important but not urgent becomes a down spiral that is difficult to break.

5.7.7 Take personal responsibility

Table 5.23 gives the stages a manager, or any person, can go through when something goes wrong. The problem with all of these stages is that they add no value, they contribute nothing toward solving the problem and preventing its recurrence. Yet these behaviors are extremely common in many organizations. The alternative to these non-value-added behaviors is taking personal responsibility. It starts with the manager, with the manager accepting (taking) responsibility for whatever has happened and deciding how to proceed optimally.

What Deming's theory of management provides is a system for everyone to take personal responsibility. When the manager has taken personal responsibility, the space is opened for everyone else to also take personal responsibility. Focusing on (1) the system and related processes and (2) the data facilitates the process of taking personal responsibility.

5.7.8 Have a passion for learning

High-performance electronics industry managers have a passion for learning. They are committed to lifelong learning, having the attitude they are never too old to learn. A passion for learning includes:

- Increasing their subject matter expertise
- Making sure they know what is going on in their industry
- Enhancing their capacity to manage effectively
- Inspiring others to have a passion for learning

5.8 The Deming Prize

The Deming Application Prize [35] is an annual award presented to a company or division of a company that has achieved distinctive performance improvements through the application of company wide quality control (CWQC). See Chap. 2 for more information about CWQC.

CWQC is a set of *systematic activities* carried out by the *entire organization* to effectively and efficiently achieve *company objectives* and *provide products and services* with a level of *quality* that satisfies *customers,* at the appropriate time and price. These terms are further explained in Table 5.24.

The Deming Prize Guide for Overseas Companies 1996 provides two checklists to use as reference material for organizations that promote quality control and that challenge for the Deming Application Prize. Table 5.25 gives the checking points for the organization, and Table 5.26 lists the checking points for their senior executives.

The Deming Prize Guide for Overseas Companies 1996 also summarizes results that have been achieved by companies that have applied for the Deming Application Prize. These results are recapped below:

1. *Quality stabilization and improvement.* With improved reliability, safety, and quality that meet society's needs, many companies have acquired a world class reputation.
2. *Productivity improvement / cost reduction.* Reduced development and design troubles, decreased part defects, lower manufacturing defects and rework, etc. have been achieved. Productivity has also been enhanced by improving production control systems and worker-hour reduction activities.
3. *Expanded sales.* A systematic approach to satisfying customers' requirements, strengthening management system for distribution of new products, and deploying TQM into sales activities has contributed to capturing new customers and to preventing out-of-stock problems.
4. *Increased profits.* Item 3, expanded sales, and item 2, improved productivity/reduced cost, work together to increase profits.

TABLE 5.24 Definitions of CWQC

Systematic activities	Organized activities that involve everyone at all levels and all parts of the company. Such activities are led by management and guided by establishing appropriate quality strategies and policies.
Carried out by the entire organization to effectively and efficiently achieve	It is necessary to use appropriate scientific methods, including statistical techniques and to repeatedly rotate the management cycle of PDCA to maintain and improve the quality of jobs and human resources that are involved in the process of "provide" as mentioned below. In maintaining and improving the quality of jobs, CWQC activities address not only the quality of products and services, but also cost, quantity and delivery, safety, and motivation. In maintaining and improving human resources, not only education and training but also self-development and mutual development are important.
Company objectives	Refer to securing appropriate profit for the long term. Also, they include contributing to the happiness and satisfaction of customers, society, and employees.
Provide	Refers to activities from producing "products and services" to handing them off to customers, including survey, research, planning, development, design, product preparation, purchasing, manufacturing, installation, inspection, order taking, sales and marketing, maintenance, after-sales services, and disposal and recycling after usage.
Products and services	Includes manufactured products (finished products and parts and materials), system products, software, energy, information and all other benefits that are provided to customers.
Quality	Refers to usefulness (both functional and psychological), reliability, and safety. Also in defining quality, influence on the third parties, society, the environment and future generations must be considered.
Customers	Includes not only buyers but also users, consumers, and beneficiaries.

SOURCE: Reprinted from Ref. 35 with permission.

TABLE 5.25 The Deming Application Prize Checklist

Policies	1. Quality and quality control policies and their place in overall business management 2. Clarity of policies (targets and priority measures) 3. Methods and processes for establishing policies 4. Relationship of policies to long- and short-term plans 5. Communication (deployment) of policies, and grasp and management of achieving policies 6. Executives' and managers' leadership
Organization	1. Appropriateness of the organizational structure for quality control and status of employee involvement 2. Clarity of authority and responsibility 3. Status of interdepartmental coordination 4. Status of committee and project team activities 5. Status of staff activities 6. Relationships with associated companies (group companies, vendors, contractors, sales companies, etc.)
Information	1. Appropriateness of collecting and communicating external information 2. Appropriateness of collecting and communicating internal information 3. Status of applying statistical techniques to data analysis 4. Appropriateness of information retention 5. Status of utilizing information 6. Status of utilizing computers for data processing
Standardization	1. Appropriateness of the system of standards (standard operating procedures) 2. Procedures for establishing, revising, and abolishing standards 3. Actual performance in establishing, revising, and abolishing standards 4. Contents of standards 5. Status of utilizing and adhering to standards 6. Status of systematically developing, accumulating, handing down, and utilizing technologies
Human resources development	1. Education and training plans and their results 2. Status of quality consciousness, consciousness of managing jobs, utilization, and understanding of quality control 3. Status of supporting and motivating self-development and self-realization 4. Status of understanding and utilizing statistical concepts and methods 5. Status of QC circle development and improvement suggestions 6. Status of supporting the development of human resources in associated companies

SOURCE: Reprinted from Ref. 35 with permission.

TABLE 5.25 The Deming Application Prize Checklist (*Continued*)

Quality assurance activities	1. Status of managing the quality assurance system 2. Status of quality control diagnosis 3. Status of new product and technology development (including quality analysis, quality deployment, and design review activities) 4. Status of process control 5. Status of process analysis and process improvement (including process capability studies) 6. Status of inspection, quality evaluation, and quality audit 7. Status of managing production equipment, measuring instruments, and vendors 8. Status of packaging, storage, transportation, sales, and service activities 9. Grasping and responding to product usage, disposal, recovery, and recycling 10. Status of quality assurance 11. Grasping of the status of customer satisfaction 12. Status of assuring reliability, safety, product liability, and environmental protection
Maintenance/control activities	1. Rotation of management (PDCA) cycle 2. Methods for determining control items and their levels 3. In-control situations (status of utilizing control charts and other tools) 4. Status of taking temporary and permanent measures 5. Status of operating management systems for cost, quantity, delivery, etc. 6. Relationship of quality assurance system to other operating management systems
Improvement activities	1. Methods for selecting themes (important problems and priority issues) 2. Linkage of analytical methods and intrinsic technology 3. Status of utilizing statistical methods for analysis 4. Utilization of analysis results 5. Status of confirming improvement results and transferring them to maintenance/control activities 6. Contribution of quality circle activities
Effects	1. Tangible effects (such as quality, delivery, cost, profit, safety, and environment) 2. Intangible effects 3. Methods for measuring and grasping effects 4. Customer satisfaction and employee satisfaction 5. Influence on associated companies 6. Influence on local and international communities
Future plans	1. Status of grasping current situations 2. Future plans for improving problems 3. Projection of changes in social environment and customer requirements and future plans based on these projected changes 4. Relationships among management philosophy, vision, and long-term plans 5. Continuity of quality control activities 6. Concreteness of future plans

TABLE 5.26 The Deming Application Prize Checklist for Senior Executives

Understanding	1. Are the objectives of quality control introduction and promotion and enthusiasm clearly understood? 2. How well do they understand quality control, quality assurance, reliability, product liability, etc.? 3. How well do they understand the importance of the statistical way of thinking and the application of quality control techniques? 4. How well do they understand quality circle activities? 5. How well do they understand the relationship of quality control and the concepts and methods of other management activities? 6. How enthusiastic are they in promoting quality control? How well are they exercising leadership? 7. How well do they understand the status and the characteristics of their company's quality and quality control?
Policies	1. How are quality and quality control policies established? Where and how do these policies stand in relation to overall management? 2. How are these policies related to short- and long-term plans? 3. How are these policies deployed throughout the company? 4. How do they grasp the status of policy achievement? Are they taking appropriate corrective actions when needed? 5. How do they grasp priority quality issues (priority business issues)? Do they make effective use of diagnostic methods such as top management diagnosis? 6. How well are targets and priority measures aligned with policies? 7. What kind of policies do they employ for establishing cooperative relationships with associated companies?
Organization and human resources	1. How is the company organized and managed to effectively and efficiently practice quality control? 2. How are the authorities and responsibilities established? 3. Is the allocation of human resources suitable? 4. How do they strive to make employees happy and satisfied? 5. How do they grasp and evaluate employees' capability and motivation? 6. How do they strive for interdepartmental cooperation? How do they utilize committees and project teams? 7. How do they relate to associated companies?
Human resources development	1. How clear is the philosophy for hiring, developing, and utilizing human resources? 2. How appropriate are the employee education and training plans? Are the necessary budget and time allocated? 3. How do they communicate the policies for quality control education and training, and how do they grasp the status achieving their policies?

SOURCE: Reprinted from Ref. 35 with permission.

TABLE 5.26 The Deming Application Prize Checklist for Senior Executives (*Continued*)

	4. How do they provide education and training specific to the company's business needs? 5. How well do they understand the importance of employee self-development and mutual development? How do they support this effort? 6. How do they strive to develop quality circle activities? 7. How interested are they in developing human resources in associated companies?
Implementation	1. What kind of measures do they have for the effective and efficient implementation of quality control? 2. How good is the overall coordination of quality control and other management systems? 3. How do they grasp the status of improvement in the business processes and the individual steps of these processes so as to provide products and services that satisfy the customer needs? Are they taking necessary corrective actions? 4. How well are the systems for developing new products and services, new technologies, and new markets established and managed? 5. How well are the necessary resources secured and allocated for establishing and operating management and information systems? 6. How do they grasp the effects and contributions of quality control to the improvement of business performance? 7. How do they evaluate their employees' efforts?
Corporate social responsibilities	1. Is the company structured to ensure appropriate profit for a long time? 2. How well do they regard employee well-being (wage levels, working hours, etc.)? 3. How well do they regard employee self-realization? 4. How well do they strive for coexistence and coprosperity with associated companies? 5. How well does the company contribute to the local community? 6. How well does the company exert efforts to protect the environment? 7. How well does the company positively impact the international community?
Future vision and future plans	1. How do they assure the continuity of quality control? 2. How do they anticipate and cope with changes in surrounding business environment and progress in science and technology? 3. How do they grasp and cope with changes in customer requirements? 4. How do they consider their employees and help them achieve happiness and satisfaction? 5. How do they consider and manage relationships with associated companies? 6. How do they plan for the future to cope with the items above? 7. How do they utilize quality control to achieve the future plans?

5. *Thorough implementation of management plans/business plans.* Policy management, one of the TQM methods, has been widely used to efficiently achieve management plans/business plans that center around profit plans.
6. *Realization of top management's dream.* By establishing the high goal of winning the Deming Prize and by promoting TQM with the examination date in mind, many companies are able to achieve at once many things they had wanted to accomplish for years.
7. *QC by total participation and improvement of the organizational constitution.* During the process of the Deming Prize examination, companies have noted that sectionalism breaks down, interdepartmental communication improves, barriers between departments are removed, and a sense of unity is instilled.
8. *Heightened motivation to manage and improve as well as to promote standardization.* Employees improve understanding of scientific management and ability to understand the work situation and thus are more capable of coming up with improvement ideas.
9. *Converged large power from the bottom of the organization and enhanced morale.* First-line supervisors and employees acquire simple managerial and analytical skills, become more interested in their own work, and acquire the ability to autonomously manage and improve their day-to-day tasks in a logical manner.
10. *Establishment of various management systems and the total management system.* Quality assurance and other management systems for new product development, profit, cost, and vendors are established along with many other cross-functional management systems. All these different systems then become connected through the total management system.

5.9 Conclusion: Nothing Left but to Do It

This chapter first described Deming's theory of management. Flowing from the theory are 16 specific operating principles for management. The theory and the principles were then translated into the seven action steps for high-performance electronics industry managers. Common management practices were reviewed in the light of Deming's theory with suggestions for alternative actions. More about leadership and other key concepts for success were outlined before the chapter closes with the call to action—there is nothing left but to do it.

We have covered:

- Deming's theory of management
- Operating principles for management
- Translating the theory into action
- A review of common management practices
- More about leadership
- Other key concepts for success

There is nothing left but to do it. And, by applying plan–do–study–act, continually improving how you do it.

References

1. Reprinted from, *The New Economics for Industry, Government, Education,* by permission of Massachusetts Institute of Technology, and the W. Edwards Deming Institute. Published by MIT, Center for Advanced Educational Services, Cambridge, MA 02139. Copyright 1986 by The W. Edwards Deming Institute.
2. Garvin, David A., *Managing Quality: The Strategic and Competitive Edge,* The Free Press, New York, 1988, pp. 180–181.
3. Reprinted from *Out of the Crisis,* by W. Edwards Deming by permission of MIT and The W. Edwards Deming Institute. Published by MIT, Center for Advanced Educational Services, Cambridge, MA 02139. Copyright 1986 by The W. Edwards Deming Institute.
4. Statements by Dr. Deming on numerous occasions to his graduate management classes at New York University.
5. Deming, 1993, p. 53.
6. Deming, 1993, back cover.
7. Dr. Deming called this his second theorem during presentations of his four-day seminar, "Quality, Productivity, and Competitive Position."
8. Deming, 1993, p. 96.
9. Deming, 1993, p. 35.
10. Deming, 1993, p. 50.
11. Senge, Peter M., *The Fifth Discipline,* Doubleday, New York, 1990, p. 63.
12. Dr. Deming so stated on numerous occasions to his graduate management classes at New York University.
13. Deming, 1993, p. 33.
14. Ackoff, Russell L., *The Democratic Corporation,* Oxford University Press, New York, 1994, pp. 206–207.
15. Deming, 1993, p. 53.
16. Shewhart, Walter A., *Statistical Method from the Viewpoint of Quality Control,* Dover Publications, Mineola, NY, 1986, p. 23.
17. Deming, 1993, p. 212.
18. Deming, 1993, p. 182.
19. Drucker, Peter, *Fortune Magazine,* December 20, 1996, p. 56.
20. Deming, 1986, p. 85.
21. Deming, 1986, p. 59.
22. Deming, 1993, p. 94.
23. Deming, 1986, pp. 23, 24.
24. Reichheld, Frederick F., *The Loyalty Effect,* Harvard Business School Press, Boston, 1996, p. *vii.*
25. Deming, 1993, p. 35.

26. Statements by Dr. Deming on numerous occasions to his graduate management classes at New York University.
27. Deming, 1993, p. 22.
28. "The Deming of America," video, Petty Consulting Productions, Cincinnati, 1991.
29. Deming, 1993, p. 17.
30. Deming, 1993, p. 33.
31. Deming, 1993, back cover.
32. Deming, 1993, p. 119.
33. Deming, 1993, p. 119.
34. Deming, 1993, pp. 119, 120.
35. The Deming Prize Committee, *The Deming Prize Guide for Overseas Companies,* Union of Japanese Scientists and Engineers, Tokyo, 1996.

Chapter

6

Total Quality Management

Larry Coleman
Director of Total Quality Systems (TQS)
The Boeing Company

6.1 Definition of Total Quality Management

An interesting thing about Total Quality Management (TQM) is that there are as many definitions as there are definers. Sashkin and Kiser write "TQM means that the organization's culture is defined by and supports the constant attainment of customer satisfaction through an integrated system of tools, techniques, and training. This involves the continuous improvement of organizational processes, resulting in high quality products and services." [1] The official Department of Defense definition is: "Total Quality Management is both a philosophy and a set of guiding principles that represent the foundation of a continuously improving organization. TQM is the application of quantitative methods and human resources to improve the material and services supplied to an organization, all the processes within an organization, and the degree to which the needs of the customer are met, now and in the future." [2] On the other hand, Tom Peters writes that "Those who put all new management ideas under the TQM banner are making a big mistake. TQM is not about life writ large, it's about products that work unerringly." [3]

We can glean some common threads from the above definitions. TQM is a system; it focuses on customers; it utilizes a continuous improvement philosophy; it applies quantitative methods. The bottom line is products that work unerringly. This is the heart of electronics—quality and reliability. This is why an electronics company must practice TQM principles in order to be successful in today's global marketplace. Embracing TQM as a concept is not difficult. Implementing TQM as a practice is another matter. It requires leadership from the top, training throughout, and the never-ending application of continuous process improvement methods and tools in all facets of the enterprise. Continuous improvement implies change and that change will engage our entire work force. In deference to Mr. Peters, the issue is not about "life writ large" but how management ideas are appropriately integrated into a TQM system. The key is the Total Quality Management system architecture. There is probably no single TQM definition because organizations require within their systems the elements that complement their particular culture. There is no "one size fits all." There is, however, a general model that can indicate how the various elements of a Total Quality Management system might be integrated. Within that generalized model we can place the tools and techniques that best fit a particular organization. However, before we introduce a model, let's take a look at the genesis of this management philosophy we call TQM.

6.2 Background

The United States of America has developed so much of the world's industrial technologies, processes, and tools and yet has failed to retain many of the resulting markets. Why is this? The answer is not simple since there are multiple factors involved including economics, cultures, attitudes, politics, and competition. In the book *Made in America* [4], a study by a team of MIT scientists, engineers, and economists, it was concluded that six interrelated patterns of behavior best characterize the evidence:

- Outdated strategies
- Short-term horizons
- Technological weaknesses in development and production
- Neglect of human resources
- Government and industry at cross-purposes

Regarding technological weakness, the book goes on to delineate that "American companies evidently find it difficult to design simple, reliable, mass-producible products; they often fail to pay enough attention at the design stage to the likely quality of the manufactured product; their product development times are excessively long; they pay insufficient attention to manufacturing processes; they take a reactive rather than a preventive approach to problem solving; and they tend to under exploit the potential of continuous improvement in product and processes. [5]

There have been numerous books and articles written on improving quality and productivity. As the search for *the* solution continues, we have created an entire vocabulary of acronyms which identify the tools and methods for the quick fix to industrial ills. We can control our processes with SPC, plan production using MRP, maintain low inventories with JIT, develop new products with IPTs, understand our customers' needs with QFD, improve productivity with SMTs, reduce variability with DOEs, and the list goes on.

In the late 1980s was born the acronym TQM (Total Quality Management). TQM was an attempt on the part of the Department of Defense to recapture the United States' eroding industrial base by addressing what was perceived to be the root cause—the quality of its products and services. Although TQM was a new term, the philosophy and principles behind it were not. Its origins go back to the early 1950s during the postwar restoration of Japan and its industries. The State Department sent to Japan a physicist and statistician by the name of Dr. W. Edwards Deming to develop sampling techniques for surveys of housing, nutrition, employment, agriculture, and fisheries. Dr. Deming's background had included working at Western Electric's legendary Hawthorne plant in Chicago circa 1924, where the management methods there would help shape his future thinking regarding his management philosophy. In 1927 Dr. Deming accepted a position with the U.S. Department of Agriculture. While at the department of agriculture, he was introduced to Walter A. Shewhart, who was a statistician at Bell Telephone Laboratories in New York. Shewhart had been working on a technique to bring industrial processes into what he called *statistical control.* Dr. Deming recognized the value of Shewhart's theory and continued to work with him for several years. Shewhart's work further influenced the basis for Deming's management philosophy.

The statistical methods of Shewhart, refined by Deming were, in the meantime, being recognized by the U.S. government. Subsequently, Dr. Deming was asked by the Census Bureau to apply these methods to the 1940 census where the techniques were further refined. In 1941 Dr. Deming began teaching statistical quality control methods to people engaged in wartime production and procurement, thus during the war effort there was a national emphasis on quality control. Unfortunately, as the war ended and the demand for consumer goods rose, American management soon abandoned statistical quality control as requiring too much effort. "By 1949, Dr. Deming says mournfully, 'there was nothing—not even smoke.'" [6]

6.2.1 Dr. Deming, Japan, and TQM

After the Second World War, Japan lay in virtual ruin. Its industries had been decimated and morale was at a low ebb. In 1947 Dr. Deming was recruited by the Supreme Command for the Allied Powers to help prepare for the 1951 Japanese census. During his tenure there, he made an effort to learn about the Japanese culture. He spent time attending cultural events and visiting areas of interest. He invited statisticians he had met during his studies to be his guest at various times. He in turn was invited to their social events, thus a relationship began to develop between Dr. Deming and his peers in Japan.

During this same period of time, an organization of Japanese engineers called the Union of Japanese Scientists and Engineers (JUSE) was being formed for the purpose of aiding in the reconstruction of their country. These engineers had obtained books and papers on Shewhart's work and they were captivated by his theories. They became aware that Dr. Deming had worked with Shewhart. Since several of the engineers knew Dr. Deming and his abilities, they thought he could help them gain further insight into the application of Shewhart's theories toward their recovery effort. Subsequently, Dr. Deming was invited by JUSE to deliver a lecture to their research workers, plant managers, and engineers on quality control methods.

In June, 1950, Dr. Deming delivered the first of a dozen lectures to standing-room-only crowds. Mary Walton writes: "The response was gratifying, but Dr. Deming nevertheless was troubled by his experience in the United States, where Statistical Quality Control had flourished for such a brief period. Midway through the first lecture, he would later say, he was overcome with a sense of déjà vu. He was not talking to the right people. Enthusiasm for statistical techniques would burn out in Japan as it had at home unless he could somehow reach the people in charge." [7] This reinforced another key principle in Deming's management philosophy. Unless you can obtain the commitment of the people in charge of the organization, and enlist their involvement in creating change, little will happen.

Dr. Deming did meet with Japan's chief executives (the Keidanren) and told them that if they wanted to survive as a nation, they had to produce quality goods that met their customers' needs. If they applied the quality tools they had been taught, they could do that. Dr. Deming emphasized that they needed to work with their suppliers and their processes to provide for the understood needs of their customers. In addition, they should strive to continuously improve their products and processes. This is, in fact, the essence of TQM. Japanese management took Dr. Deming very seriously and diligently applied the principles he taught. Today, one of the most prestigious awards that can be received in Japan is the Deming Prize. It is awarded to an individual for accomplishment in statistical theory and to companies for accomplishments in statistical applications. Additional information concerning Dr. Deming and his management techniques, as well as information on the Deming Prize, may be found in Chap. 5.

Certainly no one would argue that Japan has become a dominant force in the world economy. Their manufactured goods are considered to be the best in virtually every market they have addressed. As a result, they have gained a significant if not dominant market share in consumer electronics, optics, automotive, steel, ship building, and semiconductors, to name a few. In a very large part, the disciplines of Total Quality Management played a significant role in the postwar development of Japan and their stature in the global marketplace.

6.2.2 The TQM gurus

It would be incorrect to single out *only* Dr. Deming as the major influence in the TQM movement. Although Deming's fifth point, to *improve constantly and forever the system of production and services,* is the driver for continuous process improvement (CPI) and the heart of TQM, there have been many other significant contributors.

One of Dr. Deming's contemporaries was Joseph Juran. Juran also had his roots at Western Electric, along with Deming and Shewhart. Juran also spent time in Japan lecturing soon after Deming. Juran has established the Juran Institute which promotes quality systems worldwide. "Juran's greatest contribution has been in defining and teaching how to create customer-oriented organizational systems." [8]

A Japanese disciple of Deming, Kaoru Ishikawa, is another major influence on TQM. Kaoru's father, Ichiro, was the leader of the Japanese Keidanren, the executive council that had invited Deming to speak in 1951. Kaoru is credited with developing the fishbone diagram. This simple graphical diagram has been identified as one of the seven quality tools and is used extensively in problem solving. It is also referred to as the Ishikawa diagram. He has written two significant publications [9] that have promulgated the use of quality tools and techniques in Japan and the United States.

A third significant contributor to the TQM movement is Armand V. Feigenbaum. Feigenbaum is credited with coining the term *total quality control.* His book by the same title [10] was an early classic in the concept of a systems approach to quality management.

We should also mention Phil Crosby, who wrote the classic book *Quality Is Free* which began turning the heads of corporate leadership in recognizing the cost benefits of sound preventive practices, and William Conway, one of America's early Deming disciples who emphasizes productivity improvement through waste reduction using Deming's principles.

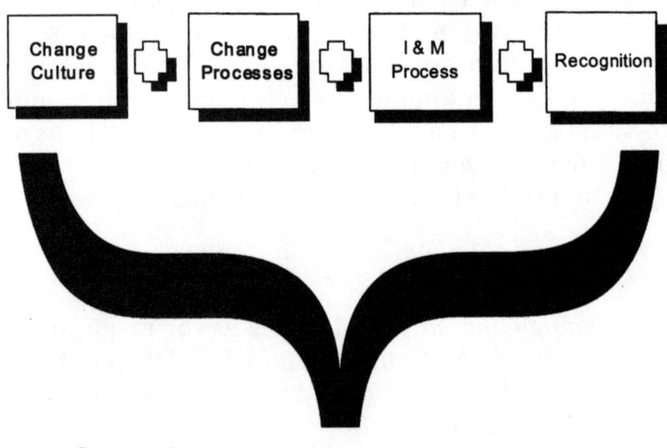

Figure 6.1 An effective TQM system.

6.3 TQM Model

The model shown in Fig. 6.1 indicates four key elements of a TQM system:

- Culture change
- Change processes
- Implementation and measurement processes
- Recognition processes

The most difficult element to deal with in the concept of a systems model is the cultural element. The other three elements are processes and their related tools. Culture, on the other hand, is fundamentally the values and behaviors of the organization. This doesn't seem to fit neatly into a TQM system model concept, and yet it is the most critical element of TQM. We will deal with some specific attributes of culture when we cover that element. We will then pursue it further when we take a look at change processes. Implementation and measurement are the processes we use to assure that the changes in our organizational systems are yielding favorable results. Recognition processes are those methods used to reinforce the organizational behavior that will result in the culture being sought.

All of the elements of the model shown in Fig. 6.1 are necessary for an effective TQM system. As new tools and methods are developed or enhanced that support TQM, they can easily be integrated into the elements of the model without the risk of being misunderstood as initiatives outside of the system. One of the advantages of having an integrated systems approach to quality is that you avoid the "program of the month" syndrome that so often frustrates attempts at incorporating TQM into organizations. On the other hand, back to Tom Peters' statement, we could end up with life writ too large. In other words, TQM is not the *organizational system,* but a part of it. Elements such as strategy and structure are probably best dealt with outside the TQM system but on the other hand, must be considered when dealing with change, which is at the heart of TQM. When we address change processes we will consider the impact of change to the organization and make a strong argument that the impact on the organization as a whole must be integral to any alteration within the organizational elements.

6.3.1 Culture

Undoubtedly the most complex element of the TQM system is changing the culture. To begin with, what do we mean by culture? If we take a pure dictionary approach to the definition, it would read "the totality of socially transmitted behavior patterns, arts, beliefs, institutions, and all other products of human work and thought." [11] Culture is what we are. It is manifested in part by behavior, i.e., what we do, and by products, i.e., what we produce. Dealing with culture change in organizations is no easy task. Michael Porter stated that "Change is an unnatural act, particularly in successful companies; powerful forces are at work to avoid it at all costs." [12] It is generally thought that change will occur only as the result of a significant emotional event. This is often true of our personal lives. We do not change lifestyles, for example, unless we are faced with a serious medical or family problem and if we choose not to change, the consequences are often tragic. Unfortunately, it is often too late to change when the change is driven by such an event. At best, organizations are required to spend inordinate amounts of resources in order to effect the needed change in a crisis. It would be best to incorporate cultural changes in the organization as part of an ongoing strategy to implement the organization's vision. Now we have introduced two additional factors, vision and strategy. Suffice it to say, we are not going to delve into these topics other than to state that for TQM to be successful in an organization, it must be an integral part of the vision and strategy of the organization.

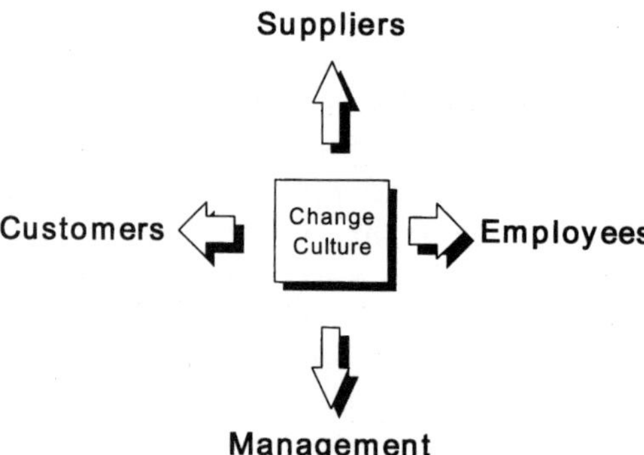

Figure 6.2 Cultural aspects.

The primary cultural aspects of TQM are shown in Fig. 6.2. They include:

- How we view our customers
- How we view our suppliers
- How we manage
- How we view our employees

Suppliers and customers. It should be clearly evident to anyone involved in any business that two of the key factors in the success of that business are the front end and the back end—the external suppliers and customers respectively.

Let us first deal with external customers. First and foremost, TQM organizations are focused on meeting the needs and expectations of their customers. They are concerned not only with whether their product or service meet specifications but also whether those specifications meet their customers' needs. They do not *rely on* slogans or gimmicks. One of Deming's management principles is to eliminate slogans, exhortations, and targets for the work force. The same thought can be applied to customers. Producers can have Madison Avenue place all the gloss possible on their products and services, but if they don't meet customer needs, the customer is not going to continue to buy. Quality organizations go to great lengths to understand their customers by obtaining feedback, and acting on that information. TQM organizations are bringing their customers into the new product development very early in that process. During the design of the Boeing 777, Boeing's key customers took part in defining the product by making inputs regarding their needs in an aircraft. Many organizations are relying on tools such as the House of Quality [13] to understand key product parameters, rank their importance to the customer, evaluate the relationships between parameters, and deploy the requirements into the manufacturing processes and ultimately into the products. Others use surveys, marketing research, and focus groups. The bottom line is that quality organizations listen to the voice of the customer and act on what that voice is telling them.

In the TQM philosophy, the concept of customers and suppliers also extends to within the organization. When we view a generalized process flow (Fig. 6.3), we can consider the inputs to each step as being from a supplier and the output to a customer. In other words, whomever we provide our product or service to is our customer and whoever supplies us a product or service is our supplier. This creates an interesting dilemma when we look at customer/supplier relationships both externally and internally, namely you are always both. As a matter of fact, two parties, teams, departments, or organizations are both to each other. For example, party A is a "supplier" to party B; when party B sends party A a document specifying the goods or services they want to procure, is not party A the customer, having received an input from party B? This relational dilemma has brought a whole new perspective to organizations operating in a TQM environment and has bred a new term, *supplier partnering*. Regardless of the contractual relationship between two parties, the quality view is that customers and suppliers must work in a collaborative manner to assure that the needs and expectations of both are met.

Management. Management in the organization makes or breaks the change to a TQM culture. Unequivocally, without the commitment and involvement of the top leader, TQM efforts will, at best, be suboptimal. The reason for this is quite simple; someone needs to lead the effort. There is a natural tendency to resist change, and, unless there

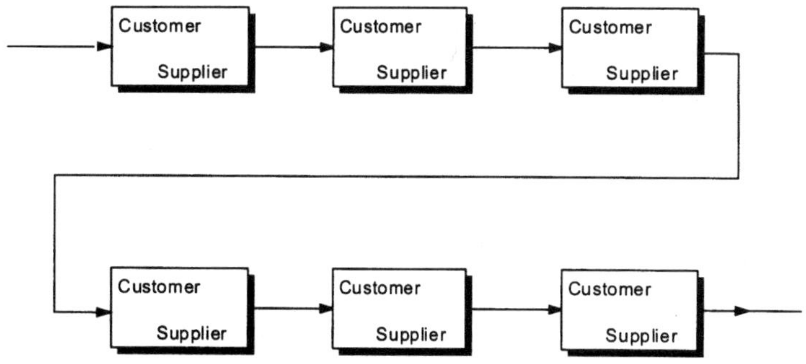

Figure 6.3 A different concept of organization.

is a leader with a vision and a passion for assuring that change, it simply cannot sustain itself. Creating and sustaining a total quality culture requires the efforts of the entire management team. We will discuss how that is accomplished when we address change processes. *What* needs to change is the topic of this section. There are four significant principles that managers must understand:

- Everyone in the organization is responsible for quality
- Management is responsible for the "system" that enables the organization to perform
- There is a significant difference between managing and leading—and we need leaders
- People are the organization's most valuable resources

Everyone responsible for quality. In a TQM organization, a primary tenet is that each individual in the organization is responsible for the quality of his or her output. This is simply restating that the needs of the customer are paramount. We have done a major disservice to ourselves in organizations by not placing the responsibility and accountability for quality onto each individual. We have developed an inspection mentality. We have created huge infrastructures for the sole purpose of inspecting. As a result, we do interesting things like having numerous signature "approvals." In a real sense, approvals are a form of inspection. The irony is the more approvals, the more likely there will be errors because no one is responsible. In fact, everyone will assume the previous approver inspected adequately, or the next one will. As a result, cycle time and productivity suffer and the product is likely to be of poorer quality than if no one had approved it. The originator must be held accountable for the quality.

The administrative assistant should not rely on the boss to find the typos, the drafter should not rely on the checker to find errors, the assembler should not rely on the inspector to detect faulty workmanship, the engineer should not rely on the test system to make up for inadequate design margins. A fundamental responsibility of management is to create an environment where the producers of goods and services are responsible for the quality of their output.

Management and the system. Management is responsible for providing the environment that will assure quality products and services. There is a saying often quoted, "How can I soar like an eagle when I work with a bunch of turkeys?" This could be restated, "How can I produce a quality output within an unquality system?" Dr. Juran and Dr. Deming have maintained that at least 85 percent of problems are due to causes in the system (a composite of all the organization's processes) and fewer than

15 percent are caused by (under the control of) the workers. Regardless of the ratio, the system must sustain a TQM environment. The system is owned by, and the responsibility of, management. The implication is quite clear; if we want real improvement, we need to work on the system, and the onus is on management. If we blame the employees rather than the system, we violate another of Deming's management principles, "Drive out fear." Employees learn quickly not to surface problems if the result is to shoot the messenger. One of the best ways to improve the system is to train and empower employees to make changes to the system's processes and subsequently take ownership of those processes.

Management versus leadership. Another cultural shift in organizations is breaking out of the plan-and-control paradigm and beginning to developing leadership characteristics in managers. Simply stated, management needs to lead the cultural change. Table 6.1 shows John P. Kotter's summary of the processes of leadership and management. [14]

As can be seen from the definitions in Table 6.1, management processes aren't bad. They produce order in an organization. In TQM organizations, we need more than order. Kotter writes:

TABLE 6.1 Management and Leadership

Management	Leadership
1. *Planning and budgeting*—setting targets or goals for the future, typically for the next month or year; establishing detailed steps for achieving those targets, steps that might include timetables and guidelines; and then allocating the resources to accomplish those plans.	1. *Establishing direction*—developing a vision of the future, often the distant future, along with strategies for producing the changes needed to achieve that vision.
2. *Organizing and staffing*—establishing an organizational structure and set of jobs for accomplishing plan requirements, staffing those jobs with qualified individuals, communicating the plan to those people, delegating responsibility for carrying out the plan, and establishing systems to monitor implementation.	2. *Aligning people*—communicating the direction to those whose cooperation may be needed so as to create coalitions that understand the vision and that are committed to its achievement.
3. *Controlling and problem solving*—monitoring results versus plan in some detail both formally and informally, by means of reports, meetings, etc.; identifying deviations, which are usually called *problems*; and then planning and organizing to solve the problems.	3. *Motivating and inspiring*—keeping people moving in the right direction despite major political, bureaucratic, and resource barriers to change by appealing to very basic, but often untapped, human needs, values, and emotions.

SOURCE: Reprinted from Ref. 14 with permission.

> Leadership is very different. It does not produce consistency and order, as the word itself implies; it produces movement....This does not mean that management is never associated with change; in tandem with effective leadership, it can help produce a more orderly change process. Nor does this mean that leadership is never associated with order; to the contrary. In tandem with effective management, an effective leadership process can help produce the changes necessary to bring a chaotic situation under control. But leadership by itself never keeps an operation on time and on budget year after year. And management by itself never creates significant useful change. [15]

The organization emulates the values and beliefs of the management. Managers are often the role models. Many an employee has been heard saying, "If I were running this place, I'd do it differently." Yet when employees have the opportunity to "run the place" they do it the same as the boss did. Why? Because it worked for the boss, who set the norm and established the culture. Sashkin and Kiser write: "It is employees' direct experience of the patterns of management behavior and action that defines the values and beliefs that make up the culture...." [16] Note that the words are *behavior* and *action*. Employees do what management does. They generally do not respond to exhortations or slogans, they respond to action. We are reminded of another of Deming's principles: Eliminate slogans, exhortations, and targets for the work force.

So what we mean by TQM leadership is "Follow me!" not "Do it, I'm behind you all the way." It is asking "What can I do to help, what can I do to keep us moving in the right direction?"

TQM and employees. There is no question that management establishes the culture of the organization. When it comes to maintaining the current culture or implementing the changes necessary to establish a new culture, the employees make it happen. Management who have the vision and management can lead, but leaders need followers. Management does not implement; employees implement. Without employees' ongoing involvement in continually improving processes, products, and services, there is no TQM.

How can TQM organizations best enlist the commitment and involvement of employees? Not by slogans and exhortations, not by fear, not by control, but by involving them. Notwithstanding the need for leadership, employee involvement may be the single most important element in the implementation of a TQM culture. Edward Lawler, writes " If major change requires employee acceptance for the change to be implemented, an involvement oriented approach to management probably is superior both in terms of speed with which the change can be implemented and the quality of implementation." [17]

Employee involvement will often be in the form of teams. These teams can be ad hoc problem-solving or process improvement teams, moderate-term project teams such as integrated product development teams, or long-term product teams such as manufacturing, accounting, and procurement. The type of employee involvement depends on the type, interdependency, and complexity of work. In some cases, individual work may be appropriate but in either case, involvement is a key ingredient. [18] Changing to a TQM culture often means a change in the organization's structure. Hierarchical and bureaucratic organizations are not conducive to high employee involvement. High involvement requires the authority to act. It requires the empowerment of employees.

The word *empowerment* is the subject of countless books and articles and, like any term that becomes a "buzzword," it is misunderstood. In its most basic form it simply means to invest with power. In an organization this means to invest with authority and responsibility. Power is the ability or capacity to perform or act effectively. [19] In traditional bureaucratic organizations, power is held toward the top of the organizational pyramid. At the bottom of the organization is the knowledge of the day-to-day operation, as shown in Fig. 6.4. Operational knowledge is with the engineers, assemblers, test technicians, sales people, accountants, and buyers—the people who make it happen. The people doing the day-to-day work know more about that work than any level of management. The paradox is that the further

Empowerment Means
Shared Power

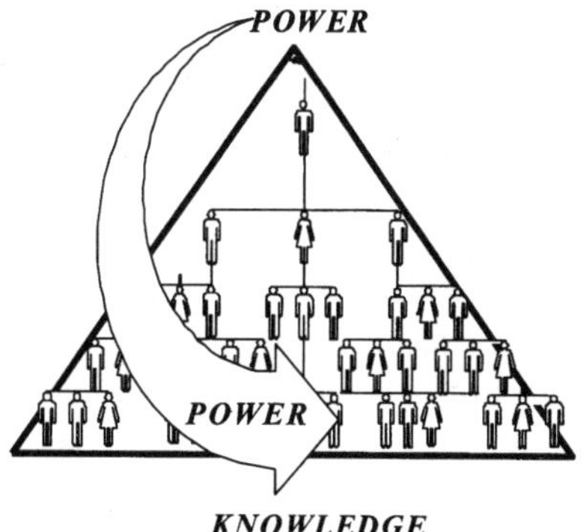

Figure 6.4 Empowerment.

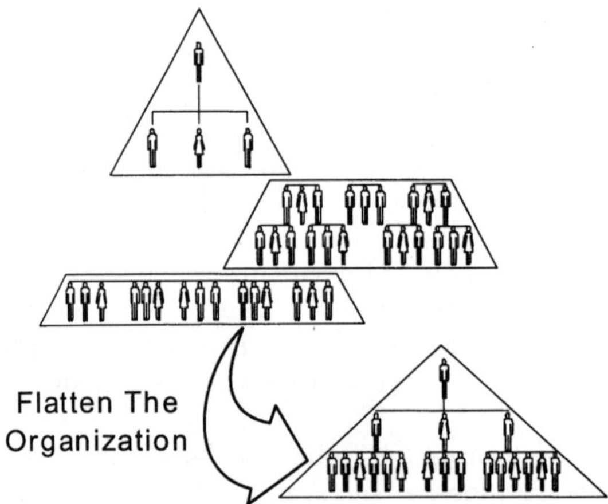

Figure 6.5 One option: eliminate levels of management.

toward the top, the more the capacity to act, but the action is at the bottom. What is needed is to get the power closer to where the work is. There have been some pretty radical things done to try to move the power and knowledge closer together. One common ploy is to rapidly restructure organizations under the umbrella of TQM. It is announced that "We will become a team-based organization. Beginning next week all our work groups will be called teams and our supervisors will be called facilitators." Another "quick fix" is to merely eliminate levels of management (Fig. 6.5). This gets the power closer to the work. Unfortunately, if not properly planned and orchestrated, it can wreak havoc on an organization. One of the primary tenets of organizational redesign is to involve the employees through a systematic process, for example, the sociotechnical systems approach [20] or the business process reengineering approach. [21]

The structure shown in Fig. 6.5, albeit leaner in terms of levels of management, nonetheless remains a pyramidal, hierarchical structure. Since empowerment is a critical factor in TQM organizations, we need to break out of the bureaucratic structure paradigm. One way to do that is quit looking at organization as pyramids. Let's look at the organization from a product team perspective:

- Product teams have front-to-back responsibility for their product. A product in this sense is whatever they produce—hardware, software, specifications, services, or concepts. The team is responsible and accountable for a product that meets their customer's needs and expectations. The customer is the internal or external entity that receives the product.

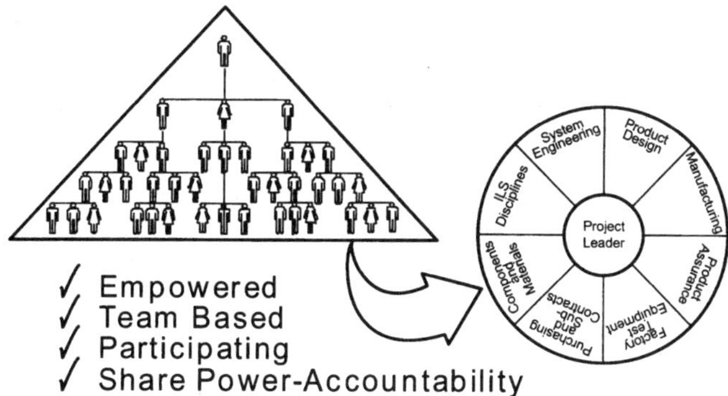

Figure 6.6 A different concept of organization.

- Teams require the latitude to use their creativity and innovation. In most organizations, teams do not have carte blanche. The team is provided boundaries from management placing limitations on their empowerment. The boundaries, however, must be established so they do not overly constrain the team. Boundaries are like playing areas in athletic events. Football, for example, would not be a viable competitive sport if the field were 20 yd long and 10 yd wide.
- Teams are multidisciplined since the team is responsible for the product front to back. In the production of an electronic assembly, for example, there would be purchasing, material control, assembly, and testing to name a few.
- Responsibilities on teams are shared, the team is responsible for meeting their customer needs.
- Team leadership is a facilitating role rather than a direct control role.

When we look at all the above characteristics of a team, the symbol of a circle as shown in Fig. 6.6 may well represent how we should view the new organizational paradigm.

Depending on the team's product, the circle can be divided into appropriate segments. Figure 6.7 depicts an electronics assembly team and an integrated product development team. This view of a TQM organization helps eliminate the stereotype of the direct and control structure and promotes the concept of a collaborative structure with shared responsibility. The ideal evolution of teams in TQM organizations would be to self-management. There is no universally agreed upon definition for *self-managing.* In some organizations, self-managing means that a team runs its enterprise like a minibusiness and is empowered accordingly. In a general sense, self-managing

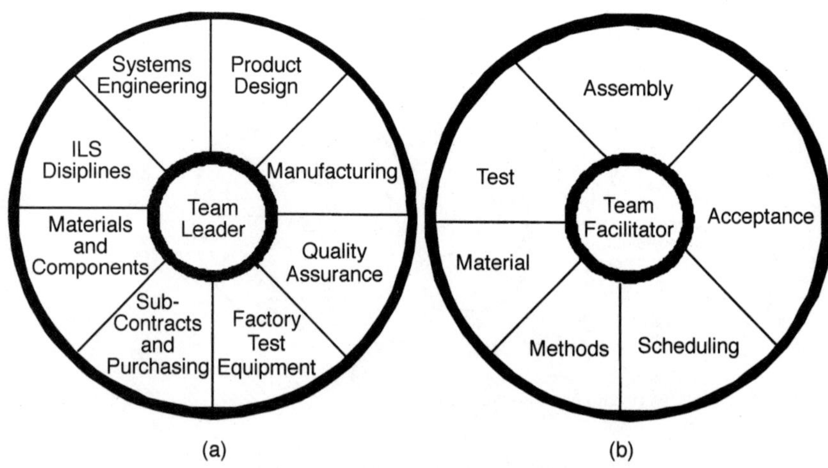

Figure 6.7 Team models. (a) Integrated product development team, (b) electronics assembly team.

means taking on many of the responsibilities that have been traditionally management's. As a minimum, a self-managing team should be involved in planning, goal setting, data gathering and analysis, problem solving, team administration such as meeting administration, overtime, vacations, and budgets. Members of self-managing teams become business owners and stakeholders.

6.3.2 Change processes

When an organization is successful at providing the leadership to begin inculcating a TQM culture, the next step is to begin the process of organizational change. We will address two change processes that should be considered when implementing a TQM system. The first is a strategic organizational change process and the second is a tactical change process.

The strategic change process must be led by senior management. It is at the heart of creating an organization that is aligned with a TQM strategy and requires decision making at the highest levels of the organization. The process must consider all elements of the organizational system. One such model, shown in Fig. 6.8, based on the McKinsey 7-S Model [22], has been used successfully. Regardless of the model that is used, a structured and methodical approach to change is necessary. The organization must be viewed as a complex system requiring well-thought-out change strategies.

The organizational system model indicates that organizations consist of eight key elements. When change is made to one element of the organization, the impact of that change must be considered relative to the

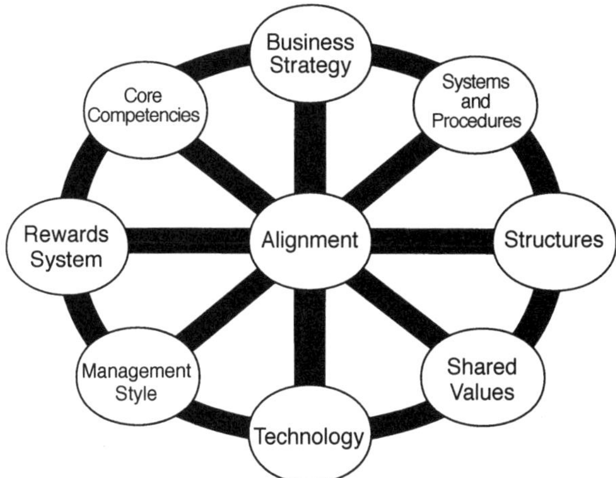

Figure 6.8 Organizational system model.

other elements. In other words, the organization must be looked at holistically. For example, if the strategy is to create a team-based TQM organization, it is imperative that the reward systems, policies and procedures, management style, values, structure, etc., be in alignment. Too often, companies work only on structure (i.e., creating teams) and do not consider that increased employee involvement requires such things as revised compensation systems. If teams are to self-manage, the management style must be enabling, using coaching and counseling versus directing and controlling. Policies and procedures in bureaucratic organizations often create operational boundaries that unduly constrain employees. They are unable to innovate and create better ways of performing their task and servicing their customers.

The process for using the organizational system model in creating strategic change is shown in Fig. 6.9. The leaders of the organization must develop an overall strategy. Based on this strategic plan, future states for each of the system elements are developed. Priorities are established for the changes desired. Teams are formed to address the priorities. Team membership is obtained from a diagonal slice of the organization. These teams characterize the current state of the organization. The teams then identify the changes needed in the organization to bridge the gap between the current state and the future state. Included in this step is a review of all the elements of the model to assure alignment. Once the required changes are identified, an action plan is developed. This plan requires a significant length of time to implement, as well as a serious commitment of organizational resources. This process then becomes a routine part of the organiza-

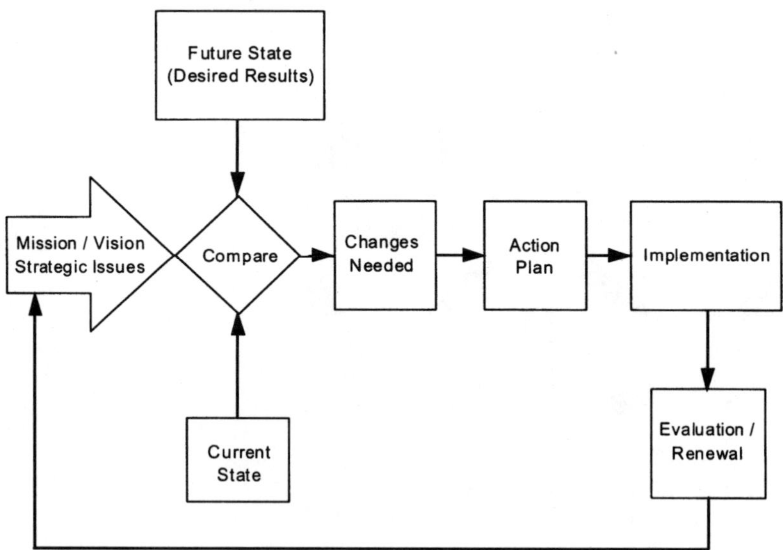

Figure 6.9 Strategic change process.

tion's strategic planning process, always under evaluation and renewal, thus creating an organization that is, at least, considering the possibility of continually reinventing itself—a learning organization.

The second change process a TQM organization requires is a method for improving and optimizing the subprocesses of the organization. We will refer to it as the tactical change process. There are innumerable models available to aid organizations in their quest for continuous process improvement. Contained within these models are the requirements to learn and apply a plethora of analytical and problem-solving tools. Statistical process control (SPC) is discussed in Chap. 3. Others will be addressed in more detail here to provide the reader a context for their application. The model for CPI introduced here is a relatively simple model, as shown in Fig. 6.10. It is based on one used by Hewlett-Packard.

Before we embark on process improvement, let's develop a simple definition of a process. Like all elements of organizational systems, we can add degrees of complexity ad nauseam. For our purposes, Fig. 6.11 provides a basic process model that depicts the essential elements of a process. The rectangle labeled *process* can be as simple as a single activity or as complex as an entire business process. Every process has an input and an output. For the purposes of CPI, we will generalize the input as being provided by a supplier and the output as going to a customer. Suppliers and customers can be either internal, external, or both.

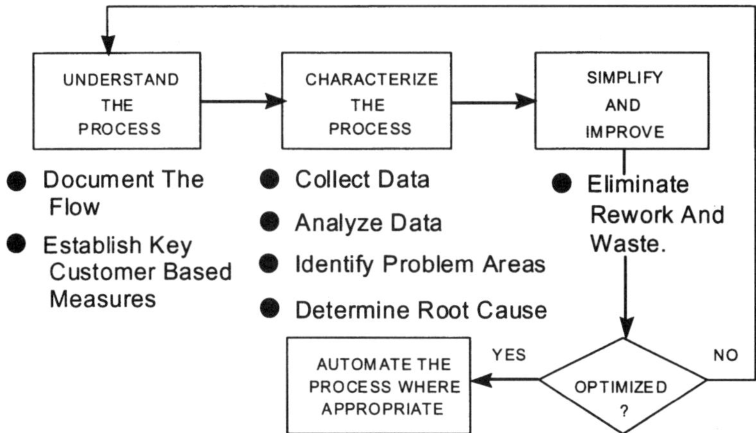

Figure 6.10 Continuous process improvement model.

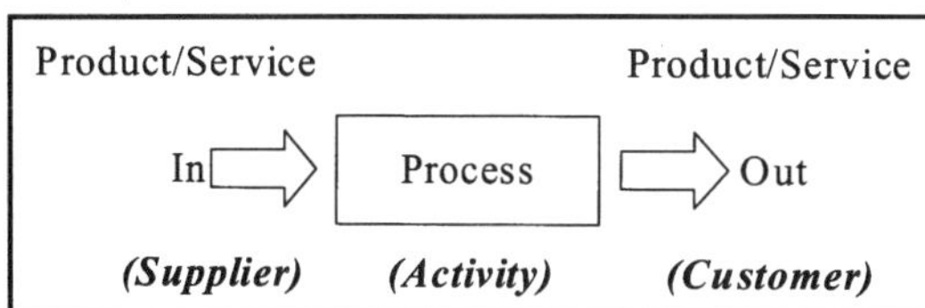

Figure 6.11 Basic process model.

CPI activities are very effectively accomplished by employee teams. For the sake of brevity we will refer to these teams as PATs (process action teams). The PAT should include representatives from the supplier, the customer, and the process owner. The PAT's membership should be 5 to 6 folks maximum if possible. One of the principles of team problem solving is synergism—the outcome of the combination of the team members being more effective than the sum of the individuals. In order to obtain synergy, it is important that team members be able to interact openly and freely. Managing group interaction in large teams is difficult at best. Another important aspect of teams is the decision-making process. Ideally consensus decisions are sought, and again, consensus is difficult in a large group. Table 6.2, from Peter Scholte's *Team Handbook,* [23] provides a few pointers on consensus decisions.

There are several preliminary activities that take place before a PAT jumps into process improvement. The most obvious is to identify the process that requires improvement—"the problem." There are three general ways a PAT discovers a process improvement opportunity:

TABLE 6.2 Consensus

Consensus is

- Finding a proposal acceptable enough that all members can support it; no member opposes it.

Is not

- A unanimous vote—a consensus may not represent everyone's first priorities.
- A majority vote—in a majority vote, only the majority get something they are happy with; people in the minority may get something they don't want at all.
- Everyone totally satisfied.

Requires

- Time.
- Active participation by all group members.
- Skill in communication; listening, conflict resolution, discussion facilitation.
- Creative thinking and open-mindedness.

SOURCE: Reprinted from Ref. 23 with permission.

1. Management requests it.
2. Data collection and analysis indicate it.
3. Brainstorming ideas generate it, i.e., "how to make things work better around here."

Once the problem is identified, a team leader is selected. This is often the process owner. Team membership is then identified. The team should have as its initial task setting ground rules for themselves. Ground rules establish the behavioral norms for the team. These include such things as decision making, conflict management, participation/communication, etc. The team then develops a mission. The mission includes such things as the task, the deliverables, the scope of the effort, and meeting time and frequency. In addition, the team should develop a time-phased activity chart (Gantt chart). Once this is accomplished, the team is prepared to embark on the process improvement task. Figure 6.12 *a, b,* and *c* is a detailed view of the continuous process improvement model. It shows each step broken down into purpose, key activity, and tool/technique. These tools and techniques can easily be adapted by an electronics company to aid in continuous process improvement. The process shown has been used in an electronics environment with a high degree of success. The figure can easily be adapted to the methodologies that have been found to be successful for your particular business. What the reader needs to bear in mind is that there are no silver bullets when working with CPI tools. The idea is to be very pragmatic when it comes to problem solving and use what works. There are several important elements to problem solving that are recommended:

STEP	PURPOSE	KEY ACTIVITY	TOOL / TECHNIQUE
1. Document the flow	o Understand process steps and sequences o Provide basis for establishing measures o Know customer supplier relations	o Identify customers o Identify products and services o Flow chart the process	o Flow chart o Brainstorm
2. Establish Key Customer Based Measures	o Understand customer needs and expectations o Determine appropriate measures	o Review process o Communicate with customers o Evaluate existing measures o Install new measures o Develop measurement methodology	o Survey o Brainstorm o Nominal Group Technique o Flow Chart o Multivoting
3. Collect and Record Data CHECKSHEET	o Establish benchmark of current performance o Provide information on reasons for nonconformances o Observe process behavior for signals	o Collect data and reasons for nonconformances o Record data on checksheets, charts and graphs	o Control Chart o Run Chart o Survey o Simple Graphs o Checksheet

Figure 6.12a Detailed improvement process, steps 1 to 3.

- A systematic and disciplined approach
- A team effort (several heads are better than one)
- The metrics (charts, graphs, etc.) be displayed so the team has ownership of the problem and the subsequent improvement
- Regular management reviews

On the last point, we go back to culture. The management reviews are not for the purpose of managing. They are for the purpose of involvement and communication. When the leadership is involved and aware of the activities of the employees in their process improvement endeavors, it has a tremendous synergistic effect on the process and the organization. The leadership and the employees share a common feeling of problem ownership and the euphoria of achieving

STEP	PURPOSE	KEY ACTIVITY	TOOL / TECHNIQUE
4. Analyze Data 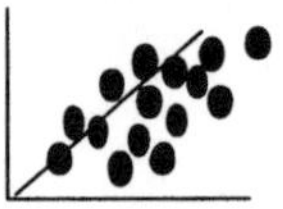	o Determine process Stability o Determine Process Capability	o Review & Summarize Data o Compare to Requirements o Calculate Process Capability Index	o Control Chart o Histogram o Scatter Diagram
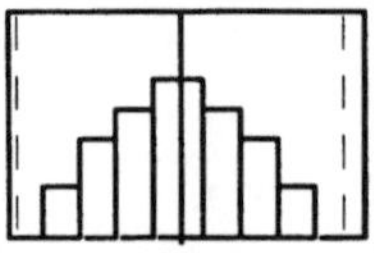	o Validate Measures	o Determine Causal Relations	o Brainstorming o Run Chart o Multivoting
5. Determine Problem Areas 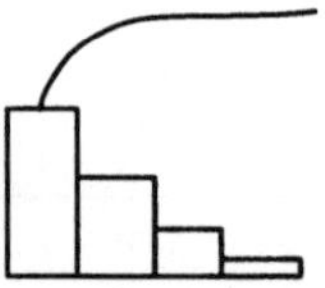	o Identify Areas in Process Causing Problems o Identify Process Simplification Opportunities	o Rank in Order Problem Areas o Correlate With Process o Determine Most Probable Cause(s)	o Brainstorming o Pareto Chart o Nominal Group Technique o Flowchart o Multivoting

Figure 6.12*b* Detailed improvement process, steps 4 and 5.

improvements. The insights as a result of the dialog that accompanies these sorts of gatherings are remarkable. Not only is there a mutual understanding but the solutions are often more optimized as the perspectives of the two elements are openly shared. There are many ways that reviews can be orchestrated. It can be integrated into the implementation and measurement process that we address in the next section, or it can be a separate effort such as a management CPI council that has the organizational charter to oversee CPI activities. Depending on the size of the organization, these councils can be at functional and business levels, at the top organizational level or both. The key is to have our leadership lead.

The first and most important step in process improvement is to understand the process. The best way to do this is to have the team construct a flowchart or process map. It is not likely a team can make viable improvements to a process they do not have common understanding of. This activity alone will surface many interesting opportunities that had probably not been considered previously. Once the process is understood, key measurements are developed. These are based on customer needs and internal controls. Baseline data is then

STEP	PURPOSE	KEY ACTIVITY	TOOL / TECHNIQUE
6. Determine Root Cause	o Determine the Root Cause(s) of a Specific Problem	o Identify All Probable Causes o Identify Potential Subprocess Problems o Decide on Cause Factor o Verify Cause	o Brainstorm o Nominal Group Technique o Cause & Effect Diag. o Experiment o Survey o Multivoting
7. Eliminate Redundancies, Rework and Waste Continuous Improvement	o Achieve Improved Level of Process Performance	o Establish Objectives o Identify Solutions o Implement Plan o Continue Process	o Brainstorm o Nominal Group Technique o Control Chart o Run Chart o Pareto Chart o Flow Chart

Figure 6.12*c* Detailed improvement process, steps 6 and 7.

collected and used to characterize the current process performance. This is important in order to understand the current capability of the process as well as to provide a reference for process improvement. Many of the statistical tools referred to in Chap. 3 are used during this phase. One tool used for collecting attributes data is a checksheet. The checksheet is used to record both the quantity of nonconforming items and the observed causes of these nonconformances. For example, a nonconforming printed wiring board assembly may be caused by excess solder, insufficient solder, solder voids, dewetting, damaged components, incorrect component installation, etc. Figure 6.13 illustrates a typical checksheet.

The checksheet is a valuable tool for determining the frequency of occurrence for each cause factor. This data can be transcribed onto a bar chart arranged in the order of the number of occurrences of each factor. This type of chart is called a *Pareto chart,* named after the Italian economist Vifredo Pareto, who hypothesized that roughly 80

PWA CHECKSHEET

Operator________________ Date__________________

Part No.	Serial No.	Defect code	Qty	Defect code	Qty	Defect code	Qty

Figure 6.13 Checksheet example.

percent of wealth is distributed in 20 percent of the population, or what he termed the "vital few." This principle has been determined to have universal application in many systems and was dubbed by Juran as the Pareto principle. An application of the principle in an electronic circuit card assembly would be that 80 percent of the nonconforming assemblies are the result of 20 percent of the possible causes. The Pareto chart is then used as a tool for deciding on which factors to address when trying to improve a process. In other words, getting the most bang for the buck. Figure 6.14 illustrates a Pareto chart.

Once data is collected and analyzed, the team will identify the possible causes of the problem and then verify those causes. It is important that the process be constructed to ferret out the root cause of the problem. One very useful tool for this activity is the fishbone diagram. The fishbone diagram is known also as a *cause-and-effect* diagram or an *Ishikawa diagram* after its developer, Kaoru Ishikawa. It is fundamentally a structured method of recording cause factors as they are suggested in a team brainstorming session. A box is drawn and labeled with the

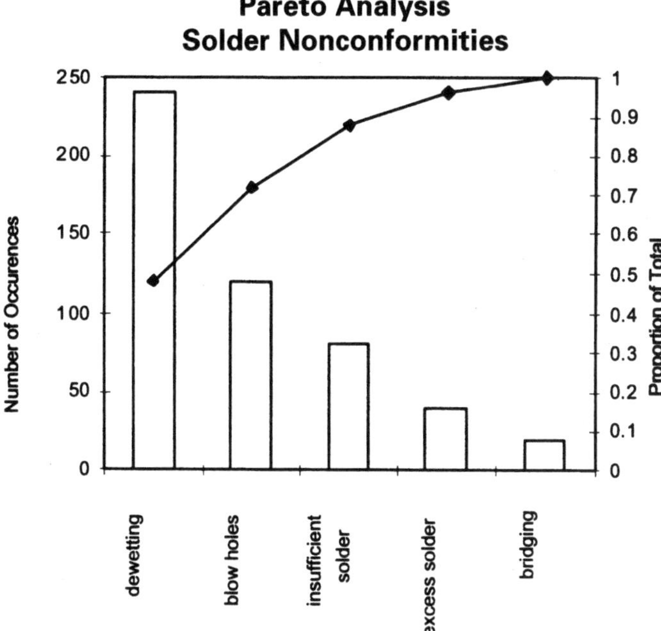

Figure 6.14 Pareto chart example.

problem statement. A line is drawn from the box and four main branches of the fishbone are drawn and labeled with the four generalized cause categories of manpower, machines, methods, and material. Virtually all root causes of problems can be placed into one of these "cause factors." Labeling the branches helps assure that the team considers these factors when identifying root cause. The team brainstorms every possible cause of the problem being considered. One of the techniques used to get to root cause is to always question why until no answer is evident. For each why, a branch is drawn to the previous cause. Figure 6.15 shows a fishbone diagram for printed wiring assemblies being rejected for dewetting. Another technique for determining root cause is a tree diagram where each cause factor continues to be broken down until the root cause is identified. Figure 6.16 illustrates a tree diagram for the same problem. Regardless of the technique used, it must be recognized that the result is, in effect, the opinions of team members. Those opinions must be validated as fact, which requires some further data gathering such as a test or experiment. After the cause is verified, solutions must be developed, and an implementation plan initiated.

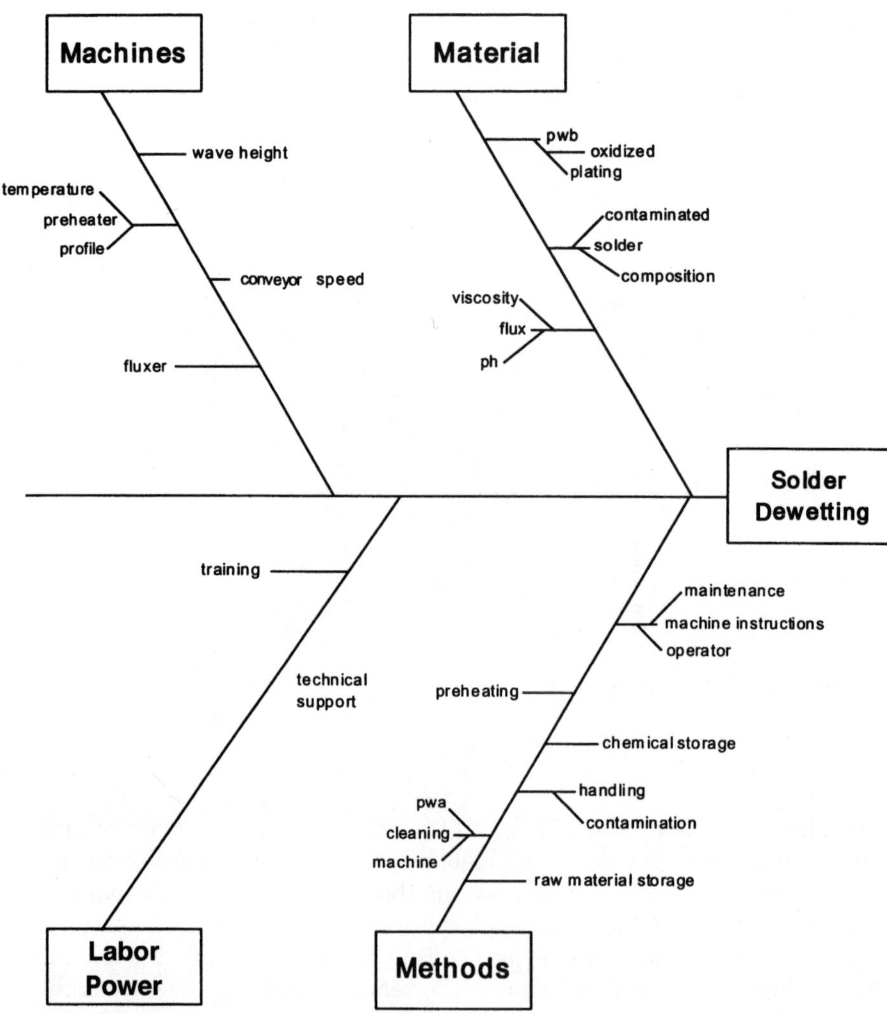

Figure 6.15 Fishbone diagram example.

6.3.3 Implementation and Measurement (I&M) Processes

In order to institutionalize continuous improvement in a TQM organization, there must be method for evaluating the results of the changes and for stimulating ongoing change. This method must consider both the financial and operational aspects of the organization (Fig. 6.17). It must also be tied to business goals and the strategic direction. Too often, each element of an organization is heading off in its own direction without coordinating its efforts with the other ele-

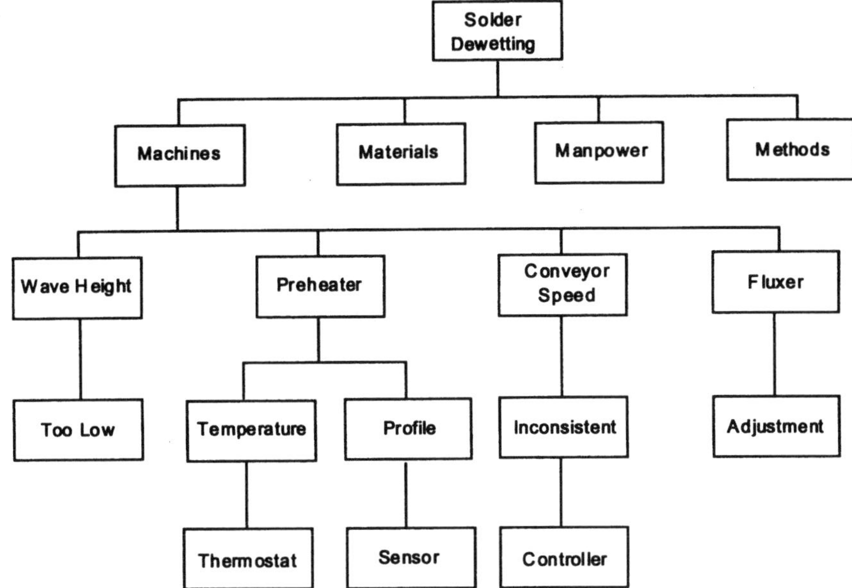

Figure 6.16 Tree diagram.

Figure 6.17 Implementation and measurement process.

ments. The result, at best, is suboptimization, but too often these activities can be counterproductive. One way an organization can assure that its well-intended TQM program does not fall into this "change for change's sake" syndrome is to develop an organization-wide implementation and measurement process. The essential principles of this process are

1. It is coordinated between all major elements of the organization
2. It promotes ongoing process improvement
3. It measures both financial and operational results
4. It is part of an integrated planning process.
5. It is regularly reviewed

Principle 1. The TQM organization communicates horizontally across the organization. This is true because it has a constancy of purpose, Deming's first point. An indication of constancy of purpose is all elements of the organization head down the same path with the same destination in mind. With an implementation and measurement process, the improvement efforts of all elements are compared to assure the resources are available to complete the tasks, to assure all the tasks support the organization's objectives, and to prevent duplication of effort. This process has the advantage of making all the elements of the organization aware of improvements and the opportunity to utilize them. It is also a very powerful supplement and catalyst to the ongoing grass roots process improvement activities being conducted by the PATs.

Principle 2. The properly designed implementation and measurement process is a stimulus for institutionalizing continuous improvement. The implementation portion consists of developing strategies that support the organization's short-term and long-term plans, planning the implementation, and tracking the progress. The method for developing the operational improvement strategies should be both top-down and bottom-up. We are reminded of the principle that the people closest to the work understand it better than anyone. It follows then that they should have a significant input to operational change strategies. On the other hand, there are change drivers coming from the strategic and business planning side of the organization that also need to be input. The inputs need to be in such a form that down-select decisions can be made. Table 6.3 is a format that meets that criterion. These inputs are distilled and prioritized through a collaborative effort between the operational and business elements of the organization. Once priorities

TABLE 6.3 Strategy description

Type of strategy:	Describe the general strategy type, e.g., reduce accounts payable cycle time.
Strategy statement:	State briefly what the strategy will accomplish.
Purpose/benefit:	Describe the purpose of the project and the benefit to the organization. If it is in support of a specific strategic initiative, so state.
Resources required:	Describe any internal and external resources required and their associated costs. This includes financial resources.
Key interfaces:	Describe internal and external interfaces required.

are established, the specific projects are flowed back to the originators for implementation. Typically, an implementation plan in the form of a Gantt chart is developed. This provides the discipline necessary to assure that the major tasks are identified, responsibilities are assigned, and completion times are committed. Table 6.4 is an example

TABLE 6.4 Strategy Action Plan

Process Owner:

Strategy Type:

Team Leader

Strategy Statement:

		FY 96												FISCAL YEAR			
		1 ST QUARTER			2 ND QUARTER			3 RD QUARTER			4 TH QUARTER						
KEY ACTION(S)	RESP	O	N	D	J	F	M	A	M	J	J	A	S	97	98	99	00
TASK 1	PERSON 1		↑														
TASK 2	PERSON 2			↑													
TASK 3	PERSON 1				▲				↑								
TASK 4	PERSON 3			▲				↑									
TASK 5	PERSON 2						↑		↑								

of such a plan. The implementation progress is reviewed and tracked by senior management to assure the resources are properly allocated and to reinforce their support of ongoing change.

Principle 3. In a TQM environment, we look at financials as results, not ends. The reality, however, is that the people that invest in our organizations, are, for the most part, bottom-line-oriented. Unless there is an overwhelmingly compelling reason for them to do so, investors don't buy "The check is in the mail" or "Trust me." For that reason, we need to be certain we have a balanced set of metrics in our implementation and measurement process. [24] We need to be certain we understand the financial ramifications of implementing or not implementing system improvements. Employees need to understand the cause-and-effect relationship between how processes function and the balance sheet. Electronics assemblers should be able to equate cycle time with return on assets. Accounts payable clerks should understand how their process affects discounts. Engineers should understand the impact of design changes on inventory. When employees are asked to be involved in changing the organization, they become business partners and need the information necessary to make business decisions.

Principle 4. The implementation and measurement system is an integral part of the organization's planning process. It is the implementing mechanism for the strategic and operating plans. It is unfortunate, but true, that some organizations create magnificent plans but cannot implement them effectively simply because they have not involved the implementers.

Principle 5. There is a lot of truth in the adage that we do best that which the boss watches. It is a way of authenticating what is important. For this reason, the implementation and measurement process should have regular and formal reviews by the senior management of the organization. This has several benefits: It provides employee teams an opportunity to spotlight their hard work; it gives senior management direct visibility of the impact of change processes; it reinforces the importance of change to the organization. If staged properly, with representatives of all elements of the organization at the reviews, this again promotes horizontal communication.

6.3.4 Recognition

Organizational cultures are fickle, or, perhaps better stated, "adaptive." The values and behaviors of the employees will adapt to those of the management. Two of management's key actions are *what* they recognize and *how* they recognize employees. Recognition takes many forms, i.e., tangible (both remunerative and nonremunerative) and intangible. In a TQM organization, recognition is a vehicle to reinforce the behaviors which the organization is promulgating. In the case of financial rewards, it's putting your money where your mouth is.

This section is not a treatise on human behavior or Maslow's hierarchy of needs, it is simply an observation. If you recognize and reward certain behaviors (and better yet, as a manager, model them), your odds are pretty good that the behavior will continue. With all due respect to Dr. Deming [25], we might even mention that during an employee's performance evaluation, they should be evaluated on whether they are demonstrating and encouraging the principles of TQM. For most organizations, there are significant changes in behavior that must take place in order for TQM to be effective. Again, what the boss watches (and values) we do well.

From our model, it can be observed that recognition is one of the four major factors in promoting continuous improvement. It follows then that a TQM organization should have a strategy for employee recognition. Communication is an essential element of recognition and therefore communication media become an integral element of recognition programs.

Many TQM organizations hold regular employee meetings to allow senior management to communicate directly with employees. These meetings are ideal settings for employee and team recognition. Regular team meetings and staff meetings also provide a means of communicating recognition. These forums are excellent vehicles for honoring groups and individuals in front of their peers. Newsletters or company newspapers are effective vehicles for recognition. Newsletters can dedicate a section to promoting TQM, recognizing individuals and teams, and highlighting significant achievements as a result of utilizing CPI disciplines.

Strategically located bulletin boards with current business information, metrics, or displays provide a highly visible means of communicating to both employees and visitors. Recognition displays can consist of team and employee photos prominently exhibited. When you tour a world class facility such as Miliken, you observe numerous displays emphasizing employee involvement. WalMart continually recognizes employees by featuring them in advertisements and throughout the stores. A General Electric division uses employees' photographs in its advertising calendars. When you observe these forms of recognition, you immediately conclude that the organization values its people, and the people usually confirm that conclusion.

Many companies use electronic media. Scrolling electronic signs with up-to-the-minute information are popular. Businesses today have local-area networks, which are an ideal communication medium. The Internet can provide Web pages for communicating everything from financial performance to current business news to job opportunities. This is especially useful for larger organizations with geographically dislocated facilities. Intranets are becoming popular for coupling Internet technology with local-area networks. With today's technology, there is no operational reason why employees cannot have virtually all the information necessary to make informed decisions. In a very real sense, information is recognition. We are saying to our employees that we want them to be informed and value their involvement

As was mentioned above, recognition takes many forms. In an organization promoting high employee involvement, recognition should have a bias toward teams. Regardless of the method used to decide on candidates, a well-thought-out reward system must be designed to reinforce the behaviors and values that are the basis of the TQM organization. Recognition can include areas such as leadership, entrepreneurship, and significant achievement. Individual recognition is important but the success of individuals, whether because of their leadership or task accomplishment, is rarely the result of working in a vacuum. One method of creating a team bias is encouraging teams to nominate individuals for recognition. Achievement awards can reinforce teaming by having teams nominate themselves. Teams can be recognized for creating change by the innovative application of TQM tools or for a notable accomplishment. Regardless of the recognition, it is important that it be communicated to the organization to reinforce high-performance behaviors.

Rewards associated with recognition can run the gamut from social rewards to a meaningful portion of an individual's wage. Social rewards such as a handshake with a "job well done," a token remuneration such as a Savings Bond, a pizza party, or a trophy are all appropriate when they are consistent with the performance being recognized. Rewards that contribute significantly to an employee's wage are becoming increasingly more popular. These reward systems fall into two basic categories: skill-based pay and performance-based pay. According to Lawler, "The two most critical features of an organization's pay system are the degree to which it is tied to performance and how it determines the worth of an individual employee." [26] Skill-based pay systems focus on people and their value to the organization. They reward employees on the skills and knowledge they have, based on predefined needs of the organization. This system requires that training and assessment be an integral part of the system. Performance-based pay focuses on organizational performance and the competitive environment. According to Lawler: "Financial rewards are vital to a proper balance of power, information, knowledge, and rewards in an organization. If such rewards are missing, individuals have no financial accountability for how they use the information, knowledge, and power they are given to improve organizational performance." [27] He goes on to say that lack of financial rewards can create major equity issues which can harm the organization's culture. One popular form of pay for performance is called *gainsharing*. In gainsharing, a baseline of performance is established and the employees are rewarded according to gains in that performance. These performance gains are quantified according to predefined formulas and the resulting financial gains are shared with the employees. Gainsharing plans are popular because they promote employee participation and ownership. For additional thoughts on performance-based pay see Chap. 5.

6.4 Additional Thoughts

Total quality management in today's electronics industry is most likely not a system that will give it any overwhelming advantage. It is fast becoming a norm. Simply put, it is probably essential for survival. See Sec. 3.3, on TQM and teams. In May of 1991, a GAO study of 20 Baldrige finalists concluded that implementing a TQM system was a common factor in their selection (see Chap. 9). [28]

We have described TQM not as a program but as a system. It is integral to the effective operation of any enterprise in today's global marketplace. The way it has been presented provides a context for the raft of tools and processes that are available for organizational change and improvement. The TQM system has several essential elements requiring leadership from the highest level of the organization and a culture that has a bias for change, These elements include change processes, implementation and measurement, and rewards and recognition. Your organization is encouraged to develop a TQM system that best fits your vision and strategic intent. You are encouraged to measure your progress in your journey toward TQM maturity. Table 6.5 is an example of a matrix that could be used. [29] You are encouraged to use this format and tailor it to your needs. You could also develop a time line using the same criteria in order to develop a plan for implementing and tracking progress toward your maturity as a TQM organization. We hesitate to say that at a certain time you

TABLE 6.5 TQM Maturity Assessment

Maturity	Culture				Change processes			Implementation and measurement	
	Customer	Management commitment	Training	Employee involvement	CPI	Organization	Recognition	Tools*	Metrics
5	Supplier of choice. Customer desires a long-term relationship.	All management involved in TQM.	*X*% of every employee's hours used for training.	Self-managed teams utilized to make decisions within boundaries.	CPI used routinely to improve all business processes.	Organization infrastructure supporting teams is evident.	Some form of success sharing is in place.	All employees routinely utilize problem-solving tools.	Entire organization displays and utilizes metrics for improvement.
4	Customer involved in design process.	Evidence of top management commitment to TQM.	Employee training is evident but not evenly distributed.	Teams involved in decision making.	CPI used routinely to improve manufacturing/service processes.	Organization structure promoting teams is evident.	Routine employee/team recognition is conducted throughout the organization.	Manufacturing /service employees routinely utilize problem-solving tools.	Manufacturing /service elements routinely display and utilize metrics.
3	Customer feedback used in decision making.	Vision-implementing plan in place.	Formal training plan for the organization developed.	Employee teams utilized.	CPI used periodically to improve some business processes.	Management and employees involved in organizational system change.	Reward/recognition conducted sporadically.	All employees are trained in the use of basic problem-solving tools.	Some evidence of metrics being used to monitor and improve processes.
2	Customer feedback solicited and analyzed.	Vision including TQM developed.	Training needs for the organization identified.	Management requests employee input but makes decisions.	Process improvement model/method adopted.	Organizational change process adopted.	Reward/recognition program developed.	Manufacturing /service employees are trained in the use of problem-solving tools.	Value of metrics as a management tool understood.
1	Customer rating unknown or presumed.	No vision exists.	No formal training program.	Traditional hierarchical organization.	No process improvement program in place.	No organizational change process in place.	No reward/recognition process in place.	Problem-solving tools are not formally taught.	No evidence of metrics being used to monitor and improve processes.

*Tools include, as appropriate, statistical process control, design of experiments, quality function deployment, seven management tools, and any other problem-solving and variation reduction techniques.

need to be at a certain place in your journey toward organizational excellence. There are many factors to consider, not the least of which is changing the values and beliefs of the organization (the culture). Suffice it to say that a plan is needed—the journey of a thousand miles begins with the first step; that first step depends on where you are as an organization.

Regardless of the model used or the structure of your plan, TQM should be approached holistically with the participation of your entire organization. When a TQM system is properly orchestrated, you will never be surprised at the results but you will be continually amazed.

References

1. Sashkin, Marshall, and Kenneth J. Kiser, *Putting Total Quality Management to Work,* Berrett-Koehler Publishers, San Francisco, 1993.
2. William B. Scott, "TQM Expected to Boost Productivity, Ensure Survival of U.S. Industry," *Aviation Week and Space Technology,* December 4, 1989.
3. Tom Peters, *The Pursuit of WOW!, Every Person's Guide to Topsy-Turvy Times,* Vintage Books, New York, 1994.
4. Dertouzos, Michael L., Richard K. Lester, and Robert M. Solow, *Made in America, Regaining the Productive Edge,* The MIT Press, Cambridge, MA, 1989.
5. Ibid.
6. Walton, Mary, *The Deming Management Method,* Perigee Books, New York, 1986.
7. Ibid.
8. See Ref. 1.
9. The two publications are *Guide to Quality Control,* UNIPUB, White Plains, NY, 1986, and *What is Total Quality Control? The Japanese Way,* Prentice-Hall, Englewood Cliffs, NJ, 1985.
10. Feigenbaum, Armand V., *Total Quality Control: Engineering and Management,* McGraw-Hill, New York, 1961.
11. *The American Heritage Dictionary of the English Language,* 3d ed., Houghton Mifflin, 1992.
12. Porter, Michael E., "The Competitive Advantage of Nations," *Harvard Business Review,* March–April 1990.
13. King, Bob, *Better Designs in Half the Time: Implementing QFD Quality Function Deployment in America,* Goal/QPC, Methuen, MA, 1989.
14. Kotter, John P., *A Force for Change, How Leadership Differs from Management,* The Free Press, New York, 1990.
15. Ibid.
16. See Ref. 1.
17. Lawler, Edward E., III, *The Ultimate Advantage, Creating the High Involvement Organization,* Jossey-Bass Publishers, San Francisco, 1992.
18. See Ref. 17 for information on high-involvement teams.
19. See Ref. 11.
20. Passmore, William A., *Designing Effective Organizations,* John Wiley & Sons, New York, 1988.
21. Hammer, Michael, and James Champy, *Reengineering the Corporation,* Harper Collins Publishers, New York, 1993.
22. Waterman, Robert H., Jr., Thomas J. Peters, and Julien R. Phillips, "Structure Is Not Organization," *Business Horizons,* June 1980.
23. Scholtes, Peter R., *The Team Handbook,* Joiner Associates, Madison, WI, 1988.
24. For an interesting presentation on this subject, see Kaplan, Robert S., and David P. Norton, "The Balanced Scorecard—Measures That Drive Performance," *Harvard Business Review,* January–February 1992.

25. Dr. Deming professed that American management suffered seven deadly diseases. Disease number 3 is *evaluation of performance, merit rating, or annual review.* In addition, his point number 8 (of Deming's 14 points) is to *drive out fear.* It is argued that performance evaluations promote fear. It could be argued that the context (primarily quotas) of both principles above are different from modifying organizational behavior but we'll let you draw your own conclusion. Two excellent books that delve into Deming's principles are Mary Walton's, previously cited, and William W. Scherkenbach, *The Deming Route to Quality and Productivity,* Mercury Press/Fairchild Publications, Rockville, MD, 1988.
26. See Ref. 17.
27. Ibid.
28. GAO Report, "U.S. Companies Improve Performance through Quality Efforts," GAO/NSIAD 91-190, May 1991.
29. Adapted from DOD 5000.51G, *Total Quality Management Guide,* US DOD 8/89, p. 22.

Chapter

7

Subcontractor/Supplier Quality

William A. Kirsanoff

Procurement Quality Engineer
The Boeing Company

7.1 Introduction

Sources of supply have played an important, sometimes decisive, role in modern history. As the United States expanded westward, the new towns depended on the dry goods store; the dry goods store depended on the stage lines to transport the goods from suppliers in the east. Later, still rural America counted on the Sears catalog for a vast array of needs. During the Second World War, the Allies expended great effort to maintain supply lines from North America to Europe, even as the Axis Powers tried to disrupt that supply line. The future of the western world truly depended on the effort to maintain the connection from supplier to end user. In the process-oriented electronics industry, selection, coordination, and management of suppliers can be as vital to a business as the convoy ships were to Europe (Fig. 7.1). This chapter explores the key factors in successfully managing supplier quality, and building supplier relationships.

Today, in the electronics world, supplier relationships are critical. Electronic system companies are dependent on component and printed wiring suppliers for the success or failure of their products. Contract manufacturing in electronics is growing at tremendous rates. From 1994 to 1995 contract manufacturing increased 51 percent, reaching $59 billion in 1996, and is expected to increase by 82 percent by the year 2000, according to electronic industry sources. [1] The interfaces between customers and suppliers in this supply chain will be vital to the quality of the product procured. Good relationships will grow, others will fail, and only electronics manufacturers with good quality systems and good supplier relationships will succeed.

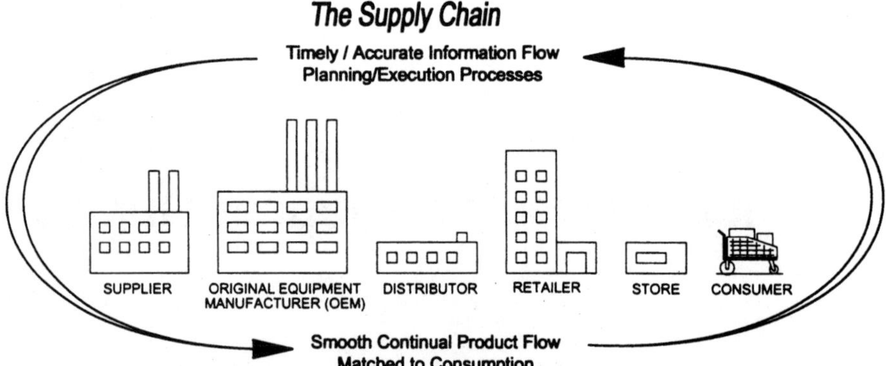

Figure 7.1 Linking the virtual enterprise. (*Adapted from Managing Automation, October 1996; used with permission.*)

7.2 Supplier Selection

In most manufacturing environments, the design engineering function harbors a large store of catalogs, and suppliers are selected according to whose catalog is on top of the stack. Little attention is paid to the supplier itself until the design is committed. Sales and marketing people recognize this, and target their efforts to the designers, providing catalogs and handbooks frequently in an effort to keep their product line fresh in the designer's mind, and on top of the designer's desk. This technique has expanded to the Internet, where supplier Web pages furnish valuable information to designers, while advertising their products. While catalogs and Web pages can be helpful tools in finding sources of supply, successful supplier selection requires a little peek under the hood. One of the biggest mistakes customers make in supplier selection is to concentrate on the supplier's product and ignore the supplier's business.

Since most successful manufacturing companies concentrate on a *core business competency* with a long-term business commitment, suppliers should be selected for their core competencies and for long-term strength (Fig. 7.2). [2] This is where a team approach to supplier selection can be of great value. By using a team approach, team members can concentrate on their areas of specialization to evaluate the whole supplier. The designer may look at the product and the product's capability to support design and reliability goals. The buyer may concentrate on the pricing and delivery. The supplier quality function focuses on the suppliers' ability to control manufacturing processes and deliver product consistent with the design. Under an ISO 9000 system, this type of team approach is required. [3]

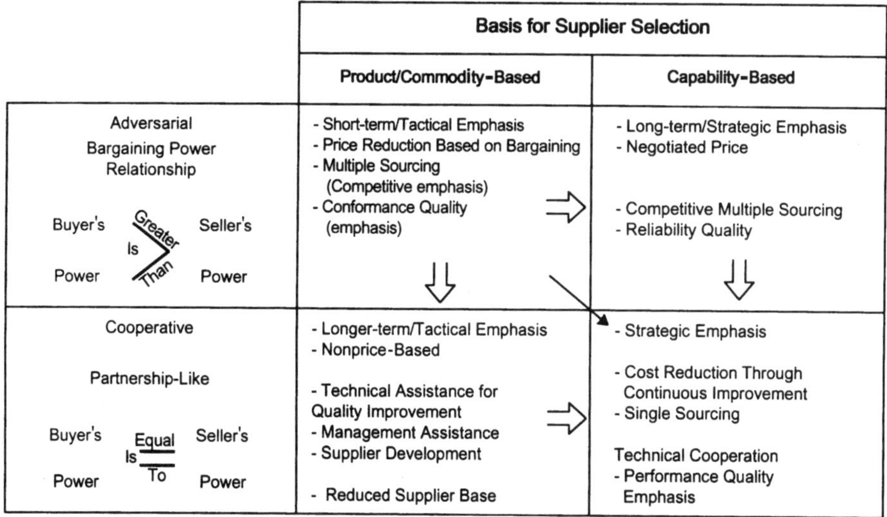

Basis for Supplier Selection
Product/Commodity-Based
Capability-Based
Adversarial
Bargaining Power
Relationship
Buyer's Power Is Greater Than Seller's Power
- Short-term/Tactical Emphasis
- Price Reduction Based on Bargaining
- Multiple Sourcing
(Competitive emphasis)
- Conformance Quality
(emphasis)
- Long-term/Strategic Emphasis
- Negotiated Price
- Competitive Multiple Sourcing
- Reliability Quality
Cooperative
Partnership-Like
Buyer's Power Is Equal To Seller's Power
- Longer-term/Tactical Emphasis
- Nonprice-Based
- Technical Assistance for Quality Improvement
- Management Assistance
- Supplier Development
- Reduced Supplier Base
- Strategic Emphasis
- Cost Reduction Through Continuous Improvement
- Single Sourcing
Technical Cooperation
- Performance Quality Emphasis

Figure 7.2 Shifting paradigm for supplier relationship. *(Reprinted from Ref. 2 with permission.)*

The team should also look at the business strength of the supplier candidate. Is the candidate a supplier that will be in business for the long term? Does the supplier have the capital equipment or the resources to obtain the capital equipment necessary to complete your contract and support their current continuing business base? Is the supplier involved in ongoing litigation or regulatory investigation that will impact your intended relationship? These are the questions frequently missed in the supplier selection process. Business information is relatively easy to obtain. Services like Dun & Bradstreet specialize in providing business information. Public companies in the United States are required to file form 10K with the Securities and Exchange Commission, and provide annual reports to stockholders. These reports are generally available for the asking, and provide a wealth of information. Privately held companies may prove more difficult to obtain business data from, but by looking at the supplier's business, many pitfalls can be avoided.

Supplier selection must also consider supplier value rather than price. Recall Dr. Deming's point 4: "End the practice of awarding business solely on the basis of price tag alone." [4] Price is certainly a critical factor of value, but it should not be the only factor. If a lower price is achieved at the expense of other critical factors, there are no savings, and it is likely there is loss. A good price for product delivered too late to use, or packaged so poorly that it arrives broken, or

TABLE 7.1 Highest Ranked Capabilities for Competing Successfully on a Worldwide Basis

In order:

- Highest perceived product quality
- Low product prices
- Product support/customer service/short lead times
- Delivery performance consistent to promise
- Product features that offer highest performance
- Ability to bring new products from concept to customer in the shortest amount of time

SOURCE: Reprinted from Ref. 5 with permission.

fails to achieve performance needs is no deal. Total supplier performance needs to be considered in evaluating costs. Is one supplier adding more value for the cost proposed than another? Consider the five most critical supplier capabilities identified by firms participating in a recent survey shown in Table 7.1. [5] Later in this chapter, the concept of best value will be discussed in more detail, and we will explore ways of measuring supplier performance and supplier rating using the best value concept.

In most cases, when a buy decision is made rather than a make decision, it is because the customer expects to find better expertise in the supplier pool than within the customer's organization. The supplier should be expert in its field. At times, the customer may have an in-house expert in the supplier's field, but the supplier has the overall organization and production infrastructure required. In any case, the buy decision indicates the customer's belief that an outside source can produce the product better or more efficiently than the customer can. To select suppliers that meet this expectation, supplier selection should focus on core business competency of the supplier in addition to addressing the other factors identified in Table 7.2.

In determining if a supplier knows the product, several factors can be reviewed. The supplier should work to some form of a standard. Work instructions should be in place to support the standard. The

TABLE 7.2 Factors in Choosing a Good Supplier

Business stability
Core competency
Past experience
Ability to meet schedule
Plant capacity
Location or virtual location capability

supplier should have good internal measures. Test data should be collected, maintained, and reviewed. Corrective action for deficiencies and a system for process improvement should be in place. The supplier should have a system for tracking products throughout the manufacturing cycle, delivery, and beyond. The supplier should show a commitment to product reliability in the field, track field returns, and use the information garnered from field returns to improve future performance. These and other pertinent metrics should be used to monitor and manage key aspects of the supplier's operation.

A supplier's core business competency may not be the same as the supplier's product. A well-known major retailer of electronic equipment can be relied on for everything from carbon composition resistors to home satellite television receiver systems. The retailers' core competency, however, is not electronics, as their advertising would indicate. The retailer's core competency is marketing! One must look beyond the surface when evaluating a supplier's core competency.

When evaluating the supplier's core competency, determine what it is the supplier does well. Consider whether what the supplier does well is what you are looking for as a customer. If what the supplier does well is what the customer needs, the supplier is a viable candidate, and can continue through the selection process. Later in this chapter, we will again discuss core competency as a criterion for supplier partnership.

The best, most capable supplier in the industry is not a good candidate for procurement if the supplier cannot meet the customer's schedule demands. At the beginning of the personal computer revolution, IBM contacted several potential suppliers of operating system software for their soon-to-be-released IBM PC. The odds-on favorite for the contract was a company that at the time made the most popular operating software for small computers. That company, however, was too busy to respond to the request for quote in a timely manner. Ultimately, the contract to develop an operating system for the IBM PC was awarded to a small start-up software house that could support IBM's schedule demands. Today, Microsoft, that small start-up company, is the world leader in PC operating systems and a software powerhouse.

The candidate supplier must have the ability to meet schedule demands. Tracking supplier performance on past contracts is one way to make that evaluation. If the customer does not have past experience with the supplier, the supplier may have documentation to demonstrate performance. A good candidate supplier could be expected to routinely track schedule performance internally for its own management needs, and could be asked to share that data with the customer. In the case of major, long-term procurements, consideration of the supplier's plant capacity may be appropriate. If a supplier is performing on time for the current customer base, but is operating near capacity, the supplier should be asked to address steps to increase plant capacity for the term of the contract.

Location can be a factor in supplier selection. It is easier to coordinate activities and schedule just-in-time shipments with a local supplier. For a contract that will entail a significant amount of development and engineering interchange, local suppliers reduce the need and expense of business travel. This is not always an alternative in a global marketplace. Local suppliers may not even be the best alternative. Because of the global scope of commerce, a support infrastructure has developed to meet the demand. Worldwide shipping and express delivery services make control of just-in-time issues as manageable across the globe as across the city. Technology and the Internet provide a viable alternative to many business trips. A supplier with an engineering issue half a world away can digitally photograph the product, tooling, or other item of discussion, type a description of the issues, and use electronic mail to send the image and text to the customer in seconds. The electronic transfer could even take place while the customer and supplier engineers are talking about the issue over the phone.

Another alternative to travel for business conferences is teleconferencing facilities. Teleconferencing can provide two-way video and audio, as well as hookups for document transfer, either by fax or computer imaging. For businesses unable to afford dedicated facilities, many cities have teleconferencing centers available for rental. Low-cost video phones and Internet imaging systems are other alternatives to frequent business travel. While technology cannot eliminate the need for all face-to-face business meetings, telecommunications can reduce the frequency of business travel and many of the difficulties associated with a dispersed supplier base.

For compliance with para. 4.6 of International Organization for Standardization (ISO) 9001, a company needs to document its method of evaluating and selecting subcontractors on the basis of their ability to meet requirements. A database of all suppliers, by type, grade, class, or other identification should be set up and linked with the supplier rating system discussed in Sec. 7.7.

7.3 Supplier Partnership/Teaming

The TC176 committee conclude their list of eight management principles addressing customer supplier relationships: "Mutually beneficial relationships between an organization and its supplier enhance the ability of both organizations to create value." [6] While this principle applies to all supplier relationships, it is the heart of supplier partnership.

The historical relationship between customer and supplier in the United States has been adversarial. The customer would pit potential suppliers against each other to put downward pressure on pricing. This resulted in large supplier bases for customers to manage, with two, three, or more suppliers providing the same products. There was little loyalty of suppliers to customers, or of customers to suppliers. The lowest price won the business, not long-term relationships. This adversarial relationship resulted in a great deal of waste, excessive supplier administration, duplicative efforts in support of proposals, bids, and contract awards, suboptimization of products as engineering teams held each other at arms length to prevent giving one supplier an unfair advantage over another.

In the 1980s, major American companies began shrinking their supplier bases to reduce the amount of administration on the large, redundant supplier bases required. The term *supplier partner* was used for any preferential relationship with a supplier, even if the only preference was that the supplier was not one of those cut from the customer's supplier base. Few of the relationships formed, after the carnage of losing suppliers was swept away, could truly be considered partnerships. True partnerships have several important attributes not found in most customer/supplier relationships. The relationship should be truly committed to mutual benefit of both the customer and supplier, and the development of mutual trust. The partners should work together to improve their performance over time, and share in the cost savings from performance improvements. Sharing of information should be open, with agreements in place to protect each party's intellectual property rights. Technology and even personnel should be shared to optimize the overall system. The supplier should be involved early in system design, before specifications are completed. Pricing should also be open, based on actual costs, not just market forces.

It has become popular to refer to suppliers as partners, or team members. In fact, only a small segment of a manufacturer's supplier base qualifies to be considered a supplier partner. Like all partnerships, a supplier partnership includes trust, risk, risk sharing, and mutual support. In selecting a supplier partner, it is again necessary to look beyond the product line. Table 7.3 illustrates the long-term relationships that are present in true partnerships and differentiates them from a typical customer/supplier arrangement. [7]

A supplier partner should have an institutional philosophy compatible with the customer. If the customer is intent on continuous improvement and internal research and development, the supplier

TABLE 7.3 True Partnership Characteristics

- Joint sharing of purchased item and product development costs
- Complete and mutual trust between buyer and seller
- Minimal need for lengthy contractual agreements
- Sharing of cost savings resulting from performance improvement efforts
- Open sharing of information between buyer and seller
- Agreement or understanding concerning intellectual property rights
- Prices that have as their basis a supplier's true cost structure rather than "what the market will bear" pricing
- Relationships that remain positive through difficult economic circumstances
- A commitment to the mutual benefit of buyer and seller
- Longer-term contracts covering the life of an item
- A mutual commitment to joint performance improvement over time
- Sharing of technology or personnel as needed

SOURCE: Reference 7; reprinted with permission.

partner should have a system to complement and support those needs. The customer should understand what specifications and standards the supplier normally works to. The customer should understand how the supplier's business system and documentation are kept current, and may want to be notified, or even approve changes affecting the product the customer buys. Does the supplier maintain contact with key industry groups, such as the Electronic Industries Association (EIA) or Institute for Interconnecting and Packaging Electronic Circuits (IPC) to keep current with its business and business trends? Does the supplier know the product lines and measure and understand its own performance? Like all partnerships, a supplier partnership must be a two-way relationship. The customer cannot merely declare that a particular supplier is a supplier partner. The supplier should have a stake in the outcome of the partnership. Will the supplier share internal performance measures with a customer partner? Will the supplier share development efforts and research with the customer? The customer and supplier should be in the partnership for mutual benefit.

7.3.1 Criteria for partner selection

As in any supplier selection activity, the supplier partner candidate should be selected for the supplier's core competency. After determining that the supplier has a core competency focus that matches the customer's needs, consider the breadth of the supplier's capabilities. While the rare case exists where a supplier is desired for a single, focused item, more often suppliers are expected to provide a type or family of products. When comparing supplier partnership candidates, consider if one of the candidates can supply the entire line of that type of product needed to support the customer. When a single supplier has a wide breadth of product in a core competency needed by the customer, and the supplier's business is generally healthy, the supplier becomes a candidate for becoming a true supplier partner.

One mistake often made as customers seek supplier partners is to base selection of supplier partner candidates based on procurement dollars. When seeking areas for supplier partnership, customers should focus on critical competencies needed to support ongoing production demands. A critical supplier may account for a small percentage of procurement dollars, but supply materials or services vital to the strategic success of the customer.

The supplier partnership process should begin with the customer developing a strategic plan for procurement. The plan should address procurement needs over the next 5- to 10-year period. What technologies will the customer need to have supported by the supplier base? What potential suppliers exist in the current supplier base with core competencies supporting those technologies. In the rapidly developing electronics industry, technologies needed 5 years from now may not exist in the marketplace today. When this is the case, what suppliers appear to be the likely leaders when the technology does appear? By developing a strategic plan for procurement in leading edge technologies, a customer may find it prudent to invest in a potential supplier through teaming arrangements or development contracts to assure the needed technology is ready when the customer needs it.

Building partnerships is not easy. It takes effort and time before significant results are produced. Use of quality function deployment, benchmarking, or other tools can help to begin a relationship. Jointly undertaking quality improvement initiatives is often a good way to build the trust that sustains a supplier partnership. Table 7.4 identi-

TABLE 7.4 Total Quality Approach with Suppliers

- The company exhibits the highest levels of quality performance.
- The company is willing to commit the resources toward quality improvement initiatives with suppliers.
- The company has in place a system or means to identify those suppliers most likely to benefit from joint quality improvement efforts.
- The company can identify areas of performance or quality improvement at selected suppliers.
- Selected suppliers are ready, willing, and able to work closely with the company to improve quality.
- The company can define and quantify world class or best-in-class levels for different performance categories.

SOURCE: Reference 7; reprinted with permission.

fies several factors that can affect a customer's ability to undertake a joint quality improvement approach with a supplier. [8]

7.3.2 Partnership tools

One of the key tools emerging in supplier partnership is supply chain management through the Internet. With the advent of greater Internet accessibility and improved security and software tools, the Internet has moved from the exclusive domain of large corporations and research centers to become a resource for businesses of every size. What started as a marketing tool, much like an electronic catalog, is now a secure two-way communication tool. Customers can perform their own stock checks and place orders through the Internet. Technical information can be exchanged. Computer-aided design and manufacturing (CAD/CAM) files can be exchanged between customer and supplier engineering groups, eliminating dual engineering drawings, one produced by the customer, one by the supplier. Customer and supplier share the same tools and the same data files.

The use of increased electronic connectivity can result in improved customer service with lower costs. Customers can make routine inquiries themselves, keeping customer service representatives available for higher-value or more complex tasks. Package delivery services now offer self-service package tracking through the Internet. Catalog companies use electronic mail to respond to customer inquiries for more information or special orders. Retailers and distributors place their orders directly into manufacturer's computer systems. Campbell Soup Company of Camden, NJ, uses an Internet-accessible customer service system for managing the order life cycle. In an article published in *Managing Automation,* Ron Ferner, a Campbell Soup vice president noted that the system "will help us dramatically decrease our order cycle times, improve invoice accuracy, increase control of trade fund and reduce customer deductions—factors that will contribute to annual cost savings in excess of $18 million." [9]

Supplier partnership and supply chain management are closely tied to trust, shared resources, and shared information. Communication is a critical element in supplier relationships. In an article published in *Managing Automation,* John Lischefska, vice president for business development for SynQuest, said it well: "the supply chain is not really a chain at all but a network with interaction and feedback and a flow of information and material that is not hierarchical." [10] The use of computer technology to improve communication and shared access to information and technical resources makes it easier than ever to facilitate close working relationships in a global marketplace.

7.4 Supplier Quality Assessments

Many companies use a system of supplier quality audits, surveys, and supplier process audits as part of their supplier selection and supplier management process. These assessments take on many forms, and can vary with the nature of the procurement or planned procurement. Many businesses accept third party certification to recognized standards, such as the ISO 9000 series, as sufficient to assure the supplier has an effective business system in place. Sometimes, a questionnaire, or "desktop survey," is used, where the supplier is asked to score itself in various areas of quality system compliance. On-site audits or surveys are frequently used, and take several forms as well. On-site process audits are frequently used as supplements to a customer's acceptance of a third party certification. With the process-oriented electronics industry, supplier use of process metrics, statistical process control (SPC), and variability reduction methods are included in supplier assessment. In some cases, and particularly for more critical suppliers, a combination of methods may be more appropriate than any one method alone.

In selecting the method of supplier quality system approval to be used, there are several factors to consider. What kind of supplier is needed? A basic supplier that is used for simple, readily available commercial products may be accepted to a much lower standard than a supplier producing a unique, technically challenging product (Fig. 7.3). Is the supplier candidate a new supplier, or a supplier with a current and continuing relationship with the customer? The prime customer may flow down specific requirements, such as ISO 9000 certification, or advanced quality practices like those required for QS

Figure 7.3 Quality management system framework.

9000 certification for the automotive industry, or unique standards for manned spaceflight–rated devices that must be addressed. The customer's ability to verify product compliance through inspection may be a factor. Many products cannot be verified by inspection; in this case, the supplier must control the manufacturing process to assure the end product meets customer requirements. A review of the supplier's quality system needs to assure that appropriate safeguards and controls are in place.

7.4.1 First party assessment

Assessments are frequently categorized by the relationship between the assessor and the organization being assessed (Fig. 7.4). A *first party assessment* is a self-assessment where the organization uses its own people to perform the assessment, report the results, and verify and approve any corrective actions. First party assessments are an important part of any effective quality management system, and a basic requirement of ISO 9000 systems. Assessments may be carried out by a specialized audit function within the organization, or by cross-functional teams staffed with people temporarily assigned to the audit function. For a formal assessment, staffing the audit team with people from the function being audited should be avoided. People tend to form a scotoma, or blind spot, to system deficiencies they deal

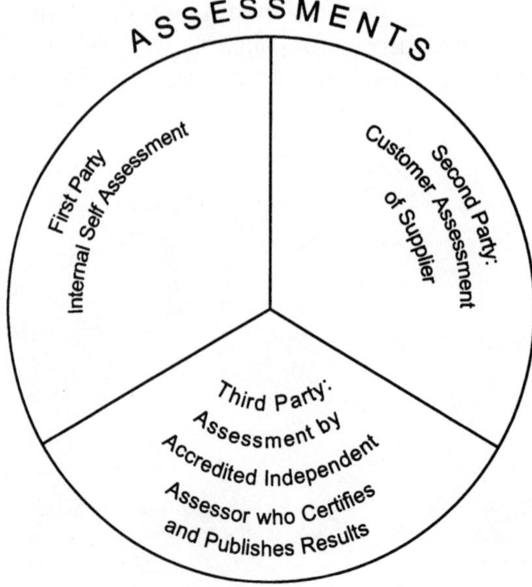

Figure 7.4 Quality assessments.

with every day, and so miss the opportunity to bring the deficiencies to the fore in an audit, where they can be corrected. That is not to say self-audits within a function are not good. Self-audits within a function can be a good tool for keeping disciplines in shape, but should not be considered substitutes for an independent assessment.

Desktop surveys are one form of first party surveys. They are attractive to many customers because of their low cost. A checklist built around the customer's requirements is sent to the supplier for the supplier to complete, making a desktop survey a first party assessment to a customer's checklist. The only cost is postage. While desktop surveys can be useful when completed faithfully by the supplier, they have severe limits. ISO certification from an accredited registrar is much better than a desktop survey. Once the desktop checklist has been sent to the supplier, there is no control over who performs the assessment. While many people would assume a quality checklist would be sent to the quality manager, or at least the quality department for completion, many times the sales representative completes these checklists. The checklist should, of course, be completed by the person with the best understanding of the supplier's quality system, whomever that may be. Another major drawback of desktop surveys is interpretation. The supplier may have a very different interpretation of some questionnaire issues than the customer does. If this is the case, the results of the desktop survey would not be valid. The rarest situation is also the most dangerous, and that is the supplier that intentionally misrepresents its system on the desktop survey. While intentional misrepresentation is rare, it does happen, and any company that intentionally misrepresents itself is not a company one wants as a supplier.

7.4.2 Second party assessment

An assessment of a supplier by a customer is called a *second party assessment.* The customer performs the assessment against the customer's requirements, documents findings and observations, and approves corrective actions. While a good second party assessment can be valuable to both the supplier and customer, these assessments can have large costs to both parties. Customers are faced with sending assessors or assessor teams out on business travel to effect the assessments, and suppliers find themselves hosting these assessment teams with increased frequency as their customer base increases. When supplier personnel are supporting a second party assessment team, other duties are not supported and jobs don't get done. This can create a significant hardship for smaller suppliers that cannot afford to increase staffing for such activities. This is one of the key reasons for the concept of *third party assessments* for assessment and approval of a supplier's basic quality management system.

7.4.3 Third party assessment

Third party assessments started to become popular in the 1980s as a way to control the cost of supplier quality management system assessments and as a way to minimize the number of assessments performed at a given supplier. Many suppliers found that they were hosting surveys from one customer or another several times a month the year around. This cost the supplier precious resources in personnel to support the survey and respond to findings. Customers found they too were expending tremendous resources sending assessors across the country and around the world.

Industry groups formed in some areas to share supplier approvals. One of the first was the aircraft/aerospace industry group known as CASE, Coordinated Aerospace Supplier Evaluation. Organizations like CASE could publish joint listings of suppliers approved by their members, and other member companies could choose to accept the listing as evidence of an acceptable quality system. Companies that participated saved costs by sharing these results, and members agreed to perform a minimum amount of surveys each year to carry their share of the load. This arrangement had two key drawbacks, however. Since each member company performed surveys and audits to its own standards, and trained its surveyors and auditors to its own standards, there was no consistency in the assessment results. Additionally, antitrust and libel concerns prevented these organizations from publishing results of unacceptable assessments. Member companies found they were surveying the same unacceptable suppliers, and still duplicating efforts. While this type of organization was a step in the right direction, the result was still less than optimum.

Another approach was a true third party certification. The idea was that a recognized independent organization would perform the survey and place its own stamp of approval on the supplier's system. Customers would, hopefully, accept this certification of acceptability in lieu of performing their own surveys. One of the first drawbacks of this system was cost. Historically, customers absorbed the cost of sending their auditors and surveyors to suppliers, conducting the reviews, and internally publishing the results. The suppliers bore the expenses of providing a contact person to accompany the auditor or surveyor, and responding to any findings. In the early attempts at third party approvals, two methods were attempted to pay for the certifications. In one approach, a group of customers would be identified as using the same supplier, and they would be asked to share part of the cost of the

TABLE 7.5 Benefits of ISO 9000 Registration

- In some markets, registration is a legally mandated requirement.
- Customers are more receptive to partnerships.
- Many companies are requiring certification from their suppliers.
- ISO 9000 allows smaller companies to compete with large ones, as it can ensure "approved vendor" status with some customers.
- Customers have more confidence in the supplier's quality management system.
- Preventive techniques are in place to foster continuous improvement.
- Forces discipline to ensure that well-documented procedures are in place.
- Increased operational awareness by all employees of the quality system.
- Personnel are adequately trained.
- May help gain a market edge.

review. The supplier would also pay a fee to the third party. Many suppliers found the fees requested to be quite high, and the acceptance of the third party certification was low. Suppliers did not see the promised benefit of fewer reviews, and were asked to pay for reviews their customers previously absorbed the costs for. Customers had difficulty accepting the certifications because they had different internal and contractual standards they were expected to adhere to, and the third party certification did not assure all aspects of these different standards. In addition, each third party organization operated to its own standards, and certification from one organization did not have the same meaning as certification from another.

It soon became clear that the major roadblock to third party systems was a universally accepted standard. The advent of ISO 9000 addressed this key issue. With the formation of the European Union, and increased globalization of manufacturing and the marketplace, standards organizations from each of the major industrial nations agreed to adopt a single set of standards for quality systems. Included in the structure would be a system to accredit certifying organizations and certify individual auditors to a single standard. Today, ISO certification is rapidly becoming a basic requirement for suppliers (Table 7.5). Often, ISO 9000 certification is considered sufficient to assure a supplier has an acceptable quality system in place. In other cases, ISO 9000 certification is the first step, followed by a customer assessment focusing only on the customer's unique requirements. The automotive QS 9000 and aerospace standards for space-rated parts are good examples of customer-unique requirements that may be assessed after ISO 9000 certification. Some customers also require the use of advanced quality system tools for continuous improvement within the supplier's ISO system, such as quality function deployment or SPC/variability reduction systems. As we examine the various supplier evaluation methods below, we will consider how they may be applied in an ISO 9000 environment.

7.4.4 On-site assessment

On-site assessments, whether second or third party, take on many forms. Some of these assessment methods have developed from military procurement practices as implemented by major aerospace and defense contractors during the space race and cold war years. In the traditional lexicon, a *quality system survey* is usually defined as a complete review of the supplier's documented quality business system. A survey usually concentrates on the company's procedures. A *quality system audit* generally concentrates on key areas of the supplier's quality system, and should review both the company's procedures and verify effective application of the procedures.

More recent developments focus more on manufacturing processes. A *manufacturing process survey* is an overall review of a particular manufacturing line, and is oriented to the hardware rather than the procedure, although documentation is still a key factor. In the electronics industry, process surveys can have a place of great importance. A process capability study is an intensive analysis of a particular manufacturing process to determine the limits of the process. Process capability studies are frequently performed as first party assessments, and results may be requested by the customer.

An on-site second party assessment has definite advantages over desktop surveys. The customer is in control of the assessor, and can better influence the assessor's interpretations of the standard guiding the assessment. During the assessment, the assessor has the opportunity to assure that the supplier understands the customer's interpretation of the standard. The assessor has more freedom to focus on any anomalies that may be detected during the assessment. On-site assessments are, obviously, more effective in detecting a supplier that is misrepresenting its capabilities. Perhaps most important, the assessor has the opportunity to personify the customer in the minds of the supplier's personnel, putting a face on an otherwise impersonal entity.

In some cases, a customer may hire a third party organization to provide an assessor for a supplier assessment. These situations should not be confused with third party assessments. In this situation, the third party representative is not an independent third party, but rather an agent of the customer. The assessment remains a second party assessment. The customer may choose to hire the third party because of a lack of sufficient in-house personnel, a temporary need for expertise in a given area, or inconvenient location. The customer may still have the opportunity to coordinate interpretation issues with the assessor and identify particular areas of concern for the customer, but overall has less control over the assessor than the customer would have with an in-house employee.

In third party assessments, such as the ISO 9000 certification discussed in Chap. 4, the assessor is truly independent, and generally has no direct contact with the customer at all. The customer's acceptance of the third party assessor's results are based on the third party assessors accreditation and reputation. When a supplier considers entering a relationship with a third party registrar, it is important to look in to the registrar's accreditation and the registrar's reputation with the supplier's customer base. There is nothing that requires a customer or potential customer to accept a particular registrar's certification. If a registrar were to develop a reputation for lax standards while somehow retaining accreditation, customers may choose not to honor that registrar's certification. Some customers may also require the registrar be accredited by a particular authority. While most U.S. companies and many companies worldwide recognize accreditation by the Registrar Accreditation Board (RAB), some customers may require use of a registrar accredited by their home country's accreditation authority.

7.4.5 Quality system survey

Many companies have systems that require that quality system surveys be initially performed prior to or shortly after a contract award, and periodically afterwards. Since the traditional survey takes place before work on the contract is in process, the survey generally concentrates on the quality business system capability rather than implementation on the production line. Where a similar line is in place at the supplier, however, verifying implementation of the quality system on that line should be included in the survey. Under quality function deployment (QFD), process capability must be assessed and understood in advance of contracting.

The initial survey performed to verify the supplier's quality system has the capability to support the customer's requirements. Under an ISO 9000 system, the supplier is required to assess itself for compliance to contract requirements before accepting a contract. Many customers perform follow-up surveys periodically to assure the system originally approved is still in place and in use, and that it continues to support the customer's requirements. The frequency of the follow-up surveys varies. Many companies prefer to schedule surveys on an annual basis. Other options include reviewing performance data before performing a follow-up survey: If the data indicates the supplier's system is still performing effectively, the follow-up survey may be waived for the next period. In customer/supplier relationships that include exchange of SPC data and variability reduction goals, follow-up survey frequency is often reduced, or eliminated in favor of process management techniques. Customers usually limit the number of times a follow-up survey may be waived for good performance unless some other method of periodic assessment is in place.

While follow-up surveys frequently review the complete supplier quality system in the same manner as the initial survey, another approach is to select key segments of the system for follow-up review. The segments reviewed are then tracked against the total body of requirements so that over time, all segments are revisited. This approach is really an audit of the quality system following initial system approval. The audit approach can be more cost-effective where the basic system appears sound. Areas of particular concern can be targeted for the follow-up surveys, but care should be taken to cover all system areas within a few cycles. If a particular area is considered highly critical, such as internal audits, or a repetitive problem, that area may be reviewed at each cycle, along with a mix of other system areas. This is the approach used by most ISO registrars: periodic full system surveys with system audits during the time between.

7.4.6 Quality system audit

Quality system audits can be used by themselves, but since the system is not baselined by the customer prior to beginning the audits, there is greater risk to this approach. The risk may be mitigated when the customer has confidence in the overall system through other means, such as approval by an industry group like CASE. If this approach is selected, the audit segments should be arranged to cover the most critical areas first.

7.4.7 Qualified manufacturers list

One third party system unique to the microelectronics industry is the *qualified manufacturers list* (QML). In the late 1980s and early 1990s, the United States government was looking for ways to lower the cost of military procurement. One key cost driver identified was the use of specifications that told suppliers *how* to achieve desired performance requirements. These "how to" specifications assured the government a consistent, reliable product from suppliers approved and listed on a *qualified parts list* (QPL), but the system was inflexible. This inflexibility made it costly for suppliers to take advantage of new technologies to reduce cost without impacting performance or reliability requirements. Additionally, the system was slow to provide

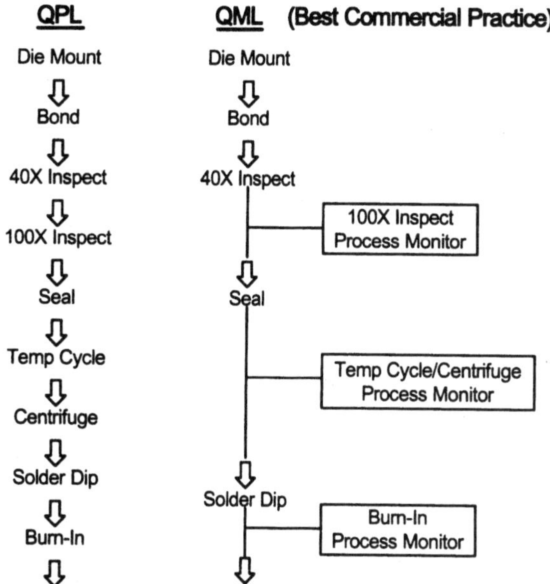

Figure 7.5 Flow comparison.

for new technology devices that system manufacturers desired to lower their own costs.

With the release of MIL-PRF-38535 in March 1995, MIL-PRF-38534 in August of the same year, and MIL-PRF-19500 in January 1996, a major change occurred in the way the United States government would control suppliers of military microelectronics. These documents, for hybrid microcircuits, integrated circuits, and discrete semiconductor devices, are *performance specifications,* establishing verification requirements that devices are required to be *capable* of meeting. Suppliers are encouraged to utilize best commercial practices and reduce or eliminate non-value-added tests and inspections (Fig. 7.5), provided that the devices remain capable of performing to defined requirements.

Under a QML program, the supplier's manufacturing line for a given technology is *baselined.* The baseline defines the flow of verifications and tests the supplier uses to assure the devices are capable of meeting performance requirements. Once the baseline is approved, the supplier may make changes with approval of the *qualifying agency,* usually the Defense Supply Center, Columbus (DSCC), formerly known as the Defense Electronics Supply Center (DESC), or a government contractor acting as an agent of DESC, or through a *tech-*

nology review board. Once a manufacturer's line is certified, a change control system coupled with a system of first and third party verification audits is maintained to assure the system continues to provide devices compliant with performance requirements while allowing the flexibility to take advantage of new methods and technologies.

If a technology review board (TRB) is used, as a minimum it must consist of representatives of the manufacturer's design, material procurement, assembly, test, reliability, and quality functions. The TRB is responsible for overseeing the manufacturer's qualified lines, the quality management program, audit program, and quality improvement program. The board is responsible for approving changes to the process baseline, monitoring and analyzing process data, taking necessary corrective action, and maintaining records. The board reports the status of the quality management program to the qualifying activity. The use of a TRB allows a great deal of autonomy for the manufacturer, but the TRB is charged with significant responsibilities that must be carried out with diligence for the QML line to maintain certification.

7.4.8 Process certification

While the requirements of the QML process are far more flexible and responsive to manufacturer-driven process optimization and technology advancement than its qualified parts list predecessor, some manufacturers have found their product areas are not covered under the technologies addressed by current QML specifications. Additionally, commercial enterprises that do not supply product to the military or government aerospace programs desired a standard system for manufacturing line certification that had many of the same attributes of QML, but with a broader application. In June of 1992, the Electronic Industries Association (EIA) attempted to fill this need with their ANSI/EIA 599, *National Electronic Process Certification Standard.*

The 11-page ANSI/EIA 599 attempts to provide broad guidance for electronic process certification. The standard is intended to be applied to suppliers already approved to ISO 9001 or ISO 9002, and focuses on certification of the process, not qualification of the product. The EIA may have been too successful in its endeavor to develop a broad standard. Many feel the standard is too broad for meaningful implementation. In July of 1996, the EIA first addressed this concern with the release of a companion to ANSI/EIA 599, EIA 681, *Assessment Guide for Process Certification.* There is also an effort under way to release a revised standard, ANSI/EIA 599-1, that adds more detail and structure to the original concept.

While EIA 681 is billed as an assessment guide, the combination of ANSI/EIA 599 and EIA 681 comes much closer to a commercial implementation of the QML process than ANSI/EIA 599 alone. As one would expect, the assessment guide focuses primarily on the conduct of the assessment and definitions of the various types of assessment. In addition to these assessment guidelines, EIA 681 addresses device qualification and decertification processes, two key items missing from ANSI/EIA 599.

The marketplace, in 1997, still has not embraced the EIA standard and guide for process certification, but as awareness of the documents increases, that may change. One area that may be holding the marketplace back from using the standards is the lack of a designated certifying authority. While QML certification flows from the Defense Supply Center, Columbus, the EIA documents leave the identity of the "certifying authority" intentionally open, so that any commercial entity can make use of them. If the documents do find a home in the marketplace, it will likely start with second party assessments and certifications well before a third party registrar can be found to champion the cause.

7.4.9 Process survey

While third party process certification is still in its infancy, more traditional second party manufacturing process surveys are increasing in use, and the EIA 681 guide is quite useful here. Manufacturing process surveys have been used for many years to validate controls on *special processes,* those processes that affect product characteristics that cannot be inspected or easily tested, like plating of hybrid packages, covers, device headers, and lead frames, or anodizing of assembly chassis. It has long been acknowledged that special processes must be controlled to assure the process results, rather than use inspection to verify results. Today, there is greater recognition that most manufacturing operations can be assured through process controls, and that by assuring process controls, manufacturing yields and product reliability both improve.

Second party manufacturing process surveys generally have two goals: (1) verification that the supplier has appropriate controls and documentation of its process and (2) familiarizing the customer with the supplier's process. A simplified example of a process survey checklist is shown in Table 7.6. An actual checklist should also include key characteristics to be verified for the actual process being surveyed. For a manufacturing process survey, the surveyor should have a good understanding of manufacturing processes and process controls.

TABLE 7.6 Sample Manufacturing Process Checklist

- ☐ Process flow defined and documented
- ☐ Staffing plans in place
- ☐ Appropriate process metrics identified
- ☐ Metrics current
- ☐ Line proofing completed
- ☐ Control limits defined
- ☐ Test equipment in place, certified, and calibrated
- ☐ Maintenance schedule defined and read
- ☐ Training plans in place, records maintained
- ☐ Capability analysis complete
- ☐ SPC system set up
- ☐ Material replenishment system in place

Ideally, the surveyor has familiarity with the specific type of process involved, although perhaps not the exact implementation used by the supplier. Frequently, a member of the customer's engineering staff is included in a manufacturing process survey, bringing specific experience to the team. Having engineering participation in a process survey also helps develop the customer's expectations for the product. This is particularly important for specialized processes or custom products where processing of the item has or potentially has a great effect on the customer's processes. When an engineer chooses an item for use, the engineer may assume some elements of the supplier's manufacturing process. Since different engineers may approach the same challenges differently, the assumptions may not be valid, and those assumptions may affect later processes. With a better understanding of the supplier's actual processes, the overall system may be better optimized. For example, a mold release used by the supplier may be an unexpected source of contamination for the customer, or the supplier may be spending a great deal of effort to maintain a characteristic that has no effect on the customer's use. Concurrent engineering teams make use of manufacturing process surveys with process familiarization specifically in mind.

When a known process is to be surveyed, well-defined checklists should be prepared in advance. The survey should consider critical process parameters, controls for those parameters, and documentation for the process and controls. There is often a tendency on the part of a customer to want to specify *how* a supplier's process is to be controlled. Care should be taken to avoid this unless there is an overriding reason that affects the customer's product. Many industry standards exist that suppliers may adopt into their systems. These standards, such as those from the Electronics Industries Association or Institute for Interconnecting and Packaging Electronic Circuits (IPC) should be identified and available when used.

A process survey should assure that the process is controlled, but recognize that there may be more than one way to achieve that control. For example, a rinse bath following an etching operation must have pure, circulating water to avoid contamination. The method used to purify the water and monitor the purity must be valid, but there may be several alternatives that are acceptable.

A process survey should verify that the supplier process is well documented. ISO 9000 para. 4.9 establishes several requirements for documenting and controlling processes. The documentation must include process control factors, process measurement methodology, and defined process control limits. The supplier must also document the process qualification technique. After determining the supplier has the process and process controls well documented, the surveyor should look at the processing area and verify the process and controls are in use as described.

Many supplier processes are considered proprietary to the supplier, and may be trade secrets. This can make verification of proper documentation and controls more of a challenge. In most cases, unless the customer is also a competitor in a similar product line, a supplier will allow the customer to review the process and documentation in place, although the customer is generally not permitted to retain copies of any materials. In practice, it is rarely necessary for the customer to retain supplier processing documentation, and retaining supplier proprietary documents carries risk. A customer is required to protect a supplier's proprietary information the customer is exposed to. In some cases, customers are required to sign proprietary information agreements before the supplier will permit access to the process.

In the case of a proprietary process that is considered a trade secret, the supplier may not permit any customer review of the process. In these cases, the customer should take steps to assure the process is documented, controlled, and repeatable in the best manner possible without violating the supplier's secrecy requirements.

Process capability analysis is covered in detail in Chap. 3. A process capability survey of the supplier's process is a necessary step in any supplier management program based on statistical process control and variability reduction. Monitoring the process capability of a supplier's trade secret process is one option to obtaining process oversight without causing the supplier to release trade secret information. Process capability analysis is also used in supplier partnership arrangements that include variability reduction goals, and in supplier certification activities, discussed later in this chapter.

7.5 Flowing Down Requirements

Most suppliers want to perform well, supporting their customers, and reaping the rewards of repeat business. If the customer has selected suppliers with appropriate care, those few dishonest supplier candidates will have been weeded out before a contract has been let. The question then remains: Why do some contracts seem to go so wrong? In most cases, the answer is poor communication, frequently resulting in incomplete understanding of requirements definition.

The process of flowing requirements from customer to supplier, and verifying that the supplier understood what the customer meant, can be far more complex than it appears on the surface. Let us consider ordering a simple cup of coffee. Not too many years ago, a customer in Boston ordering a cup of coffee "regular" would receive coffee with creamer and sugar. The same order in New York would result in a serving of black coffee with sugar. The specification "regular" had different meanings in these two regions, even though New York and Boston are in adjacent states. Moving across the continent to California, the customer ordering coffee "regular" would most likely get a cup of plain black coffee, or perhaps just a blank stare: the specification "regular" had no meaning at all applied to coffee there. The rich texture of colloquial expression may be fading from common language today, but it is still a strong factor in industry. Words do not have the same meaning to all industries. The specification *hardened* applied to steel does not mean the same as *hardened* applied to an electronic component. Care in communicating contract flowdowns is absolutely critical in achieving desired results.

The act of communication can be looked at with the following model (Fig. 7.6): Sender's initial thought or concept exists only in the

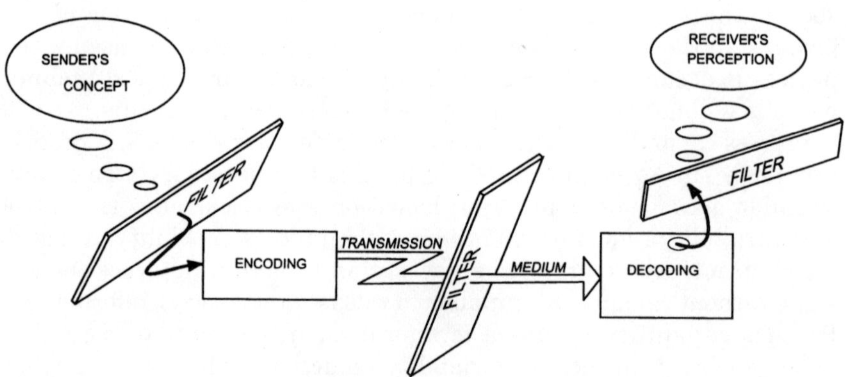

Figure 7.6 Open-loop communication model.

sender's mind. It is the goal of communication to relate this to the receiver. First, the thought is encoded—converted into language or pictures, for example. The type of encoding may be partially determined by the transmission medium to be used. People sitting across each other at a table would likely use spoken language to communicate, along with visual cues we call *body language*. Other times, we use written language, and in the case of engineering and product specifications, often diagrams and detailed drawings. The receiver decodes the information into its own perception of the sender's original thought. The receiver's perception is not a perfect duplicate of the sender's thought. As we saw in the "coffee regular" example above, language is not a perfect way to encode thoughts. This is one reason pictures or diagrams may be added. Also, there are aspects of communication called *filters* that affect how a thought is perceived. Words that have special meaning in one industry, but not in another for example, become filters, disrupting clear communication at times, while improving communication at other times.

If we consider a requirements document as flowing from a designer to a buyer, to a sales representative, to an engineer, to a production planner, to a worker in manufacturing before the item is produced, it is amazing the result ever resembles the initial concept. At each stage, the communication model above, with all its filters and flaws, was applied. For successful transmission of a concept, the communication model adds one more element, a critical one: feedback.

Feedback is a mechanism that allows the receiver to transmit its perception of the initial concept back to the sender, completing the loop. A communication model including feedback is referred to as *closed-loop communication* (Fig. 7.7). We see closed-loop communica-

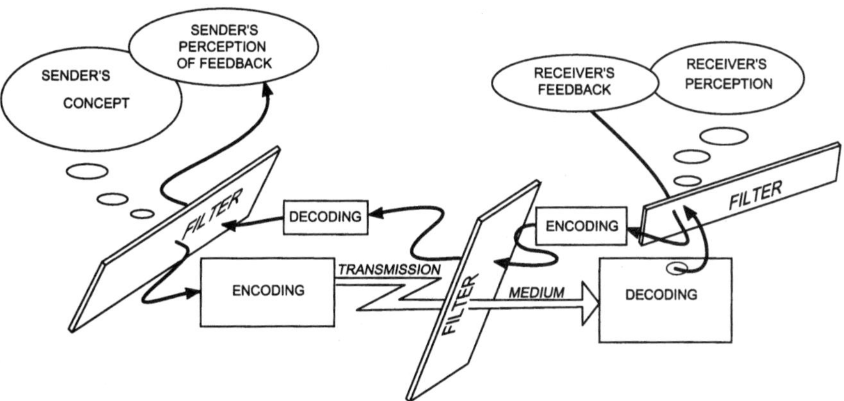

Figure 7.7 Closed-loop communication model.

tion on board Naval vessels between the captain and the helmsman; for example, the captain orders: "All ahead full!" to which the helmsman replies: "All ahead full, aye!" The more critical or complex an instruction, the more important closed-loop communication is. When ordering a number 343 widget from a catalog, ensuring that the buyer and seller are using the same version of the catalog and the same part number may be all the feedback necessary. In specifying a custom-designed part, feedback must be raised to higher levels, and include more members of the buying and selling team. When a performance specification is flowed down to a supplier, the supplier's design documents for meeting the performance specification should be reviewed and approved by the customer.

7.5.1 Feedback in procurement

Several systems have developed to support feedback during performance of a critical procurement. First we will discuss two formal conference forums frequently used by companies to support the need for feedback in understanding requirements: preaward and postaward conferences. These are formal procurement conferences, and require participation of the customer representative and supplier representative responsible for contracting, usually the Buyer's and supplier's sales or contracts representative. Other participants depend on the type of requirements raising concerns for the procurement. Typically, the team will include as a minimum a customer quality representative and an engineering representative. Large companies making major procurements frequently have specific program representatives that support pre- and postaward conferences. See Teams in Chap. 6.

Preaward conferences are used to provide a two-way forum where customer and supplier can work to assure a mutual understanding of requirements before a contract is agreed to. A preaward conference should be considered when a proposed procurement is complex, or performance goals or milestones are unusually challenging. Preaward conferences usually focus on proposed contract requirements, supplier capabilities, and the supplier's approach to performance on the contract after award. Under an ISO 9000 system, the supplier is required to perform an internal contract review prior to accepting the order. [11] A preaward conference can help to facilitate this review. In an environment of performance specifications (see Chap. 2 for a definition of performance/detailed specifications), the supplier develops detailed specifications to fulfill the customer's performance requirements. A preaward conference to thoroughly review and understand the customer's requirements can be critical to success.

Postaward conferences are designed to provide a forum for the customer to make sure the supplier fully understands all contract requirements prior to commencing production; even in the case of an ISO 9000 system, where the contract was reviewed prior to acceptance, a postaward conference may have value as a method for assuring feedback and a closed communication loop. In either case, when these conferences are used, key customer and supplier representatives should attend, including as a minimum the contracting agents (Buyer's and supplier's contracts or sales representative), the engineering representatives, and the quality representatives, all part of the product team (see Chap. 6).

Design reviews are critical in requirements-driven procurement. While requirements-driven procurements are preferred, since they afford the greatest degree of freedom to the supplier to optimize manufacturing, they also drive a need for the supplier to communicate their approach back to the customer. ISO 9000 requires internal design reviews under para. 4.4.6. It is often good practice to also perform design reviews with customer representatives. Once again, the object is to maintain closed-loop communication between customer and supplier. If a supplier product is critical to the customer's assembly, or has critical interfaces with other items outside the supplier's control, design reviews can facilitate the information exchange necessary to assure that contract requirements are achieved. If the development effort includes establishment of interface requirements, joint design reviews should be considered essential. In addition, as a system design matures, interfaces or performance requirements sometimes need to be adjusted to compensate for new information or failure of other components to achieve design goals. Joint design reviews provide the opportunity for customer and supplier to work together to optimize component design for system performance.

Technical interchange meetings (TIMs) are another tool that may be used to facilitate customer/supplier communication. These meetings can be informal and called at the discretion of either party, or they can be formal, scheduled events that are included in the contract. The purpose of a TIM is again to provide a forum for supplier and customer engineers and other team members to address technical issues together and assure that technical approaches used by the supplier support the customer's requirements in an efficient, effective manner. When a procurement includes a significant level of development effort, scheduled technical interchange meetings are an excellent tool to coordinate supplier and customer design experience.

Excellent business systems take every opportunity to facilitate communication between customer and supplier. Mechanisms are maintained to support informal and formal requests for information or requirements clarification, as well as to provide notification of concerns or problems. While many vehicles can be used for this type of communication, from formal contract letters or specific controlled forms to quick exchanges of electronic mail, and some companies give little thought to this type of exchange, excellent businesses develop specific systems for this use and stress their availability to the supplier. The supplier is encouraged to use the system with confidence that open communication will create a positive relationship.

7.5.2 Clear description

Whatever communication medium is used, under an ISO 9000 system, or any good business practice, purchasing documents must clearly describe the product to be supplied. Describe the item in terms that include grade or other classification. For example, military transistors may be classified as JAN, JANTX, JANTXV, or JANS depending on the level of reliability required of the device. Give a full title or name for the item, including any other related identification. The ¼-in lock washer you need may not be just any ¼-in lock washer. If you need a ¼-in stainless-steel internal star lock washer per your company's specification LWSHR-1234, revision D, you need to state it in your purchase order. If the buyer requires a specific quality system standard be followed by the seller, that too must be included, for example: "Supplier must be certified to ANSI/ASQC Q9002-1994." [12]

7.6 Assessing Product Compliance with Requirements

Assessment of supplier product compliance with customer requirements historically means inspection. While inspection remains one method of product assessment, inspection alone is usually inefficient and costly. Under advanced quality systems, other assessment methods are used to complement and, frequently, replace inspection. In all but the rarest of cases, the supplier is required to assess the compliance of a product prior to shipment to the customer. With this in mind, the customer should focus on assuring the supplier is effectively performing product compliance assessment. This section will address both the traditional receiving and source inspection concepts of product assessment and some alternatives.

7.6.1 Importance of early detection

It is important to recognize that the cost impact of defective product increases substantially as the product progresses further through the

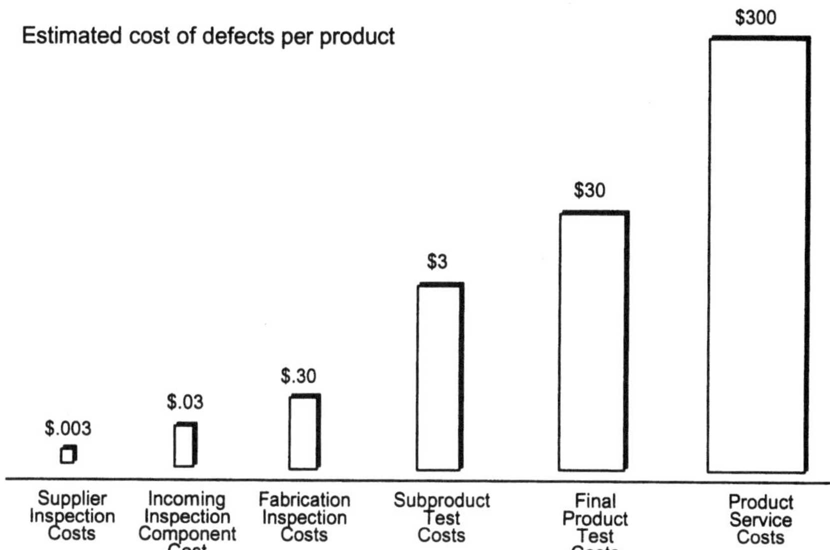

Figure 7.8 Escalation of cost of errors down the production line. (*From Ref. 13; used with permission.*)

manufacturing process. Every effort needs to be made to detect defects at the earliest possible point. The chart shown in Fig. 7.8 graphically depicts the cost of defects at various points in the manufacturing process for electronic assemblies. [13] It is clearly far more cost-effective to detect defects at the supplier, where the cost per defect is less than one cent, than to detect a defect at final test, where the cost is $30 or more per defect. Additionally, schedule recovery is significantly easier when defective product is discovered earlier in the manufacturing process. Less time has been wasted manufacturing and testing the defective product, and less time is required to replace the product.

Another important consideration is the effect of supplier defects on product yield through final test. For many years, electronic components were typically accepted to lot sampling criteria designed to assure 99 percent good product. In many manufacturing environments, 99 percent good product feeding the line is not good enough. It is not unusual for a module to use 50 to 100 pieces of some components, such as resistors, diodes, and inductors. If these components are accepted into stock as 99 percent acceptable, every module built using the components could fail in test, resulting in costly rework and additional testing. Table 7.7 addresses the effect of supplier defects on

TABLE 7.7 Effect of Parts Defects on Board Yield

ppm defective	Effect on printed wiring board yield (at 100 parts/board), %
5000	61
3000	75
2000	82
1000	90
100	99

ppm = parts per million.

product yield. Clearly, product defects need to be measured in parts per million (ppm), or even parts per billion (ppb) to avoid failures in module or assembly test.

Inspection of supplier product alone cannot assure defect levels in the ppm or ppb range. That does not mean that inspection does not have a place in supplier management and supplier product assurance. Inspection is a tool, and a good tool, provided the limitations are recognized.

7.6.2 Receiving inspection and certification

Traditional receiving inspection is effective as a coarse screen. Product damaged in shipment and incorrect or mismarked product are examples of things effectively detected in a traditional receiving inspection function. A good receiving inspection function also reviews certification and test documents supplied with product. These documents form part of the communication feedback loop previously discussed. A certification from the supplier, for example, stating that the product supplied contains no ozone-depleting chemicals, demonstrates that the supplier recognized the requirement for no ozone-depleting chemicals existed in the contract, and provides evidence for recourse if the requirement is not met. Test data can be similarly used to assure that the supplier recognized special test parameters required that differ from a supplier's catalog product.

Certifications can become more important when items are purchased through distributors, but specific restrictions need to be controlled. In some cases, a customer wants to limit procurement of a common item to a specific manufacturer's product. A requirement that the distributor certify the origin of the product provides evidence of the understanding of the requirement, and is a vehicle to verify that the manufacturer is indeed the one specified by the customer.

Certifications of chemical analysis, or material characteristics cannot always be taken at face value. In the 1980s widespread fraud was found in the distribution of grade 8 threaded fasteners. Some distributors were substituting lesser materials and certifying them as grade 8. Grade 8 fasteners are used in critical, high-strength applications, and failure of these fasteners in many cases could be life-threatening. These fraudulent fasteners found their way into many critical assemblies in all aspects of industry, from automotive parts to aircraft and aerospace products. The fraud was initially detected through two paths: review of field failures and periodic testing of items against certifications. The fraudulent certifications themselves became critical evidence in the prosecution and conviction of the people responsible for the fraud.

Periodic verification of chemical analysis and certifications can detect more benign errors too, and should be used as part of a supplier control program. In one case, a supplier of clad wire was required to certify the absence of antimony, and routinely did so. When the customer performed an independent analysis of the wire, antimony was found. Investigation revealed that antimony was present in a braze material used in the cladding process, and therefore was deposited within the wire. In this case, the error was not intentional on the part of the supplier, but the certification was in error, and the material did not comply with contractual requirements. Through the years, mistakes in handling and storage of lead wire, weld wire, and raw stock have often lead to erroneous certification of analysis, and have been detected by periodic independent verification.

More sophisticated receiving inspection functions can measure or retest critical product characteristics and track results well enough to determine product substitutions or major changes in the supplier's process. In most cases, this can be performed with limited testing of key parameters on a sample of items received in each lot. This type of receiving inspection and test, coupled with supplier statistical process control data, can be used to track a supplier's process rather well. In an extreme case, the customer can track the supplier's processes better than the supplier.

One large corporation actually retested and characterized all active electronic components in receiving inspection, and maintained running process charts of product performance against their internal product needs. The customer then worked with the suppliers to optimize the supplier's process for those characteristics the customer desired. In some cases, the products were not unique for the customer, but rather the supplier's standard catalog product lines. The customer was able to show the supplier that variability reduction coupled with a shift in parameter means could optimize the product for the one customer and still be compliant to the supplier's catalog description. This customer was using sophisticated receiving inspection data to control the supplier's process as a strategic tool to in effect obtain custom product without specifying custom product and paying premium prices. To make that approach work, the customer had to be a large consumer of the product in question. The customer maintained higher test efficiency than the supplier base because the test operation's costs were spread across several product lines. Had the customer required each of its suppliers to perform similar testing and evaluation, the cost would have been multiplied by each supplier, and, of course, the costs would have been returned to the customer as special test charges. Once the supplier processes were optimized for the customer's requirements, routine testing at receiving was eliminated. Only periodic sample testing was performed to verify continued process control.

7.6.3 Verification and validation

There are several factors to address when selecting verification methods. Volume, device complexity, criticality, past experience, and type of procurement all come into play in making the decision. A small number of items may not warrant much verification if the application is not critical; verifying the item is what was ordered and its basic functionality may be sufficient. On the other hand, a single complex and critical item may warrant extensive verification before acceptance. Items manufactured at high volume using defined processes, as many electronic components are, are best verified by using process control. If the item is considered for delivery supporting a just-in-time (JIT) program, process control is a must. Only after process controls are in use by the supplier and advanced quality techniques are in place should an item be selected for JIT procurement. JIT procurement has no margin for rejected lots caused by out-of-control processes, or delay through the inspection function. JIT is discussed in Chap. 2.

In a requirements-driven environment, the supplier's detail specifications should include verification and validation steps to assure that all the customer's requirements are fulfilled by the supplier's product. In some cases, there may be specific qualification testing, and/or the verification may take the form of a first article inspection, where each requirement is verified on the first production item or first lot. Validation, per ISO 9001, para. 4.4.8, is generally review of the verification of all functional requirements—a review of the functional configuration audit (FCA) and physical configuration audit (PCA) and the acceptance test passed on the first production item.

Verifications and validations are required as part of an ISO 9001 quality management system to verify that the customer's performance requirements have been met. After a process or design change, a first article verification of those attributes affected by the change can be performed, and the results compared to the initial first article records. This methodology can also be used by a customer as a periodic audit to assure that unauthorized process changes have not been implemented, intentionally or inadvertently, by a supplier.

7.6.4 Destructive parts analysis

Another method useful for assuring process integrity as well as workmanship on critical products is a destructive parts analysis (DPA). Since DPA requires consuming one or more devices, and the process of DPA can be costly, it is generally used for more critical applications, such as space-rated or automotive safety-rated items. DPA allows verification of many workmanship areas that are not fully verified through other inspection or test methods. For example, by sectioning a bond area, the exact condition of the intermetallic area can be assessed to assure the bond process is performing as expected. Voiding, cracking, or contamination hidden from view by internal components during x-ray of potted devices can be seen in DPA. DPA should be considered whenever a product has critical features that are hidden from other forms of assessment.

A related form of verification is a tear-down analysis. A tear-down analysis is similar to a DPA, but performed on a larger assembly. A tear-down analysis is frequently but not always destructive, since in many cases the parts may be reassembled to a compliant condition. Once again, the primary reason for a tear-down analysis is to perform a detailed workmanship verification. Often, a tear-down analysis is used during qualification to determine what degradation may have occurred from environmental testing.

7.6.5 Assessment

When statistical process control techniques, variability reduction, or other advanced quality techniques are used to assure compliance, there should be specific agreements for reporting and reviewing data. See Chap. 3 for more information on advanced quality concepts. If there are specific variability reduction goals, progress toward those goals should be planned and results tracked on a regular basis. Many customers and suppliers make variability reduction agreements, but fail to follow through by monitoring progress. This failure not only can result in missed goals, but also could result in the customer not being apprised of deteriorating control of the line. Many companies make the mistaken assumption that when advanced quality techniques are applied, assessment activity is eliminated. Of course, this is not the case. The techniques of assessment shift to more efficient and usually more effective methods, and there is usually a reduction to the amount of time spent in assessment, but assessment is not eliminated.

7.6.6 Automated testing

The vast majority of electronic components and "black box" assemblies are tested using automated test equipment. When this is the case, good practice dictates that the customer should review the test software and software configuration controls. Problems in this area are frequently seen when a component is procured that is a special selection or otherwise similar but not identical to the supplier's standard catalog product. People have come to rely on computer-controlled equipment and the image of accuracy such equipment has developed. What is forgotten is that the best, most accurate test station created is only as good as the software written by a person interpreting the customer's requirements. Simple errors, typographical errors, transpositions, slipped decimal places, can result in automated, high-volume testing to completely wrong limits or with wrong conditions. Test software should never be used for acceptance without performance of a verification to requirements and configuration controls preventing unauthorized changes to the software.

Another manifestation of automated test equipment problems can be seen when qualification or low rate production programs use manual test bench setups, and high-volume production is shifted to automated equipment without sufficient review of the automated system. Automated test systems frequently use different techniques to apply test conditions than a technician may use on a test station. Instead of applying power for a test for several minutes while a technician takes several readings, the more efficient automated system may apply power for only a few seconds while the data is gathered. While in theory the same test is performed, the results may not be the same. When an approved test system is migrated to automated equipment, it is good practice to perform correlation verifications between the systems. Test one item on the manual system, record the results, and retest the same item in the same environment using the automated system. In comparing the readings from the two tests, look for differences that exceed expected test repeatability. The automated system should not be certified for use until any differences are understood and accounted for in acceptance criteria.

7.6.7 Source inspection

Source inspection is a method used since the beginning of merchant trade for product acceptance. Source inspection is merely a product appraisal performed before the product is released from the supplier facility for shipment to the customer. In an inspection-based acceptance system, source inspection is frequently used when special tooling or test equipment available only at the supplier's facility is necessary for inspection. Another reason for inspection at the source is for in-process assessments, such as a visual inspection prior to sealing a hybrid or integrated circuit or before potting a potted assembly. Since the supplier is responsible for assuring product compliance, the source inspection is usually a repeat of effort already performed by the supplier. Traditional source inspection is usually an unnecessary redundancy. Customer source activity can be effective, however, if a broader perspective is applied. This broader scope activity is often referred to as *source surveillance.*

Source surveillance activities are generally performed by a quality engineer rather than an inspector. Surveillance activities may include some traditional inspection activities, but the greater benefit comes from process monitoring and supplier coordination. Rather than appraise the product, source surveillance appraises the supplier's assurance system. When advanced quality techniques are in place, the customer's source surveillance representative is usually responsible for reviewing the supplier's process control data and monitoring progress toward variability reduction goals. Where an integrated team of Engineering, Purchasing, and Quality Assurance, sometimes called a product team, is used for supplier management, the source surveillance representative is often charged with all routine supplier management tasks that require customer representation at the supplier facility. These extend beyond traditional "quality" tasks to include, for example, review of schedule milestones or gathering engineering data. This elimination of functional barriers brought about through the use of integrated teams can be a source of cost savings. Where a functionally segregated organization may send a quality representative for supplier product assessment, a buyer to review schedules, and an engineer to coordinate technical issues, an integrated team can often rely on one team member for many supplier interface activities.

7.7 Measurement and Rating Systems

There are several ways to look at suppliers. One good tool to use is a perceptual mapping chart. First look at your own business: What kind of a supplier is your business, as seen by your customers? Next,

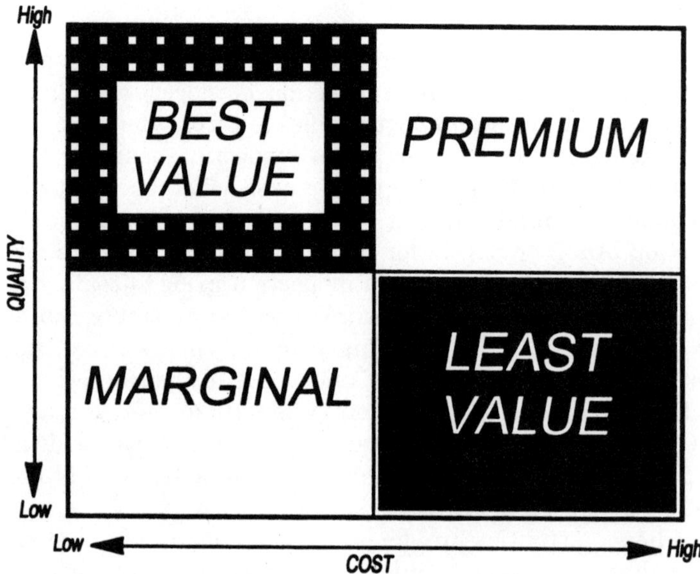

Figure 7.9 Perceptual mapping of suppliers.

consider how your suppliers seem to you. On the chart shown in Fig. 7.9, two factors are compared: quality and cost.

In the case where both Quality and Cost is high, shown as the upper right quadrant of the chart, the supplier is judged to be a "premium" supplier. Luxury cars are often marketed as being in this position. The marketing approach is to convince the buyer that the extra quality is worth the extra cost. In the lower right quadrant, the supplier is seen as high cost and low quality. At first, it may not seem that a supplier can have a business with low-quality, high-cost product, but look at the many fads that cycle through retailers each year. Many of these items are indeed low quality with a high cost. An additional key attribute, however, is that they seldom remain on the market for long. Looking at the lower left quadrant of the chart, we see the opposite of the premium supplier: low quality and low cost. In general, these suppliers are seen as having the least value to a manufacturer. Low-cost souvenir items often fall into this category, and many manufacturers are successful in that niche, but it is a niche market: small and specialized. The remaining quadrant is the region of "best value" suppliers. These suppliers are seen as high quality and low cost. This is the region claimed by the broadest range of successful suppliers. We all like to believe we are buying high quality at a low cost. A rating system should provide evidence to support or refute this popular claim.

It is important to note that this is a perceptual mapping chart. In this measurement, the perception, or belief, of the person conducting the measurement is what counts. This perception may not be, and often is not, objective fact. Most supplier selection begins with the buyer's perception of the seller, but it should not end there. Ultimately, supplier selection should always be made with objective data to support the decision. The amount and nature of the data will depend on the kind of supplier being selected. If the supplier is critical to the buyer's success, as much information as possible should be collected and assessed during the selection process. If the supplier has little impact on the supplier's business, much less evaluation may be needed.

Many manufacturers group their suppliers by their criticality to the manufacturer's system. Critical suppliers, whose product is of strategic importance to the manufacturer, are one group. These strategic suppliers should be selected with the most care, and managed with great attention. Strategic suppliers often have one-of-a-kind products, or have unique capabilities that set them apart. In many cases, the need for their product or service makes the selection automatic, but the need to have objective data is not eliminated by the fact that the decision to use the supplier has been made. Effective management decisions require knowledge of the supplier and the supplier's business, and suppliers must be managed.

After supplier selection, the need for measurement continues. Supplier performance must be tracked to support management of each supplier and of the overall supplier base. There are many approaches to supplier performance measurement, and we will explore several, but in all cases, effective supplier measurement must always be based on objective data driven by the terms agreed to by both buyer and seller in the contract. Those agreed-to terms may be stated directly in the body of a formal contract, on the face of a purchase order, or within documents referenced in those documents and specifically made a part of the contract through the reference. As different measurement approaches are discussed, we will return to this concept of objective measurement to agreed criteria.

In most manufacturing companies, small and large, supplier rejections at receiving inspection are measured. Most often, this measurement takes the form of a percent of lots or pieces accepted versus the percent rejected. This is a simple and straightforward measure, and if based on agreed criteria, a good measurement to use. Some slightly more sophisticated systems will measure both lot and piece part rejections through a formula to derive a score.

7.7.1 Assigning responsibility

What most of these systems fail to consider is responsibility for the rejection. Most of these simple systems are responsibility-blind, and the bias is to assume that any rejection is the fault of the supplier. Not only may this bias be unfair to the supplier, it may cause corrective action activity to be misguided and fail to address the root cause of the problem. It is important that every rejection be objectively reviewed and responsibility be placed appropriately with either the supplier or customer. If the rejection is caused by a customer drawing error, or a failure of a desired parameter not included in the procurement contract or specification, the responsibility for the rejection lies with the customer, and the rejection should not negatively impact the supplier's rating.

Measuring acceptance of supplier product through receiving inspection is usually not enough in today's electronic manufacturing business. If supplier product is inspected at the source, prior to shipment, the results of those source inspections need to be included in the measurement system. There is also the issue of certified suppliers or ship-to-stock suppliers, where no receiving inspection, or reduced skip-lot inspection may occur. Other measurements must be used to track the performance of these suppliers.

A system for collecting supplier part failure data from the manufacturing floor is important for all good supplier performance measurement, and vital for measuring suppliers not subjected to receiving or source inspection. Part rejection information from manufacturing must also be reviewed for responsibility, and this review may be more difficult than the review in receiving. One consideration, often neglected in manufacturing rejection systems, is components removed in rework that are not or may not be defective, but are replaced as part of an overall troubleshooting procedure. Care must be used in reviewing raw parts removal data to ensure that the reason for removal is addressed. For example, a rework procedure for a bad power transistor may require removal of the resistor and five resistors suspected of overstress due to the transistor failure. When parts removal data is reviewed, the resistors will appear as a larger problem than the transistors because of the large number removed, when in fact, the resistors are not the problem at all. The data collection system should be designed to identify the reason for part removal so that corrective action can be directed appropriately.

7.7.2 Delivery

Of course, measuring acceptance rates and functionality of supplier product is only part of the supplier's overall performance. Delivery is a factor of increasing importance in today's environment. With more emphasis on asset management, reduced warehouse stock, and just-in-time shipment, delivery performance is critical. Once again, it is important that the measurement be based on agreed criteria as defined in the contract. If the contract is written with a "will ship" commitment, the measurement system must be based on the shipment date, not the date of receipt. Some systems track an "in-stock" or acceptance date, but delays in the buyer's dock-to-stock cycle can then negatively affect measured performance. It is usually best to base contracts and measurement systems on "on-dock" dates. It is easier to track performance to an on-dock date, and the buyer has more flexibility in prioritizing the dock-to-stock cycle.

7.7.3 Supplier rating and incentive program

The supplier measurement system preferred by the author was originally developed as a cost-based system by Rockwell International Corporation in 1984. The system is called SRIP (pronounced *shrip*), for supplier rating and incentive program. The system was developed to combine several delivery and quality performance factors into a single rating number that represents the cost of nonconformance with contractual requirements. Since its inception, Rockwell's SRIP program has been widely copied in both commercial and military/aerospace companies, public utilities, and in some military procurement organizations. In 1990, *Purchasing* magazine speculated that the SRIP program could be a standard for industry. [14]

The advantage of SRIP is that it is a comprehensive system, taking into account delivery performance as well as quality performance factors. The system can be adapted to the factors measured within any business system. While setting up a measurement system based on SRIP takes some work up front, maintenance is mostly transparent to the business system, since SRIP is driven by normal business activity.

The basic premise of SRIP is to measure nonproductive costs from past procurement with a supplier and apply them as a prediction of future performance. A formula (Fig. 7.10) calculates the nonproduc-

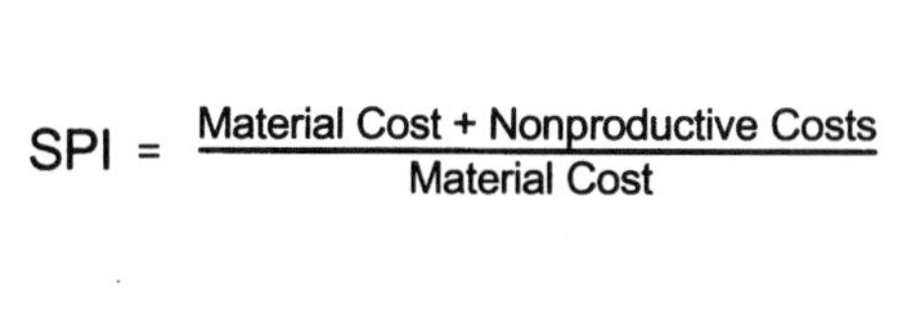

$$SPI = \frac{\text{Material Cost + Nonproductive Costs}}{\text{Material Cost}}$$

Figure 7.10 The basic SRIP formula for supplier performance index (SPI).

TABLE 7.8 Nonproductive Events in SRIP

Nonproductive Events Are Additive	
Quality assurance events	Schedule events
Source rejection	Interim undershipment
Resubmittal to inspection	Early receipt
Return to supplier	Overshipment
Rework at customer	Late receipt (late receipt is a daily assessment until on schedule or closed)
Issuance of a corrective action request	
Material review action	
Latent defects (shop floor)	

tive costs against material costs to derive a score that can be used either as a simple scoring factor or as a bid multiplier to adjust bids to better reflect predicted true cost.

The first step in developing a SRIP rating system is to determine the factors to measure. In determining these factors, consider two key things: (1) What measurable supplier performance factors drive your manufacturing performance? (2) Which of these performance factors do you currently track in your business system? In many cases, companies already have systems in place to track the factors they find most important. Table 7.8 lists several factors that may be tracked in a SRIP-type rating system.

The next step is to determine the weights for each negative factor. In the original SRIP system, the weights were standard costs calculated after a time study of each of the tracked events. Table 7.9 provides an example of how these costs could be determined. In practice, the weights can be derived in any manner provided they are applied equitably. In the table, each action has been determined to have a standard time. That standard time is multiplied by the appropriate labor rate to determine the standard cost. The activities involved in a rated event are totaled to determine the standard cost used for the event. For example, returning a shipment to a supplier may include the time in receiving inspection to document the rejection, time for a quality engineer to review the rejection, the buyer's time to process the request for returning the material, debiting the supplier, and adjusting the purchase order to reopen requirements, time for manufacturing to replan schedules, and time to repackage and actually ship the parts back to the supplier. When one examines the activities involved, a simple "return to supplier" is not so simple.

TABLE 7.9 Cost Factor Development

Cost at $50 per Worker-Hour

Function	Worker-hours per occurrence	Return to supplier	Accept with repair	Accept non-conforming material	Material late	Material early	Excess material	Short material
Receiving inspection	0.3	$15	$15	$15	$15	$15	—	—
MRB* review	0.9	45	45	45	—	—	—	—
Technical analysis	1.0	50	50	50	—	—	50	50
MRB disposition	0.6	30	30	30	—	—	—	—
Production reschedules	2.1	105	105	—	105	—	—	105
Inspection	0.3	—	15	—	—	—	—	—
Packaging	0.4	20	—	—	—	—	—	—
Shipping documentation	0.2	10	—	—	—	—	—	—
Inventory/carrying costs		—	—	—	—	60	20	—
Nonproductive cost per occurrence		$275	$260	$140	$120	$75	$70	$155

*MRB = Material Review Board

	Supplier A	Supplier B	Supplier C
Quoted price	$1000.00	$1050.00	$1025.00
Times (X) SPI	1.450	1.230	1.086
Total projected cost	$1450.00	$1291.50	$1113.15

- Supplier C is selected for award, even though the quoted price is not the low bid.
- Experience indicates supplier C is the lowest-*cost* supplier. The lowest *cost* is the best buy.
- New suppliers or suppliers with insufficient data for a valid SPI are weighted at their commodity group average.

Figure 7.11 Comparison of three bids.

The period of time to accumulate data needs to be determined. The time period should be long enough to collect a reasonable base of data, and should be compatible with other measurement systems used. In most companies, data is tracked for 1 year. How often the ratings, referred to as the *supplier performance index* (SPI) are calculated is somewhat dependent on the volume of receiving activity. As a general rule, the SPIs should be calculated at least monthly, but not less than weekly. If the ratings are calculated too frequently, it becomes confusing.

Once the data is collected for each supplier, it becomes a simple matter to perform the calculations and derive the ratings. A "perfect" supplier would have an SPI of 1 according to this formula, and the SPI would increase to something greater than 1 as performance declines. The original SRIP concept used the ratings as a multiplier for competitive bids, awarding the contract to the qualified supplier with the lowest adjusted cost after applying the SPI multiplier (Fig. 7.11). The ratings can also be used to rank potential suppliers, the lowest SPI ranking first.

When there is a large, diversified supplier base, it is usually helpful to group suppliers by broad product types to provide effective benchmark comparison. With a smaller supplier base, or a supplier base that is fairly homogeneous, grouping categories of suppliers may not be necessary. If suppliers are grouped in categories that are too narrow, some categories may include only one or two suppliers, which is not desirable.

One of the benefits of a SRIP-type rating program is that it can be used as a tool to reduce an overly large supplier base. SRIP works in two ways to help reduce the supplier base. As the rating system is used for contract awards, poor-performing suppliers will stop receiving contracts and naturally fall off the active supplier rolls. The other

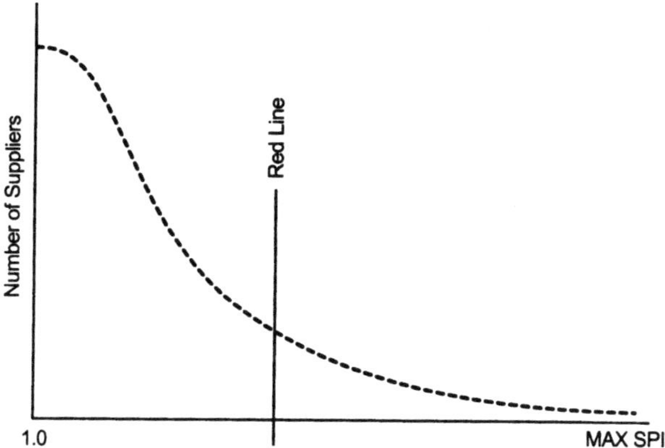

Figure 7.12 Selecting a SRIP redline.

factor in the program that aids in reducing the supplier base is referred to as the *redline*. As data from an existing supplier base is compiled and analyzed using SRIP, it is typical to discover that a plot of the SPI numbers takes the form of half a bell curve, with a long tail (Fig. 7.12). A point along the curve is selected to be a cutoff. Suppliers with an SPI above that point are ineligible for bid or award of new business. Any exceptions, such as a sole source supplier, should require independent review either by management or a peer review team. Since continuous improvement in the supplier base drives SPI numbers downward, the redline should be reevaluated from time to time to remain valid.

In cases where the supplier base is diversified, it may be helpful to set different redlines for each commodity group. It is often the case that some categories of supplier as a whole perform poorly when compared to other categories. When a redline is selected to be meaningful for the total supplier population, it may exclude an entire category.

One drawback to the simplified SRIP formula shown earlier is a natural bias in favor of suppliers that provide costlier product lots. A supplier that delivers $10,000 worth of product each month can absorb a $500 nonproductive cost with little effect. A supplier delivering $1000 worth of product each month would be hard hit by the same $500 nonproductive cost. To address this problem, a lot normalization factor Q was developed (Fig. 7.13). The Q factor adjusts the total nonproductive costs for each supplier to compensate for these lot differences. The expanded SPI formula including Q is shown in Fig. 7.14. Figure 7.15 shows a comparison of two supplier SPIs, calculated without and with Q.

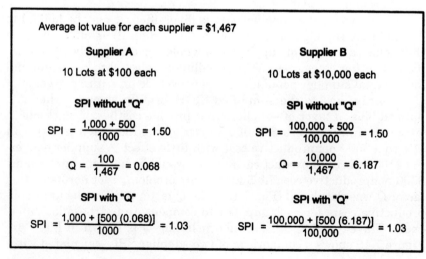

$$\text{"Q"} = \frac{\text{Average Lot Value for Individual Supplier}}{\text{Average Lot Value for all Suppliers}}$$

Figure 7.13 What is *Q*?

$$\text{SPI} = \frac{\text{Material Cost} + [\text{Nonproductive Costs (Q)}]}{\text{Material Cost}}$$

Figure 7.14 SRIP formula.

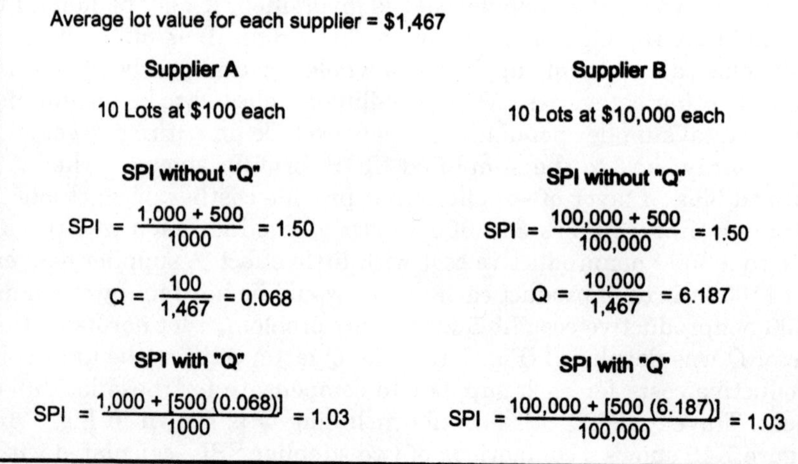

Average lot value for each supplier = $1,467
Supplier A
10 Lots at $100 each
SPI without "Q"
SPI = 1,000 + 500 / 1000 = 1.50
Q = 100 / 1,467 = 0.068
SPI with "Q"
SPI = 1,000 + [500 (0.068)] / 1000 = 1.03
Supplier B
10 Lots at $10,000 each
SPI without "Q"
SPI = 100,000 + 500 / 100,000 = 1.50
Q = 10,000 / 1,467 = 6.187
SPI with "Q"
SPI = 100,000 + [500 (6.187)] / 100,000 = 1.03

Figure 7.15 An example of how *Q* works.

Another use for the *Q* factor is to improve the resolution of the SPI rating. As a SRIP program matures and suppliers improve, the significant digits of the SPI shift further to the right. Adding a constant in front of the *Q* factor inflates all the nonproductive costs proportionally to move the significant digits back to the left.

No supplier rating system can be effective in improving supplier performance unless the supplier knows it is being rated, what its rating is, and how that rating compares to other suppliers. Using SRIP, this is accomplished by providing each supplier with a regular report with its own SPI, and the average SPI for the commodity group, or for all suppliers if commodity groups are not used. A SRIP report should also identify what nonproductive events were charged against the supplier for the current rating period. For other systems, a similar level of information appropriate to that system should be provided to all suppliers on a regular basis. Suppliers should be given the opportunity to have their data reviewed if they feel there are inaccuracies, and corrected if inaccuracies are discovered during the review.

Communicating performance measures back to the supplier is another form of feedback in closed-loop communication. A good supplier will take corrective action when performance measures point to a need for improvement. Communication with the supplier is the most important factor in a rating system and must be maintained in a clear, professional manner at all times.

7.8 Supplier Certification

Supplier certification is a special type of supplier partnership. While all certified suppliers should be considered supplier partners, not all supplier partners should be certified. The idea behind supplier certification is to move supplier product directly from the receiving dock to stock, or even the manufacturing floor, without additional receiving inspection. This relationship requires a high degree of confidence in the supplier's ability to provide consistently compliant product, and trust that the supplier will maintain the necessary processes to assure continued compliance. Since there is no routine assessment during the receiving cycle, the customer should maintain a traceability system capable of recalling any product released into the system that may subsequently be determined to be suspect.

Although many certified supplier programs are initiated as cost reduction activities, cost reduction is a poor motivation for beginning a certified supplier program. During the initial phases, supplier certification can actually increase costs to the customer. The benefits of supplier certification primarily lie in reducing cycle time delays caused by redundant inspection operations at source and receiving, and the delays caused by material holds pending inspection by the customer. Additional benefits are achieved by the shift from an inspection-based product assessment to a metrics-based process assessment. The costs of the inspection operations is generally shifted to other oversight activities, such as SPC data review and periodic supplier and product audits. As the certified supplier program matures, there are cost savings, sometimes substantial savings, but the investment must be made in the beginning, and oversight must not be removed as much as shifted.

There are many supplier certification programs in use today. Most of them can be identified as belonging to one of three basic groups: (1) delegated inspection programs, (2) SPC/variability reduction VR programs, or (3) combination programs, containing elements of both delegated inspection and SPC/VR programs. In all cases, suppliers are first screened to selection criteria, evaluating the supplier's product acceptance by the customer's receiving and source inspection, delivery performance, and volume of continuing business with the customer. The customer then approaches the supplier with a proposal for certification. As with any good customer/supplier relationship, the supplier certification relationship should be mutually beneficial for both customer and supplier. The supplier must willingly agree to certification for the process to continue.

7.8.1 Delegated inspection

Delegated inspection programs generally consist of selecting one or two supplier employees for training as the customer's agents for product acceptance and process monitoring. The customer provides the delegated inspector with training in areas including the inspector's responsibilities as an agent of the customer, completion of the customer's inspection and assessment records, and product acceptance or release methods. After completion of the training program, the delegated inspector accepts authorized product on behalf of the customer, and usually is required to conduct periodic process assessments on behalf of the customer, with results forwarded to the customer. Since the delegated inspector is placed in the difficult position of representing the customer while employed by the supplier, it is critical that the program be fully supported by the supplier's management and that the supplier assure the delegated inspector's independence and authority.

7.8.2 SPC/variability reduction

SPC/variability reduction programs are based on a consistent flow of product from lines under statistical process control. In these programs, the supplier and customer agree to monitor the process through regular exchange of SPC data. These programs usually include mutually agreed to variability reduction goals for the supplier to achieve. Suppliers are normally required to send SPC data for each delivery lot with the product. The customer reviews the data as part of the acceptance process.

Even though SPC can be applied to short run production, SPC based supplier certification programs are best suited for longer, stable process runs. Fortunately for the electronics industry, this type of process is fairly common. Even though a particular product may not be in constant production, the processes are the same for a variety of products, and many are maintained under process controls and variability reduction programs.

7.8.3 Combination programs

The third common family of certified supplier programs combine elements of each of the others. SPC and variability reduction goals are established for some if not all processes, and a delegated customer representative is selected and trained to perform reviews and assessments for the customer. This approach serves to strengthen each of the other approaches. It adds objective use of SPC data to the delegated inspector programs, and it assures a specific review of lot data by a person trained by the customer prior to shipment that may not occur under an SPC-based program without a delegated inspector.

No matter which form a certified supplier program may take, most have common elements in the selection process. Supplier data is usually reviewed for historical cost, delivery, and acceptance performance. A subjective evaluation is generally performed regarding the supplier's responsiveness to customer inquiries, customer service requests, and special requirements. The supplier is then approached with a proposal to become a certified supplier. If the supplier agrees, the supplier's management structure and quality management system is reviewed to customer criteria.

7.8.4 Additional elements in certification

While the above elements are indeed important, and need to be addressed during the selection process, many programs miss other elements that are vital to an effective certified supplier program: the customer's strategic needs, supplier core competencies to meet those needs (and thus future business potential for the supplier in the relationship), and the supplier's business stability.

Many supplier certification programs make the mistake of becoming supplier recognition programs. Suppliers that have performed well in the past are "certified" by the customer as a way of saying "Thank you for an outstanding job." As important as supplier recognition is, it should never be confused with supplier certification. Certifying a supplier whose period of contract performance will end in the near term without potential future business is a waste of time, money, and effort for both parties, and in the long run will not be appreciated even as supplier recognition. Supplier recognition focuses on past performance; supplier certification should focus on future business strategies.

The customer's strategic plan for procurement, discussed in Sec. 7.2, should include core competencies needed for supplier partners and certified suppliers. While supplier partners may be needed for all core competencies identified in the strategic plan, not every supplier partnership will warrant supplier certification. Maturity of the supplier and the supplier's product line must be considered, as well as the volume of planned procurement and the cost of more traditional assurance methods compared to the cost of properly mounting and maintaining a certified supplier effort.

Once target core competencies have been identified for supplier certification, the customer's supplier base can be reviewed for certification candidates. Historical data can be reviewed for delivery and acceptance targets as set by the customer's selection criteria. Cost data for outstanding performers can be compared to that of other suppliers in the same technology fields, and business analysis should be performed using available public records in advance of contacting the supplier with proposal for certification. The customer should make an effort to utilize every information source available in the screening process. Many large corporations have several divisions that use the same suppliers, yet they fail to consult with each other before approaching supplier certification or supplier recognition decisions. This lack of internal communication can lead to embarrassing results. Imagine approaching a supplier with a certification proposal at the same time a sister division is in dispute with the same supplier.

When developing supplier selection criteria for certification, aim high. Many programs set goals too low, and then get just what they aim for. It is common to see three-tiered supplier certification programs (Table 7.10). The lowest tier, which we will refer to as a Level III certified supplier, is frequently little more than a supplier recognition program. Suppliers are selected through a screening process that includes reduced performance goals in delivery and acceptability, and may be asked to show "intent" to use continuous improvement to reach the next level goals. Level III certified suppliers in this type of program are rarely afforded any reduced oversight, and in fact often

TABLE 7.10 Typical Certified Supplier Levels

Level III	Level II	Level I
▪ 95% product acceptance ▪ 95% on-time delivery ▪ Intent to use continuous improvement ▪ Supplier is considered in customer's future planning ▪ Supplier agrees to pursue certification boards	▪ 98% product acceptance ▪ 98% on-time delivery ▪ ISO 9000 compliance ▪ Minimal use of advanced quality techniques to manage manufacturing ▪ Mutually determined variability reduction goals ▪ Supplier is involved in customer's planning ▪ Slightly reduced oversight ▪ Supplier managed primarily through metrics	▪ 100% product acceptance ▪ 100% on-time delivery ▪ ISO 9000 certification ▪ Uses advanced quality techniques to manage business ▪ Mutually determined variability reduction goals ▪ Joint planning of future strategies ▪ Reduced oversight ▪ Supplier managed through metrics
Recognition	Certification	

are asked to provide more process data in addition to the normal assurance activities performed prior to "certification." Where this level is used as an incentive for improvement with a true intent for eventual certification, it can be a positive step. Care should be taken that a Level III certification is not awarded as a "pat on the back" to a supplier that will not be in the customer's future strategic plan.

A Level II supplier under a three-tiered approach is usually screened to slightly more stringent criteria, and awarded slightly reduced oversight. Initial use of process metrics is usually required from a Level II certified supplier, including SPC of some if not all processes, and C_{pk} variability reduction goals for at least "critical" processes. Programs also routinely require the supplier to demonstrate a "philosophy of continuous improvement" for supplier certification. Many programs additionally require ISO 9000 compliance, perhaps with an intent and schedule for actual ISO 9000 certification within a given time limit to be awarded certified supplier status at this level.

In a three-tiered approach, it is at Level I that a supplier moves from a certified supplier in training to a true certified supplier. This is usually the first time the supplier is required to maintain 100 percent on-time and acceptance performance. Supplier assurance methodology shifts from product assessment to process assessment.

Supplier process data is a routine requirement, with mutually agreed variability reduction goals. Critical processes may be required to maintain tightened variability reduction standards, for example, maintenance of a $C_{pk}>1.33$. A Level I certified supplier may be expected to show evidence of SPC use, or other valid performance metrics in nonmanufacturing business areas as well as on the manufacturing floor. If ISO 9000 certification is not already a requirement for supplier selection, it is frequently a requirement for a Level I certified supplier.

Once a supplier is certified, it is important to maintain supplier oversight. Most supplier certification programs include a first party assessment element, requiring the supplier to maintain a system of internal audits. The customer should review the audit program and assure that all critical areas of the supplier's quality management system and manufacturing process are reviewed. Suppliers may be asked to send audit reports to the customer, or maintain records for customer review. Whatever method is used, the customer should review the records and assure that any findings are closed out with positive corrective action that addresses the root cause of the finding.

The customer should also maintain a process to review certified supplier SPC and variability reduction data that may be required, and work with the supplier to maintain variability reduction goals. Many customers periodically perform product audits, including destructive physical analysis or assembly tear-down inspections to assure continued product compliance from certified suppliers. The data from the product audits may be compared with the supplier's SPC data to assess the validity of the supplied data to the product produced.

Supplier certification is an effective system for reducing dock-to-stock cycle time and shifting from inspection-based product assurance to process-based product assurance. While several approaches to supplier certification exist, there are common elements that apply to most systems. In all cases, supplier certification should never be allowed to become supplier abandonment. Any successful supplier certification program requires that supplier oversight not be eliminated, but shifted to more effective process control methods.

7.9 World Class

World class manufacturers must have world class suppliers. Not too long ago, a high-quality supplier was almost guaranteed business success. In the world class competition, being a high-quality supplier is no longer enough. High quality is the entry requirement;

other performance factors lead the way to success: a long-term vision to exploit core competencies, effective development and use of technology to bring new products to market, the ability to bring new products to market faster and more efficiently than competitors, strategic supplier relationships, outstanding customer service and support, these have become the discriminators that set world class apart from the rest.

World class companies (see Table 7.11 on next page) must focus on core competencies, both within their own organization and in selecting suppliers. [15] The days of adversarial supplier management are gone; supplier relationships must be mutually beneficial. Supplier management in world class companies begins with a strategic plan that includes long-term supplier relationships that leverage resources. World class companies maintain constant communication with their supplier chain, using standardized systems wherever possible to reduce duplication of activities between customer and supplier. The use of ISO 9000 as a business system standard is a starting point, complemented by use of advanced quality techniques to reduce variability and eliminate waste. Suppliers are given the freedom to develop innovative methods for achieving requirements and optimizing manufacturing activities. The manufacturing process is controlled and monitored through the use of metrics, and the controlled process assures product that fulfills expectations. World class companies dominate their markets by staying on the leading edge of their technologies, developed in partnerships that reward the company and the company's supplier chain. To judge where your company is in developing world class quality, the assessment in Table 7.12 on p. 439 is provided.

References

1. "Industry News," *Circuits Assembly,* November 1996, p. 8.
2. Carter, Joseph R., and Ram Narasiman, *Purchasing and Supply Management: Future Directions and Trends,* Center for Advanced Purchasing Studies, Tempe, AZ, 1995, p. 91.
3. See ISO 9001, para. 4.4.3.
4. Reprinted from *Out of the Crisis* by W. Edwards Deming, by permission of MIT and The W. Edwards Deming Institute. Published by MIT, Center for Advanced Educational Services, Cambridge, MA 02139. Copyright 1986 by The W. Edwards Deming Institute.
5. Monczka, Robert M., and Robert J. Trent, *Purchasing and Sourcing Strategy: Trends and Implications,* Center for Advanced Purchasing Studies, Tempe, AZ, 1995, p. 19.
6. TC176 Committee from the Internet, QMP, address: http://www.wineasy.se/qmp/about.html, Nov. 24, 1996. Source document cited as ISO TC 176/SC2/WG15/N125. Will be available from ANSI when printed.
7. Monczka and Trent, p. 42.
8. Monczka and Trent, p. 43.
9. Schultz, George, "The Future of Supply Chain Management," *Managing Automation,* October 1996, p. 51.

TABLE 7.11 World Class Purchasing

Characteristics	Old style	World class approach
Executive leadership/commitment	▪ Keep out of trouble—low profile	▪ Executive committee support for integration across company and corporate plans
Strategic positioning organization	▪ Lower-level plant focus ▪ Reports to manufacturing manager ▪ No strategic relevance	▪ External/internal customer focus ▪ Matrix management ▪ High-level positioning—second, third, or fourth level
Functional leadership	▪ None corporatewide ▪ Limited vision of role	▪ Companywide customer-focused leadership ▪ Establishes integrated visions ▪ Works at results and processes ▪ Drives supply/supplier management strategies companywide
Functional/horizontal integration	▪ None	▪ Cross-functional, cross-location teaming ▪ Part of the technology, manufacturing, and strategic planning process
Supply-base strategy	▪ None ▪ Short-term reactive ▪ Multiple suppliers	▪ Quality driven ▪ Design standardization ▪ Concurrent engineering ▪ Supply-base optimization ▪ Commercial strategy emerging ▪ Horizontal/vertical supplier strategies
Supplier management	▪ Reactive	▪ Focused supplier development ▪ Joint performance improvement efforts ▪ Value focused ▪ Total cost improvement ▪ Supplier benchmarking
Measurement	▪ Delivery	▪ Customer oriented ▪ Total value/cost focused ▪ Benchmarking against best in class
Systems	▪ Plant focus ▪ Material flow	▪ Global databases ▪ Historical performance data ▪ Strategy ▪ Extensive electronic development/interface: EDI/CAD/CAM, etc.

source: Reprinted from Ref. 15 with permission.

TABLE 7.12 Purchasing Self Assessment

ELEMENT	Not in effect	Elementary stage	Needs improvement	Strong	Excellent
A. Supplier Selection	1	2	3	4	5
1. Integrated supplier management team is used to select and manage suppliers.					
2. Supplier's business stability is included in selection criteria.					
3. Supplier core competency is actively evaluated during supplier selection.					
4. A corps of preferred suppliers is identified and evaluated for selection before unknown suppliers are considered.					
5. Supplier selection criteria include quality performance, schedule performance, and technical performance.					
B. Supplier Partnership/Teaming					
1. Strategic purchasing plan is in place and used to identify current and future supplier and supplier development needs.					
2. Mutually beneficial partnerships are developed with suppliers to enhance the ability of both organizations to create value.					
3. Information is shared openly with supplier partners; agreements are in place to protect customer's and supplier's intellectual property rights.					
4. Customer and supplier participate in joint quality improvement initiatives.					
5. Technology is used appropriately to provide "virtual collocation" when supplier partners are not located close to the customer.					

Total score:
220–275
Excellent progress—on the way to world class
165–220
On the way—keep working
110–165
Just beginning; plan, schedule, reallocate resources
<110
Hire consultant/reorganize

TABLE 7.12 Purchasing Self Assessment (Continued)

ELEMENT	Not in effect	Elementary stage	Needs improvement	Strong	Excellent
C. Supplier Quality Assessment	1	2	3	4	5
1. All suppliers are certified to appropriate international standards by accredited registrars.					
2. Supplier quality systems are assessed and found acceptable before final supplier selection.					
3. Customer assures that suppliers perform effective first party assessments on their system and follow through with corrective action when necessary.					
4. Second party assessments, when used, are process-focused to avoid redundant assessment activities.					
5. Customer assessments focus on requirements without directing specific methods for fulfilling requirements.					

TABLE 7.12 Purchasing Self Assessment (Continued)

ELEMENT	Not in effect	Elementary stage	Needs improvement	Strong	Excellent
	1	2	3	4	5
D. Flowing down Requirements					
1. The closed-loop communication model is understood and feedback methods are used to assure mutual understanding of issues and requirements.					
2. Requirement-based contracting is used, giving maximum freedom for the supplier to develop methods for fulfilling requirements.					
3. Purchase orders contain clear and complete requirements information including all information referenced in ISO 9001 para. 4.6.3.					
4. Review activities appropriate for the complexity and criticality of the procurement are used to assure understanding of requirements.					
5. Pre/postaward conferences, design reviews, and other interchange meetings, when used, include all appropriate supplier management team members.					
E. Assessing Product Compliance with Requirements					
1. Customer assures that supplier's detail specifications include verification and validation steps for compliance to all requirements.					
2. Customer source activities focus on surveillance rather than inspection, assuring the supplier is performing effective product assessment to requirements.					
3. Once supplier processes are controlled to assure compliance to requirements, customer testing/inspection is reduced to periodic samples to assure continued process control.					
4. Just-in-time procurements are not put in place before process controls and advanced quality techniques are in place to assure compliance at the supplier.					
5. Test software used for acceptance is reviewed against customer requirements for use.					

TABLE 7.12 Purchasing Self Assessment (Continued)

ELEMENT	Not in effect	Elementary stage	Needs improvement	Strong	Excellent
F. Measurement and Rating Systems	1	2	3	4	5
1. Comprehensive supplier measurement is used that includes quality and schedule performance.					
2. Measurement system uses objective, verifiable data.					
3. Measurement system includes failure data from manufacturing and end users.					
4. Measurement system includes regular performance feedback to the supplier.					
5. Measurement system includes requiring corrective action when performance degrades.					
G. Supplier Certification					
1. Supplier certification is reserved for suppliers identified in the strategic procurement plan for long-term relationship.					
2. Certified supplier selection criteria is defined and includes quality and delivery performance, core competency, customer support, and future business potential.					
3. Supplier certification is a mutually beneficial activity.					
4. Supplier certification program includes continuous improvement/variability reduction goals.					
5. Following certification, a program of supplier monitoring assures continued supplier performance to selection criteria.					

TABLE 7.12 Purchasing Self Assessment (Continued)

ELEMENT	Not in effect	Elementary stage	Needs improvement	Strong	Excellent
H. World Class Purchasing	1	2	3	4	5
1. Executive leadership is involved in purchasing for integration across the company.					
2. The focus is on internal customer satisfaction.					
3. The focus is on external customer satisfaction.					
4. The Integrated visions documented and on view in the company					
5. Cross-functional teams are in place.					
6. Your company is quality-driven					
7. Design standardization is completed					
8. There is focused supplier development activities					
9. Your company is best-value–focused					
10. New product development time is reduced.					
11. Supplier benchmarking is in place.					
12. Global databases are in place.					
13. Historical performance data is in place					
14. Purchasing/quality strategies are defined and documented.					
15. Electronic interfaces established:					
a. Electronic data interchange (EDI)					
b. Electronic purchase orders					
c. CAD/CAM					
d. Videoconference capability					
e. Use of the Internet					

10. Schultz, p. 48.
11. See ISO 9001 para. 4.3.2.
12. See ISO 9001 para. 4.6.3.
13. Cole, Robert E., "Improving Product Quality through Continuous Feedback," *Management Review,* October, 1983, p. 10, based on data provided by Weston Ison, General Electric Co.
14. Stundza, Tom, "Can Supplier Ratings Be Standardized," *Purchasing,* November 8, 1990, pp. 60–63.
15. Monczka and Trent, p. 72–73.

Chapter

8

Benchmarking

Gerald J. Borie, Consultant

Past International President
Society of Reliability Engineers

8.1 Definition

Benchmarking has been practiced since before the beginning of recorded history. It, like so many other things, is not new. Only the term is somewhat new in its current context. And so it is true that benchmarking, as probably practiced by cave dwellers, continues to crop up in history in ever more sophisticated forms. It is a good example of an art form evolving into a science before our very eyes without the subject's awareness over centuries or millennia.

The concept of comparing things is not new. It is something that humans have been doing since the first cave dwellers compared their war clubs to the neighbors' war clubs while considering an attack. Many manufacturing businesses employ this same technique today. However, the degree of sophistication has not changed much since the day of the cave dwellers. A more analytical approach to benchmarking is required if it is to meet the needs of today's complicated industrial and business practices, especially in the electronics industry.

The art turned science is very simple, that is, noncomplex in its theory. The complexity, and therefore its risks, lie in the execution of the task.

We often hear that common phrase, "It's as simple as apple pie." Benchmarking is a process and as such must be carefully followed and, as in any value-added activity, preparation time must be expended to succeed. Webster's dictionary defines a benchmark as "something that serves as a standard by which others may be measured." [1] By that definition we establish two main or key points: there is the "standard" and there are "others." An expanded definition is:

> A point of reference describing a specific process or subprocess attribute, that serves as a standard by which you measure the health and vitality of that process or subprocess. It is used for purposes of comparison against world leaders in that process or subprocess so that one may plan to attain world class status through continual improvement.

Benchmarking is defined by David T. Kearns, CEO of Xerox Corporation, as "the continuous process of measuring products, services, and practices against the toughest competitors or those companies recognized as leaders." [2]

The word *benchmarking* originally was a land surveyor's term. In that context, a benchmark was a distinctive mark made on a rock, building, or wall, and it was used as a reference point in determining the position or altitude in topographical surveys and tidal observation. Today a benchmark is a sighting point from which to make measurements; a standard against which others could be measured. [3]

The essence of benchmarking is measuring, managing, and satisfying customer requirements and expectations, assessing your strengths and weaknesses, finding and studying the best practices wherever you find them, and adapting what you learn to your circumstances. Benchmarking is simply the comparison of one thing to another to determine the differences or similarities and to quantify these differences. Most times, the comparison is against leading or world class companies (see Fig. 8.1).

Benchmarking is commonly performed in order to make a decision. This path is commonly chosen because it makes the task of making decisions so much easier. Benchmarking can play a vital role in making sound decisions because it is fact-based. Care must be taken executing, analyzing, and reporting the outcomes of benchmarking so that the data will lead to real-world improvement. In electronic man-

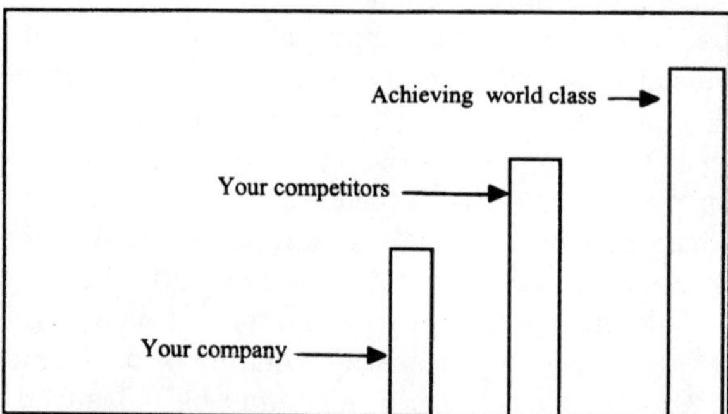

Figure 8.1 Benchmark to move ahead.

ufacturing, the task of benchmarking must be viewed in the context of the process under investigation. Simple processes do not require complex efforts. Fit the activity to the complexity of the process. Benchmarking is directly related to process improvement.

The purpose of benchmarking is to provide a target for improving the performance of the organization. The benchmark targets improvement of the process outputs or the performing of the actual process. Benchmarking brings a focus on customer-driven project management improvement efforts by emphasizing desired outcomes. It also nurtures wholesome competition by creating the desire to be the best. Benchmarking provides a common focus to hold the organization together by measuring critical areas and analyzing these critical areas against the best. This targeting of the best reinforces continuous improvement by keeping everyone aiming at a long-term objective. Identifying a target requires knowledge of process capabilities and a clear vision of the spread between the two. Figure 8.2 shows one method of plotting the spread for a multistep process.

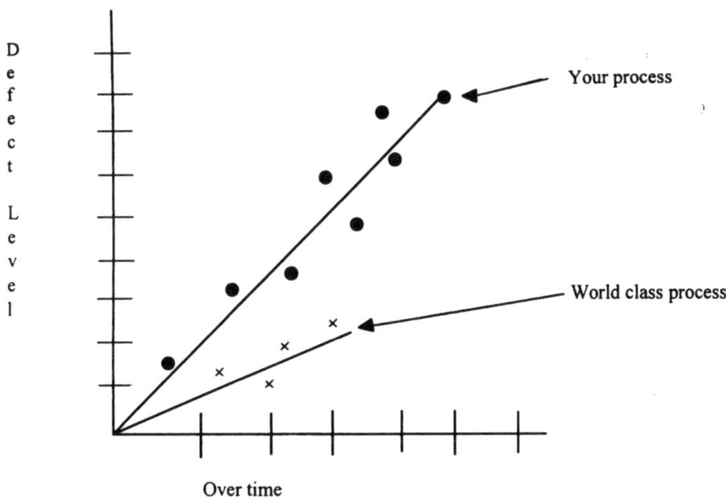

In today's electronic manufacturing environment defect levels are measured in parts per million (PPM) at the component level.

Figure 8.2 Visualizing multistep process variation.

8.2 History

As has been alluded to in the previous section, benchmarking has been around for a very long time. Let us look at a very early example of this art form turned science. Picture, if you will, a time in evolution when people lived in caves and used only rudimentary tools; a time when speech was nothing more than grunts and groans, and might was recognized as right. If for any number of real or imaginary reasons cave person A had a score to settle with cave person B, they most likely settled their differences with the use of physical force. Using their weapons of choice or availability (rocks, clubs, etc.), our cave persons selected their weapons and planned an attack strategy. Did they dash at each other in blind rage, or did they move with measured caution in order to establish just what they would face in combat? The survivor probably did the latter.

Who was the very first benchmarker? Possibly the second person to light a fire. The second fire starter watched the first fire starter and then borrowed the practice.

Have you ever heard the saying "Discretion is the better part of valor"? Most combatants would like some sort of intelligence about the adversary prior to taking up a challenge, especially a physical confrontation. This is where benchmarking came into play. No, not in the sophisticated form we know today, but the idea was the same. It was and still is the tool of choice to make an informed decision, assessment, or comparison of the balance of power.

Benchmarking gives wise practitioners the advantage of knowing where they stand with respect to their competition. This is true in all competitions, be they in business or warfare. If people don't know where they are, and they don't know where exactly they want to go, it doesn't matter what path they take to get there because they still won't know where they are. The foregoing is only a good plan for failure.

Does the term *balance of power* sound familiar? If you had grown up during the years of the cold war or had lived through the years when Ronald Reagan was President of the United States, the phrase would have been all too familiar. The United States was locked in an ideological conflict with the then Union of Soviet Socialist Republics, commonly referred to as Russia. At this point you are probably wondering, why the history lesson? What is its relevance to our subject matter? Well, the linchpin to the U.S. decision-making process was information that was arrived at by the process of benchmarking. See Fig. 8.3. No, they did not call it that; rather, they just lumped it under the title *strategic information analysis.* The same was true of the Soviets. Both sides needed to know where they were with respect to

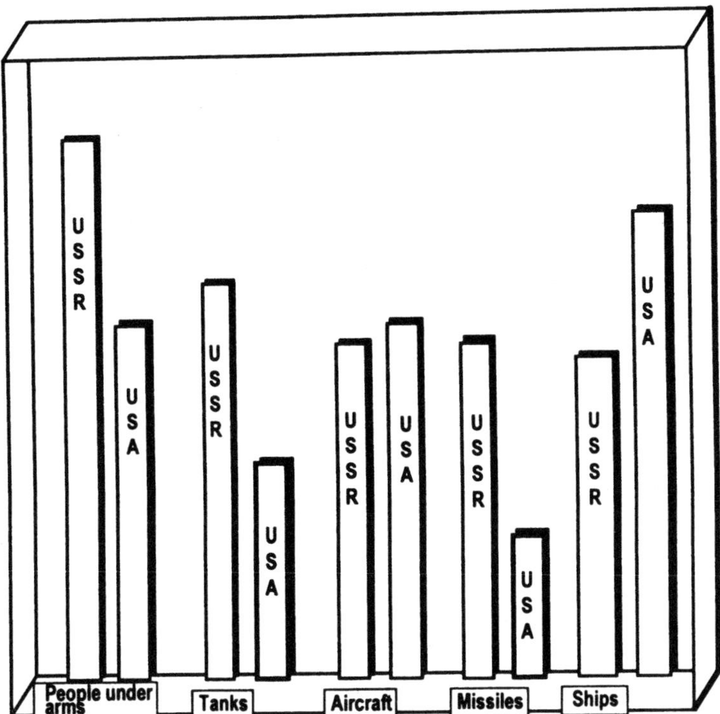

Figure 8.3 Example balance of power between the United States and the Soviet Union in the 1970s.

the other, in order not to miscalculate and make the fatal mistake from which there was no retreat. No one really wanted to start World War III, and no one wanted to be the second strongest if hostilities broke out.

This is just common sense. But what was world class power? In order to do this data collecting and analysis, both sides had specialized agencies. The United States had the Central Intelligence Agency (CIA) and the Soviets had the KGB. Both were users of the benchmarking technology even though they were unaware of the fact. Figure 8.3 shows what some of these analyses probably looked like. Historians will be the first to point out that the CIA was not always correct in its conclusions as a result of its benchmarking activities. Perhaps if they had had the advantage of *today's* sophistication in the 1960s through the 1980s, they could have done a better job, notwithstanding the use of false or misleading data that both sides employed. This example is not intended to demean their efforts nor the end results. Proper technique, as we will discuss later in this chapter, evolved from the lessons learned from the cave person right through current time.

If you will once more place yourself in the cave person's situation, picture yourself facing a giant of a warrior who has a club that makes yours look like a toothpick. Is this when discretion becomes the better part of valor? Should you turn and run? Would it not have been much better to find out what you were up against prior to the shock of the battlefield? Benchmarking, properly executed, would have told you that the odds of defeating the warrior may have been too great to overcome with the assets at your disposal, and that the plans you had made, based on gut feelings rather than facts, were in error.

North American quality programs of the 1970s and early 1980s provide a good example of the dangers associated with being internally focused and approaching performance improvements from a historical perspective. Plans for improving acceptable quality levels (AQL) by 8 to 10 percent per year were commonplace during those years. Quality performance in the 96 to 98 percent range was considered excellent, and a 99 percent level was viewed as an unreasonable expectation. Meanwhile, Japanese competitors were measuring quality performance in terms of a few hundred parts per million (ppm). (*Note:* As a point of reference, 99 percent is equivalent to 10,000 defects per million.)

From a historical perspective, 10 percent year-after-year improvements appeared reasonable. But from a competitive standpoint, North American companies were far behind the Japanese and losing ground. Worst of all, they were oblivious to what was happening. Misplaced expectations about productivity, product costs, product reliability, customer service, and other vital measures contributed to significant loss of market share in many industries. Fortunately, a few enlightened companies recognized that the traditional, evolutionary approach to change would not yield the dramatic improvements required to regain competitive standing in the global marketplace. An externally focused approach, one based on best practices—regardless of industry—would be required. Xerox and Motorola were two U.S. companies that pioneered this external approach. Both recognized the potential of benchmarking as a driver of continuous improvement and responded to the competitive challenge from abroad with the result that both firms regained market share from Japanese rivals.

Benchmarking was instrumental in providing the much-needed wake-up call to Xerox in the early 1980s and led to its tossing out many traditional practices for managing its business. Benchmarking pointed the way to quantum improvements in product quality, reliability, cost, time to market, and total customer satisfaction. The changes were so positive that Xerox applied for and won the Malcolm Baldrige National Quality Award in 1989. Today, benchmarking is fully integrated into the Xerox management process. Complete information on the Baldrige Award may be found in Chap. 9. For companies seeking the Malcolm Baldrige Award, the application guidelines specify benchmarking as a mandatory management process. A recent survey indicated that approximately 40 percent of the Fortune 800 companies are using the benchmarking process to drive improvements in customer satisfaction. [4]

Xerox introduced the concept of benchmarking to present-day corporations in 1979. In no time, it seemed that the idea of advancing corporate and organizational improvement by collecting and adapting the best practices of others started what some call a new quality science.

Today, many organizations have dedicated benchmarking departments led by managers who specialize in it. Benchmarking is an easy process that any organization can learn and inexpensively use. While some companies report spending more than $1 million annually on benchmarking, smaller organizations spend far less. A recent International Benchmarking Clearinghouse (IBC) membership survey finds that 1 to 2 days of training in benchmarking should be adequate preparation for most people. More than 80 percent of the respondents say benchmarking study leaders need two or less days of training. [5]

Private industry is not alone in the use of benchmarking as a tool for improvement. NASA facilities, such as the Dryden Research Center at Edwards Air Force Base, have recognized the value that can be gained in Research and Development (R&D). The United States Postal Service is another example of a quasi-governmental service organization using the technique in competition against commercial companies such as Fed Ex, United Parcel Service (UPS), and others. The process has no limitations. The more complex the business or process, the more you need a defined process to achieve the desired world class status.

8.3 Using and Understanding Benchmarking

Benchmarking continues its presence as one of the top five processes that companies look to, to succeed, according to The Benchmarking Exchange. Since resources are becoming more affordable and accessible, even small to medium businesses are now getting in the loop. [6] No business is too small or too large for benchmarking.

Many people in the electronic manufacturing industry have heard the term *benchmarking*. This is really not surprising. This industry is one of the fastest evolving in terms of technology, even as it is faced with outside pressures and downsizing. Only the fittest will survive this shake-out. World class manufacturers will still be there when the dust settles. Therefore, striving to be world class is a matter of survival. Benchmarking is one of the tools that will help you get there.

One of the problems in using this technique is that it is easily misused and misapplied. Let us explore what benchmarking is not. It is not uncommon for a person to measure something and label it as a benchmark. An example of such a measurement could be an individual's weight.

A measurement of one of your company parameters which was made yesterday and compared to the same parameter today is not a benchmark because the key ingredient of "other" is not present. If, however you were comparing a measurement from Company A to a similar measurement from Company B, you would be benchmarking, in our context and your definition.

Measurement, according to Webster's, is "a figure, extent, or amount obtained by measuring, i.e., a basis or standard of comparison." [7] A measurement may be made in an attempt to track the success rate of losing weight. No single measurement is a benchmark, nor is such a measurement a goal. A goal is where you want to be at the end of the weight loss activity. The first measurement is nothing more or less than the first measurement. In order to be a benchmark (B), it must be a comparison between two or more identical or similar processes (P) or activities ($P = B$). Even if you weigh yourself every day and keep exact records of your weight loss attempt, and you make comparisons to the days preceding the latest measurement, you are still not benchmarking. You are tracking your progress toward a preestablished goal. The goal may have been established as a result of using benchmarking; it may lead you to where you want to be, but the fact remains that it is not benchmarking; it is still just a measurement. This does not degrade its need nor its usefulness. In fact, measurement is the only factual way to know where you are with respect to the goal you have set using benchmarking as a means of comparison against some subset of the universe of things of interest to you.

Benchmarking is only as successful as the quality of the information generated. It's essential to develop quantitative and qualitative measures with two characteristics [8]:

1. The measure genuinely *reflects the best performances on a sustained basis*. One-time super performances are not as full of impact as a continuous record of achievement. "Genuine" is key

here, because many people are skeptical when substantially better performances are cited. They'll immediately challenge measures that do not seem authentic.

2. The factors have a bearing on quality, customer satisfaction, profitability, and other areas that contribute to the performance and individual/organizational success.

The steps for developing effective measures are as follows [9]:

Measure the right things

Ensure that a measure is understood and accepted by its users

Set measures that are easy to work with

Include target performance levels; benchmark high leverage areas

Feed back results, quickly and clearly

Reward exceptional performance

Benchmarking can be better understood if we use a real-world business example. Let us suppose that company A finds that its cost of doing business is making it noncompetitive. It may attempt to utilize benchmarking for some or all of its processes ($B = P$) against those of other companies whom they believe do a particular process better than they. *Better* in this case equates to lower cost, more reliability, and shorter flow time. (We will not get into how that decision is made at this point in the chapter.) When the comparison is made against the same process or subprocess of companies B, C, and D and subjected to detailed analysis, the results are benchmarks! These benchmarks are then carefully ranked with respect to their world class position for that process or subprocess. Remember, a company may be world class for one or more processes and yet not be a world class company. Now that there exists an ordered ranking of the process, company A can now make an informed decision where in that ranking they wish to be.

Being in first place is not always in the best interests of a company at a given time. There are economic and other short-term and long-term management plans which will or may mitigate against a decision to invest available assets for this needed process enhancement. Benchmarking is to understand and lead.

Each of the following steps must be carried out by top management in implementing benchmarking [10]:

- Learn the mechanism of benchmarking, what it is, and how it can become a "driver" to continuous improvement.

- Create an environment in which the status quo is unacceptable and the standards of performance are best in class.
- Provide for the education of all functional managers in the techniques of benchmarking.
- Establish benchmarking teams and initialize pilots in each major functional area. Establish checkpoint reviews to ensure adherence to the process.
- Establish an internal network of benchmarkers to promote, facilitate, and improve the benchmarking process.
- Integrate benchmarkers into the formal operating and strategic planning processes of the company.

Benchmarking need not be conducted with companies identical to yours. In many cases this type of approach will end in failure. This result is predictable since these companies are your competition. As such they may not see it as advisable to help you be as good or better than they. There will be exceptions among more enlightened companies. There is no attempt to stop you from trying, only an attempt to forewarn you that the outcome of getting your direct competition to benchmark with you may not be too successful. There is, however, a broad avenue open for your travel down the benchmarking lane. Remembering that any process or subprocess is broken down to its smallest elemental nature, you will find that many enterprises are involved in the same fundamental process. In order to realize this windfall of data, it will be necessary to look outside your industry. For example, if you are looking at the processes of warehousing and kitting, all or almost all manufacturers are involved in the same processes. There in no compelling reason to restrict your efforts to only the electronics industry. In fact, it would be very healthy to look at a mixture of businesses. This will allow you to look at an entirely different developmental trail that leads to the same juncture in the road. An analysis of such data could result in some real eye-opening facts.

Benchmarking against only one company can lead to a trap, even if you are sure that the company is world class. For the sake of statistical sanity, you need to check at least two or three companies at a minimum. This is not to say that there is any particular validity to two or three companies. However, in our industry there are not many who would argue with a sample of three for a task such as this. (If there is a purist in your audience who insists that for real statistical validity and assessing risk you need to have a larger number of trials, and even if you believe he or she is correct, let the people who have to pay for this level of correctness handle the situation. If senior management eventually agrees with the purist, you can always point to the cost and say you were just trying to keep costs down.) Small sample size always introduces error. However, you do not have a random sample and, as such, the risk is already greatly minimized.

At the core of benchmarking is the concept of learning and sharing. By comparing work practices with others, you may gain valuable information that you can adapt to your own situation. That is why going outside your own industry often results in startling revelations.

You'll find best practices not only in your industry, but in places you may have thought had no relationship to you whatsoever. Moreover, you can benchmark just about everything, from machine downtime to employee overtime to delivery time. Every company must grapple with these issues, and every manager can learn from others' experiences. Table 8.1 shows an example of wildly varying companies in the same process.

Benchmarking is a useful quality tool that will help your company continually improve its processes by learning how others do it. To benchmark, you first evaluate your own operation's processes to identify weaknesses and strengths; then you must identify, study, and adapt from others who may be doing it better.

You can attribute some of the recent popularity of benchmarking to the Malcolm Baldrige National Quality Award, which requires all company entries to benchmark. Another, perhaps more powerful reason to benchmark is to keep up with world class competition.

Benchmarking is most often used to identify the best organizational practices, but you can also use it to improve performance by studying what your competitors know and the rate at which they learn it.

TABLE 8.1 Industries with Similar Processes to Electronic Manufacturing

Process	Types of companies
Kitting	Vending machines Auto parts stores Supermarkets/grocery stores
Assembly	Automotive companies Appliance companies
Just in time	Automotive companies Freight companies (Federal Express/UPS) Furniture companies
Inspection/text	Medical devices Cosmetic manufacturer Nuclear power facilities

Research shows that people like to know where they stand with managers, with the company as a whole, and with their contemporaries. You can benchmark with others in your own company to take advantage of information easily available, or you can benchmark with individuals in other companies to give yourself and others a more complete picture of what is considered the best.

To be able to learn wherever you find a better way of doing things means you will have to fight the biggest obstacle to organizational progress: self-satisfaction. It takes a great deal of ego suppression and an open mind to look closely at how you do what you do and ask, "But how can we be *better?*"

Goodlow Suttler, a general manager at Analog Devices, was one of 28 people who created the Center for Quality Management in Boston. Suttler says he had thought that Analog operated well until he saw how American Baldrige Award winners and Japanese Deming Award winners operated. "The practices we saw in both Japan and the United States were incredibly motivating and moving. After I came back, I had enough to propel myself for several years." [11] One of the great values of benchmarking is that if you learn nothing else, at least you've taken a good, hard look at how you do business.

Sometimes benchmarking can also limit the thinking of a creative electronics company. If you only look to "copy" world class companies, you may limit yourself, for the real world class companies of the twenty-first century may need to invent things that the world class companies of the 1990s do not yet know. The computer, automation, and artificial intelligence field, including robotics, is moving so fast that it is unrealistic not to believe there are many quantum leaps to be made in this field. Creative minds need to use benchmarking from fields outside of electronics to find processes that need to be fixed in electronics, much like Xerox used L. L. Bean (the mail order clothing company) to talk about handling inventory. [12]

In Sec. 8.4 we will discuss the necessity for understanding a process in its entirety prior to benchmarking it, and you will be challenged to prove to yourself that you cannot effectively and successfully benchmark any process unless you completely understand that process or subprocess. You must also have knowledge about the shadow support processes involved in measuring and evaluating the outcome of the process under study. An example of such a shadow process may be the accounting department, with which you have never agreed as to the proper methodology of assessing cost for rework across departments. This is just one of many such shadow processes in your factory.

8.3.1 Benefits of benchmarking

Benchmarking has many benefits. Benchmarking is a major tool for quality and productivity improvement. Each employee will obtain value from the process. It provides four basic advantages [13]:

1. Benchmarking requires certain specific customer satisfaction criteria, so that you can identify and focus on best practices (means) and best performance (ends). The link to the customer is clearly seen.
2. Benchmarking becomes a powerful analytical tool in the improvement process as members of the organization compare their practices and results against those of leaders in customer satisfaction and profit.
3. Benchmarking helps establish the gap between current organizational performance and best in class. You can then set targets with a clear understanding of why the improvement is necessary.
4. Benchmarking motivates actions that result in pride of accomplishment.

According to The Benchmarking Exchange, in 1992 the average cost of one benchmarking study was greater than $80,000; in 1996, it has dropped to only $8000. [14]

Benchmarking stimulates an external focus on being competitive and provides creditable data for establishing aggressive yet attainable goals—aggressive in that we are measuring ourselves against best in class and attainable in that someone is actually performing at those levels.

What should you benchmark? The answers are

1. Those items that make up the highest percentage of fixed or variable costs
2. Anything that affects quality, costs, or cycle times
3. Processes or functions of greatest strategic importance to your organization
4. Anything you do that has the greatest room for improvement
5. Whatever you do that you can improve
6. Anything that supports the company's or your department's success
7. Any factors that separate your organization from the competition

Before you begin, you must define your benchmarking objectives. Your benchmarking study must have clear, accurate objectives based on customer requirements. Make sure you know what your customers (or potential customers) want before you do this. You should be polling customers regularly through phone surveys, mail surveys, focus group studies, site visits, or a combination of all of these methods. Successful benchmarking companies use the following criteria to help them decide on a suitable benchmarking objective:

- Is it of interest to our customers?
- Does it focus on a critical business need?
- Is it in an area where additional information could influence plans and actions?
- Is it significant in cost or key nonfinancial indicators?

Benchmarks ideally will create constant improvement and change within your organizations. Benchmarking can be used to identify areas in which you can make significant improvements by adapting or matching systems that are proven better. You can improve 12 areas of organizational activity by using benchmarking [15]:

1. *Meeting customer requirements.* By examining other organizations' successful processes, you often can get valuable information on consumer demand and responses within your industry. Adapting the best practices of others, wherever they are, will help you meet and beat customer expectations—the best way to match and surpass your competition. Best practices wouldn't exist if users didn't prefer them, and the best way to find out what customers want *and* whether you're meeting those requirements is to consistently survey.

2. *Adapting industry best practices.* Benchmarking done right (making sure that a cross section of workers and managers directly participates in benchmarking teams) will ensure consensus support and enthusiasm for changes suggested by any benchmarking study.

3. *Becoming more competitive.* Benchmarking studies challenge long-held ideas by showing gaps between your organization's perceived performance and its actual competitive performance. Leading manufacturers develop new products up to two-and-a-half times faster than the industry average, and for half the cost. The benchmarking gap is the difference between the industry average and the industry best. Benchmarking against the competitive best also saves time and costs associated with the old way of improving: trial and error.

4. *Setting relevant, realistic, and achievable goals.* Effective benchmarkers feel confident that their goals are realistic because they can link well-defined customer requirements with proven business practices. The major problems of blue-chip companies over the last few years can teach us an important lesson: Market forces can be quick and deadly and can destroy or hobble the strongest of organizations. Benchmarking helps organizations anticipate market changes and validate goals.

5. *Developing accurate measures of productivity.* By comparing your internal processes to best practices, leaders and employees get a better understanding of your company's strengths and weaknesses.

6. *Creating support and momentum for internal cultural change.* Benchmarking can sensitize your employees to the need for continual improvement in areas such as productivity growth, defect rate reduction, and control of direct and indirect costs.

7. *Setting and refining strategies.* Strategic lessons learned earlier by other companies you choose to analyze can help your organization refine strategy, predict results of possible changes, and forecast changes in your market. With benchmarking, contingency plans can be developed and implemented faster and cheaper than if developed from scratch.

8. *Warning of failure.* A benchmarking program should tell you if and when you are falling behind your competitors in cost, customer satisfaction, technology, or business processes.

9. *Testing the effectiveness of your quality program.* Benchmarking will test whether your quality initiatives and competitive strategy are sound.

10. *Reengineering.* Benchmarking is a necessity for organizations engineering or reengineering their processes and systems. Experts say reengineering without benchmarking will produce flat 8 to 10 percent improvements, not the 80 to 98 percent performance improvements often seen with radical redesign like reengineering.

11. *Promoting better problem solving.* Does benchmarking improve problem-solving ability, or is problem-solving ability necessary to benchmark? Probably both. Standard problem solving provides a framework that makes work teams more effective. Standard problem solving also prompts teams to root their analysis in empirical data, which supports management by fact, a key ingredient in developing and maintaining a quality organization.

12. *Providing an education and creativity boost.* People become accustomed to operating in certain ways. Even if those ways are harmful, most people resist changing because the old way of doing

business is so comfortable. What benchmarking does is challenge the old way.

Regular benchmarking is like cleaning out your closet. You always find some things you don't need and a few things you didn't know you had, but could use. Regular benchmarking of critical functions ensures that you and your managers and employees remain open to new ideas, evolving technologies, and changing trends.

8.3.2 Objections to benchmarking

Benchmarking makes sense, but many organizations and leaders discourage the borrowing of others' good ideas. For example, many executives spend too much time on "problem" units or individuals. A benchmarking pro would instead spend more time with top-performing units or individuals and try to understand why they are so outstanding.

Some see benchmarking as cheating or industrial espionage. The fast-learning, big-achieving companies have a "we can learn from anyone" attitude that encourages the sharing of nonproprietary ideas and systems. Others say benchmarking is nothing but copycatting, a system that leaves no room for improvement; but the most successful benchmarkers know that you don't adopt, you *adapt* processes and systems to your unique business. Edison may get credit for the light bulb, but it was the research of the nameless person who came before him that Edison used in his experiments. Still others are afraid to benchmark because they don't want to expose their weaknesses to other world class standards.

8.3.3 Convincing the boss

Should you benchmark? Ask the following questions to see if you should benchmark:

1. Can your organization afford to stop improving?
2. Can your organization afford to stop learning?
3. Can your organization afford to stop competing for its position in the marketplace?

If you answered no to any of these questions, you should benchmark.

By now you should have a good idea of what benchmarking is, who has benchmarked, and what the benefits and criticisms of benchmarking are. Let's see if you can convince your boss of the need to benchmark. Most of us try to convince others to do something by stressing why it is important to us. That is a major mistake. Your boss wants to know why he or she should benchmark. What is in it for the company? Follow these steps [16]:

Step 1. Write a clear, concise, to-the-point statement of the problem. The attitude of most people you try to persuade will be, "What do you want me to do? Why should I do it?" Make sure the problem you come up with is a problem that affects the target of your persuasive effort, his or her unit, or the company. Be brief.

Step 2. Describe for your target the specific, tangible, negative effects the problem has on him or her. How much does the problem cost in money, time, production, or quality? You have to know your target well to be able to pinpoint the most effective negative effects to use. Use more than one in your argument: Where one may not sell, a second or third may.

Brainstorm a list of five negative, tangible effects the problem has on your target. Don't be afraid to get suggestions from others. Then use no more than the top three.

Step 3. Describe your solutions to the problem. Offer more than one alternative for a target who may need convincing or is hostile to your proposition. People like to be presented with alternatives, but sometimes one solution is enough.

Step 4. Describe all the specific, tangible positive results that will come to your target, his or her unit, or the company by using your solution. How much money or time is saved? How much will quality improve? How much market share will the company gain? Brainstorm a list of five positive, tangible effects the solution will have on your target. Don't be afraid to get suggestions from others. Then use no more than the top three.

Now you have a structure for a logical argument to persuade anybody in your organization that your company needs to benchmark. Use this structure, write out your argument (do not think you can keep it in your head) and practice! This exercise won't guarantee 100 percent success in persuading others, but it will get you a lot closer than trying to persuade others with arguments based on what you want.

8.4 The Benchmarking Process

A complete understanding of a process to be benchmarked is essential. Likewise, it is important to understand the benchmarking process. There are four main types of benchmarking, and then a 15-step process for doing benchmarking. It is important to understand which type of benchmarking and then to follow the process that is defined. Not following the entire 15-step process will lead to benchmarking that does not meet the strategic purpose.

8.4.1 Types of benchmarking

First, let's define types of benchmarking. There are four main types of benchmarking: internal, competitive, functional, and world class.

Internal benchmarking. This type of benchmarking looks inside the organization for similar processes and units that seem to do it better. This is the first type that any organization does. It helps to first know your internal processes, so look within all your units. This is also the fastest and cheapest. It is also easiest to manage, since both sides of the benchmark work for the same company. It is least threatening. However, it can be the most difficult if units benchmarked are in an internally competitive environment. This may happen at large conglomerates where sister divisions with similar processes benchmark each other. The key benefit is that you look at your internal processes. You must understand your own processes before you can benchmark against others.

Competitive benchmarking. In this kind of benchmarking the organization looks at competitors and examines their processes. This type of benchmarking seeks other institutions that are performing better than the customer-driven, project management organization. When these processes are found, the competitor's performance is compared to the customer-driven project management organization.

However, many organizations are reluctant to share trade secrets with direct competitors and that makes this kind of benchmarking difficult. At times, competitors may try to mislead the benchmark partner. In the electronic industry, the Electronics Industry Association and publications like *Industry Week* and other electronics magazines provide actual data about electronic competitors, as do some of the world class books by Schonberger. Public (government) records (e.g., Department of Defense public information and Department of Commerce) and marketing data from competitors can provide information directly about processes. Electronics industry journals also give extensive information. With the increase of business information on the Internet, this can be another good source of information.

If your company is in competition for products or services, it is necessary for your organization to know about your competitors, and gather data publicly. After some experience in benchmarking, an organization can benchmark with a competitor on specific processes and agree not to show any proprietary data. It is good if both companies have similar management philosophies, and a win-win philosophy for both companies must be established as part of the benchmarking objectives prior to beginning.

Functional benchmarking. This type of benchmarking examines any outside or inside activity that is functionally like the process under examination. This allows benchmarking outside your industry to look at how others handle similar functions. This type of benchmarking is useful for coming up with functional approaches that allow you to overtake your competitors. It is very popular for accounting, finance, human relations, and information systems departments to use this type of benchmarking, but in this changing world of quality assurance, this author suggests that it would be good for companies to exchange approaches to quality assurance in different philosophies, especially in the advent of ISO 9000.

World class benchmarking. This type of benchmarking is best for organizations that have experience in benchmarking and know how to apply what they've learned to improve their own processes. This type of benchmarking involves finding those industries that are known to be world class. There is much literature on world class manufacturing (WCM) in which companies can be selected for possible benchmarking. The list in Chap. 2, World Class Manufacturing, might be a good place to start. Past Baldrige winners are listed in Chap. 9. With this type of benchmarking, the team should begin by understanding the literature on WCM and world class quality, as stated in Chaps. 2 and 3.

8.4.2 Benchmarking steps

The process of benchmarking (see Fig. 8.4) is defined sufficiently to facilitate the benchmarking process. It is very concentrated in the front end, as an organization needs to understand itself and have good plans, including metrics, to do good benchmarking. And, like any process involving change, the benchmarking process shows that once plans are implemented, they must be verified. If procedures are changed, or actual process documentation is changed, perhaps the verification will come with an independent audit function; but the estimated levels must be audited to ensure that the strategic reason for benchmarking this specific process was established.

The benchmarking process begins as a result of strategic planning. A complete understanding is needed of all the areas to meet total customer satisfaction, as outlined in the beginning of this chapter. The

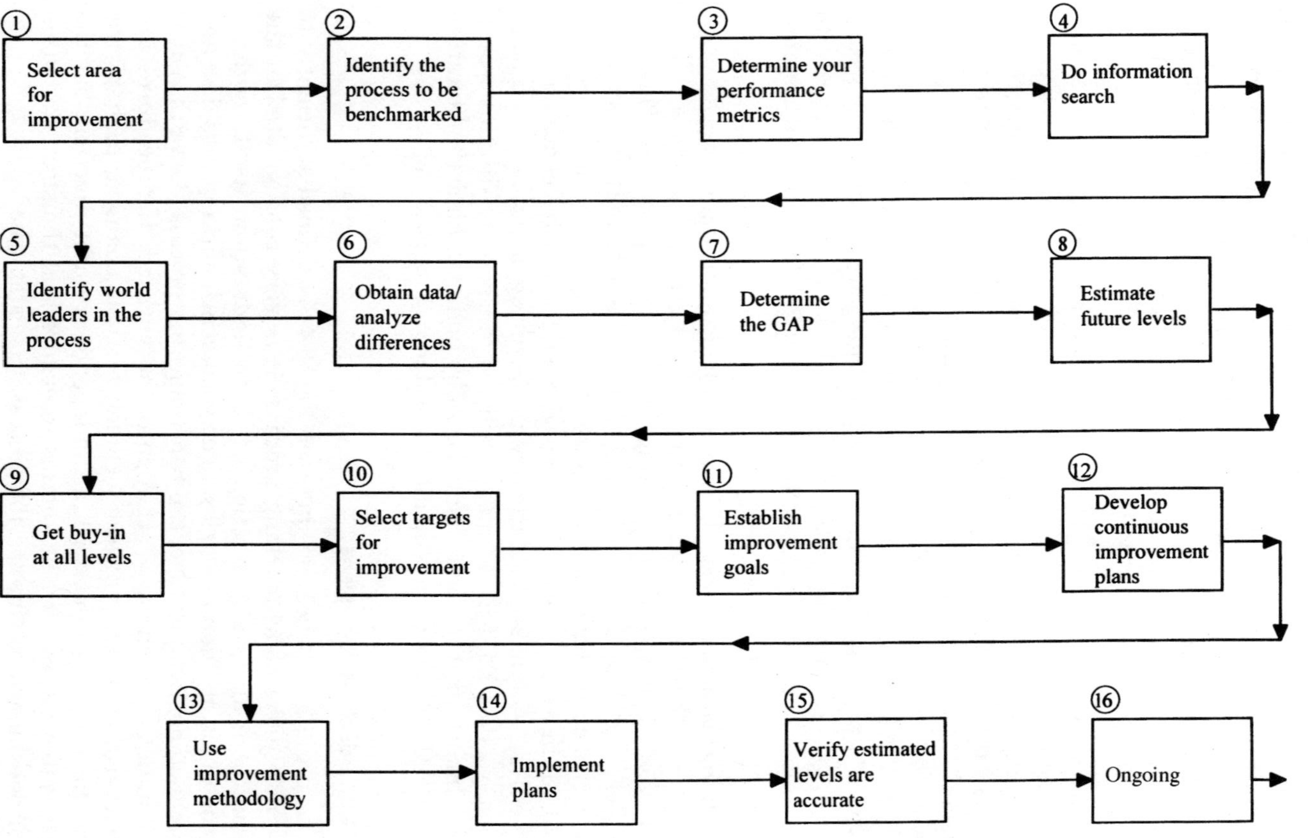

Figure 8.4 The benchmarking process.

mission outcome of strategic planning provides the focal point for benchmarks for the organization. In our example, the mission is as follows:

> Our mission is to be the leading organization seeking continuous improvement focused on total customer satisfaction by delivering continuously improving, value-added, results-oriented, customer-satisfying products and services.

Explanation of the specific steps follows.

Step 1: Select critical areas for improvement The first step involves listing the areas considered significant for success of the mission, namely the customer needs and expectations, deliverables to meet the customers' specification, and the internal processes to satisfy the requirements. Quality function deployment phase 1, outlined in Chap. 3, provides an excellent tool for listening to the "voice of the customer."

The organization may decide to benchmark all of the areas critical to customer satisfaction at this stage. During other stages in the Total Quality Management improvement methodology, the team may select other processes to benchmark. See Table 8.2.

Step 2: Identify the exact process to be benchmarked. This step consists of identifying the process to be benchmarked. On the surface this seems simple enough. However, this process requires some serious consideration and prioritization. In today's complex business world, we are involved with hundreds of thousands of processes that make our product or service available to the customer. Any attempt to measure every process and/or subprocess is not only inefficient but extremely costly and absolutely unwarranted.

What is needed is to define all of the processes that your activity comprises. This, in turn, requires detailed flow diagrams of your processes. Then differentiate between those processes which are the vital signs (major indicators) and those that are background. The trick here is to define those processes which truly define the "true health and vitality" of your business. To help in this process, the process owners must be involved. To determine the vital signs, it is often useful to use the human body as an analogy. When we think of the health and vitality of the human body, which is made up of countless processes, we think of the vital signs (i.e., heart rate, pulse rate, blood pressure, and temperature). When you equate the pulse points or vital signs of your process to that of the human body, you must ask yourself, "Is this the heart rate, or pulse rate, or blood pressure, or temperature of my process, or is it one of the background processes?" Not that the background processes are unimportant, but when looking for the vital sign, we need to identify

TABLE 8.2 Examples of Processes to Benchmark

Company Level	
1. Product line breadth and depth	6. Image
2. Feature sets available	7. Perceived value
3. Product performance	8. Impact on environmental stress
4. Product consistency	9. Overall customer satisfaction
5. Service level	
Specific Internal Processes	
1. Reliability	18. In-circuit test
2. Maintainability	19. Functional test
3. Life-cycle cost	20. Cleaning processes
4. Ease of use	21. Conformal coating
5. Attractiveness	22. Statistical process control
6. Warranty	23. Variability reduction
7. Wave solder	24. Dock to stock
8. Infrared (IR) solder	25. Just in time
9. Vapor phase solder	26. Kanban
10. Autoinsertion	27. Electrostatic discharge methods
11. Semiauto insertion	28. Crimping methods
12. Kitting	29. Bonding methods
13. Inventory control	30. Rework methods
14. Planning	31. Test processes
15. Scheduling	32. Receiving inspection
16. Document control	
17. Work instructions	

SOURCE: From Ref. 17 with permission.

that point which tells us something important about the health and vitality of the process under consideration. Again using the human body as an analogy, the doctor would not routinely order blood tests unless one of the vital signs indicated that there was something wrong. Therefore in this first of the next 10 steps, it is vital that all of the processes be defined in detail and that the processes' owners identify the "vital signs" for purposes of benchmarking. Table 8.3 provides some

TABLE 8.3 Examples of Benchmarking Parameters Applicable for Electronic Manufacturing

	Productivity
Output labor hour or employee	Output/packaging costs
Output raw material(s)	Output/process water use
Output/energy use	Output/maintenance hour or cost
Output/capital employed	Output/work-in-process inventory
	Timeliness
Percent on time to truck	Cycle time of main product (order to delivery)
Percent on time from supplier	Cycle time to main product (plant start to truck)
	Quality
Scrap of raw material	Work in process due to quality failure
Rework	Scrap of packaging
First-pass yield	Unplanned equipment downtime
Customer return rate	Documentation accuracy
Misdelivery	Quality-related receivable, overdue
Incomplete delivery	
	Other
OSHA reportables	Computer utilization
Employee turnovers	Vehicle utilization
Absenteeism	Backlog
Process equipment utilization	Overtime

SOURCE: American Productivity and Quality Center. From Ref. 13; used with permission.

good examples. This, then, is the process or subprocess you will measure and attempt to compare (i.e., benchmark) to one or more world class or best-in-class organizations. An example of this can be seen in the following excerpt relating to Dow-Corning registration when examining ISO registration [17]:

> The final key thrust (in deciding whether to get registered to ISO 9000) was benchmarking to see where Dow-Corning stood in relation to its competition. While Dow-Corning felt that it had a competitive advantage even prior to registration, it needed to obtain objective evidence. Dow-Corning used benchmarking to judge its performance against the best of the best and concentrated on performance and differentiation variables to assess the degree of its customers' satisfaction. These benchmarking variables included product line breadth and depth, the feature sets available, product performance, product consistency, service level, image, value, impact on environmental stress, and overall customer satisfaction. The results of the benchmarking study were displayed graphically and showed the relative performance of Dow-Corning to its competition and to the best in class.

Step 3: Determine your exact performance (metrics). During this step the selected critical areas are measured. To do so, metrics must be developed. The metrics development process is described in Chap. 3 and Secs. 8.6 and 8.10, and metrics are determined for the benchmark prior to step 6.

Step 4: Determine where to get benchmark information. Since the benchmarking information becomes the target, getting the right information is the most important aspect of benchmarking. The sources of information for process performance measurements are numerous. The only real source of a benchmark for performing an actual process is the process-performing organization.

Some sources of benchmarking data include the following:

- Computer databases
- Industry publications
- Professional society publications
- Company annual reports and publications
- Conferences, seminars, and workshops
- Other organizations within the same professional association
- Consultants
- Site visits
- Internet
- Benchmarking process engineering (PE)
- The Web site http://www.benchmet.com (highly recommended by the quality industry)
- World class suppliers

Step 5: Identifying world class. Identifying world class or best in class is quite tricky. In most segments of our economy there are no available data to identify a world class leader. There are, of course, the Fortune 500, Baldrige, *Industry Week* Best, and similar lists, but for the process you are considering, there is no proof that *for that process* the identified company is world class. Therefore, you need to have a specific methodology to determine what enterprises actually have that world class process as part of their activity (remember, not necessarily in your industry).

What process are you benchmarking? Now that you have a list of companies or agencies which you believe (based on your industry knowledge) are or may be world leaders in the process under consideration, it is now time to quantify your initial intuitive beliefs. Again, using a chart similar to Fig. 8.3, list the processes against which you would measure world class standing. Note that some of them show "big picture" benchmarking [18] and some show internal processes that can be benchmarked.

Step 6: Obtaining data. This step is actually relatively simple. However, there are some roadblocks and at least one pitfall. Let us discuss the pitfall first. As in any comparison of data, it is paramount that the data be exactly the same. This apples-to-apples comparison must be made prior to the exchange of data. Again, it is not uncommon once the data is exchanged and compared to find that parameter differences may be accounted for by subtle differences in the data collection or analysis systems.

Now that the apples-to-apples issue is understood and the potential benchmarking candidates have been identified, all that remains is to actually make contact with the identified world class process owner. There are several ways this can be done. Actual application of this process has shown that the most effective method is to use personal contacts, if they exist.

If this is not an available avenue, then a choice must be made between two equally effective methodologies. The first is to send an introductory letter to a sufficiently senior person at the identified potential benchmark partner (i.e., a director or vice president) describing your efforts and the benefits to both parties and requesting that your letter be directed to the process owners in the organization. Or you may start the process with a telephone call, followed by a letter. Industry experience has shown that you should contact management.

If reluctance to exchange proprietary data is encountered, it can be overcome by requesting normalized data or convincing the potential benchmarking partner that the data is actually not proprietary. One good way to get off a stalemate situation is to volunteer to send some sample data for review. In no case should you exchange any data that your company would consider improper for release. Therefore some ground rules need to be established to safeguard such sensitive data.

Now it is time to analyze the data. When you are sure that the data are totally comparable, you can proceed to the next step.

Step 7: Gap analysis. This step involves actually assessing the difference between your measure of the parameter of the process under consideration and the data you have collected from the other sources. The difference between measurements is known as the gap analysis (Fig. 8.5). This value can be either positive or negative. A danger at this point is to draw a final conclusion from too few samples or data sampled over too short a period of time. As an example, if you have data from only one other source and the gap analysis shows that your parameter is positive (i.e., better than your benchmarking partners), the tendency exists to assume that you are world class. Therefore more than one set of independent data is highly desirable. There is no hard and fast rule for determining the optimal number of benchmarking partners. The author believes the number is somewhere between 3 and 6, based on empirical analysis. Likewise, the time period should be between 6 and 12 months. Both of these rules of thumb are of course driven by the process under consideration.

Set benchmarks for performance. Since gaps between current performance and customer expectations will be part of the assessment, the organization can establish targets to define where it must be to truly gain a competitive advantage in the eyes of the customer, so when it sets targets (generally around the best-in-class benchmarks for quality, productivity, costs, and service), it can establish measurement systems to evaluate progress toward those marks. Beating those marks becomes the compelling rationale for the TQM process and for motivating the organization toward improvement. [19]

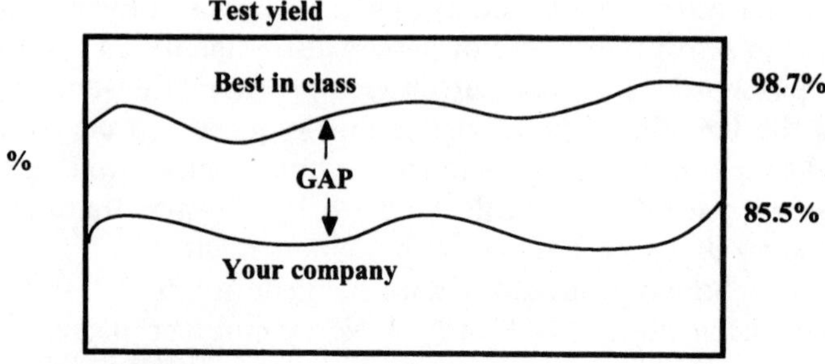

Figure 8.5 Gap analysis on first time test yields.

Step 8: Estimate future levels. With the gap now defined, plans for continuous improvement can be made. However, it is strongly suggested that analysis not be started prior to determining what change may occur to the database over the interval of time your plan for improvement may take. As an example, if your gap analysis placed you behind the best in class by 10 percent and you set your goal to increase by 18 percent over the next 2 years so that you can be best in class, you may have a rude awakening at the conclusion of that period. You cannot expect everyone else to be standing still during the time interval when you are getting better. An assessment needs to be made on the potential growth in your benchmarking partner's parameter. Once this probable growth is estimated, you are now ready to go to the next step.

The benchmarked target may take several years to achieve, depending on the current performance and capability of the organization. The organization establishes a plan to achieve the benchmark. For example, in year 1, the organization aims for improvement over its current performance. In year 2, it seeks to be competitive. In year 3, it targets best-in-organization and strives for class performance goals. In year 4, it achieves the world-class status.

Step 9: Get buy-in. Prior to setting specific improvement goals, it is necessary to obtain management and process owner acceptance of the measurement data as well as the vital signs being benchmarked. Without this buy-in, improvement efforts face a very low probability of success.

Actually the buy-in starts in step 1, when the process is chosen and the vital signs are identified. It is necessary, however, to maintain this early buy-in at step 10 to assure maximum support for step 11.

Benchmarking studies are most effective when performed by the process owners with the full support of the managers who have a stake in the results—namely, those responsible for the function being studied in another company and those who have the authority to make changes in operational processes. This creates management endorsement and leadership. Using the data generated by careful assessment, management can reach consensus on issues more quickly and establish priorities and tactics to accomplish their TQM strategy. The first three critical fundamentals become much easier to execute under these conditions.

Step 10: Select target benchmarks and set improvement goals. During this step, the electronics organization selects the target benchmarks to meet its mission. A thorough understanding of long-range goals and the current process development must be defined, prior to selecting benchmarking. As part of the decision process, it is important to determine which benchmarks take a priority position in the queue for improvement. Each organization has priorities and needs and must make its own decision tailored to its needs and its customers' needs. The only word of caution posted by the author is to remember, as a good business principle, that the value of improvement without associated payback is very questionable except in extreme cases. Each organization must come to grips with what the author considers a truism: No organization, no matter how world class it may be considered, is world class in each and every one of its processes. Don't push for improvement just to be number 1 unless it makes good business sense.

Step 11: Establish improvement goals. Once buy-in has been established and the goals set, the next step is to plan the actual improvement effort. There is nothing unique to planning for this effort with respect to benchmarking. The only advice is to keep validating step 8 during this process. If the original assessment of future levels was incorrect then your goals and/or plans may have to be modified to coincide with the more accurate data as it is developed.

Step 11 documents the needed resources. Missing ingredients quickly become apparent. Leaders cannot expect employees to make the progress TQM can generate without providing tools and resources. True leaders provide resources when they see the facts and the opportunities awaiting them.

Step 12: Be diligent; develop continuous improvement plan. Benchmarking is a continuous process. To be effective, benchmarking should be integrated into the formal operating and strategic planning processes. It should be a normal part of everyday business done by individuals in decision-making positions or those gathering or generating data for the purposes of decision making. This includes all deliberations that form a part of strategic planning.

Step 13: Use improvement methodology to achieve desired performance. The benchmarking and improvement process must be continuous. The organization must establish a continuous improvement system to achieve the target.

Step 14: Implement plans. As simple as it sounds, this is the area where many efforts fall apart. It is similar to the problem of the salesperson who lays all the groundwork and then fails to ask for the order. Poor implementation will result in poor results, regardless of how well steps 1 to 13 have been accomplished. It is suggested that a structure be developed for this overall effort (which has one person clearly in charge), to drive the effort to a successful completion. At completion, establish a steering group with representatives from each operating group reporting to a single process (benchmarking) manager. Each of these representatives is the single point of contact (SPOC) for its vice president and is totally responsible for all of the effort within that organization.

Step 15: Verification. The final step, that of verifying that the levels chosen (i.e., step 8/15) are correct, is actually an ongoing activity. Once benchmarking data exchanges are initiated, it is important to continue to keep the information flowing. This data should be continually reviewed to assure that the goals are correct. If the original goals were established on insufficient data or during a period of out-of-control process activity, or conversely at the only point when the process was in control, you may be chasing a ghost target. The only way to assure that this has not happened is to continue exchanging and analyzing data during the entire improvement process, or at least most of it. The frequency of data exchanges can be modified after this first round of improvement activity is completed. It is important to remember that benchmarking is an ongoing process; it is directly tied to, and should drive, all continuous process improvement efforts.

8.5 Processes and Process Understanding

A complete understanding of a process to be benchmarked is one of the keys to success. In fact, any business that does not have a documented set of detailed processes to ensure that standard processes are in place will not get to world class. The software industry, through the Software Engineering Institute (SEI) at Carnegie Mellon University, has moved the software industry ahead by insisting that all software processes be documented and standardized. ISO (not only the 9000 quality series, but almost all ISO documents that deal with getting work done) stresses the development of standard processes. ISO certification cannot be obtained without this documentation. And all industries, especially electronics, believe that all critical processes must be documented down to the lowest level of detail, so that the "recipe" will not be lost just by changing personnel. As we move into the twenty-first century, two events enforce this: the buying and selling of companies, and the fact that a generation is retiring that has not done a good job of mentoring the next generation in many areas. In all functional areas of the electronics company, the processes need to be defined and documented and have metrics collected against them. By documenting the processes in all functions, we understand their complexity, and what interfaces affect the process. The next anecdote illustrates this clearly.

The author was benchmarking at a large airframe manufacturer's electronic facility. After laying all of the necessary groundwork for the visit, including getting agreement on the "vital signs" (to be discussed later) to be benchmarked and submitting new data prior to the visit for our partner's familiarization, I was still astounded by some of the data as we started the comparisons. The working sessions were disrupted by an apparent disagreement between the data and the normalized cost data that they had agreed to share. My team of experts in all of the subprocesses were stumped by the apparently unexplained dollar deltas. What was even more surprising was that our processes were almost identical. After discussing this mystery during the 6-h meeting we decided to head home and try to better understand the seemingly disparate data. We wanted to plan the next step with our highly respected partner and to see if we could resolve this mystery prior to contacting them again with benchmarking as the subject. Obviously, if we could not correlate and understand the physical and the cost data, then there would be no hope for any improvement plan, given the fact that they may have valuable data contained within the puzzle. In frustration I discussed the problem with several of my coworkers over coffee. The only person to pose a question for which I had no answer worked in the finance directorate. It was then that I first realized that a process is much more encompassing than I had previously envisioned. In fact, finance is part of every process, as it is the ultimate presenter of the cost of any process.

A second meeting between the two companies was established with both partners having their finance departments in attendance. This meeting lasted only 2 h. The whole problem was that the technical experts we both brought to the table the first time were not trained in finance, nor were they expected to be. When we finally compared apples to apples, the engineering data and the financial data now went together like a hand and a glove. Truly knowing a process is more than just knowing the design or the flow on the manufacturing floor—a lesson that will not be soon forgotten by the author.

The real lesson to be learned here is not new, nor is it complicated. You cannot build on a foundation that is incapable of supporting the data it is intended to produce. Without a proper foundation, in this case a full understanding of the process, the data gathering and the analysis phase of benchmarking will not stand the test of time and management scrutiny. Just how does one get this magical understanding of a process? The answer is by flowcharting. Simple words, but a somewhat complicated process in itself. It takes time and the assistance of a number of people, experts in the process.

8.5.1 Process flow diagram

The process owners are the key to success. Without their help there will be no prerequisite foundation of information on which to build a successful benchmarking effort. To prove this let me reinforce the thought that the flowcharting process is not simple. Invite some process owners who think they really know a process to try this supposedly very simple task. Have them record the steps of tying a shoelace. This should be very easy, since most of us learned to do this prior to first grade. Give them 10 or 18 min to record the process. This record should consist of a number of discrete, simple statements in bullet format. Now that this simple task is completed, pick someone who was not involved in the recording of the process and instruct them to follow the steps with great precision. No deviations are allowed. When this challenge is taken, it normally results in two distinct conclusions. First, it is a good way to get everyone to laugh. Secondly, it takes a little of the cockiness from the people.

The combination of process flow diagrams and process owners has the sweet smell of success. All that is needed is to tie in the data that describes the process and to record the results. Sounds simple.

You start by selecting one of your processes that you have reason to believe will or can benefit, viewed in the light of benchmarking. Frequently such an effort starts by isolating and recording all of the processes in your factory. If this exists and you can verify this, you can skip this step. The next step is to decide on some discriminator or combination of discriminators to rank the overall burden each process contributes to the cost of doing business. This list of discriminators could include the defect rate, scrap rate, touch labor content, and any other task that makes sense to you and your team members. Conventional wisdom would instinctively attack the largest numbers first. Caution needs to be exercised and a little common sense applied. Some processes, even though costly, are very cost-effective and should not be picked just because they are large as compared to other processes. Cost versus quality versus criticality needs to be considered. One discriminator by itself does not indicate a poor process that must be worked. It could be a true indicator, but, again, caution needs to be exercised. When the process of discrimination is complete, we have arrived at a point where graphical description of the data should be undertaken using some sort of diagram to depict all of your data in a Pareto chart. This will make the order of your flowcharting task evident to everyone and will, in most instances, forestall any question about why your team started where they did.

Why take the time to develop all of these process flow diagrams? Remember that the goal is to have a detailed picture of how your company spends its assets in search of profits. If you hope to benchmark with another company that you have reason to believe is better at doing the same thing as you, you will need to be able to discuss each step in infinite detail and the value that it adds to the product as a function of its cost.

You need to define the process under study from the top down, in industry-common terms. You cannot start from the bottom up. The first step is simply to define the subject. Let us assume that the process of interest in our case is moving product from final build to final test to the sell-off room. The next step would be to flow diagram this process. Figure 8.6 shows how this is done.

As simple as this may seem, it is vital to get into recording all the data, as all future analysis flows from this simple model. It is also important to remember that you may be working as part of a team and that not each team member is at the exact same level of knowledge concerning the process.

This flow diagram works as a leveler of knowledge as well as the logical starting point. Of equal importance is the fact that you will most probably be required to make management presentations. The management team reviewing your presentation will more likely be receptive if they understand the process you are benchmarking. Frequent status reviews are usually more effective than one or two large reviews. You and the team will find it necessary to keep management sold on your task if you are to have any hope of final approval for the changes that are suggested. If you feed your reviewers too much data at one time, they will reach overload and you may not reach success.

Once your foundation is firmly in place, it is now time to expand the initial simple process flow diagram. Taking one subprocess or block at a time, form a set of the subprocesses that form the next higher level. Start with the first or the most logical process starting point. This second level may be serial, parallel, or a mixture of both that will describe the higher level. Using Fig. 8.6 as an example, you would start at the left-hand end of the flow and repeat the process for the next two blocks. Do not start the second block prior to completing the first block. There is no need to make the process any more complicated than necessary. We

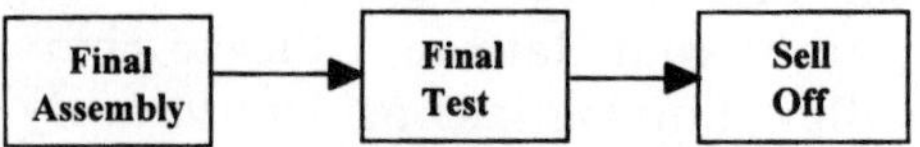

Figure 8.6 Flow diagram of moving product from final assembly to final test to sell-off room.

will start with the first block of Fig. 8.6, the *final assembly* block. In most flow diagrams it would not be logical to start in the middle of a flow. The danger in a nonlogical start is that it can lead to nonlogical assumptions, missed operations, and/or the exclusion of important interface steps. This approach can also result in missing redundant non-value-added processes. These are best recognized when an orderly flow is followed. An example of this is an inspection at the end of a sub-process and the repeat inspection at the beginning of the next process.

At this point in the process, the process owners' worth to the team becomes crystal clear to all of the team members. This is when you will find out about the little things that make the process what it is (good or bad). The value added by the process owners cannot normally be found recorded in any of the official process documentation. These are the little secrets that are lost when an employee terminates from a company. These bits of knowledge may be found on sticky notes and scraps of paper in the work area, in toolboxes, in notebooks, and in the dark recesses of the biological interface's mind. They will mean little or nothing to anyone but the recorder of the data. Now is the time to start listing all of the steps (processes) that make up the final assembly. We do this before trying to draw the blocks. It is best done in an open atmosphere such as would be found in a brainstorming session. What follows is the result of this exercise. It is not important in Table 8.4 whether this example is complete or not; only the first block is shown to demonstrate the output of the technique.

There are probably more steps involved than the 10 we recorded, and the steps would be highly influenced by the way your factory works. This demonstration of how a simple block expands will be true for most, if not all, processes you deal with in the real world. Now it is time to take our sample 10 steps and draw a flow diagram that pictures what really happens on the assembly floor. When we made the list, we did not care whether the individual step was a series operation or a parallel

TABLE 8.4 Steps in Final Assembly

1. Gather all subassemblies
2. Verify against parts list
3. Obtain the chassis
4. Check previous assemblies for completeness
5. Assemble subcomponents
6. Check for fit and alignment
7. Connect all leads and cables
8. Inspect your work for completeness and quality
9. Complete all paperwork
10. Move the product to the next process

operation. As we expand this block, we pay very close attention to what the process owners tell us. We do not try to perfect the flow at this time. Our task is to record what takes place. There will be plenty of time to remove non-value-added activities and to change the order in which things are done. Figure 8.7 depicts this visualization.

Now that we have agreement among the team members that Fig. 8.7 fairly and accurately represents the process of final build, we discuss the actual procedure on the floor and the existence of non-value-added tasks. We have not even started to see what other companies do and already we are starting to perfect our own process. This figure contains what may be considered as non-value-added tasks. Can you recognize them? Ask yourself who should be performing the duplicate task? Going through this complicated process yields a number of worthwhile results, as shown in Table 8.5.

When you choose your benchmarking partners you will need all of this data plus that for the next level down. This third-level data gets into the nitty-gritty of how tasks are to be performed. It is normally as far down as you ever have to go in benchmarking.

8.5.2 Expansion of blocks

Now let us examine one tiny piece of data from Fig. 8.7, the block labeled *assemble & verify.* Within this assembly process are a number of subprocesses that make up this task. Once again we would witness the multiplication of blocks in a third-level diagram. We would have the same outcome as when we went from the first level to the second level. The difference would be that the multiplier would be greater in this step then it was in the first attempt (i.e., going from level 1 to 2). We will not attempt to draw level 3 in this text, as the technique does not differ between levels. What we have shown is that the data looks like a pyramid in shape, starting with a simple small flow at the top and growing ever larger as you move down. Figure 8.8 depicts this expansion of data.

Some of the tasks are shown in tabular form to give the reader some feel for just how this expansion takes place. When you are recording a process, you cannot take the liberty of skipping the recording of any step (even though we have done so in this chapter to save space and time). Remember $B = P$: you must draw every block, no matter what level you are at. A single missed block can harm your benchmarking efforts. Table 8.6 shows the value to be gained and the tools developed when you go to the third level. It is important to remember that "To know the process is to improve the process."

In summary, what we have in this example is a first-level partial process flow diagram of three blocks which expands to 10 blocks at the second level, and for one of these 10 blocks we have seven more blocks at

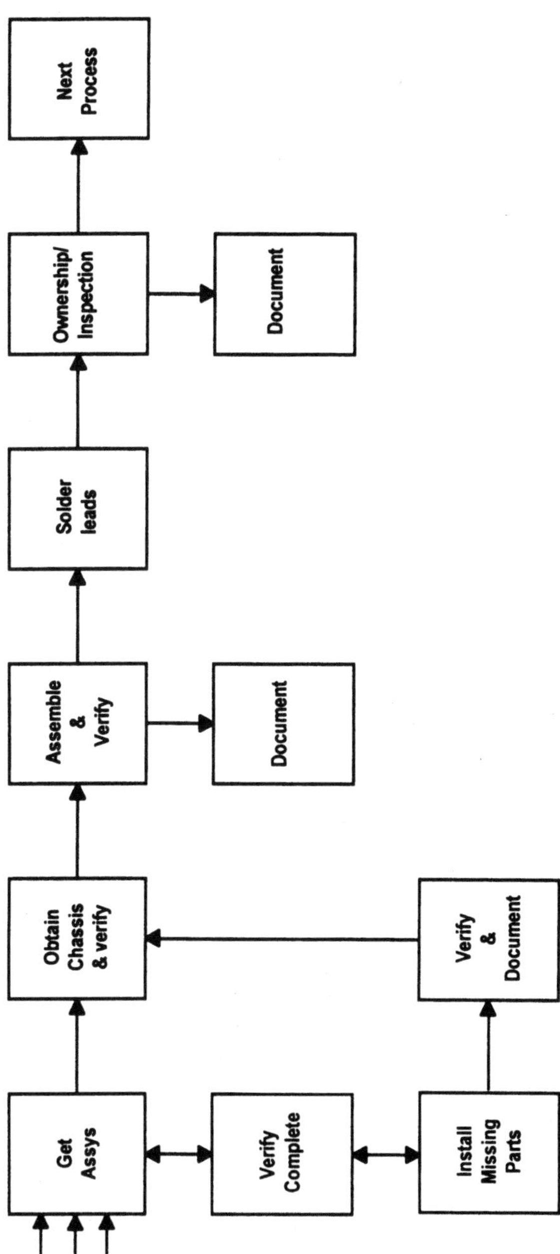

Figure 8.7 The subprocess of final assembly.

TABLE 8.5 Yield of the Process

1. You and the team learn the real process
2. You have captured the secrets that make it work
3. You spot redundant work
4. You detect non-value-added tasks
5. You get a chance to look for better ways of doing the task

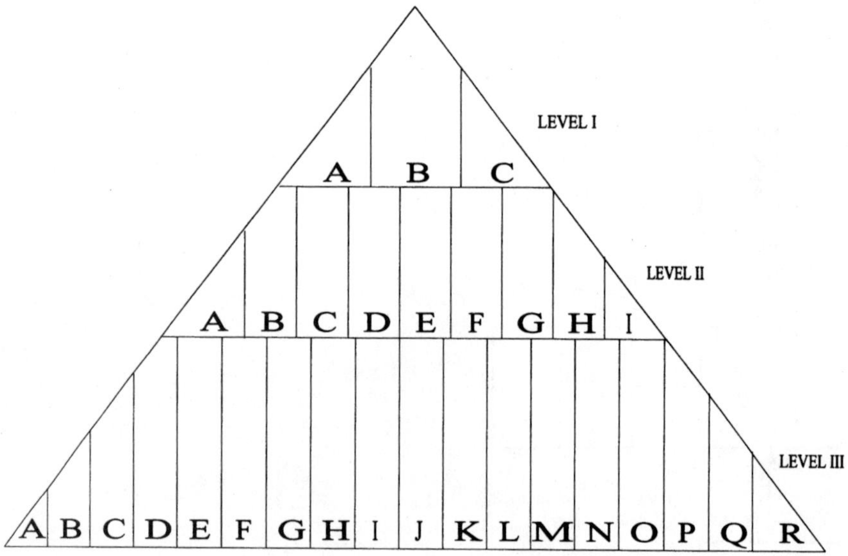

Figure 8.8 Process definition from the top down (the pyramid model).

TABLE 8.6 Process Third Level

1. Verify that you have all of the parts prior to starting
2. Notify Quality and Manufacturing if any parts are missing
3. Do you have all of the necessary mounting hardware?
4. If not, notify Quality and Engineering of the deficiency
5. Secure subassemblies and parts with mounting hardware
6. Notify Manufacturing and Quality of any deviations or difficulties that arise
7. Connect all wires and cables (solder or plug-in)
8. Notify Manufacturing and Quality of any difficulties during this step (wire has no strain relief, etc.)

level 3 (not shown). Looking back at Fig. 8.7, we can appreciate just how true the pyramid model is for this process flow diagram. What we have looked at in level 1 is one block of a total of n blocks. It should be quite evident at this time that one person should not be expected to know all of these facts. This simple truth is compounded when multiple people perform the same task and each is capable of introducing variability.

There is no stopping this manual practice, even without automation (and in low-volume production, automation may not be cost-effective). This is one more reason that supports the need for a team approach, including all of the process owners. We need not pull them all off the shop floor at one time and stop production, but this does not mitigate the need to listen to their collective voice as we do for our customers.

8.5.3 Gathering data for blocks

Now that you have built process flow diagrams for all three levels of the process, you are ready to start arranging the logic of the flow and start gathering cost or efficiency data for all of the individual blocks. The same scenario must be followed for the other two blocks of the partial process flow diagram with which we started. In the real world, a process at this level may be this simple or it may be orders of magnitude more complicated. You can start this analysis task by asking the team a set of simple questions designed to establish a dialog of inquiry that has no boundaries and no off-limit areas. Such a set of sample questions appears in Table 8.7.

There is probably a logical end to the questions you can ask. The suggestion is to have a baseline below which you never go. You can add questions based on the process involved, but there needs to be some logical bottom line that all processes must answer to before benchmarking.

All of the questions must be answered with fact. Emotional responses are taboo. Ask the contributor to go back and dig up fact to support the statement, and the team will be happy to include the contributor's input. If your team were to accept emotion in place of fact, you could answer all of the questions with a simple "no" and consider

TABLE 8.7 Dialog for Inquiry

1. Is there a simpler way?
2. Is there a faster way?
3. Is there a cheaper way?
4. Is there a more reliable way?
5. Is there a more repeatable way?
6. Is the technology we currently use state-of-the-art?
7. Why do we use the materials we do?

your job complete. Please remember that is not what the company is paying you to do. You are expected to be critical of every step against the world class manufacturers that you will soon identify. Have your house in order before you show it to others. One of the advantages to having scrubbed your flow diagrams prior to presentation is to avoid the embarrassment of having your partner do it for you in public.

Do not form any hard and fast conclusions yet. This is one of the pitfalls that looms in the shadow of the very team that has put together the suggestions. Your team may have one of the original design team members on it, and care must be taken to assure that this member does not unfairly influence the outcome. This is the reason that no suggestion or question is ever disregarded or ignored. There is no telling how this designer will react when his part of the design is being, for all practical purposes, second-guessed without the original pressures imposed on the design process. The designer needs to be part of the team and needs to understand that the team is collectively working to make the next revision of the design even better than the last. If the designer reacts in a personal way and takes the questions or suggestions as a personal attack, then the team (or the team leader, if there is one), must seriously consider asking that member to retire or to work within the framework of the team. This matter must be attended to quickly, because it will eat up time and will be harder to resolve as time progresses. This is how continual improvement results. Being concerned about criticizing a friend's design is to do a disservice to both the company and you.

You are expected to take every step at the third level and list all of the alternatives for each task and subtask. For instance, "Are separate screws, washers, lock washers, and nuts required at each attachment point? What other technologies exist to accomplish the same task?" Make a list of these alternatives for further review. Table 8.8 represents such a list.

TABLE 8.8 Alternative Technologies

1. Adhesives
2. Captive hardware
3. Stakes
4. Tabs
5. Clips
6. Springs
7. Digital
8. Analog
9. Interconnect
10. Etc.

Now you are well on your way to making benchmarking itself a value-added task. In preparing to ask outside your company for facts on identical or similar processes, you have, in some cases, reengineered your own process.

Task the team members who are experts in the alternative attachment ideas to prepare rough order-of-magnitude costs for each one, as well as the cost of the current design scheme in use. At this point do not solicit or accept recommendations for any of the design alternatives. The same is true for negative factors. There will be plenty of time to do this after you visit your benchmarking partners. It would be counterproductive to form opinions prior to seeing, for yourself, what other people do and to discover their rationale. If this task seems somewhat burdensome, it will become apparent shortly that this is one of the least expensive improvements you can make. In most cases, a reasonable amount of time has passed since the original development of the design requirements. Many things may have changed in this interval: acceptance of new technologies, new materials, new design concepts, and even techniques once considered unacceptable, for whatever reason. We just cannot pass up an opportunity to make our product world class, no matter from what quarter the suggestion may originate.

The data needs to become a permanent part of the design file. It also needs to be given wide distribution to all members of the design team; it will be of only limited value locked away in the design file. The team has the responsibility to see that both tasks are carried out in a timely fashion. We are addressing not only working on redesigns, but making sure that the designers of new equipment have the same data available to them. What a waste it would be if they had no access to this data; yet, surprisingly, this is not uncommon.

8.6 Vital Signs

Every process has its own set of vital signs. These vital signs of a process speak to the wellness of that process, just as they do of the person who interacts with the machines. If you understand what they indicate, they can quickly lead you to detect weaknesses without the need to always perform lengthy and costly testing. If you choose to disregard these vital signs, you will reach the same conclusion but with a lot of time and cost for non-value-added testing. This is not to say that testing to verify the vital signs is not in order. We are cautioning you to avoid reinventing the wheel.

When you understand your own vital signs, the transition to the manufacturing or design process will seem old hat. Why even do this step? There are two reasons. First, we do not want to go out and benchmark a sick process; and second, we want to be able to spot a sick process at our partner's facility. There is little value in making comparisons that could lead to false positives or false negatives. We must avoid the trap of garbage in, garbage out. Our goal is to obtain substantiated fact in order to make meaningful decisions which hopefully draw us nearer to being world class. If you benchmark your process against a sick process and do not recognize this fact, you look so good that you allow yourself to be lulled into thinking there is no value going any further. You stop your task and certify that you are already world class.

In the human example, we can consider some or all of the vital signs in Table 8.9. These are things that your doctor or nurse looks at when you first walk into the office. Some of the observations are so subtle or ordinary they do not even pique your curiosity.

You are now ready to embark on developing a similar list for your process. When that list is finally completed, you will be fully armed with the data you need to benchmark against other companies. Prior to setting off on your benchmarking journey, you should send your partners a list of your vital signs so that they may become acquainted with the data and you need only discuss the "deltas," or differences. This will save a great deal of time and money. By developing these vital signs, you also avoid comparing parameters that have little effect on the process.

Let us now build our list of process vital signs. There is no single correct or universal list. Some of the measures may be universal, and some vital signs may be related to common processes. We have selected the automobile as a common product for which building the vital signs should be easier because of our almost universal familiarity with the product. In addition, to show that vital signs fit just about any process or product, we will also build a list for a computer and for a process, the automatic teller machine (ATM) process for banking (Table 8.10). Each area shows processes and possible metrics to benchmark.

TABLE 8.9 Human Vital Signs

Temperature	Indicates the presence of infection
Blood pressure	Indicates condition of circulatory system
Heart	Indicates several things, including disease
Lungs	Indicates congestion and disease
Skin tone	Indicates liver malfunctioning
Pulse	Indicates pressure and rate of blood circulation
Eyes	Indicate a number of conditions, including disease

TABLE 8.10 Product/Process Vital Signs

Product: computer	Metric
Processor chip	Pentium: processes data
Random-access memory (RAM)	Fast access to data
Hard disk	Millions of bytes; stores data "32-64 mg"
Floppy disk	Stores data/size important
CD	Stores data
Speed	Depends on RAM
Size	Desktop or laptop
Weight	Especially for laptop, 7–10 pounds
Screen	Black and white or color
Size	Needs of user, 13/17/21 in
Resolution	Dots per inch
Keyboard ergonomics	Usability
Product:auto	**Metric**
Fuel	How much, what grade, miles per gallon
Oil	How much, what kind
Engine	How powerful
Muffler	How reliable
Ride	Comfortable
Steering	Easy, small ratio
Tires	Reliable, expensive, safe
Paint	Attractive, chip-proof
Cost	Inexpensive, expensive
Power	Depends on need
Warranty	1/3/5 year
Process: Automatic teller machine (ATM)	**Metric—think about time, and combining steps**
Insert card	1. What if your fingerprint combined first three steps?
Machine responds	2. Is your ATM always operational?
Person puts in personal identification number (PIN)	3. Is the response time on the items always within a second or two?
Machine gives options	
Person selects options	
Person waits for machine to react	
Machine asks for next option	
Person selects	
When complete, person picks up cash	
Machine produces receipt, returns card	

Each of the vital signs in the lists is useful only if each one has an associated numerical value or range and a unit of measure. For instance, although 18 is a numerical value, it means nothing until we add the unit of measure. The unit of measure could be pounds or inches or any other quantity. There is no set list. However, if we are to take the subjective out of our deliberations and focus only on the objective, then units of measure must be present in order to benchmark with your partners.

Let us take the product example and try to identify the hard measurements associated with each of the vital signs in our example. In some instances we will be obliged to convert the aesthetic or emotional into a metric and a unit of measure where none now exists. In this latter case you may find that there are no universal units of measure in benchmarking. To your surprise, you may find that others either ignore measures or have their own measure, or that two people have arrived at the same measure independently.

In discussing fuel, number one in our auto example, the first question that logically crosses our collective mind is fuel economy. The accepted way of defining it is in terms of volume per unit of distance. In the United States this would take the form of miles per gallon (MPG). There are several things we could say about MPG. We could elect to utilize the U.S. government mandatory MPG test numbers that appear on the price sticker attached to all new cars. We could also use data gathered from actual users in typical driving, or we could use expectations as the number or range of numbers. This brings up an interesting point. When looking at this vital sign for an economy car, we would be shocked if it were listed as 17 MPG. So this means that the measure and the measurements must be specific to the process and the product. In electronics, volts and amperes would be appropriate where felt and nodes would not.

If in our benchmarking we saw a value of 12 MPG for the same vital sign associated with a well-known luxury car made in England and used by the Royal Family, we would not be shocked by the low value, although it is quite a bit lower than the 17 MPG for the economy car. The only surprise would be that anyone would pick such a high number for such a large and heavy luxury car. I am sure no one has decided not to buy this luxury car on the basis of its consumption of fuel. We should consider whether this is a valid vital sign or just a measure of the product, and then include it when it is and exclude it when it is not.

This same process is carried out for each of the vital signs. Your team may have some interesting discussions settling on the final list of values and units of measure for the list of vital signs. Expect this to occur. It is natural. Different people have different perspectives based

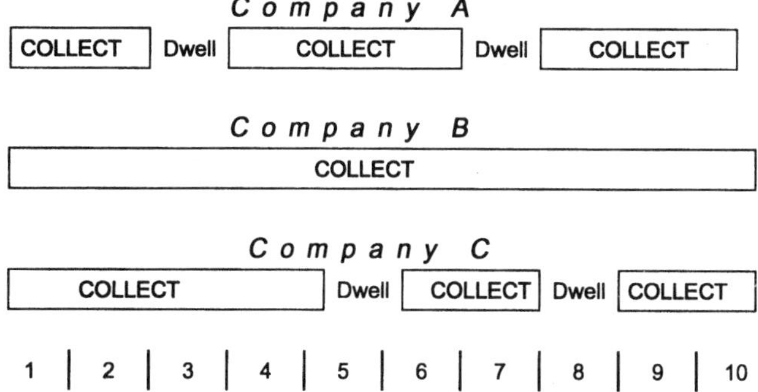

Figure 8.9 How different companies collect data.

on their individual backgrounds, which may consist of widely varying life experiences. Do not attempt to stifle this discourse. In fact, you should encourage this exchange. The most common unit of measure may no longer be appropriate (if it ever was). This free flow of ideas will be most helpful when your team tries to tie down an aesthetic value with a numerical tag that is measurable and repeatable and one you will be able to sell to your benchmarking partners. This will be slightly more taxing than selling the rest of your list, which contains industry-accepted units of measure.

Some of the companies you select for your partners may not collect data at the same level as does your company. It may surprise you to find out that some of your partners do not even collect a parameter which you consider a vital sign. This could have occurred for many reasons. The obvious reasons are that they do not consider it a vital sign, or they do not think it adds value to their control, or they never thought of collecting the data, or they collect the data somewhere in the company but the person or people you are communicating with are unaware of its existence. In such a case, tread lightly. Do not embarrass your new partner. Explain how or why your company uses the data. This may be enough to prompt them to look into its value and the ease of collecting the same data. If you end up on the receiving end, listen carefully and ask meaningful questions. After all, this is why you are benchmarking in the first place. See Fig. 8.9 above for an illustration of how different companies collect data.

8.7 Ethical Data Gathering

Many people gather data. Some are called pollsters, others have the title of consultants, and still others belong to a select group known as spies. Of this last group there are many subgroups. There is no intent in this chapter to discuss these subgroups nor to discuss the correctness or their political necessity. Many people gather data for a living. We all gather data every day in order to make decisions. We will not be discussing this casual form of data gathering that goes on every day. Some do it in the open and some do it clandestinely; some do it legally and some do it illegally, as in the Watergate scandal. What we are telling you here and now—and forevermore—is that there is no place in benchmarking for the illegal. We are not telling you to do anything against the law. If you are requested to perform such a deed, you know the law and the penalties and you know the odds against your getting another professional job, especially if you are convicted of a felony.

What benchmarking encourages is the use of all public data plus whatever private data exists in the public domain. You can get information from the vast files of the government, from your benchmarking partner, from a stockbroker, from trade publications, and through your network of associates at trade shows and symposia.

Let us look at two data-gathering scenarios. One is quite legal and the other at best is highly questionable and probably is illegal. Legally, you can observe the quantity of trash being discarded by a competitor, from a public access area, to get some feel for the level of activity at that site. There are other things equally legal that you can do. You can observe normal operations including the number of shifts being run and the number of vehicles per shift. All of these data are available to anyone who takes the time to look. You could ask questions of suppliers with respect to the activity of the company of interest. You could also ask those suppliers how much business you have to give them in order to get the same price consideration. Again, this is all legal.

What would be questionable to downright illegal would be to enter the private property of a company and rummage through their trash to count quantities and to get names of suppliers. Equally illegal, or at least unethical, would be to go through their trash and read discarded correspondence, even if you physically do not trespass. Obvious illegal acts, such as breaking and entering or bribery, are totally against ethical and legal behavior and can never be sanctioned or excused by claiming it is required by benchmarking.

So much public data is available that you probably cannot use it all. It would be a waste of time for benchmarking to spend its precious assets collecting all data, but the data it needs are a subset of all data that exist. Decide what specific data you need and go after it. If you go overboard and collect too much, you have to either analyze it and not use some of it, or cull the data after gathering it, and that is a waste of money.

Leave the questionable, cloak-and-dagger data gathering to the alphabet agencies (CIA, FBI, NSA, AIA, NTA, KGB, etc.). They are paid for the task, trained, experienced, and chartered to do it and take the risks. You are not!

A few companies, but not many, will not discuss any of this data without a legally executed nondisclosure agreement. Some will decline to discuss one or more parameters, claiming that they are trade secrets. You could present to them the idea of normalizing such data and ranking the comparison that way. In any case, do not push your point too far. If you do, you may find yourself talking to yourself and lose a potentially good benchmarking partner. Loss of your chosen partner could result in the loss of that one piece of data that you really needed to make you world class or to maintain your lofty position at the top of the heap.

Some people will discuss nothing with you. They will decline your offer, seeing some machiavellian plot to steal their secrets for use against them, or to give it to their competition. If you feel strongly about making this company one of your partners, not all is lost. Carefully search in your own company for someone who has had contact with that company at another level of management. If this fails, request that your president consider sending a letter to their president requesting that they benchmark with your company. Probably the most effective way to get your president to endorse this move is to have a final copy of the letter in your hand as you ask. This letter should be the product of your team's efforts. If you want to impress your president, have a computer disk with you on which you can make real-time changes he or she may want to make prior to signing the letter. It is always much better to leave the room with your business completed than to have to make a second visit. Remember, the president's time is very limited and expensive.

8.7.1 The ethics of benchmarking

How do you make sure you and your partners gain, not lose, from benchmarking? The International Benchmarking Clearinghouse (IBC) of the American Productivity and Quality Center has established a code of conduct for benchmarkers that includes many of the following:

- Obey the law. Don't do anything that might imply restraint of trade, bid rigging, price fixing, or acquisition of trade secrets. If you agree to share proprietary information, make sure all parties sign a nondisclosure agreement.

- Be prepared from the start. Don't waste your benchmarking partners' time by learning on their time.
- Be willing to give partners the same kind of information you want. Make sure all partners know what to expect to avoid later misunderstandings. Don't assume anything.
- Respect confidentiality. Treat any information you gain from benchmarking as confidential unless you get your partner's written permission.
- Keep benchmarking information inside your organization to improve the operational processes of your organization. Don't use the data in your advertising or promotional efforts.
- Follow the chain of command. Don't first approach the business unit; seek permission from the senior management of the parent company.
- Unless given permission, don't share benchmarking partners' names with others.
- Be honest.

8.8 Data Analysis

After your team has expended so much effort to gather the benchmarking data, do not lose sight of your objective. Two common errors are made at this place in the effort:

1. Using the data without analysis
2. Analyzing for things your task does not require.

Using raw (that is, unscreened and/or uncensored), poorly understood, or confusing data will result in conclusions which reflect those characteristics. Your team could just as easily have concocted false data based on the conclusions you want to reach if your goal is to prove you want to discredit benchmarking or to prove you can cheat the system. As an example of raw data: If one observes a factory working around the clock but with a half-empty parking lot, a number of logical additional questions need to be asked in order to understand the observations. Here are some typical questions you could ask yourself:

1. Is the company into carpooling?
2. Have they recently gone through process restructuring?
3. Has the company recently automated their production line?
4. Have they narrowed or flattened their core business or realigned their profit-generating products?

5. Are they subleasing some of their idle floor space?
6. What does their trash say about their purchases?

You would develop your questions around your data. If you fail to question your data before you use it, you will fall into one of those traps we always warn you about. Do not ask why half the data is below average. Wait for one of your vice presidents to ask the question during a review. Then you will learn that silence is the better part of valor. Keep quiet and maintain a blank stare and wait for one of his peers to answer that question. You and the team will have plenty of time to snicker later. For now, remember that data can be manipulated to tell almost any story. Someone in your audience will certainly challenge your data as well as your conclusions. Be prepared to explain why the data can be misleading if they are not associated with other related data; or why the question, although excellent, missed the goal that the team is working toward. Be prepared to answer any negative or positive question with the same enthusiasm. Never put down any sincere questioner. Do not be afraid to interject a little humor. It can be useful to get things going, especially after a touchy situation. Never go into a review without a dry run with the team. Have them pose any question, no matter how silly it seems to you. Do not limit the review to the data you used; include data you excluded and your conclusions and recommendations. Have at least one person play the devil's advocate. Better to fumble in a dry run than before your management. During this type of dry run, do not let the majority rule. A way-out question then could save you some real problems later. It is not unusual that only a single person at the dry run or at the real review will be able to see something that everyone else missed. Do not let it be tabled just because only one person recognizes a disconnect. It is easy to be shouted down, especially near the end of the dry run when everyone is tired and wants to get on to something else. The reason you have free thinkers on your team in the first place is for the unique background and expertise they bring to the team. Even the smallest voice must be given the opportunity to plead the case. You must give this question the same thorough consideration and respect as any other question. Do not turn off any team member, especially during a dry run.

The data, as you and the team had carefully decided to use it early in your planning, must stand the test of friend and foe alike. Valid data properly collected, carefully analyzed, and expertly presented by you and the team will have no problem withstanding this test. Benchmarking helps change the course of a company or, at very least, helps it maintain its world class position. This activity does not happen in isolation from other activities going on in your company, but it certainly carries great weight.

Make all of the hard work you have done really count. Do not look for economies with your data analysis. Do not skimp with your work when it comes to the data analysis. If you are still uncomfortable with the quantity or the source of your data, do not rush to conclusions. Request that management allow a prereview. Voice your concern, and ask statistical assistance if this is what you need. Make suggestions to bring the study to a proper conclusion by adding another benchmarking partner, if you need confirmation of your data. Your concern could be on either side of confidence. Your data may be too pessimistic for you to believe, or the conclusion that you are perfect does not feel right in the pit of your stomach. Make management part of the decision to go one step further, if that is what it takes. To your surprise, they may say that they do not think there is undue risk in your work. Remember that your data is never complete or representative until the team believes it is. Do not try to present half a story to management. It will not be a service to your company and it will not help either of you in the long run. You have come too far to let poor or incomplete data analysis ruin your team's hard work. Your company deserves only the best!

8.9 Building a Team for Success

One can never stress too often, or with too much emphasis, that the results of the benchmarking effort can be no better than the team that did the groundwork. When this group of experts and subexperts is being assembled, there are a few ground rules to consider. Be wary of the persons who are being pushed onto your team because their managers want to get rid of them, even for a short time. Be aware of the person, known as "the lifer," who fights all change. Both of these types can and will impede your success. On the brighter side, select members who are known to have demonstrated knowledge of the subject area. Select people who are aware of the industry as a whole and its trends. Pick people with a winning attitude and a "can do" mentality. Accept only people who firmly believe in continual improvement; if any member of this team does not really believe, you need to rethink their inclusion. Your team must consist of people who "walk their talk" or practice what they preach.

Any benchmarking initiative should start with education. Consultants who have had hands-on benchmarking experience can be especially helpful in providing education as well as guidance on the actual benchmarking study.

Then the team needs to be trained in the benchmarking process. According to The Benchmarking Exchange, in 1992, to have someone within an organization trained and in contact with potential benchmarking partners cost $20,000 per year; in 1996, the cost dropped to $780 per year. [20]

The team must remember that when they set their goals for improvement, they need to set their sights to be better than the current world class they wish to lead. The company you are looking at overtaking also wants to remain preeminent in their field. They will be looking to improve. Remember that you are looking at an elusive, moving target. You need to anticipate where it will be and plan to be in front of it. If you aim at the target, you will end up behind the target.

Make sure your team members understand consensus teamwork. If potential members do not, you will have to bring them up to speed. The key is to get everyone to say that he or she can accept the decision of the team. If you can't do this, the team cannot move past this point. Facilitation may be in order. Team consensus prevents the "I told you so" or the "I warned you" second guessing. This trait is often seen in people who cannot take risk. It is also part of the makeup of a person who is into status-quo management. If it turns out that this is the disposition of a team member, the team needs to get that person turned around, if you want to maintain forward motion. That behavior is unacceptable and must be stated as such! The best of all worlds is to have a team in which all members have had formal team training. Figure 8.10 illustrates world class goal setting.

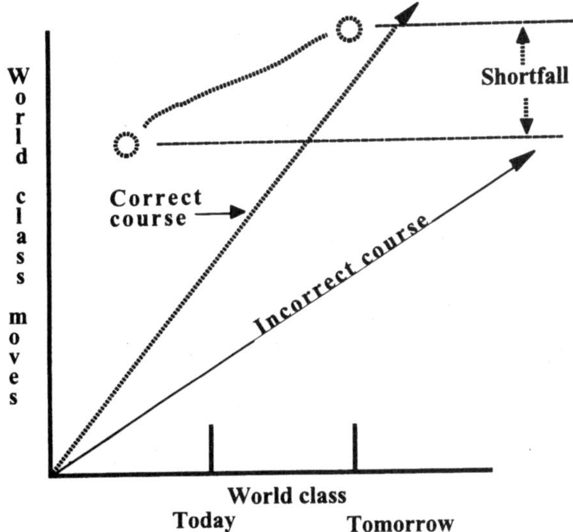

Figure 8.10 Goal setting versus a moving target.

If you feel that your benchmarking team requires facilitation, make sure that the facilitator understands your team goals and supports your team's activities. This must be accomplished prior to setting foot into the first team meeting. There is little or no chance that a trained facilitator would ever walk into an ongoing team activity cold, that is, without facts. This is a sure way to turn the team members off to the idea of facilitation. Make sure the team understands that the facilitator is here only to help them surmount the current impediment. There is, once again, a note of caution. There is probably someone in your organization who does not support your team goal and its activity. It is usually someone who fears change and is in a position to be a road block to you if the occasion arises and that person can't be blamed. Be cautious if facilitation is forced on your team without your request for assistance. This is especially true if the person behind this move is suspected to not be one of your team's supporters. A team can be derailed by poor facilitation, whether intentional or unintentional. It is a sorry fact that in some organizations you are required to struggle simultaneously with both technology and dinosaurs.

Team overall compatibility is likewise very important; in fact, it may be a prerequisite for success. If you note any personality clashes, do not try to resolve them at team meetings. That approach may polarize the team, and this will spell defeat. Meet one to one with the people involved and try to work out the differences. Be firm and make it clear that the team cannot tolerate disruptive activity. You and your company are in a day-to-day battle for economic survival, and being world class is one way your team can contribute to that win-win goal. The team can make a difference, a powerful difference. Form the team for success.

And now, a little about team leadership. Autocratic leadership does not work well with consensus teamwork. However, from time to time, teams require someone to step back to focus direction and recommend some rethought. Sometimes the leader must be autocratic in the beginning to get the team to work to the correct process, and collect and document the correct data. They need someone to perform administrative functions and be the single point of contact with the whole world. There is little doubt that a team leader is required. It is good for the team to select the leader from among its ranks and, from time to time, to rotate the leadership. This person carries no more weight than anyone else on the team. Not all team members seek a leadership position, but it is good experience for every team member to have a little leadership experience so that his or her understanding of the leader's position is appreciated.

8.10 Defining the Team Goal in Terms of the Physical

This, as I have alluded to throughout this chapter, is so important that I want you to remember it as my final caution: failure to define or completely and fully define the end goal of your benchmarking team effort will probably result in an incomplete effort as far as the company's needs are concerned. This could result in just the opposite of your initial goal to help preserve and advance your company's position.

Keep clearly in the minds of all of your team members and your management staff that everything your company does consists of processes. This simple thought is often overlooked, because many of these processes are not recognized as such. The fact that these processes are associated with continual improvement comes from process improvement. Therefore, you must first define those processes that define your company. After that, you need to specify the measures that define these parameters. You will not get very far if your goal is "to get better." How much better? It will not be of much avail to your company if your goal is to "be good." Good in the eyes of whom? In the eyes of the sales force, the accountants, your customers, or your suppliers? They will all have somewhat differing definitions of "better" and "good."

What you need to do now is to define goals as hard numbers or ranges of numbers. You need to take even the aesthetic, and convert it into hard numbers. You must give the data exposure to the light of day for all to comment. If you are considering how paint or a solder joint looks to a customer, you need to know what factors make it look good to the customer. The team needs to define these measures. If they like a deep high-gloss shine to the paint, you need to consider measuring factors such as reflectivity. There are doubtless many more measures. It is your job to convert all of these nonnumeric values to numeric values. Some of these measures will really tax your capabilities. How smooth and reflective does a solder joint have to be? However, if you do not know where you are, how do you determine where you want to go, and how do you intend to measure your progress?

Look what can be achieved if you define and measure the parameters that define your product. There was a time when one of the big three auto makers was going bankrupt. No one was buying the company's cars, which were not attractively enough priced to pull buyers away from the other two builders. The company took the time to find out what people would buy, and they built that product. They defined the measures, and they introduced many new and exciting parameters into the U.S. auto market that we now consider commonplace.

They benchmarked against the world's best and discarded the no-change-is-necessary dinosaurs from within the ranks. They remade their corporation. They redefined their processes and their product. They became so successful that their turnaround is almost unbelievable. With the techniques we have given you in this chapter, we hope your company does not get that close to closing before they recognize the benefits of benchmarking.

If your goal is not to be world class, but rather to be number two in your industry or even number three, you need to define what that means. *No,* do not turn and run at this surprising revelation. You need to define the exact measures and a route to success. It may not be your company's decision to be number one, now or ever. The cost may not justify the expense of hard-to-get capital required to get there. This is just one of many reasons. There could be a master plan to make your improvements in measured, incremental steps. Technology may be in transition, and that is where your company wants to put its money. Although it may be hard for the team to accept, especially if they are dedicated employees, you may be asked to restrict your efforts. This is an incorrect interpretation. No one is asking you to sacrifice the quality of your individual or collective ability. The task, every step of it, is the same. The only difference is the numeric goal you set. You still seek the world class companies, and you still complete a professional task. The only real difference is when you make your recommendations, they are to a different employer.

Therefore, goal setting is the key factor in achieving beneficial benchmarking results. Your goal must be attainable—possibly not all at once, or even in 1 or 2 years, but achievable. If you fall into the trap of setting goals that today's technology cannot support, everyone loses and benchmarking dies as a utilitarian process at your company. If your company's current technology cannot change but your industry does have that technology, you must lay these facts before your senior management. Investment in technology may be the only way to save what you had all worked so long and hard for. You cannot leapfrog your competition with yesterday's capabilities. If you want to be in the ball game, you need the assets and the technologies if you hope to reap the rewards.

Process improvement cannot take a company past the point of perfection. That is a given. So why state the obvious? Because it is a signal for you. It is time to look forward to new, better, and cheaper ways to do things. This is the catalyst for change. Many advances are born of not being able to squeeze anything more out of the old process. Necessity is the mother of invention. Benchmarking assists you to define that need in real time and with hard numbers. A number of people in your organization will be able to start on improvements based on the data you generate. Therefore, your team should not view their activities as isolated tasks. You should have realized that you and the team form the foundation of the next generation of your company. You will have participated in a very professionally rewarding process. You will always have the knowledge that you were an active participant in success.

8.11 World Class Benchmarking

How to pick world class companies against which to benchmark your processes will be addressed here, even though we have touched some of the attributes earlier in this chapter. If you recall, we previously alluded to the difficulty of getting a direct competitor to work with you. Of course, its data would prove very interesting, but there is work attached to obtain the exact data that you may need. In industry publications of world class companies, especially books written by Schonberger, there are lists of companies. One such list is found in Chap. 3 of this book. *Industry Week* and the Baldrige Award (see Chap. 9 for past winners) both publish their annual award winners. You can get many public documents, newspapers, and articles, including company publications. Annual reports are always public documents of publicly held companies.

There exist in our society mountains of data about everything and everyone if only you know where to look. There are official government files, trade papers and magazines, company financial reports (if publicly traded), stock market analyses, and (lest we forget) the always-effective contact with other professionals in that field. The Internet is full of information. Almost every company of any size has a home page with data it wishes to share. Sometimes a simple telephone call sets you off on the right track with a single name that suddenly explodes into a valuable network. Product shows and symposia can also be a source of data. There are so many free sources of these data that it is rarely a problem obtaining data. The real problem is determining what data to use and what to disregard as nothing more than sales data of questionable value.

World class is more than just a set of numbers. It is also a perception of the consumers. Keep in mind that the customer has two aspects, being not only a good analyst of fact but also emotional. Identifying and measuring customer needs actually involves two distinct steps. First, customer needs must be determined. This almost always requires a nonstructured, exploratory approach (focus groups, interviews, analysis of complaints and suggestions). Only after the most important customer needs have been identified is it appropriate to measure perceived quality on any sort of rating scale. A common error is to obtain scaled responses on perceived quality, based on items which management thinks are important, rather than to find out directly, and in the customer's words, what the important customer needs are. [21] Another caution concerns the direction of a potential benchmarking partner. If it has been a world class company for many years but is now on the skids, do you really want to benchmark against it? That company may still be number one, but not for much longer if it does not make some fundamental changes. The call is yours. Again a reminder about the company that is not world class as a company but is suspected to have one of the best processes of interest to you. By all means go after the data and extract whatever you can that will help your company.

So what, then, is world class? There is much literature on world class from different authors' points of view. However, Schonberger seems to be the benchmark that companies need to look at (see Chap. 2). Each company, even though cited as world class, is at a different point of time in achieving the processes. By Schonberger's definition, world class is "continual and rapid improvement." [22] Therefore, all companies will be in a different place. If you have been touting yourself as world class without data and comparing your information with Baldrige, Deming, and *Industry Week*–cited "Best Plants," then you could be accused of drinking your own bathwater. If your company does not adopt the world class philosophy of continual change, you may end up going down the drain as the real world class companies leave you in their dust.

Since electronics is so varied, depending on the product, different attributes would be apparent. Let us continue to use the automobile as an example. The first thing we need to do is to develop a list of what we believe to be world class cars. This illustrates the fact that there is a perception of world class cars, which there is. Note the "what we believe." Benchmarking, you remember, is supposed to be an exercise of facts.

When the perceived world class is benchmarked process by process, there could be automobiles that exceed in certain parameters. And of course, the automobile industry is changing constantly to achieve market share.

Shortly we will apply measurements to describe our definition of what makes these cars world class. The exercise of developing this can be done with one person, but is more meaningful when performed with a group. Do not let any of the participants discuss their reasons for their classification at this time. Accept any and all data as it is given.

TABLE 8.11 Classification of Cars

World class cars	Non–world class cars
1. BMW	1. Yugo
2. Infiniti	2. Henry J
3. Mercedes-Benz	3. Corvair
4. Cadillac	4. Lada
5. Rolls-Royce	5. Fiat
6. Jaguar	6. Jaguar
7. Astin Martin	7. Ford
8. Fiat	8. Chevrolet
9. Ford	9. Peugot
10. Chevrolet	
11. Lexus	

Once this is completed, develop a second list of non-world-class cars (i.e., world's worst cars). Again, accept any answer. In a number of instances in past trials similar to this, the same car appeared in both lists. (Please remember the elephant analogy.) Just for ease of analysis, the author has supplied these lists for you in Table 8.11. Remember, there are no right or wrong answers. These lists are only starting places to begin to define what we mean by world class. When you make similar lists for your employer, remember my earlier caution: Do not assume that just because you think something is world class, it is automatically so. This decision can only be based on facts and the consensus of the team, as shown in Table 8.11. This year's list and next year's list may differ, so continually verify your data.

These lists are quite interesting because they contain duplications. You may not agree with the listings. It is not important whether you agree or not. Remember, they represent points of view. Now it is once more time to define the measures that place a car in one list or the other. When we have a complete understanding of this phase of the process, we will be able to quantify the terms we decided to use. Table 8.12 details some parameters for world class classification.

The team can and should discuss the inclusion or exclusion of each parameter. When your list has the buy-in of the team members, you will be ready to go to the next step. The important thing to remember here is that at this time we have the results of the emotional yet intellectual effort. To some of the team members, a parameter such as a sofa-smooth ride is the most valued attribute to a world class car. To others, like a race enthusiast, road handling and speed are the most

TABLE 8.12 Parameters for World Class Classification for Automobiles

1. Comfort	10. Frequency of maintenance
2. Ride	11. Resale value
3. Handling	12. Safety
4. Torque	13. Insurance costs
5. 0–60 speed	14. *Consumers Report* data
6. Braking	15. Fuel economy
7. Reputation	16. Prestige
8. Acquisition cost	17. Snob appeal
9. Cost of operation	18. Etc.

valuable attributes in the car. In electronics, it would be similar to accuracy versus dynamic range.

The next step is or can be a lot of fun for the team, at least more fun then most of the steps you have already completed. This is where "the rubber meets the road." What you now have is a combination of fact, quasi-fact, supposed knowledge, and preference. This is where it all comes neatly tied together in the bow of preconceived emotion and bias. At this time, team members can finally discuss the merits of each value and why someone would think that any particular individual would pick that attribute. Now comes the part when you, as a team, need to append to each attribute you intend to use a very specific number or range of numbers and a unit of measure that defines why it is a step above the "also ran." These values will define why these cars are truly world class in the team's estimation. To different people the same attribute can have different meanings. Let us examine the attribute of *comfort*. You can have an attribute with more than one unit of measure. Table 8.13 shows such a case. Again, we see the pyramid effect. Comfort has at least 10 attributes.

TABLE 8.13 Measures of Comfort

1. Noise
2. Climate control
3. Road isolation
4. Ride
5. Seats
6. Carpet
7. Window tint
8. Bad weather handling
9. Ease of entry
10. Entry/exit handles

TABLE 8.14 World Class Comfort Parameters with Measures

	Parameter	Value
1.	Noise	$-X$ dB
2.	Climate control	± 2°F
3.	Road isolation	Dual-mode shocks/struts
4.	Ride	95% surface isolation
5.	Seats	Leather>class 12
6.	Carpet	Twist pile $\frac{3}{8}$ nylon 50/50
7.	Window tint	Legal limit/sun blocking
8.	Bad weather handling	Adhesion>90% minimum
9.	Easy to enter	Accommodates handicapped
10.	Ease of entry/exit	Handholds at each door

This list can be as long or as short as the definition requires. It is not at all surprising that some of the items shown in Table 8.13, can themselves be further subdivided and further defined. The need to go this fine in the task depends on the team and the process or product under investigation. Again, we have our pyramid effect. At this level as well we started with one word, *comfort,* and found 10 primary definitions. If you were to go one level lower, it would expand even further. The value of this step is that you have quantified and unitized the measures you want to compare to other companies. This is the essence of benchmarking.

Using the automobile example again, and starting where Table 8.13 left off, we can now construct Table 8.14, which will record the comfort values and the units of measure we have selected for our world class car example. You are probably asking yourself, "Why all of these separate tables?" Why can't we combine some or all of them into meaningful spread sheets? That is an excellent conclusion. The only reason that I have chosen to take my data in small bites and small steps was to make it clear to you that getting ready for benchmarking is nothing more than a number of small, rather simple steps that when combined produce a tool which is much more powerful than the sum of its parts.

Please remember that in the earlier example we were dealing only with one block of a three-block flow diagram which is part of a larger process. In each step in our pyramids for comfort we have worked with only one term to keep the example simple. When your team actually performs this benchmarking groundwork, it must be done in an orderly fashion. This orderly, logical approach is the only thing that will prevent you from chasing your own tail. If it were not for computers and spreadsheet programs you would retire early trying to analyze a very large and complicated process flow for benchmarking. Now let us look at Table 8.14.

The measures in Table 8.14 are not intended to be factual; this is just an example. When you and your team build such a table, yours will contain actual values because your team will be constituted of experts in the process or product you are benchmarking. Be sure to document what you consider world class on the particular parameter and where it came from. Expect all to challenge your measures of world class. Perceptions are separate because of experiences. What was considered impossible a few years ago is considered possible now. For this reason, and for historical reference, it is of value to commit your deliberations to paper. No one, or at least very few people, will remember all of the pertinent facts of all the data building that took place when you constructed your pyramid. This record may never be used, but when it is needed, your audience—whoever they may be—will be very much impressed by your forethought and professionalism.

As a sanity/reality check, it is necessary to know what some of your best and most friendly customers think of your results. After all, you listened to the voice of these customers in determining what was important to them when you started. With the aid of your marketing people, present them with the results. Customer surveys are an excellent tool to use. Their response could turn out to be priceless and invaluable. Your customers, internally and externally, should be included whenever possible. You have already complied with the internal part with your team reviews. If your company is really aggressive and is one that accepts reasonable risk, suggest that the same data be shown to companies that are not now customers. Their reaction to your attempt to be world class may pay off in more than one way. They may be buying from your competitor despite its not being world class. Again, when one looks at exact parameters, as opposed to perceived, one result may be knowing what you need in your product or process to get the business, and you may get the business as you improve or become world class. There is a caution that needs to discussed: Be sure that no proprietary data is ever shared. When sharing one company's data, permission is needed from the company.

8.12 Planning for Success and Continual Improvement

Success is a fragile, highly transitory state. If you are trying to improve, it is a good bet that your competition is doing the same thing. If you wear horse blinders and become myopic and concentrate on making the world's best whatever, you may find yourself losing. You obviously must make sure that either the marketplace exists or that you can develop the demand. The two must go hand in hand. Part of planning for success is being better at determining where the demand (market) will be or is heading. Being there first and with the best product is not a bad game plan. This is especially true if you want to be more than just a winner. If you have that desire to be world class, to be number one and have everyone looking to emulate you, then your choice of benchmarking your tool of choice was really smart.

Your benchmarking experience and your contribution will receive recognition from on high. You will position yourself as a valuable asset, and you will probably get other challenging tasks in addition to more benchmarking tasks. The knowledge you demonstrated in defining product attributes, how best to measure them, and how best to collect and analyze that data will be tools you will use over and over again.

The successful past experience in benchmarking will serve the entire team well in the near future. Benchmarking is not a one-time activity if you want to keep ahead of the pack. Earlier we said that success is a moving target and it is subject to technologies that often leapfrog. Benchmarking is your way to keep tabs on your field of interest. It is an ongoing process. This is not to be misconstrued to mean that as soon as you finish with a specific process you start over with that process. It assumes you have more than one process you would like to improve and make more cost-effective or to eliminate. It assumes that in the electronics industry, technology is rapidly changing. Subsequent benchmarking tasks will cost less money, as you will be fully proficient in the techniques. Your team can be expanded by bringing in new talents as recruits and have these teams led by one of your original team members. Eventually teams formed solely for benchmarking will be a thing of the past. Every process owner will be a team and will be looking for ways to produce higher-quality products more rapidly and at a lower cost.

Neither benchmarking nor continual process are one-shot processes. It would be ridiculous to say continual improvement is a one-time activity. No process stands still. There will always be variables. Absolute control is absolutely impossible. People, raw materials, and operators change. The need to monitor some vital signs will always be required. They are your early warning network. The process owners must be the ones to collect and record this process data. They must also be the ones to raise the flag when something starts to go wrong. An example of continuous benchmarking is provided in the following section.

8.13 Xerox: A Case in Point

In 1979, Japan's Canon Inc. introduced a midsize copier retailing at less than $10,000, or less than it cost Xerox to make a similar machine. Xerox first assumed Canon could do this because it priced the product below fair market value to buy market share, but Xerox engineers showed that Canon could sell its product cheaper because Canon was more efficient. Xerox took more than a year to decide to change its ways to compete. Xerox decided to benchmark Canon's processes with the objective of reducing costs. It was Xerox's turnaround through benchmarking that started the movement in the United States.

From 1980 to 1988, Xerox adapted Japanese techniques to cut its unit production costs in half and slash inventory costs more than 60 percent. Since then, the Xerox share of the U.S. copier market has climbed 80 percent, to almost 18 percent.

Everything Xerox does centers on surpassing customer expectations. Its customer satisfaction measurement system, in which more than 200,000 Xerox customers have been polled every year for the last decade, resulted in the company improving its number of highly satisfied customers by more than 38 percent.

The amount of benchmarking Xerox does has greatly increased since 1984, when it benchmarked only 14 performance elements. Now more than 240 elements are benchmarked, and the ultimate target for each attribute is the level of performance a world class leader achieves, regardless of industry. [23]

Your team must leave a strong impression in the minds of management when you do your final review. This indelible impression must be that continued improvement is the key to business success and benchmarking is the tool to make that improvement have value. Their contributions are tied together so strongly that any attempt to cut one of these processes loose from the other will spell disaster.

One final admonition: Complacency is the key to mediocrity.

References

1. *Merriam-Webster's Ninth New Collegiate Dictionary,* Merriam-Webster, Springfield, MA, 1996, p. 143. (Copyright ©1996 by Merriam-Webster, Inc.)
2. Poirier, C. C., and S. J. Tokarz, *Avoiding the Pitfalls of Total Quality,* ASQC Press, Milwaukee, WI, 1996, p. 100.
3. Patterson, J. G., *Benchmarking Basics,* Crisp Publications, Menlo Park, CA, 1996, p. 4.
4. Landry, P., "Benchmarking the Best-in-Class" in Wallace, T. F., and S. J. Bennett, *World Class Manufacturing,* Omneo, Essex Junction, VT, 1994, pp. 288–289.
5. Patterson, p. 7.
6. Dolan, T., "Benchmarking Past, Present and Future," from the Internet home page of The Benchmarking Exchange, Sept. 28, 1996.
7. *Merriam-Webster's Collegiate Dictionary,* p. 736.

8. Poirier and Tokarz, p. 103.
9. Poirier and Tokarz, p. 108.
10. Landry, p. 263.
11. Patterson, p. 6.
12. Poirier and Tokarz, p. 102.
13. Poirier and Tokarz, pp. 100–102.
14. Dolan.
15. Patterson, pp. 19–24, paraphrased.
16. Patterson, pp. 26–27, paraphrased.
17. Schnoll, Les, "One World, One Standard," *Quality Progress,* April 1993, p. 38.
18. Schnoll, p. 38.
19. Poirier and Tokarz, p. 23.
20. Benchmark Exchange, http:/www.benchnet. com. Data as of 9/28/96.
21. Kordupleski, R. E., R. T. Rust, and A. J. Zahorik, "Why Improving Quality Doesn't Improve Quality (or Whatever Happened to Marketing?)," California Management Review, vol. 38, no. 3, Spring 1993, p. 88.
22. Schonberger, R. J., *World Class Manufacturing; The Lessons of Simplicity Applied,* The Free Press, New York, 1986, p. 2.
23. Patterson, p. 11.

Chapter

9

The Malcolm Baldrige Award

Striving for World Class Quality and Total Customer Satisfaction

Eugene R. Carrubba

President
The PALC Group
Past Member, Board of Examiners
The Malcolm Baldrige Award

9.1 The Linkage between World Class Quality and Total Customer Satisfaction [1]

9.1.1 Why it's important!

Buying a product from a world class electronics manufacturer gives a customer a better chance of procuring a high-quality product. The trick, though, is to determine which manufacturer, if any, within a product category of interest is world class. Fortunately, there are certain attributes that characterize a world class quality manufacturer.

In some respects, world class quality is in the eyes of the beholder, and in this case the beholder is the customer. If the customer is not satisfied with the product that has been procured, it really doesn't make any difference if the manufacturer is viewed as world class by others. It all comes down to the degree of total customer satisfaction. For example, although they may not say it quite that way, United States consumers typically view Japanese products as having world class quality. This perception is particularly true in the case of automobiles and entertainment products like television sets and VCRs.

Rightly or wrongly, the consumer's degree of satisfaction in a purchase is translated into a perception of total quality. Generally speaking, the higher the degree of satisfaction, the higher the likelihood that a customer will view a particular manufacturer as being at the world class level. And Japanese electronics products in general do provide a high degree of customer satisfaction.

9.1.2 Quality recognition

For many years in Japan, the prestigious Deming Application Prize has been awarded to companies excelling in quality. (See Chap. 5 for more information on this prize.) This award is named in honor of one of the leading quality gurus in the United States, the late Dr. W. Edwards Deming. After World War II, the Japanese went through a painful reconstruction period to rebuild and revive their economy. Initially, their products were viewed as junk. Deming arrived on the scene and set in motion a quality revolution, largely based on the use of management and statistical quality control techniques, which the Japanese manufacturers embraced. Through Deming's guidance, and the unwavering commitment of the Japanese to religiously follow his lead, Japanese products became the benchmark of high quality. In gratitude for Deming's contributions, the Japanese have given this award annually in his honor.

Japanese manufacturers have been taking away more market share from American manufacturers in many product areas, including electronics products. Quality has become a weapon of competition. As a get-well prescription, many U.S. manufacturers have been working harder to improve their product quality and image. As a result, quality levels of many U.S. products have risen dramatically in the last dozen or so years.

The United States government, eager to foster and continue this rejuvenated quality impetus, established its version of the Deming Prize, the Malcolm Baldrige National Quality Award (MBNQA). This award, signed into law in 1987, recognizes companies in the United States that excel in quality from the viewpoint of both achievement and management. The Baldrige Award was envisioned as a standard of excellence that would help U.S. companies achieve world class quality. And today, the award's criteria are widely accepted as the standard for quality excellence in business performance.

9.1.3 Some world class quality companies

As of 1996, 29 companies out of 622 applicants have received the MBNQA since its establishment in 1988. Table 9.1 lists these winners. Together with the Florida Power and Light Company, which

TABLE 9.1 MBNQA Winners, 1988–1995

1988	Motorola, Inc. Commercial Nuclear Fuel Division of Westinghouse Electric Corporation Globe Metallurgical, Incorporated
1989	Milliken & Company Xerox Corporation's Business Products Division
1990	Cadillac Motor Division IBM Rochester Federal Express Corporation Wallace Co., Incorporated
1991	Marlow Industries Solectron Corporation Zytec Corporation
1992	AT&T Network Systems Group—Transmission Systems Business Unit AT&T Universal Card Services Granite Rock Company Texas Instruments Defense Systems & Electronics Group The Ritz-Carlton Hotel Company
1993	Ames Rubber Corporation Eastman Chemical
1994	AT&T Consumer Communications Services GTE Directories Corporation Wainright Industries, Inc.
1995	Armstrong World Industries, Inc., Building Products Operations Corning Telecommunications Products Division
1996	ADAC Laboratories Dana Commercial Credit Corporation Custom Research, Inc.

won the prestigious Deming Application Prize, a first for a non-Japanese company, these 24 companies clearly are of world class stature in the United States. Certainly there are other companies in the world, such as Toyota, which are also world class from a quality and customer satisfaction perspective. One practice common to all of these companies is that they are all implementing strong and effective quality initiatives. Motorola has its 6 sigma quality program aimed at achieving virtually zero defects in everything it does; Xerox has its "leadership through quality" program on continuous improvement; and IBM has its market-driven quality program, which, like Motorola's, is directed toward cutting defects to near zero across the board. What sets these world class companies apart from the rest is that they do all the right things from a total quality management perspective in order to achieve total customer satisfaction.

9.1.4 The Baldrige Award and ISO 9000

In spite of popular misconceptions, the MBNQA criteria and ISO 9000 standards are *not* equivalent in defining world class quality. As pointed out by Curt W. Reimann, former director of quality programs at the National Institute of Standards and Technology (NIST) and Harry S. Hertz, now director of quality programs at NIST and the head of the MBNQA program, in an article, there are many important differences relative to their focus, purpose, content, and implementation. [2] These are summarized below.

The focus of the Baldrige Award program is on improving competitiveness, while ISO 9000 focuses on conformance to quality requirements. ISO 9000 furnishes a common set of requirements for ensuring conformance with established quality systems. Although ISO 9000 also stresses the importance of the customer, there is little enforcement through the implementation process employed. On the other hand, Baldrige carries the competitiveness theme in an educational sense by promoting quality awareness and the sharing of successful quality strategies and lessons learned. Furthermore, *quality* means something different as well. In Baldrige, the customer defines quality—it is customer-driven. With ISO 9000, quality is linked directly to conformity to documented requirements, and in the manner currently implemented in the United States, not necessarily the customer's definition.

Another major difference between the quality programs deals with results. Baldrige is results-oriented, with the various approaches a company uses and deploys expected to produce results—results that relate to the business, its operations and financial performance, and of course, customer satisfaction. Here, ISO 9000 falls woefully short in that it does not require evidence of results among registered companies. To rectify this lapse, ISO is incorporating changes in its planned calendar year 1999–2000 standards update to reflect the importance of results.

The robustness of the Baldrige criteria is also another difference compared with ISO 9000. The Baldrige criteria cover all operations and processes of all work units. By comparison, ISO 9001 requirements cover less than approximately 10 percent of the scope of the Baldrige criteria, and do not fully address any of the criteria items. Figure 9.1 shows the scope of coverage graphically.

Baldrige looks for alignment and integration of quality leadership, systems, and related processes, both product and nonproduct (e.g., business processes), across a company's total operations. On the other hand, ISO 9000 mainly focuses on design, manufacturing, and service processes and procedures. ISO 9000 attempts to assure process consistency, while Baldrige targets overall business improvement.

In short, the Baldrige criteria are wider in scope, are customer-oriented, and look to approaches and their deployment to produce

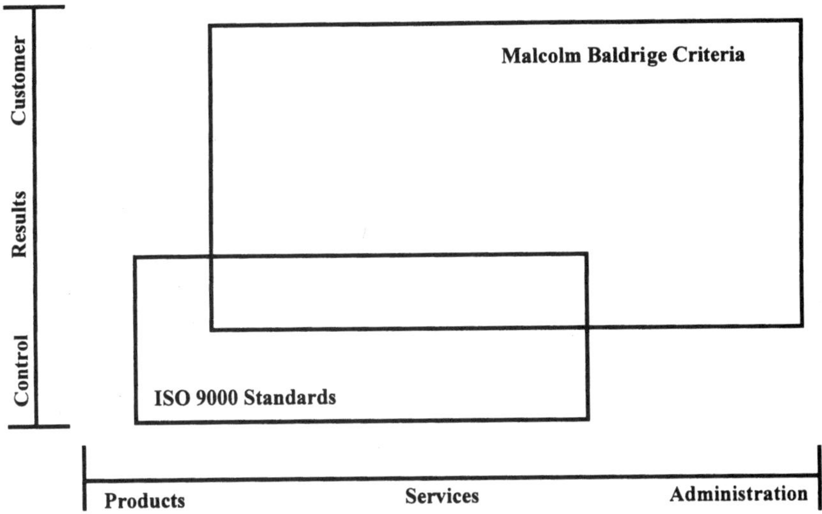

Figure 9.1 Quality system scope.

results. Certainly, ISO 9000 has value in ensuring conformance to a set of quality system requirements. However, consistent conformance to a set of requirements will not necessarily result in customer satisfaction if the customer's concerns and issues are not addressed. Few companies receive the Baldrige Award, while at the same time, thousands of companies have achieved ISO 9000 registration. This is not to imply that ISO 9000 registered companies may not have customer-focused quality systems, but the MBNQA does indeed provide a truer measure of world class quality.

9.2 Baldrige Award History and Background: Establishing a National Quality Award [3]

By the time the 1980s rolled around, quality was no longer an option for U.S. companies. It was a virtual requirement for doing business. Somehow, though, the message wasn't getting through. U.S. companies continued to lose market share to foreign competitors. U.S. companies were either ignoring the message, or didn't know where and how to start. There needed to be a way of raising awareness about quality, quality management, and quality techniques, as well as recognizing companies that have successful quality management systems. Thus, industry and government leaders looked to some sort of national quality award system as the vehicle to help regain competitiveness in the United States.

After passage was delayed by a couple of false starts, President Ronald Reagan signed Public Law 90-97, the Malcolm Baldrige National Quality Improvement Act of 1987, on August 20, 1987. This law established the annual U.S. National Quality Award. The award was named for Malcolm Baldrige, Reagan's Secretary of Commerce from 1981 until his death in a rodeo accident in July 1987. Baldrige had championed quality management and the quality improvement act, and was instrumental in drafting earlier, unsuccessful, versions of the law. The award is managed by the U.S. Department of Commerce National Institute of Standards and Technology and is administered by the American Society for Quality Control (ASQC).

The award promotes the following:

1. Quality awareness as an increasingly important element in competitiveness
2. An understanding of the requirements for quality excellence
3. The sharing of information on successful quality strategies and on the benefits derived from implementation of these strategies

The award is not given for excellence in specific products or services. Instead, it is awarded to companies for their total achievements. Up to two awards may be given annually in each of three categories: (1) manufacturing, (2) service, and (3) small businesses.

The use of the Baldrige framework and criteria have gone far beyond their intended original purpose. Over one million copies of the award criteria have been requested from and distributed by NIST. These requests are not so much for use in applying for the award, but rather for use as the basis for self-assessment and improvement. Awareness has definitely been raised! Networks have been created among individuals, companies, states, and health-care and educational institutions to share their best quality practices and success stories. Through the Baldrige Award program, quality has indeed become better understood as an essential factor in competitiveness.

9.3 Award Values, Concepts, and Framework [4]

9.3.1 Core values and key concepts

The Baldrige Award is built on a rock-solid foundation, represented by a set of core values and key concepts which allow for seamless integration of customer and company performance requirements. These core values and key concepts include:

- Customer-driven quality
- Leadership
- Continuous improvement and learning
- Employee participation and development
- Fast response
- Design quality and prevention
- Long-range view of the future
- Management by fact
- Partnership development
- Corporate responsibility and citizenship
- Results orientation

Customer-driven quality. It all starts with the customer! Don't forget: plain and simple, without customers, there is no business. A company needs to listen to its customers and understand their needs and wants. The company management system's key focus needs to be on those product and service features which are of value to its customers. Then, in order to remain competitive, the company must go further and improve upon these features to differentiate itself from the competition. It is more than simply delivering products that are defect-free or meet specifications, and resolving customer complaints. It means staying on top of new customer and market requirements, measuring customer satisfaction, being aware of new technology developments, and keeping abreast of competitors' offerings. Customer-driven quality thus becomes an important strategy aimed at retaining customers and increasing market share.

Leadership. The commitment, involvement, and visibility of the senior leadership is the catalyst that drives the systems that produce the results leading to customer satisfaction and superior business performance. Initially, they need to set the proper directions and tone through focus on the customer and on unambiguous company values and expectations. These values and expectations, however, must also be driven and reinforced by the senior leaders. They need to ensure that the right strategies, systems, and techniques are in place and being implemented to satisfy those values and expectations. Finally, because accomplishment is through people, the senior leaders must provide the appropriate motivation and training to the entire work force, including themselves.

Continuous improvement and learning. Continuous improvement is a never-ending journey. It has to be a way of life to which an organization is committed. Although continuous improvement has virtually no bounds, to be effective it should be focused on those areas that are important to the business, including customer satisfaction and retention. The continuous improvement process must be well-executed. For problems, continuous improvement should be aimed at root causes. Nevertheless, it is a worthwhile, justifiable motive for continuous improvement to simply want to do better. Thus, there will be cycles of learning involving continuous improvement based on planning, execution, and evaluation of opportunities for improvement.

Employee participation and development. As noted above, people are an important factor in a company's success. The success of the work force in improving company performance is linked tightly to the skills and motivation of the people. New skills need to be learned through education and training, and then practiced. Opportunities for personal growth through classroom and on-the-job training, as well as job rotation, need to be made easily available. With the workforce becoming increasingly diverse, education and training programs need to be tailored to fit this diversity. Appropriate recognition and career advancement opportunities should be offered to employees to provide additional motivation. Effective employee participation and development requires acquisition and use of employee-related data, and proper alignment of human resource plans with business plans.

Fast response. Fast response means more than just rapidly resolving customer complaints and issues. It also means reducing the time it takes to introduce new products and services to market. In both cases, an organization needs to examine and evaluate its processes to determine how to simplify and improve them. Typically these process improvements yield cycle time reductions, as well as improvements in quality and productivity.

Design quality and prevention. "An ounce of prevention is worth a pound of cure!" How many times have we heard this old saying? Yet, how many times have we ignored it? Generally speaking, the earlier a problem is detected and corrected in a product's life cycle, the less costly it will be. This has been proved many times over. And over. And over! Hence, catching a problem, or better still, preventing it during the design stage, is the most cost-effective approach. This philosophy applies not only to the product, but also to the processes that will be used to design, build, support, and service the product. And it extends to suppliers as well, since their participation and contributions to design quality and prevention are essential. Furthermore, with design-to-market cycle time becoming increasingly important in the competitive environment, concurrent engineering is emerging as an integral design approach.

Long-range view of the future. How about another old saw? "Fail to plan; plan to fail." That's another one we've heard any number of times. Having the vision to look into the future is important to achieving market leadership and competitiveness. But with the ever-changing business environment comes the challenge to anticipate these changes and develop appropriate strategies, plans, and resource allocations to address them. As a reinforcement to having a strong future orientation, it should be noted that long-term commitments are valued by customers, suppliers, and employees alike.

Management by fact. Too often, management decisions are based on emotion, or "gut" feel, rather than fact. Modern business management systems need to rely on relevant measurements, data, information, and analysis that are integrally linked to the company's strategy. These data should be derived from key processes, and used for improvement and assessment. Analysis of these data should be used to support decision making. Overall, these data and analyses should also be employed as inputs to planning, company performance reviews, operational improvements, and competitive comparisons. Of importance, however, is the need to establish appropriate and representative measures or indicators that lead to improved customer, operational, and financial performance.

Partnership development. Working closely in a partnership arrangement goes a long way toward helping to achieve goals. For that reason, building and maintaining partnerships is an important key to business success. These partnerships should embrace not only those key areas external to the company, but also those within it. Internal partnerships should cover relationships among company units, work teams, and labor-management. External partnerships should of course include customers and suppliers, together with educational organizations as appropriate. The external partnerships could also take the form of a strategic partnership or alliance to enhance and leverage the overall capabilities of the partners. Whatever kinds of partnerships are built, the partners must understand what the critical requirements are for success, as well as how they will communicate and track progress.

Corporate responsibility and citizenship. Corporate responsibility and citizenship, particularly as it relates to the environment, has become increasingly important of late. This corporate responsibility also includes the making of ethical business decisions, protecting the health of the public, and promoting safety. Companies need to address environmental, health, and safety issues through planning, goal setting, improvement, and use of preventive, rather than reactive, techniques. In addition to conforming to local, state, and federal regulations and requirements that impact these areas of responsibility, companies must take a leadership posture, providing education and increasing awareness of health, safety, and environmental concerns.

Results orientation. The goals of the Baldrige criteria to deliver ever-improving value to customers and improve overall company performance are oriented to results. Thus, the company's performance system needs to be focused on results as they relate to all stakeholders, including customers, suppliers, employees, stockholders, partners, the public, and the community. In order to do this well, the requirements of all the stakeholders must be understood and addressed through planning, actions, and necessary improvements.

The above core values and key concepts are embodied in seven Baldrige categories:

1.0 Leadership
2.0 Information and Analysis
3.0 Strategic Planning
4.0 Human Resource Development and Management
5.0 Process Management
6.0 Business Results
7.0 Customer Focus and Satisfaction

The categories are related and integrated into the award criteria framework shown in Fig. 9.2. The framework is composed of three essential elements: driver, system, and goal. In this framework, the senior executives of the company act as the *driver* of a *system* that achieves a two-fold *goal:* superior results from a customer/marketplace performance and business performance viewpoint. The senior executive leadership, or *driver,* (1) sets the direction; (2) creates values, goals, expectations, and systems; and (3) pursues customer, marketplace, and business performance excellence. The *system* that the senior executives *drive* includes a set of well-defined and well-designed processes for meeting the company's customer and overall performance require-

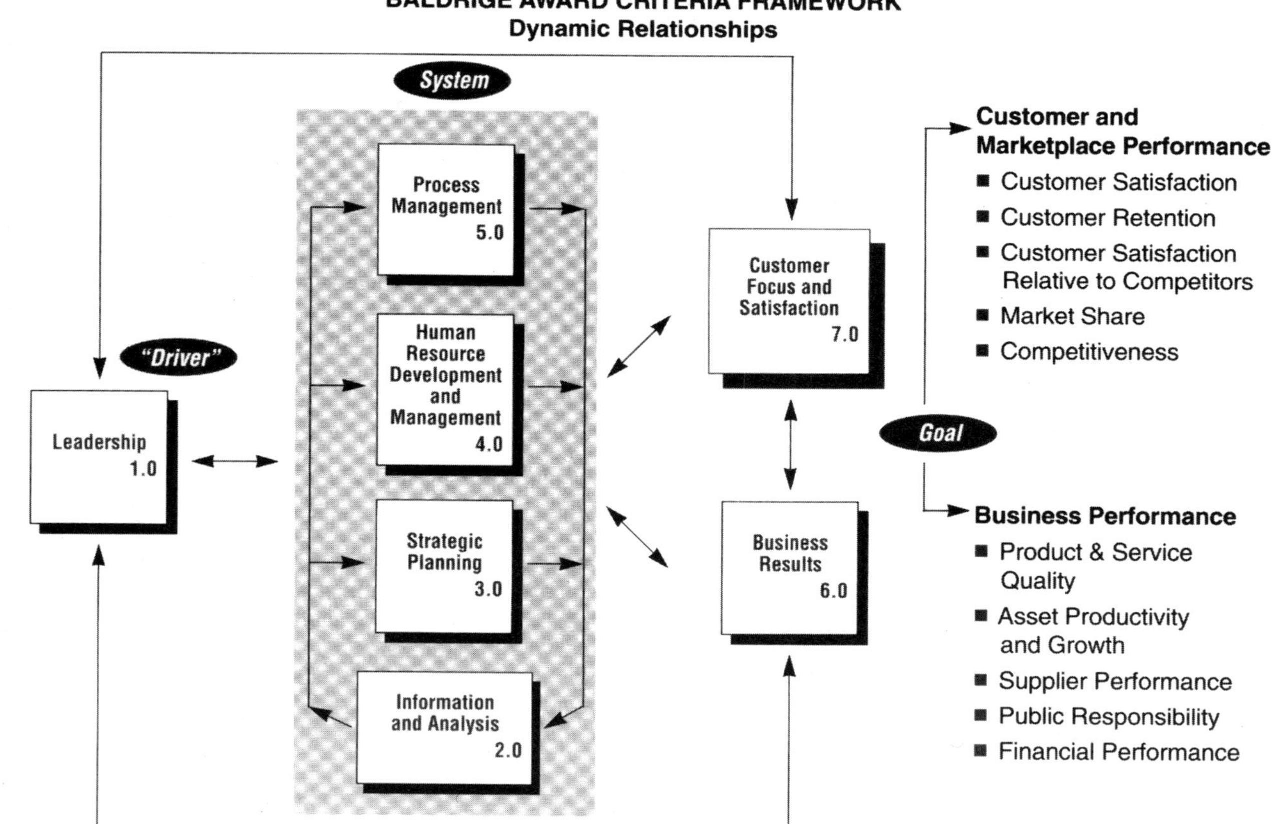

Figure 9.2 Baldrige Award criteria framework. (*Reprinted from Ref. 6 with permission.*)

ments. Finally, the dual *goal* that the leadership is striving to achieve covers (1) the delivery of ever-improving value to customers, high levels of customer satisfaction, and a strong competitive performance and (2) a wide variety of financial and nonfinancial results.

The above seven categories are further subdivided into 24 items. These items are listed in Table 9.2. Also provided in this table are the point values assigned to the various categories and items. Each item focuses on a major requirement that relates to the category in which it resides. In turn, each item consists of areas to address in the formal Baldrige Award application or during self-assessment. There are a total of 52 related areas to address in the criteria.

9.4 Award Criteria Category and Item Descriptions [5]

9.4.1 Award criteria summary

This section provides a summary of the seven Baldrige categories and the associated 24 items. It should be noted that the Baldrige Award criteria are revised annually, with changes made to the criteria items and areas to address, as appropriate to improve content and focus.

Category 1.0 Leadership. Various studies have shown that management-related issues are responsible for over 85 percent of quality problems. Leadership by the senior executives is necessary for achieving world class quality status. Therefore, the company's senior executives should endorse and commit to a strong quality value system and supportive management system to drive toward quality excellence.

Senior executives should lead by example, being committed and personally involved and visible in the quest for quality excellence and total customer satisfaction. They should have a standard of quality values demonstrated through internal policies, mission, values, and guidelines. This leadership concept should also be extended downward throughout the organization to other management levels. Finally, senior leadership should be extended beyond the workplace and into the community.

Category 2.0 Information and Analysis. Quality improvement depends greatly on data and its analysis. The selection of areas for improvement should be based on fact, not hearsay and emotion. Since quality covers a lot of areas, the data should be far-reaching to cover both product and nonproduct (i.e., administrative and support) areas. And the information should not only be collected, but also acted upon in a timely manner. It needs to evaluated and appropriate action needs to

TABLE 9.2 1996 Award Criteria—Item Listing

1996 Categories/Items		Point Values
1.0 Leadership		90
1.1 Senior Executive Leadership	45	
1.2 Leadership System and Organization	25	
1.3 Public Responsibility and Corporate Citizenship	20	
2.0 Information and Analysis		75
2.1 Management of Information and Data	20	
2.2 Competitive Comparisons and Benchmarking	15	
2.3 Analysis and Use of Company-Level Data	40	
3.0 Strategic Planning		55
3.1 Strategy Development	35	
3.2 Strategy Deployment	20	
4.0 Human Resource Development and Management		140
4.1 Human Resource Planning and Evaluation	20	
4.2 High Performance Work Systems	45	
4.3 Employee Education, Training, and Development	50	
4.4 Employee Well-Being and Satisfaction	25	
5.0 Process Management		140
5.1 Design and Introduction of Products and Services	40	
5.2 Process Management: Product and Service Production and Delivery	40	
5.3 Process Management: Support Services	30	
5.4 Management of Supplier Performance	30	
6.0 Business Results		250
6.1 Product and Service Quality Results	75	
6.2 Company Operational and Financial Results	110	
6.3 Human Resource Results	35	
6.4 Supplier Performance Results	30	
7.0 Customer Focus and Satisfaction		250
7.1 Customer and Market Knowledge	30	
7.2 Customer Relationship Management	30	
7.3 Customer Satisfaction Determination	30	
7.4 Customer Satisfaction Results	160	
TOTAL POINTS		1000

SOURCE: Reprinted from Ref. 6 with permission.

be taken. It does no good to collect reams and reams of information and data which then sit in a computer file or hard-copy file cabinet.

The data should be used to aid in management of the overall business. The company should be doing competitive evaluations and benchmarking to understand what improvements can be made to help improve its performance. The various kinds of data should be reliable and accurate and made available in a timely manner so that the results help meet the objectives of the company.

Category 3.0 Strategic Planning. An effective strategic planning process should be put in place and implemented. This process should result in plans that factor in knowledge about the customer, competition, and other key business drivers. In addition, since "to be the best is to know the best," strategic planning should also include benchmarks of the best-in-class and world class companies.

The resulting plans should address both the short-term and long-term priorities of the company. To be effective in the long run, the planning process should not remain stagnant, but rather should be dynamic and continuously reviewed and improved.

Category 4.0 Human Resource Development and Management. This criterion aims to ensure that the company's efforts are effectively focused on developing and realizing the full potential of the total work force. Tied closely to this aim is the importance of maintaining a conducive environment to achieve this end. Human resource plans should be created, implemented, and managed. All employees should be involved in achieving the company's objectives. Education and training should be aligned with these objectives.

The human resource planning should be integrated into the company's strategic planning process. The pursuit of quality should be enhanced through employee team involvement with customers and suppliers. Employee contributions to company success should be encouraged through recognition and reward systems. High employee morale should be fostered by a variety of means including employee development programs, as well as programs aimed at employee well-being and satisfaction.

Category 5.0 Process Management. A world class quality company should employ systematic approaches for total quality assurance and control of its products and services. These approaches should begin with design and introduction of products to satisfy customer needs, expectations, and requirements. They should continue into the manufacturing processes employed to build these products, as well as include the quality controls that need to be implemented to ensure that the designed-in integrity is not compromised. The products should meet the design specifications and be built by using controlled processes. In addition, since the quality of components and materials used in the product is extremely important, an effective process should be in place to assure that suppliers meet the necessary requirements.

An important element throughout should be the continuous improvement (or *kaizen,* as the Japanese call it) of processes, products, and services. Systems should document the various improvements as they provide an important "lessons learned" filing of valuable information. In addition to quality assurance and improvement processes, products, and services, these activities should be extended to suppliers and supporting services, such as sales and marketing.

Category 6.0 Business Results. The measurement and knowledge of overall business results are essential to improvement. Measurements should be made against indicators that are important to success of the business. Consequently, these measurements should cover a broad spectrum, including quality, operational, financial, human resources, administrative/support services, and supplier results.

Trends in key product and service quality measures should be reviewed and acted upon routinely. When trends go awry, the reasons should be analyzed and understood, and appropriate actions implemented to bring them back on track and prevent their recurrence. Knowing how the industry as a whole, including competitors and the best in class, are performing can also provide useful comparisons, and provide a basis for initiating further improvements.

Category 7.0 Customer Focus and Satisfaction. Last, but certainly not least, is how all this contributes to customer satisfaction. This evaluation criterion carries the greatest weight and emphasis in scoring. And well it should! One should think of this category as the grand payoff.

The company should know its customers, be responsible, and meet their expectations and requirements. At the very outset, there should be a strong knowledge of the customer's needs, expectations, and requirements (both current and future), something the Japanese do ever so well. The management of customer relationships should be considered essential. In line with the latter, it is important that standards for customer service be set high. It is also important that the company keep its commitments, such as those made in guarantees and warranties, if it is to maintain its credibility. Furthermore, timely and effective resolution of customer complaints is obviously an important aspect of total customer satisfaction.

Having done many things aimed at customer satisfaction, a company may believe that its customers are reasonably satisfied with its products. To know for sure, it should conduct surveys that take the pulse of the customer. These customer satisfaction results should be tracked religiously, and any adverse trends addressed promptly and effectively. Finally, the results of the company's customer satisfaction studies should be compared with those of industry and world leaders, as well as competitors, for important measures such as customer retention and market share growth. When it comes to total customer satisfaction, there is only one goal: to be at the top, to be the best, to be world class!

9.4.2 Detailed award criteria

Detailed descriptions of these categories and items and the related 52 areas to address, directly as extracted from the 1996 MBNQA award criteria, are provided in Table 9.3. [6]

9.4.3 Highlights of 1997 criteria changes

As noted earlier, the criteria are reviewed annually, and appropriate changes made as an important part of an ongoing process of continuous improvement. The Baldrige Program is indeed "walking the talk." In 1997, sweeping changes were made to the criteria. Even the title of the criteria document was changed from simply "Award Criteria" to "Criteria for Performance Excellence."

The 1997 criteria place greater importance on strategy-driven performance, tying company strategy to effective performance management, while addressing the needs and expectations of all stakeholders (i.e., customers, employees, stockholders, suppliers, and the public). This is reflected in the revised criteria framework, which rearranges and renumbers the categories, and offers a systems perspective that keys on customer- and market-focused strategy and action plans.

Changes have occurred throughout, at the category, item, and point value levels. Some examples of these changes follow. Category 2.0, Strategic Planning, focuses on the overall strategy planning process, as well as the company's actual strategy and its deployment. Greater emphasis is placed on organizational learning by making human resource planning an integral part of Category 2.0. What was formerly Category 7.0, Customer Focus and Satisfaction, is now Category 3.0, Customer and Market Focus. All business results are now treated in a parallel and integrated fashion, and the composite of the new Category 7.0, Business Results, now includes customer satisfaction

ABLE 9.3 1996 Award Criteria

1.0 Leadership (90 pts.)

The *Leadership* Category examines senior executives' personal leadership and involvement in creating and sustaining a customer focus, clear values and expectations, and a leadership system that promotes performance excellence. Also examined is how the values and expectations are integrated into the company's management system, including how the company addresses its public responsibilities and corporate citizenship.

1.1 Senior Executive Leadership (*45 pts.*)

Describe senior executives' leadership and personal involvement in setting directions and in developing and maintaining an effective, performance-oriented leadership system.

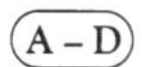

(See page 24 for a description of these symbols.)

AREAS TO ADDRESS

a. how senior executives provide effective leadership and direction in building and improving company competitiveness, performance, and capabilities. Describe how senior executives: (1) create and maintain an effective leadership system based upon clear values and high expectations; (2) create future opportunity for the company and its stakeholders, set directions, and integrate performance excellence goals; and (3) review overall company performance, capabilities, and organization.

b. how senior executives evaluate and improve the company's leadership system, including their own leadership skills

Notes:

(1) "Senior executives" means the applicant's highest-ranking official and executives reporting directly to that official.

(2) Values and expectations [1.1a(1)] should take into account needs and expectations of key stakeholders customers, employees, stockholders, suppliers and partners, the community, and the public.

(3) Review of overall company performance is addressed in 1.2b. Responses to 1.1a(3) should focus on senior executives' roles in such reviews, and their use of the reviews to set expectations and develop leadership.

(4) Evaluation of the company's leadership system (1.1b) might include assessment of executives by peers, direct reports, and/or a board of directors. It might also include use of surveys of company employees.

1.2 Leadership System and Organization (*25 pts.*)

Describe how the company's customer focus and performance expectations are integrated into the company's leadership system, management, and organization.

AREAS TO ADDRESS

a. how the company's values, expectations, and directions are integrated into its leadership system, management, and organization. Describe: (1) how the organization and its management of operations are designed to achieve companywide customer focus and commitment to high performance. Include roles and responsibilities of managers and supervisors; and (2) how values, expectations, and directions are effectively communicated and reinforced throughout the entire work force.

b. how overall company and work unit performance are reviewed. Include a description of: (1) the principal financial and nonfinancial measures used and how these measures relate to key stakeholders' primary needs and expectations; (2) how progress relative to plans is tracked; (3) how progress relative to competitors is tracked; (4) how asset productivity is determined; and (5) how review findings are used to set priorities for improvement actions.

Notes:

(1) Reviews described in 1.2b might utilize information from results Items — 6.1, 6.2, 6.3, 6.4, and 7.4 — and also might draw upon evaluations described in other Items and upon analysis (Item 2.3).

(2) Reviews might include various economic measures as well as financial ones.

(3) Assets [1.2b(4)] refers to human resources, materials, energy, capital, equipment, etc. Aggregate measures such as total factor productivity might also be used.

SOURCE: Reprinted from Ref. 6 with permission.

TABLE 9.3 1996 Award Criteria (*Continued*)

1.3 Public Responsibility and Corporate Citizenship *(20 pts.)*

Describe how the company addresses its responsibilities to the public in its performance management practices. Describe also how the company leads and contributes as a corporate citizen in its key communities.

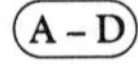

AREAS TO ADDRESS

a. how the company integrates its public responsibilities into its performance improvement efforts. Describe: (1) the risks and regulatory and other legal requirements addressed in planning and in setting operational requirements, measures, and targets; (2) how the company looks ahead to anticipate public concerns and to assess possible impacts on society of its products, services, facilities, and operations; and (3) how the company promotes legal and ethical conduct in all that it does.

b. how the company leads and contributes as a corporate citizen in its key communities. Include a brief summary of the types of leadership and involvement the company emphasizes.

Notes:

(1) Public responsibility issues (1.3a) relate to the company's impacts and possible impacts on society associated with its products, services, facilities, and operations. They include environment, health, safety, and emergency preparedness as they relate to any aspect of risk or adverse effect, whether or not these are covered under law or regulation. Health and safety of employees are not addressed in Item 1.3. Employee health and safety are covered in Item 4.4.

(2) Major public responsibility or impact areas should also be addressed in planning (Item 3.1) and in the appropriate process management Items of Category 5.0. Key results, such as environmental improvements, should be reported in Item 6.2.

(3) If the company has received sanctions under law, regulation, or contract [1.3a(3)] during the past three years, briefly describe the incident(s) and its current status. If settlements have been negotiated in lieu of potential sanctions, give an explanation. If no sanctions have been received, so indicate.

(4) The corporate citizenship issues appropriate for inclusion in 1.3b relate to efforts by the company to strengthen community services, education, health care, environment, and practices of trade or business associations. Examples of corporate citizenship appropriate for inclusion in 1.3b are:

- *influencing and helping trade and business associations to create school-to-work programs;*
- *communicating employability requirements to schools;*
- *influencing national, state, and local policies which promote education improvement;*
- *partnering with and charitable giving to schools, e.g., sharing computers and computer expertise;*
- *developing trade and business consortia to improve environmental practices;*
- *promoting volunteerism among employees;*
- *partnering with other businesses and health care providers to improve health in the local community; and*
- *influencing trade and business associations to engage in cooperative activities to improve overall U.S. global competitiveness.*

TABLE 9.3 1996 Award Criteria (*Continued*)

2.0 Information and Analysis (75 pts.)

The ***Information and Analysis*** Category examines the management and effectiveness of the use of data and information to support customer-driven performance excellence and marketplace success.

2.1 Management of Information and Data

(*20 pts.*)

Describe the company's selection and management of information and data used for strategic planning, management, and evaluation of overall performance.

AREAS TO ADDRESS

a. how information and data needed to support operations and decision making and to drive improvement of overall company performance are selected and managed. Describe: (1) the main types of data and information and how each type supports key business operations and business strategy; (2) how the company's performance measurement system is designed to achieve alignment of operations with company priorities, such as key business drivers; and (3) how key requirements such as reliability, rapid access, and rapid update are derived from user needs and how the requirements are met.

b. how the company evaluates and improves the selection, analysis, and integration of information and data, aligning them with the company's key business drivers and operations. Describe how the evaluation considers: (1) scope of information and data; (2) use and analysis of information and data to support process management and performance improvement; and (3) feedback from users of information and data.

Notes:

(1) Reliability [2.1a(3)] includes software used in information systems.

(2) User needs [2.1a(3)] should consider knowledge accumulation such as knowledge about specific customers or customer segments. User needs should also take into account changing patterns of communications associated with changes in process management, job design, and business strategy.

(3) Feedback from users [2.1b(3)] might entail formal or informal surveys, focus groups, and teams. Factors in the evaluation might include completeness, timeliness, access, update, and reliability. The evaluation might also include assessment of the information technologies used.

2.2 Competitive Comparisons and Benchmarking (*15 pts.*)

Describe the company's processes and uses of comparative information and data to support improvement of overall performance and competitive position.

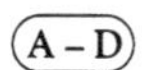

AREAS TO ADDRESS

a. how competitive comparisons and benchmarking information and data are selected and used to help drive improvement of overall company performance. Describe: (1) how needs and priorities are determined; (2) criteria for seeking appropriate information and data — from within and outside the company's industry and markets; (3) how the information and data are used to improve understanding of processes and process performance; and (4) how the information and data are used to set stretch targets and/or to encourage breakthrough approaches aligned with the company's competitive strategy.

b. how the company evaluates and improves its process for selecting and using competitive comparisons and benchmarking information and data to improve planning, overall company performance, and competitive position

Notes:

(1) Benchmarking information and data refer to processes and results that represent best practices and performance, inside or outside of the company's industry. Competitive comparisons refer to performance levels relative to direct competitors in the company's markets.

(2) Needs and priorities [2.2a(1)] should show clear linkage to the company's key business drivers.

(3) Use of benchmarking information and data within the company [2.2a(3)] might include the expectation that company units maintain awareness of related best-in-class performance to help drive improvement. This could entail education and training efforts to build capabilities.

(4) Sources of competitive comparisons and benchmarking information might include: (a) information obtained from other organizations such as customers or suppliers through sharing; (b) information obtained from the open literature; (c) testing and evaluation by the company itself; and (d) testing and evaluation by independent organizations.

(5) The evaluation (2.2b) might address a variety of factors such as the effectiveness of use of the information, adequacy of information, training in acquisition and use of information, improvement potential in company operations, and estimated rates of improvement by other organizations.

TABLE 9.3 1996 Award Criteria (*Continued*)

2.3 Analysis and Use of Company-Level Data

(40 pts.)

Describe how data related to quality, customers, and operational performance, together with relevant financial data, are analyzed to support company-level review, action, and planning.

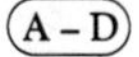

AREAS TO ADDRESS

a. how information and data from all parts of the company are integrated and analyzed to support reviews, business decisions, and planning. Describe how analysis is used to gain understanding of: (1) customers and markets; (2) operational performance and company capabilities; and (3) competitive performance.

b. how the company relates customer and market data, improvements in product/service quality, and improvements in operational performance to changes in financial and/or market indicators of performance. Describe how this information is used to set priorities for improvement actions.

Notes:

(1) Item 2.3 focuses primarily on analysis for company-level purposes, such as reviews (1.2b) and strategic planning (Item 3.1). Data for such analysis come from all parts of the company and include results reported in Items 6.1, 6.2, 6.3, 6.4, and 7.4. Other Items call for analyses of specific sets of data for special purposes. For example, the Items of Category 4.0 require analysis to determine the effectiveness of training and other human resource practices. Such special-purpose analyses should be part of the overall information base available for use in Item 2.3.

(2) Analysis includes trends, projections, cause-effect correlations, and the search for deeper understanding needed to set priorities to use resources more effectively to serve overall business objectives. Accordingly, analysis draws upon all kinds of data: operational, customer-related, financial, and economic.

(3) Examples of analysis appropriate for inclusion in 2.3a(1) are:

- *how the company's product and service quality improvement correlates with key customer indicators such as customer satisfaction, customer retention, and market share;*
- *cost/revenue implications of customer-related problems and problem resolution effectiveness; and*
- *interpretation of market share changes in terms of customer gains and losses and changes in customer satisfaction.*

(4) Examples of analysis appropriate for inclusion in 2.3a(2) are:

- *trends in improvement in key operational performance indicators such as productivity, cycle time, waste reduction, new product introduction, and defect levels;*
- *financial benefits from improved employee safety, absenteeism, and turnover;*
- *benefits and costs associated with education and training;*
- *how the company's ability to identify and meet employee requirements correlates with employee retention, motivation, and productivity;*
- *cost/revenue implications of employee-related problems and problem resolution effectiveness; and*
- *trends in individual measures of productivity such as manpower productivity.*

(5) Examples of analysis appropriate for inclusion in 2.3a(3) are:

- *working capital productivity relative to competitors;*
- *individual or aggregate measures of productivity relative to competitors;*
- *performance trends relative to competitors on key quality attributes; and*
- *cost trends relative to competitors.*

(6) Examples of analysis appropriate for inclusion in 2.3b are:

- *relationships between product/service quality and operational performance indicators and overall company financial performance trends as reflected in indicators such as operating costs, revenues, asset utilization, and value added per employee;*
- *allocation of resources among alternative improvement projects based on cost/revenue implications and improvement potential;*
- *net earnings derived from quality/operational/human resource performance improvements;*
- *comparisons among business units showing how quality and operational performance improvement affect financial performance;*
- *contributions of improvement activities to cash flow, working capital use, and shareholder value;*
- *profit impacts of customer retention;*
- *market share versus profits;*
- *trends in aggregate measures such as total factor productivity; and*
- *trends in economic and/or market indicators of value.*

ABLE 9.3 1996 Award Criteria (*Continued*)

3.0 Strategic Planning (55 pts.)

The ***Strategic Planning*** Category examines how the company sets strategic directions, and how it determines key plan requirements. Also examined is how the plan requirements are translated into an effective performance management system.

3.1 Strategy Development

(35 pts.)

Describe the company's strategic planning process for overall performance and competitive leadership for the short term and the longer term. Describe also how this process leads to the development of a basis (key business drivers) for deploying plan requirements throughout the company.

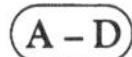

AREAS TO ADDRESS

a. how the company develops strategies and business plans to strengthen its customer-related, operational, and financial performance and its competitive position. Describe how strategy development considers: (1) customer requirements and expectations and their expected changes; (2) the competitive environment; (3) risks: financial, market, technological, and societal; (4) company capabilities—human resource, technology, research and development, and business processes—to seek new market leadership opportunities and/or to prepare for key new requirements; and (5) supplier and/or partner capabilities.

b. how strategies and plans are translated into actionable key business drivers

c. how the company evaluates and improves its strategic planning and plan deployment processes

Notes:

(1) Item 3.1 addresses overall company strategy and business plans, not specific product and service designs.

(2) Strategy and planning refer to a future-oriented basis for major business decisions, resource allocations, and companywide management. Strategy and planning, then, address both revenue growth thrusts as well as thrusts related to improving company performance. The sub-parts of 3.1a are intended to serve as an outline of key factors involved in developing a view of the future as a context for strategic planning.

(3) Customer requirements and their expected changes [3.1a(1)] might include pricing factors. That is, competitive success might depend upon achieving cost levels dictated by anticipated market prices rather than setting prices to cover costs.

(4) The purposes of projecting the competitive environment [3.1a(2)] are to detect and reduce competitive threats, to shorten reaction time, and to identify opportunities. If the company uses modeling, scenario, or other techniques to project the competitive environment, such techniques should be briefly outlined in 3.1a(2).

(5) Key business drivers are the areas of performance most critical to the company's success. (See Glossary, page 4.) The purpose of the key business drivers is to ensure that strategic planning leads to a pragmatic basis for deployment, communications, and assessment of progress. Actual key business drivers should not be described in 3.1b. Such information is requested in Item 3.2, which focuses on deployment.

(6) How the company evaluates and improves its strategic planning and plan deployment process might take into account the results of reviews (1.2b), input from work units, and projection information (3.2b). The evaluation might also take into account how well strategies and requirements are communicated and understood, and how well key measures throughout the company are aligned.

TABLE 9.3 1996 Award Criteria (*Continued*)

3.2 Strategy Deployment *(20 pts.)* Summarize the company's key business drivers and how they are deployed. Show how the company's performance projects into the future relative to competitors and key benchmarks. 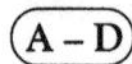 A – D	AREAS TO ADDRESS **a.** summary of the specific key business drivers derived from the company's strategic directions and how these drivers are translated into actions. Describe: (1) key performance requirements and associated operational performance measures and/or indicators and how they are deployed; (2) how the company aligns work unit and supplier and/or partner plans and targets; (3) how productivity and cycle time improvement and reduction in waste are included in plans and targets; and (4) the principal resources committed to the accomplishment of plans. Note any important distinctions between short-term plans and longer-term plans. **b.** two-to-five year projection of key measures and/or indicators of the company's customer-related and operational performance. Describe how product and/or service quality and operational performance might be expected to compare with key competitors and key benchmarks over this time period. Briefly explain the comparisons, including any estimates or assumptions made regarding the projected product and/or service quality and operational performance of competitors or changes in key benchmarks.

Notes:

(1) The focus in Item 3.2 is on the translation of the company's strategic plans, resulting from the process described in Item 3.1, to requirements for work units, suppliers, and partners. The main intent of Item 3.2 is effective alignment of short- and long-term operations with strategic directions. Although the deployment of these plans will affect products and services, design of products and services is not the focus of Item 3.2. Such design is addressed in Item 5.1.

(2) Productivity and cycle time improvement and waste reduction [3.2a(3)] might address factors such as inventories, operational complexity, work-in-process, inspection, downtime, changeover time, set-up time, and other examples of utilization of resources — materials, equipment, energy, capital, and labor.

(3) Area 3.2b addresses projected progress in improving performance and in gaining advantage relative to competitors. This projection may draw upon analysis (Item 2.3) and data reported in results Items (Category 6.0 and Item 7.4). Such projections are intended to support reviews (1.2b), evaluation of planning (3.1c), and other Items. Another purpose is to take account of the fact that competitors and benchmarks may also be improving over the time period of the projection.

(4) Projections of customer-related and operational performance (3.2b) might be expressed in terms of costs, revenues, measures of productivity, and economic indicators. Projections might also include innovation rates or other factors important to the company's competitive position.

TABLE 9.3 1996 Award Criteria (*Continued*)

4.0 Human Resource Development and Management (140 pts.)

The ***Human Resource Development and Management*** Category examines how the work force is enabled to develop and utilize its full potential, aligned with the company's performance objectives. Also examined are the company's efforts to build and maintain an environment conducive to performance excellence, full participation, and personal and organizational growth.

4.1 Human Resource Planning and Evaluation *(20 pts.)*

Describe how the company's human resource planning and evaluation are aligned with its strategic and business plans and address the development and well-being of the entire work force.

AREAS TO ADDRESS

a. how the company translates overall requirements from strategic and business planning (Category 3.0) to specific human resource plans. Summarize key human resource plans in the following areas: (1) changes in work design to improve flexibility, innovation, and rapid response; (2) employee development, education, and training; (3) changes in compensation, recognition, and benefits; and (4) recruitment, including critical skill categories and expected or planned changes in demographics of the work force. Distinguish between the short term and the longer term, as appropriate.

b. how the company evaluates and improves its human resource planning and practices and the alignment of the plans and practices with the company's strategic and business directions. Include how employee-related data and company performance data (Item 6.2) are analyzed and used: (1) to assess the development and well-being of all categories and types of employees; (2) to assess the linkage of the human resource practices to key business results; and (3) to ensure that reliable and complete human resource information is available for company planning and recruitment.

Notes:

(1) Human resource planning addresses all aspects of designing and managing human resource systems to meet the needs of both the company and the employees. Examples of human resource plan (4.1a) elements that might be part(s) of a comprehensive plan are:

- *redesign of work organizations and/or jobs to increase employee responsibility and decision making;*
- *initiatives to promote labor-management cooperation, such as partnerships with unions;*
- *creation or modification of compensation and recognition systems based on building shareholder value and/or customer satisfaction;*
- *creation or redesign of employee surveys to better assess the factors in the work climate that contribute to or inhibit high performance;*
- *prioritization of employee problems based upon potential impact on productivity;*
- *development of hiring criteria and/or standards;*
- *creation of opportunities for employees to learn and use skills that go beyond current job assignments through redesign of processes or organizations;*
- *education and training initiatives, including those that involve developmental assignments;*
- *formation of partnerships with educational institutions to develop employees or to help ensure the future supply of well-prepared employees;*
- *establishment of partnerships with other companies and/or networks to share training and/or spread job opportunities;*
- *introduction of distance learning or other technology-based learning approaches; and*
- *integration of customer and employee surveys.*

(2) "Employee-related data" (4.1b) refers to data contained in personnel records as well as data described in Items 4.2, 4.3, 4.4, and 6.3. This might include employee satisfaction data and data on turnover, absenteeism, safety, grievances, involvement, recognition, training, and information from exit interviews.

(3) "Categories of employees" [4.1b(1)] refers to the company's classification system used in its human resource practices and/or work assignments. It also includes factors such as union or bargaining unit membership. "Types of employees" takes into account other factors, such as work force diversity or demographic makeup. This includes gender, age, minorities, and the disabled.

(4) Human resource information for company planning and recruitment [4.1b(3)] might include an overall profile of strengths and weaknesses that could affect the company's capabilities to fulfill plan requirements.

TABLE 9.3 1996 Award Criteria (*Continued*)

4.2 High Performance Work Systems *(45 pts.)*

Describe how the company's work and job design and compensation and recognition approaches enable and encourage all employees to contribute effectively to achieving high performance objectives.

AREAS TO ADDRESS

a. how the company's work and job design promote high performance. Describe how work and job design: (1) create opportunities for initiative and self-directed responsibility; (2) foster flexibility and rapid response to changing requirements; and (3) ensure effective communications across functions or units that need to work together to meet customer and/or operational requirements.

b. how the company's compensation and recognition approaches for individuals and groups, including managers, reinforce the effectiveness of the work and job design

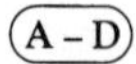

Notes:

(1) Work design refers to how employees are organized and/or organize themselves in formal and informal, temporary or longer-term units. This includes work teams, problem-solving teams, functional units, departments, self-managed or managed by supervisors. In some cases, teams might involve individuals in different locations linked via computers or conferencing technology.

Job design refers to responsibilities and tasks assigned to individuals. These responsibilities and tasks help define education and training requirements.

(2) Examples of approaches to create flexibility [4.2a(2)] in work design might include simplification of job classifications, cross training, job rotation, work layout, and work locations. It might also entail use of technology and changed flow of information to support local decision making.

(3) Compensation and recognition (4.2b) refer to all aspects of pay and reward, including promotion and bonuses. The company might use a variety of reward and recognition approaches — monetary and non-monetary, formal and informal, and individual and group.

Compensation and recognition approaches could include profit sharing and compensation based on skill building, use of new skills, and demonstrations of self-learning. The approaches could take into account the linkage to customer retention or other performance objectives.

Employee evaluations and reward and recognition approaches might include peer evaluations, including peers in teams and networks.

4.3 Employee Education, Training, and Development *(50 pts.)*

Describe how the company's education and training address company plans, including building company capabilities and contributing to employee motivation, progression, and development.

AREAS TO ADDRESS

a. how the company's education and training serve as a key vehicle in building company and employee capabilities. Describe how education and training address: (1) key performance objectives, including those related to improving customer responsiveness and enhancing high performance work units; and (2) progression and development of all employees.

b. how education and training are designed, delivered, reinforced, evaluated, and improved. Include: (1) how employees and line managers contribute to or are involved in determining specific education and training needs and designing education and training; (2) how education and training are delivered; (3) how knowledge and skills are reinforced through on-the-job application; and (4) how education and training are evaluated and improved.

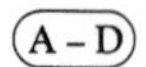

Notes:

(1) Education and training address the knowledge and skills employees need to meet their overall work objectives. This might include leadership skills, communications, teamwork, problem solving, interpreting and using data, meeting customer requirements, process analysis, process simplification, waste reduction, cycle time reduction, error-proofing, priority setting based upon cost and benefit data, and other training that affects employee effectiveness, efficiency, and safety. It might also include basic skills such as reading, writing, language, and arithmetic.

(2) Training for customer-contact (frontline) employees should address: (a) key knowledge and skills, including knowledge of products and services; (b) listening to customers; (c) soliciting comments from customers; (d) how to anticipate and handle problems or failures ("recovery"); (e) skills in customer retention; and (f) how to manage expectations.

(3) Determining specific education and training needs [4.3b(1)] might include use of company assessment or employee self-assessment to determine and/or compare skill levels for progression within the company or elsewhere. Needs determination should take into account job analysis — the types and levels of skills required — and the timeliness of training.

(4) Education and training delivery [4.3b(2)] might occur inside or outside the company and involve on-the-job, classroom, computer-based, or other types of delivery. This includes the use of developmental assignments within or outside the company to enhance employees' career opportunities and employability.

(5) How education and training are evaluated [4.3b(4)] could address: effectiveness of delivery of education and training; impact on work unit performance; and cost effectiveness of education and training alternatives.

'ABLE 9.3 1996 Award Criteria (*Continued*)

.4 Employee Well-Being and Satisfaction *(25 pts.)*

Describe how the company maintains a work environment and a work climate conducive to the well-being and development of all employees.

AREAS TO ADDRESS

a. how the company maintains a safe and healthful work environment. Include: (1) how employee well-being factors such as health, safety, and ergonomics are included in improvement activities; and (2) principal improvement requirements, measures and/or indicators, and targets for each factor relevant and important to the employees' work environment. Note any significant differences based upon differences in work environments among employee groups.

b. what services, facilities, activities, and opportunities the company makes available to employees to support their overall well-being and satisfaction and/or to enhance their work experience and development potential

c. how the company determines employee satisfaction, well-being, and motivation. Include a brief description of methods, frequency, the specific factors used in this determination, and how the information is used to improve satisfaction, well-being, and motivation. Note any important differences in methods or factors used for different categories or types of employees, as appropriate.

Notes:

(1) Examples of services, facilities, activities, and opportunities (4.4b) are: personal and career counseling; career development and employability services; recreational or cultural activities; non-work-related education; day care; special leave for family responsibilities and/or for community service; safety off the job; flexible work hours; outplacement; and retiree benefits, including extended health care. These services also might include career enhancement activities such as skills assessment, helping employees develop learning objectives and plans, and employability assessment.

(2) Examples of specific factors which might affect satisfaction, well-being, and motivation are: effective employee problem or grievance resolution; safety; employee views of leadership and management; employee development and career opportunities; employee preparation for changes in technology or work organization; work environment; workload; cooperation and teamwork; recognition; benefits; communications; job security; compensation; equality of opportunity; and capability to provide required services to customers. An effective determination is one that provides the company with actionable information for use in improvement activities.

(3) Measures and/or indicators of satisfaction, well-being, and motivation (4.4c) might include safety, absenteeism, turnover, turnover rate for customer-contact employees, grievances, strikes, worker compensation, as well as results of surveys.

(4) How satisfaction, well-being, and motivation information is used (4.4c) might involve developing priorities for addressing employee problems based on impact on productivity.

(5) Trends in key measures and/or indicators of well-being and satisfaction should be reported in Item 6.3.

TABLE 9.3 1996 Award Criteria (*Continued*)

5.0 Process Management (140 pts.)

The ***Process Management*** Category examines the key aspects of process management, including customer-focused design, product and service delivery processes, support services, and supply management involving all work units, including research and development. The Category examines how key processes are designed, effectively managed, and improved to achieve higher performance.

5.1 Design and Introduction of Products and Services

(40 pts.)

Describe how new and/or modified products and services are designed and introduced and how key production/delivery processes are designed to meet key product and service quality requirements, company operational performance requirements, and market requirements.

AREAS TO ADDRESS

a. how products, services, and production/delivery processes are designed. Describe: (1) how customer requirements are translated into product and service design requirements; (2) how product and service design requirements are translated into efficient and effective production/delivery processes, including an appropriate measurement plan; and (3) how all requirements associated with products, services, and production/delivery processes are addressed early in design by all appropriate company units, suppliers, and partners to ensure integration, coordination, and capability.

b. how product, service, and production/delivery process designs are reviewed and/or tested in detail to ensure trouble-free and rapid introduction

c. how designs and design processes are evaluated and improved to achieve better product and service quality, time to market, and production/delivery process effectiveness

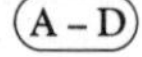

Notes:

(1) Design and introduction might address:

- *modifications and variants of existing products and services, including product and service customization;*
- *new products and services emerging from research and development or other product/service concept development;*
- *new/modified facilities to meet operational performance and/or product and service requirements; and*
- *significant redesigns of processes to improve customer focus, productivity, or both.*

Design approaches could differ appreciably depending upon the nature of the products/services — entirely new, variants, major or minor process changes, etc. If many design projects are carried out in parallel, responses to Item 5.1 should reflect how coordination of resources among projects is carried out.

(2) Applicants' responses should reflect the key requirements for their products and services. Factors that might need to be considered in design include: health; safety; long-term performance; environmental impact; "green" manufacturing; measurement capability; process capability; manufacturability; maintainability; supplier capability; and documentation.

(3) Service and manufacturing businesses should interpret product and service design requirements to include all product- and service-related requirements at all stages of production, delivery, and use.

(4) A measurement plan [5.1a(2)] should spell out what is to be measured, how and when measurements are to be made, and performance levels or standards to ensure that the results of measurements provide information to guide, monitor, control, or improve the process. This may include service standards used in customer-contact processes. The term, "measurement plan," may also include decisions about key information to collect from customers and/or employees from service encounters, transactions, etc. The actual measurement plan should not be described in Item 5.1. Such information is requested in Item 5.2.

(5) "All appropriate company units" [5.1a(3)] means those units and/or individuals who will take part in production/delivery and whose performance materially affects overall process outcome. This might include groups such as R&D, marketing, design, and product/process engineering.

ABLE 9.3 1996 Award Criteria (*Continued*)

.2 Process Management: Product and Service Production and Delivery *(40 pts.)*

Describe how the company's key product and service production/delivery processes are managed to ensure that design requirements are met and that both quality and operational performance are continuously improved.

AREAS TO ADDRESS

a. how the company maintains the performance of key production/delivery processes to ensure that such processes meet design requirements addressed in Item 5.1. Describe: (1) the key processes and their principal requirements; and (2) the measurement plan and how measurements and/or observations are used to maintain process performance.

b. how processes are evaluated and improved to improve products and services and to achieve better performance, including cycle time. Describe how each of the following is used or considered: (1) process analysis and research; (2) benchmarking; (3) use of alternative technology; and (4) information from customers of the processes — within and outside the company.

Notes:

(1) Key production/delivery processes are those most directly involved in fulfilling the principal requirements of customers — those that define the products and services.

(2) Measurement plan [5.2a(2)] is defined in Item 5.1, Note (4). Companies with specialized measurement requirements should describe how they ensure measurement effectiveness. For specialized physical, chemical, and engineering measurements, describe briefly how measurements are made traceable to national standards.

(3) The focus of 5.2a is on maintenance of process performance using measurements and/or observations to decide whether or not corrective action is needed. The nature of the corrective action depends on the process characteristics and the type of variation observed. Responses should reflect the type of process and the type of variation observed. A description should be given of how basic (root) causes of variation are determined and how corrections are made at the earliest point(s) in processes. Such correction should then minimize the likelihood of recurrence of this type of variation anywhere in the company.

(4) The focus of 5.2b is on improvement of processes — making them perform better than the original design. Better performance might include one or more of the following: operational, customer-related, and financial performance. After processes have been improved, process maintenance (5.2a) needs to adjust to the changes. Process improvement methods might utilize financial data to evaluate alternatives and set priorities.

(5) "Process analysis and research" [5.2b(1)] refers to a wide range of possible approaches for improving processes. Examples include process mapping, optimization experiments, basic and applied research, error proofing, and reviewing critical encounters between employees and customers from the point of view of customers and employees.

(6) Information from customers [5.2b(4)] might include information developed as described in Items 7.2, 7.3, and 2.3.

(7) Results of improvements in products and services and in product and service delivery processes should be reported in Items 6.1 and 6.2, as appropriate.

TABLE 9.3 1996 Award Criteria (*Continued*)

5.3 Process Management: Support Services

(30 pts.)

Describe how the company's key support service processes are designed and managed so that current requirements are met and that operational performance is continuously improved.

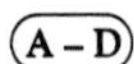

AREAS TO ADDRESS

a. how key support service processes are designed. Include: (1) how key requirements are determined or set; (2) how these requirements are translated into efficient and effective processes, including operational requirements and an appropriate measurement plan; and (3) how all requirements are addressed early in design by all appropriate company units to ensure integration, coordination, and capability.

b. how the company maintains the performance of key support service processes to ensure that such processes meet design requirements. Describe: (1) the key processes and their principal requirements; and (2) the measurement plan and how measurements are used to maintain process performance.

c. how processes are evaluated and improved to achieve better performance, including cycle time. Describe how each of the following is used or considered: (1) process analysis and research; (2) benchmarking; (3) use of alternative technology; and (4) information from customers of the processes — within and outside the company.

Notes:

(1) Support services are those that support the company's product and/or service delivery, but are not usually designed in detail with the products and services themselves because their requirements do not usually depend a great deal upon product and service characteristics. Support service design requirements usually depend significantly upon internal requirements. Support services might include finance and accounting, software services, sales, marketing, public relations, information services, supplies, personnel, legal services, plant and facilities management, research and development, and secretarial and other administrative services.

(2) The purpose of Item 5.3 is to permit applicants to highlight separately the design (5.3a), maintenance (5.3b), and improvement (5.3c) activities for processes that support the product and service design, production, and delivery processes addressed in Items 5.1 and 5.2. The support service processes included in Item 5.3 depend on the applicant's type of business and other factors. Thus, this selection should be made by the applicant. Together, Items 5.1, 5.2, 5.3, and 5.4 should cover all key operations, processes, and activities of all work units.

(3) Measurement plan [5.3a(2)] is described in Item 5.1, Note (4). Process maintenance (5.3b) is described in Item 5.2, Note (3). Process improvement (5.3c) is described in Item 5.2, Note (4).

(4) "Process analysis and research" [5.3c(1)] refers to a wide range of possible approaches for improving processes. See Item 5.2, Note (5).

(5) Information from customers [5.3c(4)] might include information developed as described in Items 7.2, 7.3, and 2.3. However, most of the information for improvement [5.3c(4)] is likely to come from "internal customers" — those within the company who use the support services.

(6) Results of improvements in support services should be reported in Item 6.2.

TABLE 9.3 1996 Award Criteria (*Continued*)

5.4 Management of Supplier Performance *(30 pts.)*

Describe how the company assures that materials, components, and services furnished by other businesses meet the company's performance requirements. Describe also the company's actions and plans to improve supplier relationships and performance.

AREAS TO ADDRESS

a. summary of the company's requirements and how they are communicated to suppliers. Include: (1) a brief summary of the principal requirements for key suppliers, the measures and/or indicators associated with these requirements, and the expected performance levels; (2) how the company determines whether or not its requirements are met by suppliers; and (3) how performance information is fed back to suppliers.

b. how the company evaluates and improves its management of supplier relationships and performance. Describe current actions and plans: (1) to improve suppliers' abilities to meet requirements; (2) to improve the company's own procurement processes, including feedback sought from suppliers and from other units within the company ("internal customers") and how such feedback is used; and (3) to minimize costs associated with inspection, test, audit, or other approaches used to track and verify supplier performance.

Notes:

(1) The term "supplier" refers to other-company providers of goods and services. The use of these goods and services may occur at any stage in the production, design, delivery, and use of the company's products and services. Thus, suppliers include businesses such as distributors, dealers, warranty repair services, transportation, contractors, and franchises as well as those that provide materials and components.

If the applicant is a unit of a larger company, and other units of that company supply goods/services, this should be included as part of Item 5.4.

The term "supplier" also refers to service suppliers such as health care, training, and education.

(2) Key suppliers [5.4a(1)] are those that provide the most important products and/or services, taking into account the criticality and volume of products and/or services involved.

(3) "Requirements" refers to the principal factors involved in the purchases: quality, delivery, and price.

(4) How requirements are communicated and how performance information is fed back might entail ongoing working relationships or partnerships with key suppliers. Such relationships and/or partnerships should be briefly described in responses.

(5) Processes for determining whether or not requirements are met [5.4a(2)] might include audits, process reviews, receiving inspection, certification, testing, and rating systems.

(6) "Actions and plans" (5.4b) might include one or more of the following: joint planning, rapid information and data exchanges, use of benchmarking and comparative information, customer-supplier teams, partnerships, training, long-term agreements, incentives, and recognition. Actions and plans might also include changes in supplier selection, leading to a reduction in the number of suppliers.

(7) Efforts to minimize costs might be backed by analyses comparing suppliers based on overall cost, taking into account quality and delivery. Analyses might also address transaction costs associated with alternative approaches to supply management.

TABLE 9.3 1996 Award Criteria (*Continued*)

6.0 Business Results (250 pts.)

The ***Business Results*** Category examines the company's performance and improvement in key business areas — product and service quality, productivity and operational effectiveness, supply quality, and financial performance indicators linked to these areas. Also examined are performance levels relative to competitors.

6.1 Product and Service Quality Results (*75 pts.*)

Summarize performance results for products and services and/or product and service offerings and results of improvement efforts, using key measures and/or indicators of such performance and improvement.

AREAS TO ADDRESS

a. current levels and trends in key measures and/or indicators of quality of products and services and/or product and service offerings. Graphs and tables should include appropriate comparative data.

Notes:

(1) Results reported in Item 6.1 should reflect performance relative to key non-price product and service requirements — those described in the Business Overview and addressed in Items 7.1, 3.1, and 5.1. The measures and/or indicators should address factors that affect customer preference — performance, timeliness, availability, and variety. Examples include defect levels, repeat services, delivery response times, and complaint levels.

(2) Data appropriate for inclusion might be based upon one or more of the following:

- *internal (company) measurements;*
- *field performance;*
- *data collected by the company or on behalf of the company through follow-ups (7.2c) or surveys of customers on product and service performance; and*
- *data collected or generated by organizations, including customers.*

Although data appropriate for inclusion are primarily based upon internal measurements and field performance, data collected by the company or other organizations through follow-ups might be included for attributes that cannot be accurately assessed through direct measurement (e.g., ease of use) or when variability in customer expectations makes the customer's perception the most meaningful indicator (e.g., courtesy).

(3) Comparative data might include industry best, best competitor, industry average, and appropriate benchmarks. Such data might be derived from independent surveys, studies, laboratory testing, or other sources.

6.2 Company Operational and Financial Results (*110 pts.*)

Summarize results of the company's operational and financial performance and performance improvement efforts using key measures and/or indicators of such performance and improvement.

AREAS TO ADDRESS

a. current levels and trends in key measures and/or indicators of company operational and financial performance. Graphs and tables should include appropriate comparative data.

Notes:

(1) Key measures and/or indicators of company operational and financial performance include the following areas:

- *productivity and other indicators of effective use of manpower, materials, energy, capital, and assets. (Aggregate measures such as total factor productivity, ROI, margin rates, operating profit rates, and working capital productivity are encouraged. Aggregate economic and/or market value measures are also appropriate.);*
- *company-specific indicators such as innovation rates, innovation effectiveness, cost reductions through innovation, and time to market;*
- *environmental improvements reflected in emissions levels, waste stream reductions, by-product use and recycling, etc. (See Item 1.3);*
- *cycle time, lead times, set-up times, and other responsiveness indicators; and*
- *process assessment results such as customer assessment or third-party assessment (such as ISO 9000).*

(2) Comparative data might include industry best, best competitor, industry average, and appropriate benchmarks.

ABLE 9.3 1996 Award Criteria (*Continued*)

.3 Human Resource Results *(35 pts.)*

Summarize human resource results, including employee development and indicators of employee well-being and satisfaction.

AREAS TO ADDRESS

a. current levels and trends in key measures and/or indicators of employee development, well-being, satisfaction, self-directed responsibility, and effectiveness. Graphs and tables should include appropriate comparative data.

Notes:

(1) Measures and/or indicators should include safety, absenteeism, turnover, and satisfaction. Comparative data might include industry best, best competitors, industry average, and appropriate benchmarks. Local or regional data on absenteeism and turnover are also appropriate. Financial measures such as worker compensation cost or turnover cost reductions are appropriate for inclusion.

(2) Measures and/or indicators of development should cover not only extent (for example, percent of employees trained or hours of training per year) but also effectiveness. Financial information such as benefit cost ratios for training is appropriate for inclusion.

(3) Examples of satisfaction factors are given in Item 4.4, Note (2).

(4) The results reported in Item 6.3 derive from activities described in the Items of Category 4.0. Results should address all categories and types of employees.

5.4 Supplier Performance Results *(30 pts.)*

Summarize results of supplier performance and performance improvement efforts using key measures and/or indicators of such performance and improvement.

AREAS TO ADDRESS

a. current levels and trends in key measures and/or indicators of supplier performance. Graphs and tables should include appropriate comparative data.

Notes:

(1) The results reported in Item 6.4 derive from activities described in Item 5.4. Results should be broken out by key supplies and/or key suppliers, as appropriate. Results should include performance of supply chains and/or results of outsourcing, if these are important to the applicant. Data should be presented using the measures and/or indicators described in 5.4a(1).

(2) Results reported should be relative to all principal requirements: quality, delivery, and price. If the company's supplier management efforts include factors such as building supplier partnerships or reducing the number of suppliers, data related to these efforts should be included in responses.

(3) Comparative data might be of several types: industry best, best competitor(s), industry average, and appropriate benchmarks.

TABLE 9.3 1996 Award Criteria (*Continued*)

7.0 Customer Focus and Satisfaction (250 pts.)

The ***Customer Focus and Satisfaction*** Category examines the company's systems for customer learning and for building and maintaining customer relationships. Also examined are levels and trends in key measures of business success — customer satisfaction and retention, market share, and satisfaction relative to competitors.

7.1 Customer and Market Knowledge *(30 pts.)*

Describe how the company determines near-term and longer-term requirements, expectations, and preferences of customers and markets, and develops listening and learning strategies to understand and anticipate needs.

AREAS TO ADDRESS

a. how the company determines current and near-term requirements and expectations of customers. Include: (1) how customer groups and/or market segments are determined and/or selected, including how customers of competitors and other potential customers are considered; (2) how information is collected, including what information is sought, frequency and methods of collection, and how objectivity and validity are ensured; (3) how specific product and service features and the relative importance of these features to customer groups or segments are determined; and (4) how other key information and data such as complaints, gains and losses of customers, and product/service performance are used to support the determination.

b. how the company addresses future requirements and expectations of customers and potential customers. Include an outline of key listening and learning strategies used

c. how the company evaluates and improves its processes for determining customer requirements, expectations, and preferences

Notes:

(1) The distinction between near-term and future depends upon many marketplace factors. The applicant's response should reflect these factors for its market(s). Methods used in 7.1a(2) and 7.1b might be the same or similar.

(2) The company's products and services might be sold to end users via other businesses such as retail stores or dealers. Thus, "customer groups" should take into account the requirements and expectations of both the end users and these other businesses.

(3) Some companies might use similar methods to determine customer requirements/expectations and customer satisfaction (Item 7.3). In such cases, cross-references should be included.

(4) Customer groups and market segments [7.1a(1)] might take into account opportunities to select or create groups and segments based upon customer- and market-related information. This might include individual customization.

(5) How information is collected [7.1a(2)] might include periodic methods such as surveys or focus groups and/or ongoing processes such as dialogs with customers.

(6) Product and service features [7.1a(3)] refer to all important characteristics and to the performance of products and services that customers experience or perceive throughout their overall purchase and ownership. The focus should be primarily on features that bear upon customer preference and repurchase loyalty — for example, those features that differentiate products and services from competing offerings. This might include price and value.

(7) Examples of listening and learning strategies (7.1b) are:
- *relationship strategies, including close integration with customers;*
- *rapid innovation and field trials of products and services to better link R&D and design to the market;*
- *close monitoring of technological, competitive, societal, environmental, economic, and demographic factors that may bear upon customer requirements, expectations, preferences, or alternatives;*
- *focus groups with demanding or leading-edge customers;*
- *training of frontline employees in customer listening;*
- *use of critical incidents to understand key service attributes from the point of view of customers and frontline employees;*
- *interviewing lost customers;*
- *won/lost analysis relative to competitors;*
- *post-transaction follow-up (see 7.2c); and*
- *analysis of major factors affecting key customers.*

(8) Examples of evaluation and factors appropriate for 7.1c are:
- *the adequacy and timeliness of the customer-related information;*
- *improvement of survey design;*
- *the best approaches for getting reliable and timely information — surveys, focus groups, customer-contact personnel, etc.;*
- *increasing and decreasing importance of product/service features among customer groups or segments; and*
- *the most effective listening/learning strategies.*

The evaluation might also be supported by company-level analysis addressed in Item 2.3.

ABLE 9.3 1996 Award Criteria (*Continued*)

.2 Customer Relationship Management (*30 pts.*)

Describe how the company provides effective management of its responses and follow-ups with customers to preserve and build relationships, to increase knowledge about specific customers and about general customer expectations, to improve company performance, and to generate ideas for new products and services.

AREAS TO ADDRESS

a. how the company provides information and easy access to enable customers to seek information and assistance, to comment, and to complain. Describe how contact management performance is measured. Include key service standards and how these standards are set, deployed, and tracked.

b. how the company ensures that formal and informal complaints and feedback received by all company units are resolved effectively and promptly. Briefly describe the complaint management process, including how it ensures effective recovery of customer confidence, how it meets customer requirements for resolution effectiveness, how it ensures that complaints received by company units are aggregated and analyzed for use throughout the company, and how it seeks to eliminate causes of complaints.

c. how the company follows up with customers on products, services, and recent transactions to determine satisfaction, to resolve problems, to seek feedback for improvement, to build relationships, and to develop ideas for new products and services

d. how the company evaluates and improves its customer relationship management. Include: (1) how service standards, including those related to access and complaint management, are improved based upon customer information; and (2) how knowledge about customers is accumulated.

Notes:

(1) Customer relationship management refers to a process, not to a company unit. However, some companies might have units which address all or most of the requirements included in this Item. Also, some of these requirements might be included among the responsibilities of frontline employees in processes described in Items 5.2 and 5.3.

(2) How the company maintains easy access for customers (7.2a) might involve close integration, electronic networks, etc.

(3) Performance measures and service standards (7.2a) apply not only to employees providing the responses to customers but also to other units within the company that make effective responses possible. Deployment needs to take into account all key points in a response chain. Examples of measures and standards are: telephonic, percentage of resolutions achieved by frontline employees, number of transfers, and resolution response time.

(4) Responses to 7.2b and 7.2c might include company processes for addressing customer complaints or comments based upon expressed or implied guarantees and warranties.

(5) The complaint management process (7.2b) might include analysis and priority setting for improvement projects based upon potential cost impact of complaints, taking into account customer retention related to resolution effectiveness. Some of the analysis requirements of Item 7.2 relate to Item 2.3.

(6) Improvement of customer relationship management (7.2d) might require training. Training for customer-contact (frontline) employees should address: (a) key knowledge and skills, including knowledge of products and services; (b) listening to customers; (c) soliciting comments from customers; (d) how to anticipate and handle problems or failures ("recovery"); (e) skills in customer retention; and (f) how to manage expectations. Such training should be described in Item 4.3.

(7) Information on trends and levels in measures and/or indicators of complaint response time, effective resolution, and percent of complaints resolved on first contact should be reported in Item 6.1.

TABLE 9.3 1996 Award Criteria (*Continued*)

7.3 Customer Satisfaction Determination *(30 pts.)*

Describe how the company determines customer satisfaction, customer repurchase intentions, and customer satisfaction relative to competitors; describe how these determination processes are evaluated and improved.

AREAS TO ADDRESS

a. how the company determines customer satisfaction. Include: (1) a brief description of processes and measurement scales used; frequency of determination; and how objectivity and validity are ensured. Indicate significant differences, if any, in processes and measurement scales for different customer groups or segments; and (2) how customer satisfaction measurements capture key information that reflects customers' likely future market behavior.

b. how customer satisfaction relative to that for competitors is determined. Describe: (1) company-based comparative studies; and (2) comparative studies or evaluations made by independent organizations and/or customers. For (1) and (2), describe how objectivity and validity of studies or evaluations are ensured.

c. how the company evaluates and improves its processes and measurement scales for determining customer satisfaction and satisfaction relative to competitors. Include how other indicators (such as gains and losses of customers) and dissatisfaction indicators (such as complaints) are used in this improvement process. Describe also how the evaluation determines the effectiveness of companywide use of customer satisfaction information and data.

Notes:

(1) Customer satisfaction measurement might include both a numerical rating scale and descriptors for each unit in the scale. An effective (actionable) customer satisfaction measurement system provides reliable information about customer ratings of specific product and service features and the relationship between these ratings and the customer's likely future market behavior — repurchase and/or positive referral. Product and service features might include overall value and price.

(2) The company's products and services might be sold to end users via other businesses such as retail stores or dealers. Thus, "customer groups" or segments should take into account these other businesses and the end users.

(3) Customer dissatisfaction indicators include complaints, claims, refunds, recalls, returns, repeat services, litigation, replacements, downgrades, repairs, warranty work, warranty costs, misshipments, and incomplete orders.

(4) Comparative studies (7.3b) might include indicators of customer dissatisfaction as well as satisfaction.

(5) Evaluation (7.3c) might take into account:

- *how well the measurement scale relates to actual customer behavior;*
- *the effectiveness of pre-survey research used in survey design;*
- *how well customer responses link to key business processes and thus provide actionable information for improvement; and*
- *how well customer responses have been translated into cost/revenue implications and thus provide actionable information for improvement priorities.*

(6) Use of data from satisfaction measurement is called for in 5.2b(4) and 5.3c(4). Such data also provide key input to analysis (Item 2.3).

7.4 Customer Satisfaction Results *(160 pts.)*

Summarize the company's customer satisfaction and dissatisfaction results using key measures and/or indicators of these results. Compare results with competitors' results.

AREAS TO ADDRESS

a. current levels and trends in key measures and/or indicators of customer satisfaction and dissatisfaction. Results should be segmented by customer groups and product and service types, as appropriate.

b. current levels and trends in key measures and/or indicators of customer satisfaction relative to competitors. Results should be segmented by customer groups and product and service types, as appropriate.

Notes:

(1) Results reported in this Item derive from methods described in Items 7.3 and 7.2.

(2) Measures and/or indicators of satisfaction relative to competitors (7.4b) should include gains and losses of customers and customer accounts to competitors as well as gains and losses in market share.

(3) Measures and/or indicators of satisfaction relative to competitors might include objective information and/or data from independent organizations, including customers. Examples include survey results, competitive awards, recognition, and ratings. Such information and data should reflect comparative satisfaction (and dissatisfaction), not comparative performance of products and services (called for in Item 6.1).

(4) Customer retention data might be used in both 7.4a and 7.4b. For example, in 7.4a, customer retention might be included as a satisfaction indicator, while in 7.4b, customer retention relative to competitors might be part of a switching analysis to determine competitive position and the factors responsible for it.

results; it has also been increased in point value (from 250 points to 450 points). Some items, or parts of items, are now combined (e.g., items 1.1 and 1.2 from the 1996 criteria have been combined into a new item 1.1, called Leadership System).

Finally, in an attempt to facilitate applicant response, changes have been made to reduce the complexity of the criteria document structure and application page limits:

	1996 criteria	1997 criteria
No. of categories	7	7
No. of items	24	20
No. of areas to address	52	30
No. of item notes	114	45
Application page limit	70	50

9.5 Evaluation and Scoring Guidelines [7]

The evaluation and scoring of MBNQA applicants is based on three dimensions. The first dimension on which an applicant is evaluated is the *approach* used to address a particular item requirement. The second evaluation dimension considers how well the applicant *deploys* that approach. Lastly, for those items that require outcomes, the third dimension looks for the approaches, and deployment of those approaches, to produce the desired *results*. A closer examination of the detailed award criteria provided in Table 9.3 indicates an appropriate coding for each item, consistent with these three dimensions: Approach/Deployment items are designated by A–D, while Results items are designated as R. The scoring guidelines are provided in Table 9.4.

When examining an *approach* item, the evaluation looks for the methods that are used to address the item requirements. Some important considerations in this case are appropriateness, effectiveness, and innovativeness of the methods the applicant has applied.

Deployment focuses on the pervasiveness of these approaches as applied by the applicant to address the item requirements. The evaluation looks for an extension of these approaches across all processes, products, nonproduct areas (i.e., administrative, service, and support), and work units.

Finally, the applicant's *results* are evaluated from the viewpoint of current performance levels and comparative levels with benchmarks such as industry leaders, competitors, and world class and best-in-class companies. In addition, results are examined for evidence of

TABLE 9.4 Scoring Guidelines

SCORE	APPROACH/DEPLOYMENT
0%	■ no systematic approach evident; anecdotal information
10% to 30%	■ beginning of a systematic approach to the primary purposes of the Item ■ early stages of a transition from reacting to problems to a general improvement orientation ■ major gaps exist in deployment that would inhibit progress in achieving the primary purposes of the Item
40% to 60%	■ a sound, systematic approach, responsive to the primary purposes of the Item ■ a fact-based improvement process in place in key areas; more emphasis is placed on improvement than on reaction to problems ■ no major gaps in deployment, though some areas or work units may be in very early stages of deployment
70% to 90%	■ a sound, systematic approach, responsive to the overall purposes of the Item ■ a fact-based improvement process is a key management tool; clear evidence of refinement and improved integration as a result of improvement cycles and analysis ■ approach is well-deployed, with no major gaps; deployment may vary in some areas or work units
100%	■ a sound, systematic approach, fully responsive to all the requirements of the Item ■ a very strong, fact-based improvement process is a key management tool; strong refinement and integration — backed by excellent analysis ■ approach is fully deployed without any significant weaknesses or gaps in any areas or work units

SCORE	RESULTS
0%	■ no results or poor results in areas reported
10% to 30%	■ early stages of developing trends; some improvements *and/or* early good performance levels in a few areas ■ results not reported for many to most areas of importance to the applicant's key business requirements
40% to 60%	■ improvement trends *and/or* good performance levels reported for many to most areas of importance to the applicant's key business requirements ■ no pattern of adverse trends *and/or* poor performance levels in areas of importance to the applicant's key business requirements ■ some trends *and/or* current performance levels — evaluated against relevant comparisons *and/or* benchmarks — show areas of strength *and/or* good to very good relative performance levels
70% to 90%	■ current performance is good to excellent in most areas of importance to the applicant's key business requirements ■ most improvement trends *and/or* performance levels are sustained ■ many to most trends *and/or* current performance levels — evaluated against relevant comparisons and/or benchmarks — show areas of leadership and very good relative performance levels
100%	■ current performance is excellent in most areas of importance to the applicant's key business requirements ■ excellent improvement trends *and/or* sustained excellent performance levels in most areas ■ strong evidence of industry and benchmark leadership demonstrated in many areas

SOURCE: Reprinted from Ref. 6 with permission.

sustained performance and linkage to key business success factors previously defined by the applicant.

9.6 Award Criteria Key Characteristics [8]

The key characteristics of the Baldrige criteria (1) are aimed at being results-oriented, (2) are purposely intended to be nonprescriptive yet comprehensive, (3) include improvement and learning cycles, (4) emphasize alignment of measures with results, and (5) are an integral part of a diagnostic system.

As indicated earlier, the Baldrige criteria are built on a framework in which the leadership drives a system that is aimed, among other things, at achieving business results. And business results in this case cover a variety of areas such as customer satisfaction, market share, product quality, financial performance, human resource development, supplier performance, and even public responsibility. With a rich set of business indicators such as these, it is important to strive for a set of balanced strategies that meet the overall business objectives. Furthermore, to achieve and sustain results, there is the need to drive for continual improvement.

Since companies may have different key business success factors and objectives, there is no specific set of tools and techniques which will achieve the results for all. Nor are there specific organizational structures and planning approaches that necessarily fit across all companies. Therefore, the focus remains on results, rather than the specific means (e.g., specific procedures, methods, and systems) by which companies achieve those results. The criteria do not spell out the specifics, but instead provide companies significant leeway in their quest of excellent business results.

Certainly the Baldrige criteria are very comprehensive. They cover the interests, needs, and expectations of all stakeholders (e.g., customers, employees, stockholders, suppliers, the community). Consequently, they include both internal and external requirements, and all the processes that relate to those requirements. In addition, they embrace product and nonproduct areas, such as services and other business support areas.

Continuous improvement is the basis for achieving sustained business results. Therefore, there is the need for continual and interrelated learning and improvement cycles. In essence, the process follows the Deming plan–do–check–act (PDCA) cycle, in which plans are developed, executed, assessed for progress, and revised based on feedback. The cycle is then repeated. Information about Deming and the development of the PDCA cycle as part of the Deming Prize may be found in Chap. 5.

The criteria enable alignment of key business factors to related measures, indicators, requirements, and goals. The improvement and learning cycles are also aligned with these overall performance objectives. Thus, the directions across the company become consistent and common in their aims.

Finally, the Baldrige criteria, when coupled with the scoring guidelines, become a two-part diagnostic system. The criteria provide the results-oriented requirements for each Baldrige item, while the scoring guidelines give the approach, deployment, and results dimensions for those same items. The assessment allows for development of an actionable plan that reflects the strengths and the opportunities for improvement for the organization.

9.7 Baldrige Award Winners and Highlights

A review of the electronics companies that have won the MBNQA reveals some interesting traits and accomplishments. Brief thumbnail sketches, summarized from the MBNQA "Profile of Winners," are provided below in alphabetical order for these winners. These summaries represent the highlights at the time each company won the award, and are not necessarily intended to portray their current situation, products, or organization.

AT&T Network Systems Group—Transmission Systems Business Unit (1992 Winner). The Transmission Systems Business Unit (TSBU), a division of AT&T, resides within the Network Systems Group. The company develops, manufactures, markets, and services systems for transporting data, voice, and images over public and private telecommunications networks. Its products include digital loop carrier systems, digital access and cross-connect systems, network multiplex equipment, and lightwave systems. TSBU is currently one of the world's largest makers of transmission systems.

TSBU is very highly customer-driven. Many of its major customers provide annual detailed, individualized report cards covering both product and service characteristics. Using this feedback, as well as measures of the quality and responsiveness of customer-support services, TSBU implements its strategic planning process to ensure alignment of performance improvement goals to customer satisfaction.

Since reliability is extremely important for transmission equipment, TSBU has focused heavily on this parameter in both design and manufacturing. As a result, the reliability of TSBU products exceeds customer expectations, enabling TSBU to be the first in its industry to offer a 5-year warranty. Furthermore, TSBU has reduced its new product development time in half and realized substantial cost savings.

TSBU's approach to quality involves a number of important activities. The company uses an iterative "policy deployment" process in its planning to link AT&T's quality principles, through TSBU's detailed objectives, to specific team-based quality improvement projects. Cross-organizational steering committees translate TSBU goals into specific quality projects required to accomplish the company's near-term and long-term goals. The company's strategic plan is communicated both to all employees and key suppliers. Organizational plans are aligned closely to the strategic plan.

TSBU has a strong team-based culture, supported by appropriate and timely training, with a very high percentage of the workforce involved in teams. Key processes are tracked through TSBU's information systems using carefully selected performance indicators which are reviewed regularly for management decision making and process improvement. Design for manufacturability is an area of focus along with design for reliability. Through the company's automation efforts, new product development times and production times have been reduced. Customer relationships are maintained through a variety of methods, including customer report cards, customer focus groups, complaint handling, technical assistance, customer surveys, and competitive evaluations. In addition, TSBU has established forward-looking programs to determine long-term customer needs and the new technology needed to meet those needs.

Corning Telecommunications Products Division (1995 Winner). The Telecommunications Products Division (TPD) of Corning was formed with a mission to manufacture optical fibers that carry large amounts of information over great distances. Today, TPD is the world's largest optical-fiber manufacturer. In spite of its market position, TPD continues to advance its technology and processes and stay ahead of its competitors.

The organization, driven by its Executive Leadership Team (ELT), is heavily focused on achieving customer satisfaction. The ELT integrates total quality principles into its overall strategic plan, "Plan to Win," which has six fundamental components: strategic direction, customer focus, formalized process management systems, continuous improvement culture, progress measurement using the Baldrige criteria, and foundation values of people, processes, and technology. TPD consistently receives high customer marks.

The many processes that the division employs in its business are characterized and documented, and maintained under a formal process management and control system. From the larger set of processes, TPD has identified core processes which are emphasized for special continuous improvement efforts. Productivity has significantly improved, allowing TPD to be the world's lowest-cost optical fiber producer, while at the same time substantially reducing defects.

Total Quality Management (TQM) is an integral part of TPD's management process. Implementing this process has indeed helped TPD develop and deliver products that are superior from a technological, cost, and quality perspective, and thus stay ahead of the competition. The ELT uses a strategy-driven, customer-focused system that melds quality into all parts of TPD's business. TPD's customer satisfaction focus is centered around a customer response system which provides the means to distill key customer requirements into continuous improvement action plans having measurable critical success factors. Its Plan to Win is executed following a four-element strategy that includes satisfying customers, competing effectively in selected markets, building affordable production capacity, and reducing manufacturing costs. To ensure successful execution of its plan, the ELT continually invests in TPD's three values: people, processes, and technology.

An integral part of the strategic planning process is the alignment of TPD's initiatives and values throughout. Organizational initiatives and values are tightly linked with short-term and long-term plans, work-group and unit goals, and individual performance objectives of all employees, including contributions made as part of teams. In fact, teamwork and interunit cooperation have become the standard modus operandi and facilitated TPD's effective, sustained improvement efforts across the organization. The results have been reflected in increasing sales, market share, and profitability, even though the price of optical fiber has dropped significantly.

IBM Rochester (1990 Winner). IBM Rochester (Minnesota) is in the business of manufacturing intermediate computer systems, including both hardware and software. Its systems are installed worldwide. They also manufacture hard disk drives.

IBM Rochester links the concept of quality directly with the customer. Customer needs and expectations weigh heavily in the design, manufacture, and delivery of its products. Customers are involved in the various product life-cycle steps through various avenues such as advisory councils. A market-driven process is used to effectively involve suppliers, business partners, and customers in delivering needed solutions.

The organization has formulated improvement plans using six critical success factors which include:

1. Better requirements definition for its product and services
2. An improved product strategy
3. A strategy of defect elimination based on 6 sigma concepts
4. Further reductions in cycle time
5. An improved education program
6. Greater involvement and ownership by employees in its strategic quality initiatives

Worldwide benchmarking is a cornerstone of IBM Rochester's continuous improvement of support processes, with numerous teams working on the identified opportunities. Quality goals become an integral part of the 5-year business plans, and are supported by improvement plans aimed at achieving the quality objectives. The organization also emphasizes education and training and makes substantial investments in this area.

IBM Rochester's efforts have paid dividends through significant increases in productivity and product reliability, with the latter contributing to an increased warranty on its products and a very low cost of ownership to its customers. In addition, product development and manufacturing cycle times have been drastically reduced. Coincident with these improvements has been an increase in share of the intermediate-computer world market.

Marlow Industries (1991 Winner). Marlow Industries is in the business of manufacturing customized thermoelectric coolers for heating, cooling, or stabilizing the temperature of electronic equipment. Most of these devices are custom-designed to meet its customer requirements. Its customers include makers of laser diodes, as well as aerospace and defense companies. Marlow has substantially increased its share of the world market for these devices by continuously improving all areas of its business by applying TQM principles.

The organization's TQM approach focuses on continuous improvement. Its TQM Council is chaired by the CEO and president, and formulates the five-year strategic business plan. Employees are encouraged to be an integral part of the continuous improvement process. Worker representatives participate in weekly council meetings. Company performance is reviewed regularly and employees are recognized for their contributions. Quality-related training is pervasive throughout the organization.

Marlow employs an extensive information system to help establish its goals and track progress against these goals. Rapid and easy access is provided to these data. The TQM system also heavily involves the organization's suppliers. Most of its suppliers have been certified to ship materials to stock without inspection.

These approaches have produced dramatic results. Manufacturing yields have improved significantly. Employee productivity has increased annually. New product design cycle times have been reduced. The cost of poor quality has been lowered significantly. On-time deliveries have been improved. Warranty times have been increased. Product prices have either been held stable or reduced. And over a 10-year period, the company has not lost a single major customer.

Motorola, Inc. (1988 Winner). Motorola develops, manufactures, and services a wide variety of communications products such as radios, pagers, cellular telephones, and modems. It is also a major producer of semiconductors. Motorola is a global company with major facilities throughout the world. In an effort to stay ahead of its competitors, Motorola pursued a successful strategy of quality improvement leading to total customer satisfaction (TCS).

Motorola was one of the first Baldrige Award winners (in the manufacturing category). This achievement came 7 years after it started a corporate program to improve the quality of its products and services tenfold. The company's goal became "Zero defects in everything we do." This was later translated into Motorola's 6 sigma initiative, or a target of no more than 3.4 defects per million products and services.

Motorola's focus is on TCS, which the company sees as key to achieving and sustaining sales and revenues far ahead of its competitors. Senior leadership and managers drive and reinforce the TCS initiative throughout the company. The rest of the company has rallied around this common goal and made significant improvements in quality, total cycle time, customer retention, and improved profitability. The company also makes an enormous investment in training its employees in acquiring and improving the skills necessary to achieve these objectives. To facilitate this companywide training initiative, Motorola established its own training center, Motorola University, with a heavy focus on quality-related training courses. Another important Motorola initiative is cycle time reduction, which covers the spectrum of processes related to design, manufacturing, sales, and delivery.

Motorola has spawned a strong, companywide team culture. At any given time, many of its employees are participants of TCS teams, aimed at improving processes which contribute to customer satisfaction. Benchmarking is used extensively by the teams to aid in their improvement efforts.

Through these various initiatives, Motorola has seen its market share grow substantially within its various business sectors. Quality defects, both in-process and delivered, have substantially decreased, as has the cost of poor quality. Cycle times have also been reduced. Customer surveys have shown a high level of satisfaction in both products and services. And employee satisfaction remains very high.

Solectron Corporation (1991 Winner). Solectron is mainly in the business of printed circuit board assembly. Other areas of its business cover the assembly of other electronic systems, subsystems, software packaging, disk duplications, custom products remanufacture, and design and testing services. Its customers include manufacturers of personal computers, workstations, disk and tape drives, and avionics, medical imaging, and telecommunication equipment.

Solectron's marketing efforts are on value-added projects that require high reliability and quality, quick turnaround, and response to change. It finds that low costs and timely delivery are associated with high quality and efficiency. The company also focuses heavily on customer satisfaction. As a result, most of its new business is additional work from established, loyal customers.

A variety of "listening" posts, such as weekly surveys of all customers, are employed by Solectron to determine how existing and prospective customers define superior performance. This direct and frequent customer feedback is used as an integral part of its continuous improvement program.

Solectron's continuous improvement efforts revolve around a team-based culture. Teams set goals which support corporate quality improvement targets. Management takes a coaching role in the overall improvement process. Two teams are established to support each customer, one working to ensure quality performance and the other focusing on on-time delivery.

Solectron's efforts have paid off. Defect rates have dropped significantly. Product rejection rates have improved markedly. On-time delivery has risen sharply. From a customer satisfaction perspective, the company has been cited by its customers with numerous awards for superior performance.

Texas Instruments Defense Systems & Electronics Group (1992 Winner). The Defense Systems & Electronics Group (TI-DSEG) is a subsidiary of Texas Instruments Inc. TI-DSEG designs and manufactures precision-guided weapons, airborne systems, infrared vision equipment and other electrooptic systems, and electronic warfare systems. The organization ranks highly as a defense electronic contractor.

TI-DSEG is using TQM to become a stronger competitor. It is on a mission of achieving 6 sigma quality and reducing product development times. Quality goals and business goals are intertwined. The organization employs a large network of teams to execute its quality strategy. Self-directed work teams possess day-to-day decision-making authority, and have the full backing of the company's executive-level Quality Improvement Team (QIT).

The QIT, led by the TI-DSEG president, drives strategic quality planning and goal setting for the company. The planning horizon includes 5-year and 10-year goals. Customer needs, obtained through various means such as formal and informal surveys, are segmented by market, and become an integral part of the entire planning process. Key customer requirements are translated into goals for product parameters, which are tracked toward customer satisfaction objectives.

The organization's TQM approach also makes valuable use of the lessons learned through its very active benchmarking program. Benchmarking results have been used to lead the way in many improvement efforts, ranging from manufacturing quality to training. The company also is pursuing a strategy of increased empowerment for both individuals and teams, complementing this strategy with enhanced information technology and diagnosis/problem-solving tools.

TI-DSEG TQM strategy has yielded significant dividends both to the company and its customers. System reliability has consistently exceeded customer specifications. TI-DSEG's manufacturing processes and techniques have been designated as "best manufacturing practices" by the Navy. Because of the company's quality improvements efforts, customer-conducted audits have been significantly reduced, as have the number of formal customer complaints. An independent customer satisfaction survey has shown TI-DSEG leading its top competitors in all customer satisfaction categories, such as product support, included in the survey. And revenues per worker have increased significantly.

Zytec Corporation (1991 Winner). Zytec manufactures power supplies for computer original equipment manufacturers. Its power supplies are also used in electronic office and testing equipment. Most of its sales are for customized power supplies. The company is also in the business of repairing cathode-ray tube monitors and power supplies, both its own as well as those of its competitors. Zytec is focused on quality, service, and value.

Zytec's quality improvement efforts are organized around Deming's 14-point concepts for managing productivity and quality. Annual surveys of its employees are conducted to track progress and assess quality commitment and satisfaction of workers. A common quality focus is assured through Zytec's interactive "Management by Planning" process that involves employees in long-term and annual improvement goal setting. This focus is further reinforced through the review and critique of the company's 5-year plans by a cross section of employees. Based on these inputs, an executive team finalizes the plans and sets broad corporate objectives. Cross-functional teams then develop appropriate supporting goals. Customers and suppliers also participate in this planning and goal setting process.

New product design and development activities are implemented through empowered, interdepartmental project teams which work closely with customers. As part of its continuous improvement thrust, Zytec ensures that its employees are well trained in analytical and problem-solving techniques. Performance at all levels is evaluated using measurable criteria. Performance improvement is furthered by an active benchmarking program which evaluates competitors' products and services, as well as the practices of recognized quality leaders in other industries.

Zytec has successfully aligned quality improvement with customer priorities, and achieved sustained results. Productivity and sales per employee have increased significantly. Manufacturing yields have improved, while manufacturing cycle times have decreased. Similarly, design cycles and product costs have also been reduced. Product quality and reliability have increased dramatically. In addition, on-time delivery has improved substantially.

9.8 Award Winners' Best Practices

Some of the best practices by the MBNQA winners can be generalized into a set of "lessons learned." These lessons are reviewed by Baldrige category in terms of key excellence indicators.

Category 1.0 Leadership. The Baldrige winners have highly visible senior leadership which focuses heavily on the customer. These senior leaders have established aggressive goals for their organization and people. They view cycle time as an important business driver. The leadership has established clear values that are easily remembered by the entire work force. Managers serve as coaches to facilitate meeting the objectives. They understand and emphasize the importance of continuous learning. The leaders serve as role models and champions for good corporate citizenship. And they realize that the fruits of their labor and of the total involved workforce doesn't happen overnight. Patience is a virtue!

Category 2.0 Information and Analysis. Management by fact is an important characteristic of the Baldrige winners. They have a quantitative orientation, with a focus on actionable data. They establish and use a variety of measures that are aligned to key business factors to gauge and drive progress. These metrics are interlinked, and measure both internal and external performance. The metrics are widely deployed throughout the entire organization and the results made accessible to the workforce. The winners typically have a strong analysis capability to enable translation of data into actions. They also routinely benchmark the best-in-class companies, both within and outside their industry.

Category 3.0 Strategic Planning. Strategic planning by the Baldrige winners integrates quality planning with business planning. Their planning horizon covers both the near term and the long term. The planning encompasses their products and services, as well as their processes. They do benchmarking derived from studies of world leaders to establish aggressive planning drivers. They use customer requirements to derive key business targets, both current and future. Furthermore, these targets are deployed to all units within the company. Recognizing the importance of its suppliers and partners, the Baldrige winners also heavily involve these stakeholders into their planning process.

Category 4.0 Human Resource Development and Management. The Baldrige winners greatly value their human assets. They view them as their internal customer, a valuable stakeholder in the success equation. Consequently, human resource planning is also integrated with its overall business planning. The winners establish comprehensive training and education programs that are both relevant and effective to its employees and its business success. They empower and cross-train their employees to be valuable to themselves, as well as the company. Recognition is extended not only to individuals, but also to teams. This focus on employees results in lower turnover, accidents, and absenteeism. In short, the winners are committed to employee satisfaction and well-being.

Category 5.0 Process Management. Process management as practiced by the Baldrige winners not only covers the spectrum of processes related to its products and services, but also extends to its administrative and support processes. Design quality is incorporated into products, services, and processes. The winners focus on cycle time and productivity, as well as quality. They also integrate timely and effective prevention, correction, and improvement activities with their daily operations across the board. And again, recognizing the importance of suppliers, the winners partner with them on key process management initiatives.

Category 6.0 Business Results. The Baldrige winners achieve outstanding results. They demonstrate a broad base of improvement trends and excellent performance across their products and services, as well as their internal operations. In particular, these excellent results are reflected in the winners' cycle time achievements and productivity. Supplier quality results show sustained improvements and excellent current levels. As indications of leadership, the winners show exemplary results when benchmarked and compared to industry leaders. From a bottom-line perspective, their business results are intimately linked to financial performance.

Category 7.0 Customer Focus and Satisfaction. It is here that the Baldrige winners really shine! They show that they really understand and have comprehensive markets that they provide their products and/or services. Markets are segmented consistently with their customer base. They focus on very proactive, rather than reactive, customer systems. They make extensive and effective use of all "listening posts," including surveys, product and service follow-ups, complaints, and customer turnover, to provide better customer relationships. Surveys go beyond current customers and extend to potential customers as well. Frontline employees who deal daily with customers are empowered to keep the customer happy, and are supported by a strategic infrastructure. Much emphasis is placed on managing customer relationships to meet their expectations. In line with this focus, significant attention is devoted to hiring, training, and maintaining the morale of these frontline employees. Finally, the Baldrige winners achieve high levels of customer satisfaction as evidenced by surveys and awards they receive from their customers.

9.9 Baldrige Total Quality Payback

9.9.1 Benefits derived from self-assessment

There are a number of important benefits to be derived by using the Baldrige framework and criteria for internal self-assessment. These may be summarized as follows:

1. Helps achieve consensus on what needs to be done to improve
2. Helps integrate the various quality management and business efforts

3. Helps lead to constancy of purpose of direction over time in the pursuit of continuous improvement
4. Helps educate top management on Total Quality Management
5. Provides an objective, credible assessment tool
6. Results in an assessment supported by facts and data
7. Necessitates comparisons to applicable benchmarks
8. Leads to action based on feedback, rather than action based on emotion
9. Provides measures of progress over time through repetitive improvement and assessment cycles
10. Focuses improvement where most needed through prioritization of findings
11. Promotes sharing internally and externally of good, effective approaches
12. Leads to a continual improvement process

9.9.2 Value of implementing a Total Quality Management system

It has been shown through a study [9] conducted by the General Accounting Office (GAO) that there is significant value to be gained in implementing a total quality management system along the lines of the Baldrige framework. The purpose of this study was to determine the impact of TQM practices on the performance of selected U.S. companies. These companies included 20 of the winners and finalists for the 1988 and 1989 MBNQA.

From this study, it was determined that each company developed tailored practices, but that the TQM systems reflected common features. Furthermore, different companies benefited over time from specific TQM practices. Most importantly, it was found that the TQM-oriented companies experienced overall corporate performance improvement in a number of key business-related success factors:

- Better employee relations were realized.
- Quality was improved.
- Greater customer satisfaction was achieved.
- Improved market share and profitability was attained.

Complete information about implementing TQM may be found in Chap. 6.

9.9.3 Stock investment performance of the award winners

The National Institute of Standards of Technology, under which the Baldrige Program Office resides, has conducted a couple of stock investment studies [10] involving the Baldrige Award winners and finalists. The first study was conducted in February 1995 and showed solid results. An update was completed approximately 1 year later to determine if these results were being sustained. The returns continue to be impressive.

The study examined the MBNQA winners, as well as those companies that received Baldrige Award site visits. The first category included 14 publicly traded companies that have won the Baldrige Award from its inception in 1988 through the 1994 award year. This category included five whole companies and nine parent companies of subsidiaries. The second category included 41 publicly traded site visit recipients from award inception through 1994.

For investment performance comparisons, the Standard & Poor (S&P) 500 Index was used. The same hypothetical dollar amount was invested in both the S&P 500 Index companies and the Baldrige winners. Similarly, the same (though a different amount than for the winners) hypothetical amount was invested in both the S&P 500 Index and the site-visit companies. Adjustments were made for stock splits and/or stock dividends, and the value on August 1, 1995 was calculated for the S&P 500 Index companies, the winners, and the site visit companies.

As demonstrated in Table 9.5, in both categories, the Baldrige companies far outperformed the S&P 500 Index companies. Performance was measured by the percent change from the amount invested in a company in the first business day in April of the year in which they won the Baldrige Award or received a site visit (or the date when they began public trading). The Baldrige award winners beat the S&P 500 Index companies by factors of 2 to 1 and 5 to 1. The Baldrige site-visit companies also outshone the S&P 500 Index companies. In fact, the investment return was reflected in a factor approaching 3 to 1 when winning companies were included in the site visit applicants.

9.10 Baldrige Self-Assessment

9.10.1 Implementing a self-assessment and improvement process

Why do a Baldrige self-assessment? One reason might be to ensure preparedness for applying for the MBNQA, or a quality award based on the Baldrige criteria. In the latter category, a number of states (e.g., Massachusetts) in the United States and countries (e.g., Finland) use the Baldrige criteria. Another reason might be to comply

TABLE 9.5 Stock Investment Performance of the MBNQA Winners

Category	Investment return, %
Publicly traded winners:	
All winners	248.7
S&P 500	58.5
Whole company winners only	279.8
S&P 500	55.7
Publicly traded site-visit applicants:	
All companies, including winners	94.5
S&P 500	40.6
Whole companies only, including winners	106.0
S&P 500	39.2
All companies, excluding winners	47.1
S&P 500	41.2
Whole companies only, excluding winners	41.4
S&P 500	38.8

SOURCE: Data from Ref. 10 with permission.

with a company's overall continuous improvement program. In fact, the many requests for the Baldrige criteria annually support this pervasive use of the criteria for improvement.

In any event, the process for implementing a Baldrige self-assessment and improvement effort involves the several steps described below. In implementing this process, it will be both appropriate and necessary to utilize personnel, either from within or external consultants, who have been trained and have experience in the Baldrige criteria.

1. Understanding the criteria. This step involves educating the various participants in the self-assessment in the Baldrige criteria. These participants should include those who will be involved both directly and indirectly in the self-assessment. Obviously, direct participants will have a need for knowledge to successfully execute the self-assessment and improvement project. On the other hand, the indirect participants will be looked on to provide the necessary support and commitment to project success. In particular, senior leaders must be included in this education process. As noted earlier, the senior leadership is the driver of the system that will produce the results. Consequently, these leaders and line managers will need to be educated in the importance of being committed, involved, and visible in the implementation of the project.

2. Training the "examiners." The individuals, or examiners, who will actually evaluate the information and data gathered as a part of the self-assessment will be required to have more in-depth training than the others involved in the project. This training will cover the criteria in more detail and teach the examiners how to evaluate the gathered information and data and how to score the results. Training will also cover how consensus is achieved among the examiners, and how a meaningful and actionable feedback report is prepared for the organization.

3. Gathering the information and data. Information and data gathering is the heart of the project. In order to evaluate the organization by the Baldrige criteria, relevant information and data will obviously need to be gathered for the various areas to address. This information and data will be gathered through a variety of means such as interviews, observations, walk-throughs, and document reviews. Organizational units should not fear providing this information and data. It is essential to the learning and improvement process that characterizes the self-assessment to "open the books" and furnish the necessary and available information and data as they pertain to the Baldrige criteria. Senior management, as well as the education and training process, should strongly emphasize the importance of this openness and discourage recrimination. Without this information and data, there will be no self-assessment, nor improvement.

4. Documenting the facts. For ease in tracking against the criteria, the facts should be documented by the organization for the examiners following the MBNQA application format defined in the Baldrige criteria. This format is embodied in a 70-page-limited document as follows:

- Business overview
  1. Provide an outline of the organization's business, indicating key business and success factors, as well as business directions.
  2. Include information on the following: basic description of the company or organization, customer requirements, supplier relationships, competitive factors, and other important factors (e.g., new business alliances, new technologies, strategy changes).

- Item responses
  1. For *approach* and *deployment* items, write responses that show what and how, systematic approaches, deployment, focus, and consistency.
  2. For *results* items, write responses that show trends, current lev-

els, comparisons with competitors, and benchmarks, relating all results to key requirements, stakeholders, and goals.

5. **Evaluating the self-assessment results.** If the company is preparing a formal application for the MBNQA, the documented facts will be submitted to NIST for examination and evaluation by Baldrige examiners. If the company intends to use the self-assessment results for improvement, it will submit the "application" to trained examiners for evaluation. The examiners will take the information and data gathered during self-assessment and documented in the organization's application, and for each item, determine strengths and areas for improvement. Based on the scoring guidelines provided, appropriate scores will be developed by each examiner at both the Baldrige item and category level. Where there is disagreement in the scores that go beyond preestablished guidelines, the examiners will proceed through a consensus process. In the MBNQA award structure, the consensus will be very exhaustive and involve as many as six to eight Baldrige examiners. In any event, the examiners will flag out as "site-visit issues" any areas that require clarification or verification before a final score is determined.

6. **Preparing the feedback report.** Documentation of the self-assessment results in the form of a feedback report by the examiners should be done with the "customer" in mind—in this case the organization. The report needs to contain sufficient particulars that allow for improvement actions. The feedback report format should be structured along the following lines:

Overall Summary

Category Summaries

Item Strengths and Areas for Improvement

Overall, Category, and Item Level Scoring

Improvement Recommendations

7. **Developing the improvement strategy.** With all the facts at hand, the next step is to lay out the improvement strategy and plan and then begin implementation. The strategy should focus on improvements that are important to success of the business. From the feedback report, a number of areas for improvement will surface. However, not all improvement areas will have the same impact on operational excellence and improved competitiveness. The areas for improvement should be attacked with some sort of prioritization scheme. One simple prioritization scheme is shown in the sample survey output of Fig. 9.3, in which the Baldrige category performance ratings (scores) are mapped against the category importance ratings. Focus for improve-

Figure 9.3 Sample survey output: importance versus performance by category.

ment is then directed at those categories that fall in the lower right-hand quadrant, where the category performance ratings are relatively low and their importance ratings are relatively high.

9.10.2 Using Deming's PDCA cycle

Another way to think about Baldrige self-assessment for internal improvement is using a modified version of the Deming PDCA cycle (Fig. 9.4), sometimes referred to as the McClaskey two-circle model. The cycle starts with inputs representing customer needs and company strategic and business plans. These inputs are fed into the right-side circle to start the self-assessment process. The process begins with planning and training, then the assessment itself, followed by a communication of the organizational strengths and improvement opportunities. The next step in the assessment cycle is the prioritization of improvement areas.

The outputs of the assessment cycle then become inputs to the improvement process. The improvement cycle (left-hand circle) repeats with planning for improvement, implementing the action plans, tracking progress, and finally keeping the improvement momentum going. The cycles then replicate themselves on some peri-

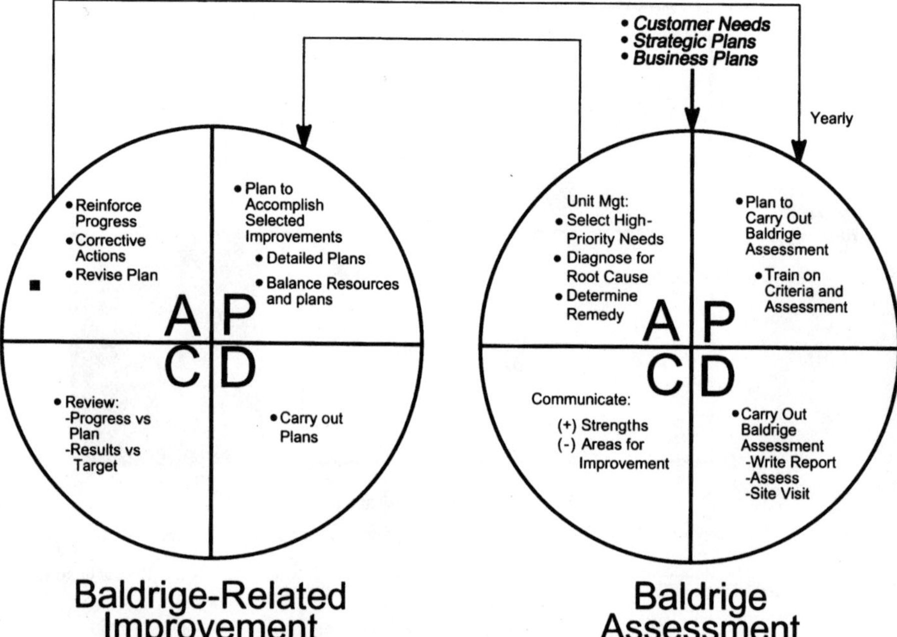

Figure 9.4 Deming PDCA cycles.

odic basis (e.g., annually). Complete information on Deming and his management concepts may be found in Chap. 5.

9.10.3 Following some useful hints

Several things can be done to ensure the effective use of the Baldrige framework for self-assessment. Some useful hints to follow during the assessment planning process are

1. Apply the award criteria without modification, at all levels—category, item, and areas to address.
2. Require that the assessed organization document the assessment results consistent with the Baldrige application format.
3. Stress the use of the internal self-assessment results and feedback as the basis for driving improvement.
4. Incorporate the results and feedback into company plans.
5. Prioritize areas for improvement with the organization's business strategies, objectives, and directions.

6. Integrate the use of the award criteria into the company's existing business and quality management efforts.
7. Avoid playing the "numbers game" with the scoring.
8. Minimize or avoid internal competition.
9. Do external, world class benchmarking.
10. Provide recognition through company awards.
11. Keep management committed and involved throughout the assessment process.

9.10.4 Barriers to self-assessment and ways to overcome them

Organizations have encountered a number of obstacles in their attempts to use the Baldrige criteria for self-assessment and improvement. Nevertheless, a familiarity with these barriers provides an opportunity to devise some ways to circumvent them.

Most often, it just isn't clear, or at least people don't want to admit, that there is a need for improvement. Consequently, when proposed, the subject of self-assessment is looked upon in disbelief or even in bewilderment. However, bringing evidence to the forefront can remarkably influence this kind of thinking. Benchmarks against world class competitors and "show-and-tell" sessions with Baldrige winners provide significant insight into how much further the organization can, or needs to, improve.

In these times of downsizing, and trying to make do with less, there is much made about the lack of time and resources to do a self-assessment. The flippant answer is that "You can't afford not to do it!" Nevertheless, there are ways to mitigate the time involved, while raising the stakes in the value of doing the self-assessment. Time can be reduced by using volunteers and/or the people who are truly interested and passionate for improvement to conduct the assessment. Simultaneously, senior management needs to reinforce the need and importance of self-assessment and improvement.

Another barrier relates to the fear of getting "bad" news from the assessment. The best way around this obstacle is to stress the need for improvement, not recrimination. Furthermore, the benefit in the scoring for use in setting a baseline should be emphasized, rather than undue focus on an absolute result.

If the training guidelines provided earlier are followed, then a shortage of trained examiners will be a nonissue. However, if it is impossible to get a cadre of people trained as examiners for the self-assessment process, then other avenues can be sought. These include using members of the past or present MBNQA Board of Examiners, or other external personnel trained and experienced in the use of the Baldrige criteria. Ideally, however, if these avenues are followed, it is beneficial to have some members of the assessed organization also receive outside training in the criteria in order to home-grow internal expertise.

Not knowing how the Baldrige criteria tie into the organization's improvement and planning efforts can be another obstacle to overcome. Communications will be very important here. So too will knowledge and awareness in how other companies couple these facets into their overall operations. A good starting point is to study the world class companies and Baldrige Award winners.

Finally, and maybe most critical as an obstacle, is a lack of commitment on the part of the organization to use a Baldrige self-assessment as the basis for improvement. As the key to circumventing this barrier, management must stay committed, involved, and visible in the process. It is management's responsibility to ensure that the assessment results are used for improvement, and that the assessment is integrally linked to the business planning cycle.

9.10.5 A real-life example—Company *X*

Immediately after winning the MBNQA, one of the early winners, a large electronics Fortune 100 manufacturing corporation, Company *X*, made a commitment to reapply in its first year of next eligibility (5 years from the date of the award). Consequently, it initiated a process to be in a state of readiness.

To prepare for reapplication, this multidivision international corporation required that all its operating groups, worldwide, conduct annual internal self-assessments. To a large degree, Company *X* would follow the modified Deming PDCA cycles described earlier.

The Corporate Quality Office began by laying out and communicating the strategy for self-assessment and improvement. This strategy was presented both to the Executive Board and the Quality Council for review and buy-in. After strategy approval, an internal Corporate Baldrige Program Office was established to lay out the detailed plans for implementation and administer the program. A program manager at the vice-president level was assigned with a small staff to support that individual in managing and administering the overall process. A timeline was developed for the project (Fig. 9.5).

Next, the awareness and understanding process started in earnest. Senior executives and Quality Council members were given a briefing on the Baldrige criteria and framework. This briefing was followed by the selection of approximately 90 individuals from the various operat-

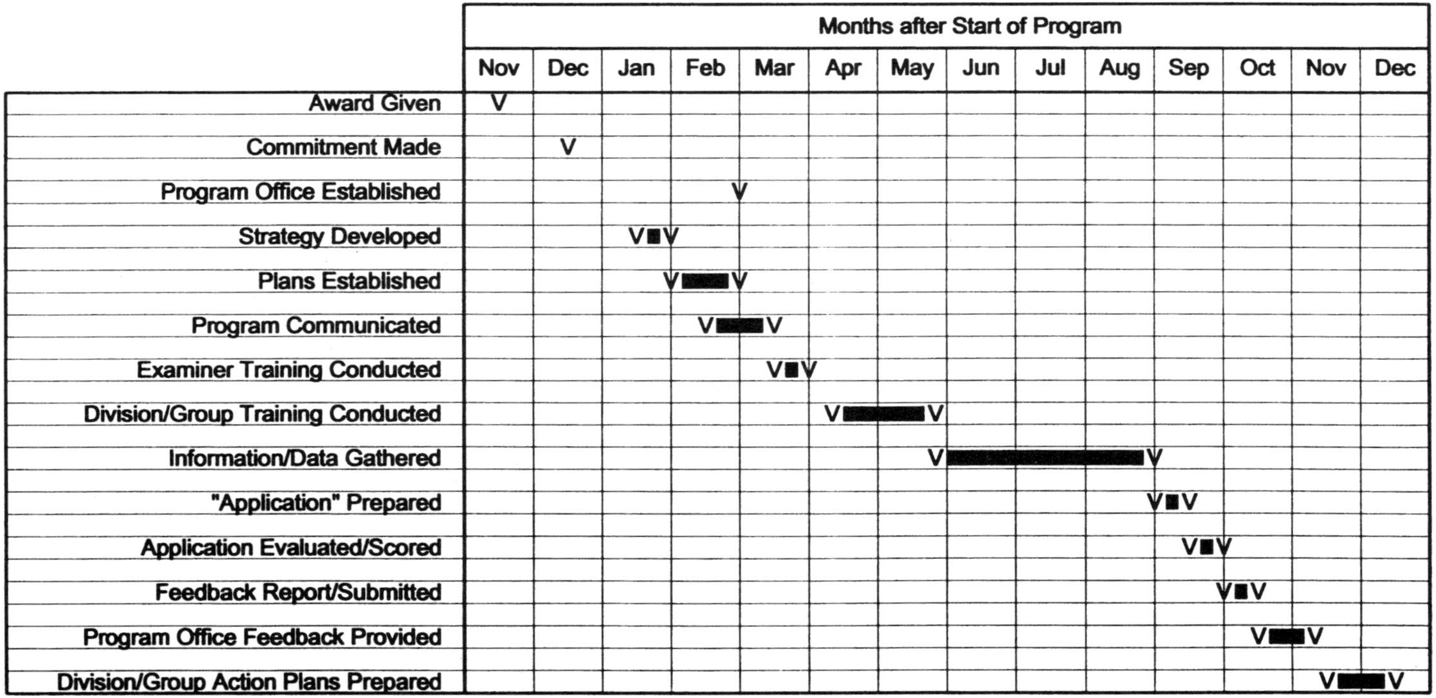

	Months after Start of Program													
	Nov	Dec	Jan	Feb	Mar	Apr	May	Jun	Jul	Aug	Sep	Oct	Nov	Dec
Award Given														
Commitment Made														
Program Office Established														
Strategy Developed														
Plans Established														
Program Communicated														
Examiner Training Conducted														
Division/Group Training Conducted														
Information/Data Gathered														
"Application" Prepared														
Application Evaluated/Scored														
Feedback Report/Submitted														
Program Office Feedback Provided														
Division/Group Action Plans Prepared														

Figure 9.5 Baldrige self-assessment timeline for Company *X*.

ing groups and divisions to receive in-depth training in the Baldrige criteria, framework, scoring, consensus process, and feedback report writing. Once trained, these individuals would assume the roles of internal examiners for the self-assessment process.

Two corporate employees, who also served as senior examiners on the MBNQA Board of Examiners, conducted the internal examiner training. The training followed the same format as that given to MBNQA examiners, lasting almost 3 days and using the same material and teaching approaches—prework case study, evening exercises, and role playing. It was very thorough indeed.

The internally trained examiners then became the trainers for the people within their operating groups and divisions. In essence, the examiner training had become a train-the-trainer course as well. At the operating group/division level, the examiners delivered executive level briefings and champion level training. In addition, they conducted Baldrige awareness sessions.

Now the fun really started! All operating groups/divisions of Company *X* were next required to perform a self-assessment of their organizations. At the conclusion of the self-assessment, the groups/divisions documented the results and submitted an "application," which adhered fully to the format and page limitations of the MBNQA application, to the internal Baldrige Program Office.

In order to execute the self-assessment, each operating group/division decided on the actual method of implementation. Establishing Baldrige category sponsors and champions was a popular approach. Following this approach, operating groups/divisions drew sponsors, or cosponsors if more convenient for load-sharing, from the senior executive ranks and assigned them to each of the seven Baldrige categories. Their job was to provide the commitment and reinforcement, through personal involvement and visibility, to sustain the project for their assigned categories. Similarly, a champion, or cochampions in some cases, was assigned to each of the Baldrige categories to provide the day-to-day coordination and management of the self-assessment—initially the information and data gathering effort and later the application preparation. In addition, the champions took on the role of operating group/division program managers to manage and report status on the overall effort for their organizations. The group/division level program manager also was the single point of contact between the operating group/division and the Baldrige Program Office.

The champions drove the information and data gathering for their assigned category. Interviews were conducted with individuals most likely to have information and data relevant to the associated

Baldrige items and areas to address. For *approach* and *deployment* items, walk-throughs were provided to gain a complete understanding of processes and how they worked. Every effort was made to verify that the activities were systematic. In addition, evidence of improvement cycles was sought. For *results* items, data were acquired in the form of trends, current levels, comparisons, and benchmarks. Only actual documented results were used.

Having gathered all the pertinent information and data, it was now time to write the application. The actual writing was done by the sponsors and champions for their assigned categories. The category page allocations were made in proportion to the scoring weights. For example, using the 1996 Baldrige criteria, scoring weights and total application page counts, Category 1.0, Leadership, would have a page allocation in the ratio of 90 points/1000 points times 70 pages, or a little over 6 pages. The program manager typically took on the job of preparing the "Business Overview" section, as well as coordinating the overall application.

The completed applications were submitted to the internal Baldrige Program Office. Teams of the internally trained examiners from the various operating groups/divisions were assigned on a confidential basis to evaluate and score each application, reach consensus, and resolve any site-visit issues. From the total group of examiners, a small select group of senior examiners was appointed. These senior examiners had the role of managing the overall application evaluation, as well as preparing the feedback report to the "applicant" (the operating group/division) based on inputs provided by the examiners at the category/item level. The feedback report also followed the format of the MBNQA feedback report, and included an overall summary, score, strengths, and areas for improvement. Since the evaluation was intended not to be prescriptive, no recommendations for improvement were provided in the feedback report.

The senior examiner for each evaluation submitted the feedback report to the Baldrige Program Office for review. After review, and compilation of relevant statistics (including relative scores), the Program Office sent the feedback report together with the scoring range of other operating groups and divisions, along to the applicant's general manager, with a copy to the group/division program manager. In addition, a summary of each group/division's scoring result, both overall and at the category level, was sent to the corporation's Executive Board for their information.

Each group/division was then required to prepare and submit an appropriate action plan of improvements, based on the findings contained in the feedback report, to the Baldrige Program Office.

Furthermore, status of actions was reviewed regularly not only by the group/division senior management, but also by the Quality Council. And then, in the spirit of the Deming PDCA cycle, the annual assessment and improvement started all over for Company *X*.

Because of the confidential nature of the MBNQA process, unless companies win the award, or voluntarily disclose the fact that they have applied for the award, no disclosure is made of their application by NIST. Therefore, in the case of Company *X*, it is unclear if it has again applied formally for the MBNQA. Nevertheless, the company is a winner by virtue of the fact that it is using a tried and proven set of criteria to maintain its world class quality status by applying a continuing learning and improvement process.

References

1. Extracted in part from Carrubba, Eugene R., and Mark E. Snyder, *You Deserve the Best—A Consumer's Guide to Product Quality and Total Customer Satisfaction,* Chap. 3, ASQC Quality Press, Milwaukee, 1993.
2. "The Malcolm Baldrige National Quality Award and ISO 9000 Registration," *ASTM Standardization News,* November 1993, pp. 42–53.
3. Extracted in part from "Who Was Malcolm Baldrige and Why Did They Name An Award After Him?—A Guide to the Malcolm Baldrige National Quality Award," U.S. Department of Commerce, Technology Administration, NIST, October 1995.
4. Paraphrased from *1996 MBNQA Award Criteria,* NIST.
5. Ibid.
6. *1996 MBNQA Award Criteria,* NIST, 1996.
7. Ibid.
8. Ibid.
9. GAO Report, "U.S. Companies Improve Performance through Quality Efforts," GAO/NSIAD-91-190, May 1991.
10. NIST *Update,* results of Baldrige Winners and Baldrige site-visited companies common stock comparison, February 1996.

Appendix

Roadmap for Quality in the Twenty-First Century

ROADMAP FOR QUALITY IN THE 21st CENTURY

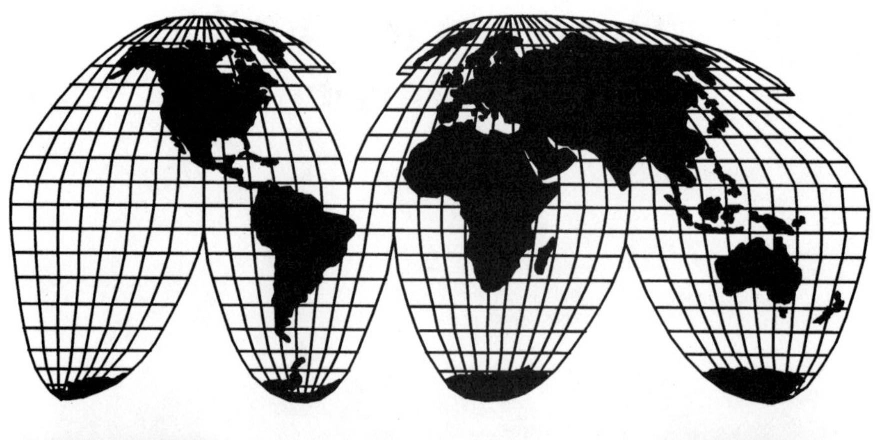

Prepared by

The Government & Industry Quality Liaison Panel

April 24, 1995

Table of Contents

TABLE OF CONTENTS
EXECUTIVE SUMMARY
INTRODUCTION
CHARTER
LEAD TEAM MEMBERSHIP

SECTION I STRATEGIC PLAN

BACKGROUND
MEMBERSHIP
VISION
STRATEGIC THRUSTS
GOAL 1. SINGLE QUALITY MANAGEMENT SYSTEM
Key Objectives
Background
Current Situation
Single Quality Management System Framework
Definitions
GOAL 2. ADVANCED QUALITY PRACTICES
Key Objectives
Background
Current Situation
GOAL 3. QUALITY MANAGEMENT SYSTEM IMPLEMENTATION AND OVERSIGHT
Key Objectives
Background
Present Situation

SECTION II TACTICAL ACTION PLAN

PURPOSE
OVERVIEW OF TASKS AND ACTIVITIES
Task 1 Develop Panel Charter, Principles, and MOU
Task 2 Define & Implement Pilot Programs
Task 3 Identify Policy/FAR Changes
Task 4 Develop DID
Task 5 Develop Training Elements
Task 6 Develop Evaluation Criteria
Task 7 Develop Evaluator Guidelines
Task 8 Develop Oversight Guidelines for Basic Quality System
Task 9 Develop Procurement Handbook/Guidelines
Task 10 Identify Advanced Quality Practices
Task 11 Develop Communication Clearinghouse
Task 12 Draft MOA to Accept Mutual Recognition
Task 13 Develop Quality Audit System Approach and Data Base

APPENDICES

ROSTER, PANEL MEMBERS/PARTICIPANTS
MEMORANDUM OF AGREEMENT
MEMORANDUM OF UNDERSTANDING

Introduction

Recent fundamental changes in the Government approach to systems development and acquisition have provided the opportunity to rethink the policies, practices and procedures which have for a long time been the cornerstones of Government acquisition. For these changes to be successful, Government and Industry organizations must work together to understand how these changes will impact each other and to develop mutually supportive practices to resolve the underlying issues. In an attempt to reach this accord, the quality community has formed the Government/Industry Quality Liaison Panel to identify and coordinate issues of mutual concern and to work together to develop the most effective solutions to problems raised.

Past practices with extensive oversight that dictated unique quality systems and detailed quality management practices have required contractors to organize and manage their facilities on a contract by contract basis. The inefficiencies of this approach have widened the rift between commercial and Government contractors and are not in keeping with acquisition reform efforts.

The Panel, as its first goal, will establish a single quality system within a facility, when desired by the contractor, capable of meeting each customer's requirements. Industry has made significant advances in the development and application of Advanced Quality Management tools, practices and processes. In order to take best advantage of these advanced concepts in the streamlined acquisition environment, source selections must solicit, judge and reward the use of advanced quality practices which have proven effective.

The Panel has, as its second goal, that the Government and Industry recognize, share and use advanced quality concepts in the requirements definition, design, manufacture and acceptance of products. Government approaches to motivate quality improvement in the past have often worked in opposition to efforts to achieve a single quality system in a contractor facility. Government activities need a consistent approach to motivate improvements in system performance and quality management. The diverse nature of the oversight community interpretation of quality system requirements has lead to confusion in Government quality expectations. In order to ensure Government activities synchronize their approach to requirements, procurement, and oversight activities, close coordination is required.

The panel has, as its third goal, the effective implementation of criteria for a baseline quality system and appropriate oversight methods. This effort will culminate in benefits to both Government and Industry. It will contribute to the national competitiveness and improve the quality and value of products and services. Although best efforts were made to identify the key objectives and strategies to facilitate and improve approaches to quality, this plan is but a beginning. It must be continually evaluated and up dated to reflect the changing realities and continual improvements in the field of quality.

Executive Summary

This plan describes a vision for quality in the 21st Century and outlines a strategy to guide Government and Industry efforts to achieve this vision. It encompasses both strategic and tactical planning as well as more detailed tasks and milestones. Leaders from participating Government agencies and Industry associations are working together as a Government/Industry Quality Liaison Panel to construct the Quality Roadmap and to champion its implementation. Government and Industry have welcomed this opportunity to advance quality practices and remove non-value added costs hidden in restrictive requirements and redundant oversight and inspection processes.

This plan identifies actions to improve the quality and value of products and services provided to the Government (better, faster, cheaper). The plan provides a framework which embraces the adoption of commercial and best industrial practices and will facilitate movement toward dual use facilities and improved international competitiveness. Its focus is to move away from the rigid quality practices of the past to a more open, flexible and effective approach to quality.

The National Performance Review is leading the transition to performance based management and recommends agencies avoid Government-unique requirements and rely more heavily on the commercial marketplace. Other key events that lead to the development of this plan were:

(1) Government acquisition reform initiatives to improve the effectiveness of acquisition, and

(2) Dual Use initiatives to assist contractors to serve both Government and commercial customers and become more internationally competitive.

CHARTER

GOVERNMENT INDUSTRY QUALITY LIAISON PANEL (G&IQLP)

I. MISSION

To enable consistent satisfaction of customer expectations through a government and industry association partnership using world-class quality processes and practices to enhance international competitiveness.

II. SCOPE

The Government Industry Quality Liaison Panel will encourage the participation of interested federal agencies and industry associations in the development and deployment of uniform quality management systems and advanced quality concepts.

III. OBJECTIVES

1) To enable use of a single quality management system within a contractor's facility that satisfies each customer's requirements;

2) To mutually recognize, share, and utilize advanced quality concepts in the requirements definition, design, manufacture and acceptance of products;

3) To establish and implement effective and efficient oversight methods.

IV. ORGANIZATION

Membership on the G&IQLP consists of representatives from federal agencies and industry associations, as identified by the Lead Team.

The G&IQLP operates under a Lead Team comprised of the designated representative of each participating federal agency and industry association. The Lead Team coordinates and plans all the activities of the G&IQLP. The Lead Team is co-chaired by one representative from the government and one representative from an industry association. The co-chairpersons are elected by the Lead Team and serve for a one year term.

The Lead Team assigns actions and establishes Task Teams, as appropriate, to achieve the objectives of the G&IQLP. Progress reports of assigned actions will be presented each G&IQLP meeting.

V. OPERATIONS

A. The Panel reflects the collective effort of all participating members. Member organization participation requires active involvement and contributions to the collaborative nature of this venture. The Panel meets a minimum of two times per year, with the Lead Team and Task Teams meeting as needed.

B. Panel member responsibilities include:

1. Participating in meetings and constructively contributing to the activities of the Panel.

2. Promoting the mission of the Panel by participating in relevant activities and promoting the objectives within their respective organizations;
3. Leading and/or serving on Panel task teams; and
4. Participating in the development and deployment of educational materials that can be used by participating organizations for training activities.

C. Lead Team responsibilities include:

1. Developing and implementing a plan for facilitating the activities of the Panel;
2. Leading, coordinating and integrating the work of Task Teams; and
3. Reporting on the status of the Panel and its projects.

D. The Co-chairpersons are responsible for:

1. Administering the operations of the Panel and the Leadership Team, including the distribution of meeting agendas, minutes, and other reports;
2. Scheduling meetings and meeting facilities; and
3. Managing interfaces with other entities.
4. Clearing information released for public distribution.

VI. INTERACTION WITH OTHER ENTITIES

Interaction with other Entities will be determined by the Lead Team.

VII. GUIDING PRINCIPLES

In seeking to fulfill our mission, the Government/Industry Quality Liaison Panel is guided by the following principles:

1. The customer judges and drives quality.
2. Senior leaders are committed and involved.
3. Improvement and learning are continuous.
4. Employee participation and development are essential.
5. Fast response is required.
6. Quality is prevention-oriented and designed-in.
7. Our view of the future is long-range and global.
8. Management is by fact and data.
9. Internal and external partnerships are essential.
10. Corporate responsibility and citizenship are the basis of action.
11. The performance system is results-oriented and value added.
12. The position of U.S. industry in the international marketplace will be enhanced.
13. Consistency is achieved with commercial and international practices.

A principle is a fixed or predetermined rule guiding behavior or policy. These principles are based on the core values (principles, standards, or qualities considered inherently worthwhile or desirable) of the 1995 Malcolm Baldrige National Quality Award criteria.

Lead Team Membership

Frank Doherty - Co-chair	**Office of the Secretary of Defense (OSD)**
Carl Schneider - Co-chair	**National Aeronautics and Space Administration**
Primus Ridgeway - Co-chair	**Electronics Industries Association**
James Parrish	**Aerospace Industries Association**
Bill Thompson	**National Security Industrial Association**
Ryan Bradley	**U.S. Air Force**
Steve French	**U.S. Army**
CAPT Murat Shekem	**U.S. Navy**
George Georgeadis	**U.S. Marine Corps**
Joel Odum	**General Services Administration**
Carol Driscoll	**Federal Aviation Agency**
CAPT Dan Kalletta	**U.S. Coast Guard**
Mike Pursley	**Maritime Administration (MARAD)**
Richard Zell	**Defense Logistics Agency/DCMC**
Maureen Breitenberg	**National Institute of Standards and Technology**
Kenneth LaSala	**National Oceanic and Atmospheric Administration**
Peter Angiola - Executive Secretary	**Office of the Secretary of Defense (OSD)**
Brian Mansir - Facilitator	**Logistics Management Institute**

Section I
Strategic Plan

Background

Two key events lead to the development of this plan: (1) Government acquisition reform initiatives to improve the effectiveness of acquisition, and (2) the drive to assist defense contractors to become capable of serving both defense as well as commercial customers and become more internationally competitive.

The first step in the formulation of this plan was the drive by Government departments and agencies to adopt the ISO-9000 series quality standards. In late 1993, representatives from DoD, NASA, GSA, and the Coast Guard joined together to develop guidelines for using the ISO standards as alternatives to the traditional military/agency standards for quality programs. Representatives of the National Security Industrial Association, and Aerospace Industrial Association were requested to form a "Quick Action Group" to provide initial industry comment on implementation guidance developed by the Government Department/Agency Team.

At the same time, industry associations were seeking improved means to communicate and cooperate with Government agencies on common quality issues. The juxtaposition of these events lead to the continuation of the Government Department and Agency efforts to develop a strategic plan for quality to support acquisition reform initiatives and to the reformation of the Industry Fast Reaction Group as a liaison panel to provide industry comment relative to the plan for responding to Government acquisition reform initiatives. As this panel formed, new initiatives emerged, such as the Secretary of Defense memorandum of June 29, 1994, "*Specifications and Standards - A New Way of Doing Business*," which was considered in the development of this plan.

Membership

The panel consists of representatives from government and industry. The government representatives are from the Department of Defense (Office of the Secretary of Defense, Army, Navy, Air Force, Defense Logistics Agency, and Defense Contract Management Command), National Aeronautics and Space Administration, Department of Transportation (Coast Guard, Federal Aviation Agency, and Maritime Administration), the General Services Administration, the Department of Commerce (National Institute of Standards and Technology, and National Oceanic and Atmospheric Administration). Industry representatives, while from individual companies, participate as representatives of the Aerospace Industries Association, Electronics Industries Association, and the National Security Industrial Association.

Vision

Consistent satisfaction of customer expectations through a Government and Industry partnership using world-class quality processes and practices.

Strategic Thrusts

At present there are three strategic thrusts or goals in this plan:

1. A single quality management system within a contractor's facility capable of meeting each customer's requirements.
2. Government and Industry recognize, share, and use advanced quality concepts in the requirements definition, design, manufacture, and acceptance of products.
3. Effective implementation of a baseline quality management system and appropriate oversight methods.

Each goal has one or more objective and associated strategies. Respective tactical plans will contain specific actions, milestones, and responsible individual(s).

This plan is to be a living document and will be revised as events dictate the need to change.

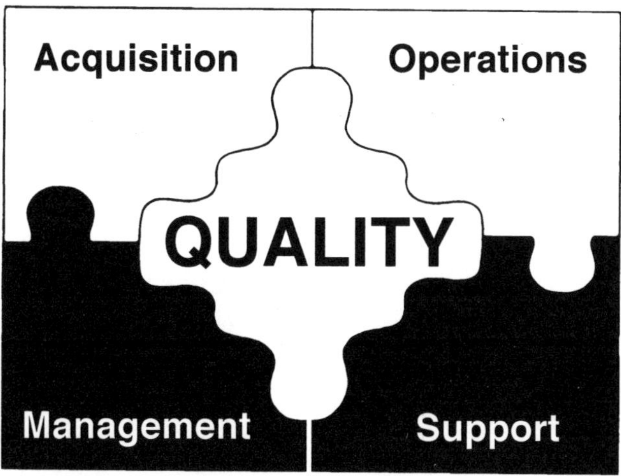

This plan recognizes the essential integrated nature of a quality management system within an enterprise and therefore touches many different aspects of business from acquisition and operations through management and support. It addresses the need for each organization to design and implement a quality management system that fits the specific requirements of the total enterprise.

GOAL 1. Single Quality Management System	***A single quality management system within a contractor's facility capable of meeting each customer's requirements.***
Key Objectives	1.1 Establish a way of doing business that enables contractors to use their basic quality management system whenever it satisfies acquisition requirements. 1.2 Create a culture that recognizes that the purpose of a quality management system is to assure efficient and effective design and production processes and the delivery of good products and services; not to constrain suppliers to a set of inflexible requirements which are monitored for sole compliance.
Background	The intent is to establish effective processes and economical methodologies that allow contractors to use their normal quality management systems whenever they meet acquisition needs, eliminating the burden of multiple and often redundant quality management system requirements. Many contractor facilities have multiple quality management systems to conform to the requirements of different customers. This situation encumbers the contractors with added cost and resources to meet all the government needs and therefore, add cost to the respective procurements. To meet future needs, the federal government must increase access to commercial technology and must facilitate the adoption by its suppliers of world-class business processes. In addition, the integration of commercial and military development and manufacturing facilitates, and the development of dual use processes and products, contribute to an expanded industrial base that is capable of meeting government needs at lower costs. Efforts to merge the federal and private sector industrial base require increased use of commercial standards and recognition of contractor quality management systems.
Current Situation	Procurement packages released by both government and contractor buying agencies/organizations include a requirement for the contractor/supplier to develop, implement, and maintain a quality management system that is compliant with stated guidance (FAA-STD-013/016, Mil-Q-9858, Mil-I-45208, NASA-HB-5300, FAR part 46, DFARS part 246, contractor equivalent documents, etc. ad infinitum). Often, redundant/overlapping specifications are listed in an apparent attempt to "assure that all bases are covered." Contractor/supplier responses are reviewed and source selection audits are performed by procuring activities based on parochial interpretations of "how those system requirements ought to be met, " with apparent disregard for consistencies between service/agencies/offices, and with little or no participation of local or resident CAS agencies and/or other buyers already doing business in the same contractor/supplier facility. In reality, almost any given contractor/supplier has only one basic comprehensive quality management system, designed to be appropriate for the products he produces and to satisfy the customers to which he delivers.

Current Situation (cont.)

This basic system may be amended as directed regarding the extent and depth of application of quality management elements to meet individually unique customer requirements, but at root remains one system with quality plans and procedures needed to meet contract requirements.

This goal recognizes that there are commercial non-complex and non-critical buys where inspection by the contractor or the standard inspection clause are appropriate.

Single Quality Management System Framework

This goal does not advocate a rigid and uniform quality management system across Government and in every contractor's facility. This plan envisions a multi-tier quality management framework. The foundation of the framework is a basic quality management system that is defined by a basic quality criteria, such as the ISO-9000 elements, and is recognized and agreed to Government-wide and across industry. This basic system would be certified or verified once annually by a Government approved auditing team. This verification of a company's basic quality management system would be recognized and accepted by all Government agencies.

Both Government and industry could augment the basic quality management system. Government could add to the basic criteria defined by the ISO-9000 elements as deemed necessary for a particular procurement. Industry, on a company by company basis, could augment their basic quality management system by the use of advanced quality concepts. Where these practices and concepts add value to the product, they would be recognized by the Government and given appropriate weight in the selection of suppliers. Certifying, verifying, or monitoring advanced quality concepts could be accomplished by a 1st, 2nd, or 3rd party agent as mutually agreed to by the Government agency and the contractor.

Specific and unique quality management system requirements could be implemented in a contractors facility where special product or service requirements exist. These unique practices could be controlled by either the contractor or the customer, as appropriate.

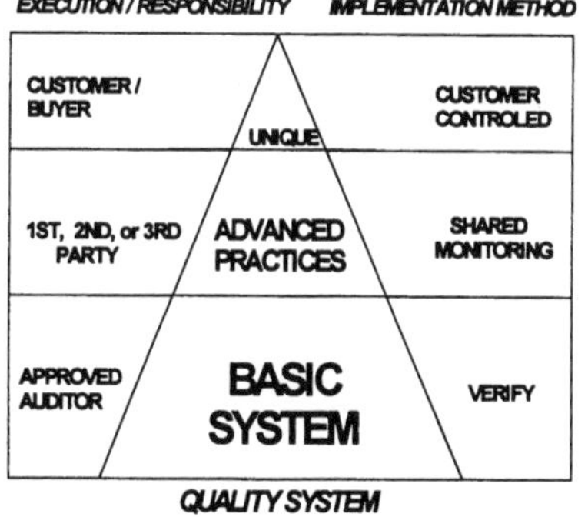

Definitions

Basic Quality management system A defined, documented, and disciplined set of practices that focus on assuring that the deliverable product (or service) conforms to the performance and configuration criteria included in the agreement between buyer and seller. The basic quality management system must provide for the appropriate controls of part and product characteristics and attributes from product design through delivery and include inspection and test criteria/methodology/data that is used to verify/validate conformance. The key consideration is the delivery of products (or services) that fall within allowable tolerances.

Advanced Quality Management System: A management commitment to excellence that is manifested in a structured, long-term approach to the continuous improvement of all processes/products. The focus is on the process(es) that produce a product or service rather than just the product itself. The employees that actually perform the process(es) are trained to use quality tools to analyze their process, streamline it, stabilize it, measure it, and then, on a continuous cycle, to reduce the variability of its characteristics or attributes toward optimizing its output. Advanced quality practices/techniques are complementary to basic quality considerations but do not supplant them.

Single Quality Management System Approach: An acquisition approach that allows a contractor to propose and demonstrate that his existing plant-wide quality management system adequately covers all the elements necessary to satisfy the program acquisition objectives as opposed to requiring him to implement a system "in compliance with a stipulated quality management system standard/specification."

Single Quality Management System: A quality system defined by the contractor for a specific facility; it is based on the elements contained in the ISO standards and is augmented by Advanced Quality Practices (engineering and manufacturing tools & techniques), as deemed appropriate for the product.

GOAL 2.
Advanced Quality Practices

To have Government and industry recognize, share, and use advanced quality concepts* in their requirements definition, design, manufacture and acceptance of products.

Key Objectives

2.1 Provide a set of defined Government expectations for value-adding advanced quality concepts.

2.2 Define techniques to give source selection credit to contractors who demonstrate value-added through the use of advanced quality concepts.

2.3 Provide recognized alternatives to inspection-based acceptance of product based on demonstrated process controls and capabilities.

2.4 Encourage continuous quality improvement through the identification, promotion, education in, and dissemination of, advanced quality concepts.

Background

Advanced quality concepts are the cumulative result of the application of more advanced and sophisticated practices in the areas of design, manufacturing, assessment, corrective action, and continuous improvement. Examples of tools and techniques utilized to implement these concepts include:

- requirements definition tools
- designing for robustness
- controlling key characteristics
- reducing variability

In order to improve competitiveness in the marketplace, both domestically and worldwide, many U.S. companies have been applying these advanced tools and techniques to a greater extent. The use of the tools and techniques such as statistical process control (SPC) and the application of other new and innovative practices has resulted in higher expectations and a "raising of the basic quality standard" by customers.

* *Advanced quality concepts are defined as those practices that are above and beyond the elements defined by the single baseline quality management system envisioned in Goal 1. Advanced quality concepts should not become contractual "how to" requirements -- rather they should be understood as potential ways for contractors to respond to stated contractual requirements.*

Current Situation

Requirements, criteria and metrics for advanced quality concepts (AQC) are frequently vague, poorly defined, or lacking altogether. There are no common benchmarks or baselines to determine if the specific application of an AQC is "value-added." When combined with lack of understanding of many AQC concepts and principles, the situation often leads to inconsistent or inappropriate application of these practices.

The current source selection system is geared toward a "minimum acceptable" quality program rather than encouraging the use of AQC to maximize value. In addition, there are no consistent means of evaluating or giving credit for the use of AQC through methods such as cost benefit analysis. With respect to product acceptance, there are no common evaluation methods to replace inspection based acceptance of products and services with acceptance based on objective evidence of process controls. While significant progress has been made in the application of advanced quality concepts, there is still a critical need for a greater understanding and education in the awareness and proper use of these concepts.

While the National Quality Award (Baldrige Award) has expressed a rediscovery of our own national interest and pride in quality, it has had limited impact on the Government in general. Few Government contractors have won the award. Even fewer Government supplier can show the results in quality, market growth or profitability to match the collective record of Baldrige winners.

Examples of advanced quality concepts are available from Ford, Boeing, and the Baldrige winners, as well as from a variety of small aggressive and innovative pockets in Government. They have not yet been marshaled into a cohesive Government wide practice of customer expectation and credit to suppliers with value-adding practices.

OAL 3. uality anagement ystem nplementation nd Oversight

Establish and implement effective and efficient oversight methods.

ey Objectives

3.1 Develop criteria for evaluation of contractor evidence of an acceptable quality management system.

3.2 Implement a mutually acceptable single evaluation process to validate the quality management system.

3.3 Redefine oversight activities.

3.4 Promote effective and efficient innovation in quality management systems.

ackground

When consensus is reached on the necessary elements of a single quality management system, methods for evaluating the effective implementation of this system and adjusting the Government oversight accordingly must be defined and agreed to by industry and Government.

In an environment where the Government procurement budgets are shrinking and are expected to continue to shrink, industry must become more cost effective to survive in this highly competitive situation. To replace lost Government procurements, most industries are seeking to expand their international presence; but, again a highly competitive situation exists which demands cost effectiveness. There is little consistency among Government agencies or within international standards in terms of quality management system requirements and evaluations thus making cost effective systems difficult to implement.

resent ituation

Current RFPs and similar procurement packages require compliance with specific quality management system documents, usually documents peculiar to the individual Government agency. To be responsive, contractors modify their systems to meet the requirements of each solicitation thus producing multiple quality management systems. Although the basic elements of the required systems are usually common, there is no recognition that a contractors system is acceptable unless it meets all the requirements of the specified standard.

Oversight of these multiple systems normally includes audits/evaluations by multiple Government agencies with little or no communication among the agencies. Additionally the oversight required by Program Offices is inconsistent, is not based on risk or performance, and does not consider results of audits/evaluations performed outside the individual Program Office.

Section II
Tactical Action
Plan

PURPOSE This tactical plan implements the Government/Industry Quality Liaison Panel Strategic Plan.

OVERVIEW OF TASKS AND ACTIVITIES

TASK DESCRIPTION	TASK LEADER	TEAM MEMBERS	MILESTONES	COMMENTS
1. Develop Panel Charter, Principles, and MOU.	Carl Schneider	Frank Doherty, Carol Driscoll, Ken LaSala,	Draft Charter - Dec. 94 Get Charter signed - May 95	Task completed. - March 95.
2. Identify Pilot Programs	Frank Doherty,	Larry Johnston, Primus Ridgeway, Rich Zell		
3. Identify Policy/ FAR Changes.	Ryan Bradley	Peter Angiola, Carol Driscoll, Frances Sullivan John Maristch Mike Pursley	DFARS Changes - Dec. 95 Complete FAR Case - July 95	DFARS Case 246 approved by DSIC. Awaiting DAR Council action. Need input on changes needed to FAR Part 46.
4. Develop DID.	Peter Angiola	Geza Pap	Obtain DID Approval - Jan. 95	Task completed. DID 81449 approved - Jan. 95.
5. Develop Training Elements.	Jo McLaughlin	Primus Ridgeway, Stephanie Strohbeck, Carol Driscoll, Roger Fisher, Michael Litchfield Leonard Courschene	Plan of Action for Training - May 95	

TASK DESCRIPTION	TASK LEADER	TEAM MEMBERS	MILESTONES	COMMENTS
6. Develop Evaluation Criteria.	Lynn Harris	E. McKenna, Michael Parke, Al Parker, Dale Cole, Wally Luther Bob Lackland,	ID Basic Elements - Jan. 95 Complete Criteria - August 95	
7. Develop Evaluator Guidelines.	Charles Cockrell	Michael Parke	Complete Guidelines - Mar 95	
8. Develop Oversight Guidelines for Basic Quality Management System.	Bill Thompson	Richard Zell, Pete Angiola, John Maristch, Dale Cole, Lindo Bradley, Eugene Moore Mike Pursley Dan Kalletta Bob Lackland,	Complete Oversight Guidance - Aug. 95	
9. Develop Procurement Package Criteria.	Michael Parke	Primus Ridgeway, Frances Sullivan, Bob Tourville, Jim Camden Richard Wojciechowski Greg Colbert	Finalize Criteria and Contract Language - Sept. 95	
10.. Identify Advanced Quality Practices Integrate Existing Voluntary Partnership Programs.	Geza Pap,	Bob Tourville, Jo McLaughlin, Dick Tracey, Steve Newman, Howard Butts, Steve French, Charlie Mercer, James Parrish, Dan Barton	ID System Practices - Feb. 95 ID Sectors - May 95 ID Scoring Criteria - Jul. 95 ID Programs, Metrics and Incentives - Sept. 95	

TASK DESCRIPTION	TASK LEADER	TEAM MEMBERS	MILESTONES	COMMENTS
11. Develop Communication Clearinghouse.	Phil Smiley	Jo McLaughlin, Ken LaSala, Maureen Breitenberg Charles Cockrell	Establish Communication Clearinghouse - Oct. 95	
12. Draft MOA to Accept Mutual Recognition.	Ryan Bradley	Frank Doherty, Larry Johnston, Chuck Bailey	Draft MOU - Jun. 95 MOU Ready - Oct. 95	
13. Develop Quality Audit System Approach and Data Base.	Carl Schneider	Earl Major		

Task 1 Develop Panel Charter, Principles, and MOU.

Problem Description	There is a need to establish guiding principles and to obtain formal recognition of the Government/Industry Panel in order to proceed with implementation of proposed actions.
Task Overview	This task will develop a Charter for the Panel: mission, guiding principles, objectives, organization, operation, etc. Then, obtain signature approval from all participating agencies and organizations.
Objective	Obtain agreement and commitment from all participating agencies and organizations to support the efforts of the panel.
Approach	• Determine charter content requirements. • Write draft charter. • Gain panel consensus on charter. • Obtain approval signatures from government and industry members.
Actions and Milestones	• Draft Charter, Principles & MOU .... • Final Charter and Principles .... • Identify required signatories .... • Develop presentation .... • Conduct Presentations .... • Obtain Approval Signatures .... • Signed MOU .... Completed April 24, 95
Products	• Panel Charter. • Memorandum of Understanding
Issues	• Obtaining high level commitment.
Team/Resources	**Carl Schneider, Frank Doherty, Ken LaSala, Carol Driscoll**

Task 2 Define & Implement Pilot Programs.

Problem Description	Implementation of new practices, such as a Single Quality System, recognition for Advanced Quality Concepts, and Reduced Government Oversight, requires a test phase to identify and correct unforeseen problems before widespread implementation.
Task Overview	To identify willing candidates for a pilot program, ensure the cooperation of Government program offices, establish a mutual criteria for success, and agree on a non-obtrusive process to monitor the outcome.
Objective	To establish models which, if successful, will lead to expanded implementation. To improve processes by adopting leading edge quality approaches. To reduce unnecessary audits and oversight To establish success criteria, and determine how to proceed with widespread implementation upon favorable outcome.
Approach	A Government/Industry Panel, comprised of members from all participating agencies, will review the proposed list and select facilities considered to be the best candidates. DLA/DCMC and program managers to agree on criteria for oversight, and for monitoring the pilot program.
Actions and Milestones	• Establish steering group .... May 95 • Scope pilot program .... May 95 • Identify candidate facilities .... July 95 • Select the candidates for initial pilot program .... Aug 95 • Define process & success criteria .... Aug 95 • Establish mutual agreements among contractors, Government program offices, and DCMC .... Oct 95 Implement pilot program .... Nov 95
Products	• Assessment of lessons learned • Recommendations relative to the utility of expanded implementation • Proposed policy and procedures to support expanded implementation • Changes/guidance relative to oversight at facilities utilizing single quality processes.
Issues	Pilot program, metrics/success criteria.
Team/Resources	**Frank Doherty, Larry Johnston, Primus Ridgeway, Rich Zell.**

Task 3 Identify Policy/FAR Changes.

Problem Description	The current language in the FAR is out of date due to recent policy changes. In addition, further changes should be proposed in order to enhance implementation of the new quality concepts.
Task Overview	This task will develop proposals to change FAR Parts 46 and 52, as well as DFARS Part 246. The purpose of these changes would be to reflect the new policies on the use of commercial standards and the concept of a single quality management system.
Objective	Identify all FAR/DAR and/or acquisition policy changes that would be required to allow contractor/subcontractor to declare a single system based on defined minimum quality management system requirements.
Approach	• Identify FAR changes needed to enable single quality management system at a contractor's facility. • Write FAR Case that implements the needed changes. • Obtain approval from the appropriate authorities of the FAR Case.
Actions and Milestones	• Identify and draft proposed FAR Case. (Case #: 90-517) • Obtain approval of FAR Case. • Identify and draft proposed changes to DODI 5000.2 • Identify changes to other policy documents.
Products	• Proposed FAR Case. • Proposed policy changes.
Issues	• Changes in Quality policy. • DoD acquisition reform (Application of Specifications & Standards)
Team/Resources	**Ryan Bradley, Peter Angiola, Carol Driscoll, Frances Sullivan, Mike Pursley, John Maristch.**

Task 4 Develop DID.

Problem Description	A data Item Description is necessary in the procurement process to obtain a quality plan that defines the contractor's quality management system.
Task Overview	This task will develop a document identifying the key elements of a quality plan; it's purpose is to guide the contractor in defining it's quality management system. This document will be listed in the DoD 5010.12-L, Acquisition Management System and Data Requirements Control List.
Objective	Identify all DID requirements that would allow contractor / subcontractor to declare a single system based on defined minimum quality management system requirements.
Approach	• Identify DID requirements needed to enable single quality management system at a contractor's facility. • Write DID that implements the needed requirements. • Obtain approval of the DID from the appropriate authorities.
Actions and Milestones	• Identify DID requirements. • Write DID. • Obtain approval of DID. (Approved Jan 95) • Submit revisions to reflect changes to ANSI/ASQC Q9000 June 95
Products	• Proposed DID and revisions.
Issues	• Compatibility with ISO documents. • Consistency with Government procurement regulations.
Team/Resources	**Peter Angiola, Geza Pap**

Task 5 Develop Training Elements.

Problem Description	While a myriad of training opportunities exist in both government and industry sectors in each of the three goal areas, there is no common language or consistent application of training opportunities. This task will seek to develop templates from which training can be framed, based on individual organizational needs, using consistent understanding of a specifically identified subject matter.
Task Overview	This task will explore, develop and evaluate the possibilities for shared training and development opportunities in government and industry sectors.
Objective	Develop and implement a common education and training program that fosters the awareness and develops skills needed to implement and achieve the objective. Establish cooperative Government / Industry training efforts.
Approach	This framework will encompass a training initiative on ISO-9000, currently under development by DCMC and AIA under the leadership of Stephanie Strohbeck (DCMC) and Roger Fisher (AIA).The approach to developing training for advanced quality concepts will be coordinated by Jo McLaughlin, based on needs identified by the Advanced Quality Concepts Working Group. A training concept for the implementation of the single quality management system and advanced quality practices will evolve based on demonstrated needs.
Actions and Milestones	ISO 9000 training team meeting Jan 24-26/95 and March 1-2/95.
Products	Government/Industry training template for quality system.
Issues	Identify training needs and opportunities in the various goal areas.
Team/Resources	**Jo McLaughlin, Primus Ridgeway, Stephanie Strohbeck, Roger Fisher, Dr. Leonard Courschene, Carol Driscoll, Michael Litchfield, Mike Pursley.**

Task 6 Develop Evaluation Criteria.

Problem Description	There is no consensus across the federal government and industry as to the criteria for evaluating the acceptability (or effective implementation) of a quality system.
Task Overview	This task will develop criteria for use by government (and industry) to evaluate contractor's basic quality management systems. The evaluation criteria will be based on ISO-9000 elements and should be recognized and agreed to Government-wide and across industry. Required actions include: (a) Review criteria used in current commercial and government quality audits. (b) Draft and coordinate proposed evaluation criteria.
Objective	Produce a set of evaluation criteria that can be used to assess a supplier's basic quality management system.
Approach	• Develop a maturity matrix to assess implementation of the basic quality system. The matrix, based on the elements of ISO 9001, can be used internally for self-assessments and by second parties during both the supplier selection process and post-award supplier evaluations. • Review available checklists from government and industry sources and determine the need for detailed evaluation criteria
Actions and Milestones	• First Draft Maturity Matrix .... Completed • Draft to Lead Team .... Completed • Review Available checklists .... Completed • Final Criteria to Lead Team .... Completed Apr 95
Products	• Maturity Matrix.
Issues	• Determine how detailed criteria should be. • Multiple interpretations of the same requirements.
Team/Resources	**Lynn Harris, Eugene McKenna, Michael Parke, Al Parker, Dale Cole, Wally Luther, Bob Lackland.**

Task 7 Develop Evaluator Guidelines.

Problem Description

Existing criteria leading to certification of auditors seem to be incomplete with respect to gaining acceptance across government agencies. In addition to acceptance of auditors themselves, there is a concern that to gain government wide acceptance of audits, some knowledge of the results of individuals audits will be required.

Task Overview

This task supports the goal to adopt a common standard for quality systems. The first concern in developing evaluator guidelines for personnel who will perform assessments under this common standard is to consider existing criteria for auditors. If we intend to apply a common quality system to suppliers then it would make sense to also use a common system for acceptance of quality auditors. The two most widely used and accepted existing criteria for auditors is that used by the Registrar Accreditation Board and that used for the Board of Examiners for the Malcolm Baldrige National Quality Award

Objective

To examine existing requirements for audit personnel to determine if common criteria exists between various independent bodies such as the Registrar Accreditation Board (RAB) or the Board of Examiners for the Malcolm Baldrige National Quality Award (MBNQA).

To determine if any common criteria that is found, is acceptable for application to evaluators which would be recognized by all government agencies

Approach

- Develop evaluator guidelines.
- Obtain Government consensus criteria for auditors based on PARs, IG audits, ACSEPs, ISOs, AIA audits, etc.
- Government/Industry Quality Liaison Panel consolidate criteria.
  - a. Auditors must be viewed as being objective, capable, and consistent.
  - b. Auditors must be certified.
    - Written examination
    - Experience
    - Demonstration of capability
  - c. Auditors must receive periodic reassessment of capability.
- Obtain Government commitment to criteria.
- May include designation of one or more Government organizations to perform all evaluations and certify all auditors.
- Achieve international recognition of validation process.

Actions and Milestones	TBD
Products	Recommendations to establish an oversight body Establish a criteria for a Government Systems Auditor. Recommendations relating to standards for certification.
Issues	• Synergy with other task groups.
Team/Resources	**Charles Cockrell, Michael Parke.**

Task 8 Develop Oversight Guidelines for Basic Quality System.

Problem Description

Industry is currently faced with redundant oversight, e.g. program office, contract technical representatives, DCAA, DCMC, NASA Centers, GSA itinerants, FAA itinerants, etc. Multiple oversight activities cause unnecessary disruptions at contractors facilities, and are a cost driver.

Task Overview

This task will determine how the criteria developed in Task #6 (Develop Evaluation Criteria) will be utilized in the source selection process (from pre-award audits on) and determine how to best reduce oversight in areas which meet the criteria. Oversight may be adjusted based on government audits, contractor self-audits, or audits conducted by other contractors.

Objective

Determine types of oversight which we can impact. Define proper level of oversight including mutual recognition of review systems by government agencies.

Approach

- Generate policies for emphasizing and promulgating alternative acceptance methodologies.
- Develop and maintain a list of recognized alternative techniques.
- Evaluate criteria used in current evaluations such as AIA contractor audits, DCMC PARs, FAA ACSEPs, Baldrige, ISOs, etc. Sort criteria into three major areas:
  a. Adequacy of documentation to describe key processes.
  b. Employee understanding of processes.
     (1) Training
     (2) Demonstration
  c. Effectiveness of implementation.
     (1) Demonstration of system effectiveness - do the procedures, tools, and people produce acceptable products?
     (2) Assurance of product conformance - is there a system in place to assure product conforms to requirements and does data substantiate that?
     (3) Provide flexibility of processes - can system accommodate variances such as DoD-HDBK-9000?
- Utilize results of past performance.
  a. Has contractor recently passed evaluation for a Government program?
  b. Does a Government organization such as DCMC have data which demonstrates contractor capability?

Actions and Milestones	• Identify types of oversight including criteria. .... 30 Apr. • Review ongoing related activities. .... 30 Aug. • Identify critical metrics for guidelines. .... 30 Apr. • Develop guidelines for oversight. .... 30 Sept. • Provide product to leadership group to obtain concurrence from government agencies to conduct pilot. .... 30 Aug. • Implement pilot including training .... Actions also include (a) consensus inputs from AIA, EIA, and NSIA, (b) Government / Industry Quality Liaison Panel integration and consolidation of industry comments, (c) Government review and approval , and (d) implementation on new and current programs.
Products	• List of types of quality related oversight. • Guidelines for oversight • Supplement to MOU.
Issues	• Ability to maintain stable team with consistent representation. • Travel funds.
Team/Resources	**Bill Thompson, Richard Zell, Peter Angiola, Dale Cole, Lindo Bradley, Eugene Moore, Mike Pursley, John Maristch, Dan Kalletta, Bob Lackland.**

Task 9 Develop Procurement Handbook/Guidelines.

Problem Description

Currently there is no guidance available on how to integrate quality and past performance in contractor source selection.

Task Overview

Existing procurement practice / language focuses quality requirements through government directives (MIL-STDS, HDBK) and contractors must demonstrate compliance of their quality management systems to these requirements.

Given varying and (at times) differing requirements from a range of government customers, the contractors ability to leverage a singular quality management system at a specific facility is greatly lessened.

Additionally, the predominant emphasis is on compliance oriented requirements versus incentivizing advanced quality.

Objective

To develop, coordinate, and deploy a means of introducing advanced quality concepts into the acquisition process such that:

- The government's requirement for a basic quality management system and program unique items are satisfied while encouraging the contractor to leverage its singular quality management system at that particular facility.
- The government's expectation for risk minimization and mitigation are addressed via the contractor's demonstrated application of advanced quality concepts.
- The government's need for oversight is adjusted according to the nature of the contract and the performance of the selected contractor.

Determine use of relevant past performance information.

Approach

Develop a set of criteria, probably in the form of general guidelines, to encourage the use of a system that uses Advanced Quality Concepts. The criteria should include general guidance describing what benefits the Government desires from the use of this system. The statement should be written in such a way that an existing quality management system using these concepts may be defined by or applied to the criteria, or a new system can easily be designed.

Develop procurement contract language, to be used in addition to, or in lieu of, the clause defined by FAR 52.246.11. The clause should ensure fairness in competition, but define the basic quality management system elements and provide incentives to the vendor to use Advanced Quality Concepts in the processes. The incentives would be dependent upon the method of contract payment to be used, but may include incentive bonuses or consideration in awards.

Approach (cont.)

Develop general procurement and contract language that enable the implementation of the basic quality elements as defined in the strategic plan. Develop a set of procurement guidelines and contract language that can be used as incentives for the proposal and implementation of Advanced Quality Concepts as defined in the strategic plan.

Actions and Milestones

The premise is to use prior work (Guide to Integrate Quality into Source Selection - dared June 1991) as a starting point, and to access material from tasks #6, and #8.

- Establish full work team representation .... Feb. 95
- Gather material from other task teams. .... Apr. 95
- Generate first draft handbook .... May 95
- Review and incorporate comments and changes .... Jul. 95
- Generate final draft .... Aug. 95
- Distribute Handbook for concurrence .... Sept. 95

Products

Handbook / Guide.

Issues

- Industry representation of work team.
- Access to government releasing organizations.

Team/Resources

Michael Parke, Primus Ridgeway, Frances Sullivan, Bob Tourville, Jim Camden, Richard Wojciechowski, Greg Colbert.

Task 10 Identify Advanced Quality Practices.

Problem Description

There are no Government processes to recognize, improve, promote, share and reward the use of value-adding advanced quality practices.

Task Overview

Current efforts to ensure compliance with a baseline quality management system must be accompanied by efforts to seek out and promote even more effective ways of integrating quality throughout the life-cycle, including requirements definition and product and process development.

This task provides the framework for identifying, promoting, and managing the evolution of advanced quality tools and implementation practices which go beyond those in the baseline quality management system. Such an effort must be integrated effectively into procurement, systems engineering, and oversight processes and must consider product complexity and cost. This will lead to increasingly higher quality products and services, more commonality between government and commercial industries and increases in global competitiveness. A system which recognizes and encourages advanced quality may also reduce the need for individual programs that promote quality improvement.

Objective

- Provide a set of defined government expectations for value adding advanced quality concepts.
- Define techniques to give source selection credit to contractors who demonstrate value-added through the use of advanced quality concepts.
- Provide recognized alternatives to inspection based acceptance of product based on demonstrated systems engineering, process controls and capabilities.
- Create voluntary partnerships that encourage continuous quality improvement through the identification, use, promotion, education, and dissemination of advanced quality concepts.

Approach

This task will review and compare existing "ISO Plus" practices (automotive industry - QS-9000, Boeing D1-9000, SEMATECH - SSQA, etc.) and process validation systems (USA - Baldrige, USA - ANSI/EIA 599, European Union - Technology Approval, etc.). The output will be a recommendation for the use of specific practices and/or system(s) and an evaluation methodology for providing credit for the use of advanced quality management systems by contractors.

Develop a common, inter-agency recognized VPP with our best contractors to: further reduce oversight; increase customer-supplier cooperation; and improve all aspects of quality (performance, cost, time).

Actions and Milestones

- Initial research on existing advanced quality tools. .... April 95
- Identify organizations to visit. .... April 95
- Team visits to contractor organizations. .... May 95
- Interim report on strategies for advanced quality. .... Aug. 95
- Compilation of existing advanced quality partnering features. May 95
- Development of advanced quality concepts syllabus. .... Aug. 95
- Initiate MOU for common, interagency recognized VPP. .... Sep. 95
- Final report. .... Sep. 95

Evaluate the best available advanced quality practices and systems such as SSQA, QSR, etc., quantifiable criteria such as Baldrige, and proven metrics to identify the customer expectation they address.

Establish a cooperative teaming framework within which these advanced quality practices can be successfully implemented.

Products

- Integrated summary of currently used advanced quality tools, implementation practices, and effectiveness metrics.
- Proposed process for continued assessment, documentation, and promotion of advanced quality tools, practices, and metrics.
- Methodology by which procurement, systems engineering, and oversight processes can provide incentives for the use of advanced quality tools and practices.
- Partnered training framework in advanced quality tools.

Issues

- Support contractor resources ($200K)
- Task team travel
- Administrative support resources ($25K)

Team/Resources

Geza Pap, Bob Tourville, Jo McLaughlin, Steve Newman, Howard Butts, Steve French, Charlie Mercer, James Parrish, Rich Zell, Dan Barton, Dick Tracey

Task 11 Develop Communication Clearinghouse.

Problem Description

Presently there is no defined management information system to disseminate quality information to Government, Industry and Academia.

Task Overview

Develop a clearinghouse mechanism for dissemination of quality information to government, industry, and academic institutions.

Objective

Establish an advanced quality practices clearing house to help promote and advance the implementation of a new quality management systems approach in government and industry.

Approach

- Establish a focal point and resource center for basic and advanced quality management system information.
- Disseminate quality management system information to all interested parties.
- Provide coordination and communication for transition to new government/industry quality management approach.
- Provide quality standards, specifications, lessons learned.
- Serve small and large businesses as well as government.

Actions and Milestones

- Identify customers and customer needs. ....
- Identify available data sources. ....
- Define clearinghouse characteristics. ....
- Define clearinghouse operating concept. ....
- Compare alternative concepts. ....
- Select best alternative. ....
- Develop clearinghouse implementation strategy. ....
- Estimate clearinghouse resource requirements. ....
- Implement clearinghouse and publicize. .... Oct. 95

Products

Fully implemented and operating clearinghouse that satisfies customer information requirements.

ssues

- What information should be disseminated by the clearinghouse?
- Who are the clearinghouse customers?
- What are the sources of information?
- What issues exist concerning copyright?
- Who clears information to be placed in clearinghouse?
- How is information to be disseminated?
- What resources will be required to implement and operate the clearinghouse?
- How should the clearinghouse be publicized?

Team/Resources

Phil Smiley, Jo McLaughlin, Ken LaSala, Maureen Breitenberg, Charles Cockrell

Task 12 Draft MOA to Accept Mutual Recognition.

Problem Description	Multiple and redundant supplier quality system audits primarily because organizations do not recognize and accept other organizations review.
Task Overview	This task will establish the basis for an agreement for mutual recognition and acceptance that validation/approval by one government agency of a contractor / subcontractor basic quality management system will be acceptable to all.
Objective	
Recognition	Achieve Government-Industry recognition that the quality management system elements included in ISO-9000 / Q 9000 defines the model that satisfies government quality management system requirements.
Agreement	Establish a Government-wide agreement to validate, accept, and mutually recognize a contractors basic quality management system based on defined baseline requirements.
Policy	Articulate to government and industry the policy to establish a single quality management system at a contractors facility that meets defined baseline government-wide quality management system requirements. Develop and issue clear and unambiguous policy statements that articulate and enable the accomplishment of the objective.
Approach	Collect and review other mutual recognition agreements as a model.
Actions and Milestones	First rough draft memorandum completed January 1995. Further action and coordination based on completion of related tasks that will provide a basis for mutual recognition.
Products	Draft Memorandum of Agreement.
Issues	This task is dependent on the completion of all other tasks.
Team/Resources	**Ryan Bradley, Frank Doherty, Larry Johnston, Chuck Bailey**

Task 13 Develop Quality Audit System Approach and Data Base.

Problem Description	Generally, audits of contractors' quality management systems are conducted independently by the respective program/project offices. If a contractor is shared by multiple projects or different government customers, there will usually be duplicative audits conducted of the same contractor with little information shared between auditing functions. This situation excessively disrupts the contractor's operations, resulting in increased costs. From the government perspective, resources are needlessly expended performing duplicative audits.
Task Overview	This task will establish a common auditing program whereby audit results obtained by one government customer will be acceptable to all government customers of that contractor. The success of this task is dependent on a single quality system existing at the contractor's facility.
Objective	The objective is to minimize disruptions of the contractors' operations with customer audits. Through sharing of audit results, government customers could eliminate the costs of duplicative audits.
Approach	Establish a working group comprised of the participating members of the Panel, to develop the framework, operating guidelines, and electronic collection and dissemination system that would be acceptable to all participants. Institute a pilot run of the integrated audit system to further refine its operation and effectiveness.
Actions and Milestones	TBD
Products	An integrated and coordinated government audit system where audit results are shared by all participating government agencies.
Issues	The primary issues in implementing a single audit system would be the establishment of mutually acceptable audit criteria, effective corrective action system, and a readily accessible data system that provides all pertinent audit results.
Team/Resources	Carl Schneider, Earl Major.

Task 14 Develop Quality Systems Training Courses.

Problem Description

There is a need to develop four courses for Level 1 through 4 based on the approved International Organization for Standardization (ISO 9000) Training Template

Task Overview

This task will develop the four courses and then obtain approval from the Government/Industry Course Review Committee.

Objective

Obtain agreement and commitment from all participating organizations and agencies to support the effort. Guidance and interpretations can be obtained from the team members.

Approach

- Read and understand the ISO 9000 Training Template
- Prepare an Activity Schedule
- Form Development and Review Committees
- Develop courses with lesson plans and instructors
- Receive Approval for the course content
- Offer courses and train-the-trainers

Actions and Milestones

- Form committees
- Approve courses
- Offer courses starting March 96 (Level 1 & 2)
July 96 (Level 3 & 4)
- Train key instructors for Government and Industry

Products

- Four courses with instructor and student materials
- Uniform approach for Quality Systems

Issues

- Obtaining high level commitment

Team/Resources

Dr Julius Hein, Jack McGovern, Roger Fisher, Hank Guck, Dale Greer, Steve Newman, Richard Zell, Jo McLaughlin

TASK 15

Problem Description

There is a need to review the four courses developed by team in task 14.

Task Overview

The team will review and approve the four courses for use by Government and Industry

---

Objective

Obtain review and approval on content and context of the courses

---

Approach

- Form Review Committee
- Furnish dates for review
- Review and approve content and context of the courses

Actions and Milestones

- Select Review Committee members
- Obtain members commitment to participate
- Clear calendar for scheduled dates
- Review and approve start 19 December 95 (Level 1 & 2)
  30 April 96 (Level 3 & 4)
- Monitor course offering

Products

- Approved courses

Issues

- Come to agreement between developers and reviewers

Team/Resources

Harry Hudson, Dale Cole, Primus Ridgeway, James Tew, Gene Barker, Chuck Packard, Karen Richter, Harold Law, Jerry Norley, Frank Doherty, Kent Nelson, Ryan Bradley, Steve French, Carl Schneider, Joel Odum, Carol Driscoll, Michael Litchfield and Robert Tourville

APPENDICES

APPENDIX A

Roster, Panel Members/Participants

Angiola, Peter, OUSD(A&T)DTSE&E/DDSE
Barnett, Dick, AMC
Butz, Jr., Howard E., AAI Corp.
Bailey, Chuck, EIA/UNISYS
Barton, Daniel, DOC/NIST
Becker, Carl, Boeing
Bradley, Lindo, USMC
Bradley, Ryan, SAF/AQXM
Bradley, Mark, USDA/AMS/LS
Breitenberg, Maureen, DOC/NIST
Butts, Carl, GM/Hughes
Camden, Jim, Hughes
Colbert, Greg, ASD/ENSP
Cole, Dale M., Martin Marietta
Cockrell, Charles E., NASA-Langley
Doherty, Frank, OUSD(A&T)DTSE&E/DDSE
Driscoll, Carol, FAA
Engbretson, Jerry, AIA/Rockwell
Fiege, Fred, DOE
Fisher, Roger L., Boeing
Fontenot, Brent, NASA-JSC
Franklin, Bob, FAA
French, Steve, SARD-DER
Gardner, Guy, NASA-QW
Georgeadis, George, USMC
Harris, Lynn, DLA-MMSL
Hudson, Harry, Hughes Aircraft
Hart, Terry, Westinghouse
Johnston, Larry, Northrup/Grumman
Jung, Dennis, NAVSEA
Kalletta, Dan, CAPT, USCG
Keller, Walter R., Army/AMCCOM (Rock Island)
Lackland, Bob, Electronic Ind-Quality Registry
LaSala, Kenneth, NOAA/DOC
Litchfield, Michael, USCG
Luther, Wally, Rockwell
Mansir, Brian E., LMI
Major, Earl, Allied Signal Aerospace
Maristch, John M., NASA-QW
Mercer, Charles R., AIA/McDonnel Douglas

Moore, Jr., Eugene A., GSA-FSS-FOA
Muzio, David, OFPP
McKenna, Eugene P., GSA-FSS
McLaughlin, Jo, NASA-U
Newman, Steve, NASA-QW
Odum, Joel, GSA
Pap, Geza, Army/SMCAR (Picatinny)
Parke, Michael, NAVSEA
Parrish, James, AIA/Sundstrand
Pursley, Mike, DOT/MARAD
Reinhard, Fred, OUSD(A&T)DP/DSPS
Ridgeway, Primus, EIA(G-43)/Rockwell
Ryskamp, J. Michael, Army/CECOM
Sapp, Rich, AIA/Lockheed
Schneider, Carl, NASA-QR
Schmitt, Robert, DLA/DCMC/AQCO
Shekem, Murat, CAPT, OSN/RDA-PI
Smiley, Phil, CAPT, OSN/RDA-PI
Smith, Edward, OSN/RDA-PI
Strohbeck, Stephanie, DLA-DCMC/AQC
Sullivan, Frances, NASA-HC
Tessier, Robert, FAA/AIR-200
Thompson, Bill, NSIA/Lockheed
Tourville, Robert M., NAVAIR, AIR-5166
Tracey, Dick, AIA(GMHE)
Walker, E. Doug, DLA-DCMC/MMSL
Wessel, Bill, NASA-Langley
Whiteley, Bob, AMCRD-IEC
Wojciechowski, Richard, OUSD(A&T)DP/DSPS
Zell, Richard, DLA/DCMC/AQCO

APPENDIX B

Memorandum of Agreement
for
Mutual Recognition of Basic Supplier Quality Management System

(Draft; 12/95)

Government & Industry Quality Liaison Panel Objectives:

> ***1) To enable use of a single quality management system, that satisfies each customer's requirements, within a supplier's facility;***
>
> ***2) To mutually recognize, share, and utilize advanced quality concepts in the requirements definition, design, manufacture and acceptance of products;***
>
> ***3) To establish and implement effective and efficient oversight methods.***

Ve, the undersigned, mutually agree that the approval of a supplier's basic Quality Management System e.g. ISO 9000 or ANSI/ASQC Q9000 Quality Management and Quality Assurance Standards), ollowing an on site evaluation by one participating signatory agency, service, or department, will be ecognized and accepted by the others as adequate evidence that the system was found to comply with pecified criteria at that time.

his agreement supports the Government & Industry Quality Liaison Panel single quality management ystem objectives and is based on the following conditions:

- The approval will state the date the supplier system was reviewed and found in compliance with basic Quality Management System criteria as defined in __________(Note: need to identify).
- The survey / audit was performed using the evaluation criteria commensurate with customer requirements and the commodity being procured.
- The survey / audit team granting approval was led by a representative who has satisfactorily completed Agency Quality Management System Training or certified lead assessor training.
- The survey / approval was performed/validated within the most recent 36 calendar months.
- Documented evidence of the review will minimize the necessity for additional Government survey / audit of the basic Quality Management System for program peculiar re-survey/re-approvals.

Appropriate agency signatures will follow below as in the 24 April MOU for G&IQLP

APPENDIX C

Memorandum of Understanding

of

April 24, 1995

GOVERNMENT

Memorandum Of Understanding
for
GOVERNMENT INDUSTRY QUALITY LIAISON PANEL

April 24, 1995

We, the undersigned, agree to actively support the mission, scope and objectives of the Government Industry Quality Liaison Panel (GIQLP).

Mission: To enable consistent satisfaction of customer expectations through a government and industry association partnership using world-class quality processes and practices to enhance international competitiveness.

Scope: The GIQLP will encourage the participation of interested federal agencies and industry associations in the development and deployment of uniform quality management systems and advanced quality concepts.

Goals:

1) To enable use of a single quality management system within a contractor's facility that satisfies each customer's requirements;

2) To mutually recognize, share, and utilize advanced quality concepts in the requirements definition, design, manufacture and acceptance of products;

3) To establish and implement effective and efficient oversight methods.

Mr. Fred Gregory, NASA
Associate Administrator

D, TSE&E OUSD(A&T)

Mr. John Burt, DOD
Director Test, Systems Engineering, & Evaluation

Mr. Blaise Durante, DOD/USAF
Deputy Assistant Secretary
Management Policy & Program Integration

RADM Thomas H. Collins, DOT/CG
Chief, Office of Acquisition

Mr. Dan Porter, DOD/NAVY
Acquisition Reform Executive

Mr. Robert E. Brown, DOC/NOAA
Director, Systems Acquisition Office

Mr. Michael Pursley, DOT/MARAD
for Mr. Joseph A. Byrne,
Director of Shipyard Revitalization

Mr. Herbert K. Fallin Jr., DOD/ARMY
OASA(RDA)
Director Assessment & Evaluation

RADM Leonard Vincent, DOD/DLA
Commander, DCMC

Dr. Belinda Collins, DOC/NIST
Director, Office of Standards Services

Mr. Dennis DeGaetano, DOT/FAA
Director of Acquisitions

Ms. Ida M. Ustad, GSA
Associate Administrator for
Acquisition Policy

INDUSTRY

Memorandum Of Understanding
for
GOVERNMENT INDUSTRY QUALITY LIAISON PANEL

April 24, 1995

We, the undersigned, agree to actively support the mission, scope and objectives of the Government Industry Quality Liaison Panel (GIQLP).

Mission: To enable consistent satisfaction of customer expectations through a government and industry association partnership using world-class quality processes and practices to enhance international competitiveness.

Scope: The GIQLP will encourage the participation of interested federal agencies and industry associations in the development and deployment of uniform quality management systems and advanced quality concepts.

Goals:

1) To enable use of a single quality management system within a contractor's facility that satisfies each customer's requirements;

2) To mutually recognize, share, and utilize advanced quality concepts in the requirements definition, design, manufacture and acceptance of products;

3) To establish and implement effective and efficient oversight methods.

Mr. Stan Siegel, AIA
Vice President, Technical Operations

Mr. Dan Heinemeier, EIA
Vice President, Government Division

ADM James R. Hogg, NSIA
President

Appendix

B

Sample Electronics Company Quality System Manual

Issue 1
January 1997

Overview

Introduction

Sample Electronics Company (SEA) Mission Statement
Control/Revision Authority
Revision Summary

4.1 Management Responsibility

4.2 Quality System

4.3 Contract Review

4.4 Design Control

4.5 Document Control

4.6 Purchasing

4.7 Control of Customer-Supplied Product

4.8 Product Identification and Traceability

4.9 Process Control

4.10 Inspection and Testing

4.11 Control of Inspection, Measuring, and Test Equipment

4.12 Inspection and Test Status

4.13 Control of Nonconforming Product

4.14 Corrective and Preventive Action

4.15 Handling, Storage, Packaging, Preservation, and Delivery

4.16 Control of Quality Records

4.17 Internal Quality Audits

4.18 Training

4.19 Servicing

4.20 Statistical Techniques

Appendix: Procedures/Processes—Compliance to ISO 9001

Overview

Sample Electronics Company, headquartered in New York City, is a public United States (U.S.) company incorporated in the state of Delaware. It is a wholly owned subsidiary of Electronics World, Inc., whose corporate offices are in Paris, France. The Company operates at three sites: New York City; Austin, Texas; and Mexico City, Mexico.

The Company, with over 100 years' experience, is the designer and manufacturer of electronics systems, subassemblies, and components for the United States and international markets. These include electronics for defense and commercial customers in the areas of circuit assemblies, control systems, weapons, navigation, and avionics systems, sensors, and telecommunication products. An organization chart is shown as Fig. QSM-1.

Introduction

This manual establishes the organization and responsibilities for quality throughout Sample Electronics, and describes the quality policies for the company operations. Both the New York and Austin, Texas, facilities do engineering design and are compliant to ISO 9001. The Mexico City facility produces electronic components and assembles product, and is compliant to ISO 9002. Sections in this manual correspond with the numbering scheme of ISO 9001.

The operation of the policies contained in this manual and their relationship with ISO 9001:1994 is defined in a series of company procedures referenced in the appendix of this document.

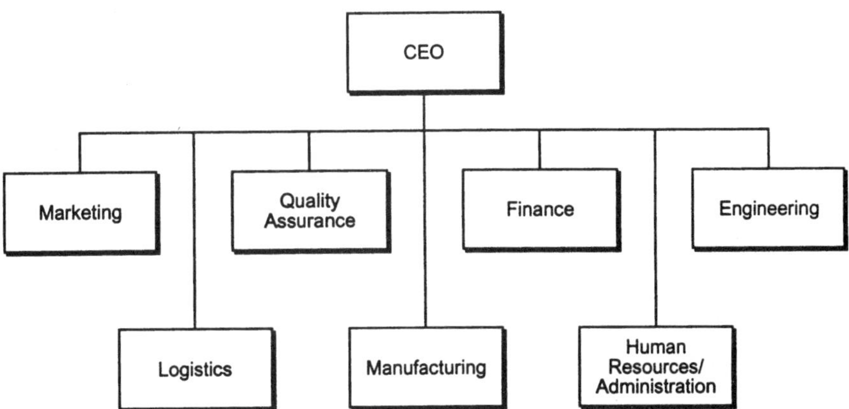

Figure QSM-1 Sample Electronics Co. organization chart.

Sample Electronics Company (SEC) Mission Statement

Sample Electronics strives to be one of the leading electronics companies in the world, by supplying our customers with quality products and services which enable them to meet their needs and enable us to achieve long-term partnerships for the future. In addition, it is our goal to fulfill our commitments to our shareholders, suppliers, and the community at large.

Control/Revision Authority

Revision approval is via the ISO management representative with concurrence by the SEC systems compliance manager, and the quality assurance director, with final authority with the chief executive officer (CEO). The Administration Department is responsible for controlling and distributing this manual in an electronic fashion on the SEC local-area network (LAN). Changes are summarized in the revision summary section.

Revision Summary

Date issued: January 1, 1997: Note changes to document on this page.

Date:	**Summary (reason for change)**

4.1 Management Responsibility

Purpose

Sample Electronics Company (SEC) ensures that the quality policy is defined and communicated to all levels of the organization.

Compliance

The quality policy of Sample Electronics Company is

> SEC's quality policy is to deliver high-quality products and services that please the customer through Total Quality Management. This policy applies to the entire company and to every employee. Our reputation as a quality company depends on us.

The company maintains a quality program applicable to the design, production, and service of its products (hardware, software, etc.). The goal of the quality program is to ensure that the needs and expectations of its customers are met or exceeded.

It is the responsibility of all members of management to ensure that their respective employees understand, implement, and maintain the quality policy. This responsibility includes: the development and implementation of procedures/processes or methods necessary to control activities affecting the quality of the product and/or service; organizational freedom and authority for verification of work affecting quality; and adequate resources to perform the work and verification activities.

By the authority delegated by the CEO, the director of quality is the ISO management representative with authority to establish, implement, and maintain a quality system in accordance with ISO 9001. The management representative reports to the CEO and the executive staff on the performance of the quality system on a quarterly basis and maintains records of the reviews, including the status of the quality objectives and any improvements to the quality system. At off-site locations, the management representative is the quality manager at each site and reports quality system status to the management representative at the New York office.

4.2 Quality System

Purpose

Sample Electronics Company (SEC) has established and maintains a documented quality system as a means of ensuring their products and services conform to specified requirements.

Compliance

Each component of this quality system is consistent with ISO 9001. The quality system is implemented through a documented set of procedures/processes, shown in Table QSM-1.

Level 1 is the quality manual. It describes the policy and methods for implementing and maintaining the quality management system. Level 2 is cross-functional procedures/processes that describe the "who, what, and when" that is defined as part of the system. Level 3 is detailed instructions, departmental procedures, and world instructions. These discuss how the quality system operations are accomplished. Level 4 is the set of forms used in the system, and the quality records that are collected and maintained.

Quality plans are prepared for use on specific contracts or projects, as required.

TABLE QSM-1 Design/Documentation Table (Sample)

Design task	Requirement	Company process	Company procedure
Design documentation	IEEE P1220/D1, December 1993	Process 4.4	SOP 1-4.4
Drawings/ specifications/ schematics	ASME Y14.24M ASME Y14.1 ASME Y14.34M	Process 4.4.1	SOP 1.4.4.1
Configuration management	MIL-STD-973	Process 4.8	SOP 1.4.8
Quality system for design	ISO 9001-1994	Process 4.1	SOP 1.4.1

The process owner of every element in the ISO 9001 standard is the responsibility of a person on the CEO's direct staff. The process owner is responsible for implementation, maintenance, corrective and preventive action, and self-audits of that element.

4.3 Contract Review

Purpose

Sample Electronics Company (SEC) has established and maintains documented procedures for the review of each contract or order received. All members of the integrated product/process team for the affected activity are part of the contract reviews, as is the executive management staff.

Compliance

Contract review is applied to the review of new and repeat work accepted under contract conditions. A review of contract/order received, with tender documents, is conducted to ensure that the requirements are adequately defined and documented, any requirements deriving from the contract/order are resolved, and the capability exists to meet defined requirements of contract/order. Amendments to contract/order are processed in the same manner.

Contract review is coordinated by the Contracts Organization (part of Finance) with all the affected employees dependent on the contract.

Records of each contract/order review are maintained by the Contracts department performing the review.

4.4 Design Control

Purpose

Sample Electronics Company (SEC) has established and maintains documented procedures to control and verify the design of the product in order to ensure that the specified requirements are met. System engineering procedures/processes are based on the *Standard for Systems Engineering,* P1220/D1, December 1993, produced by the Institute of Electrical and Electronics Engineers.

Compliance

Sample Electronics Company prepares plans for each design and development activity: Project Definition, Project Planning, Development/Environment Planning, and Verification/Validation Planning. Once the plans have been prepared, associated activities are used to ensure that the plans are being followed. Advanced quality concepts are employed during the design phase, and written into the design plans, as required. Table QSM-1 shows the design documents and what standards they were prepared from.

The organizational and technical interfaces between different groups are defined and their responsibilities are in accordance with the Integrated Product/Process Development Guide.

Design input requirements are identified, documented, and reviewed for adequacy with resolution of all incomplete, ambiguous, or conflicting requirements with those responsible. The results of any contract review activities are taken into consideration during design input.

Design output is documented and expressed to meet design input requirements, contain or make reference to acceptance criteria, and identify the characteristics of the design that are crucial to the safe and proper functioning of the product. Design output documents are reviewed before release.

Formal documented reviews are planned and conducted at appropriate stages of design with representatives of all concerned functions being involved. Design verification is performed to ensure that each design-stage output meets the design-stage input requirements. Design validation is performed to ensure that product conforms to defined user needs and/or requirements. All design changes and modifications are identified, documented, reviewed, and approved by duly authorized personnel prior to their implementation.

4.5 Document Control

Purpose

Sample Electronics Company (SEC) has established and maintains documented procedures to control its documents and data (on any form of media) that relate to the requirements of ISO 9001, including documents of external origin, as applicable.

Compliance

Sample Electronics Company reviews and approves documents and data for adequacy by duly authorized personnel prior to issue.

Document and data control procedures are established: to identify the current revision status so that the pertinent issue of the appropriate document is readily available; to preclude the use of invalid and/or obsolete documents by ensuring against unintended use or through removal from all points of issue or use; to ensure appropriate documents are available at all locations where essential operations to product quality are performed; and to ensure documents are suitably identified when they are retained for legal and/or knowledge-preservation purposes.

Most SEC documents are already in electronic media, and producers address which ones are and are not. It is the goal of SEC to have all documents in electronic media by 1999.

Changes are reviewed and approved by the functions/organizations that performed the original review and approval of documents, unless otherwise authorized. When there is a change, the nature of the change will be identified in the document, where practicable.

Pertinent background information shall be accessible as the basis for review and approval.

4.6 Purchasing

Purpose

Sample Electronics Company (SEC) has established and maintains a documented system that ensures that purchase product conforms to specified requirements.

Compliance

Sample Electronics Company Purchasing selects best-value suppliers on the basis of cost, approval status, their past performance using the supplier rating and incentive program (SRIP), and their ability to satisfy quality system requirements as defined by the contract and/or purchase order. Records of acceptable subcontractors are maintained.

Sample Electronics Company purchasing data defines the specified technical product requirements using defined quality flowdown requirements to the subcontractor to ensure the quality of the purchased product, process, or service. This may be done, in part, by quality codes or reference to other applicable technical information such as national or international standards, test methods, specifications, etc. Essential information is clearly and precisely stated in the subcontracts. Prior to release, the appropriate personnel review and approve the purchasing data. Purchasing identifies the revision status of documents controlled by Sample Electronics Company in the purchasing data.

Items procured that are too complex to verify on receipt, have a critical application, or are to be direct-shipped to the customer are source-inspected by Sample Electronics Company. This requirement is clearly stated in the purchase order to the subcontractor defining phone number and location of appropriate contacts to support the source inspection effort. Upon request, customer representatives are given the opportunity to review documents for the possible inclusion of customer source inspection. When required, the customer is afforded the right to inspect the subcontracted product at the contractor's or subcontractor's facility.

4.7 Control of Customer-Supplied Product

Purpose

Sample Electronics Company (SEC) has established and maintains procedures for the receipt, verification, handling, control, accountability, maintenance, storage, and disposition of customer-supplied product while in the custodial possession of Sample Electronics Company.

Compliance

It is Sample Electronics Company policy to develop, maintain, and follow company or customer-directed processes to document receipt of, identify, verify, protect, maintain accountability for, provide maintenance for, store, and dispose of customer property when and as directed by the customer or contract. Should the customer-furnished property become damaged, change condition, or malfunction, this condition shall be documented and reported to the customer.

Although Sample Electronics Company performs verification, the customer remains responsible for final approval on the acceptable product, since it is a customer-supplied product.

4.8 Product Identification and Traceability

Purpose

Sample Electronics Company (SEC) maintains systems for placing part numbers on engineering documentation. SEC identifies its product during all stages of production, delivery, installation, and servicing per documented procedures. Unique identification for traceability of individual products or batches is maintained in accordance with documented procedures. Identification is sufficient to provide traceability to the extent specified by individual customers.

Compliance

Sample Electronics Company Design Manual specifies the way that Engineering assigns part numbers to all products. Those part numbers identify its products during manufacturing activities. Drawings, manufacturing orders, tags, stamps, and product nameplates are utilized as a means to identify products during production, delivery, and installation. Configuration control of hardware and software is maintained during all stages of design, testing, integration, delivery, and installation.

4.9 Process Control

Purpose

Sample Electronics Company (SEC) ensures that processes which directly affect quality are identified, documented, and planned for production, installation, and servicing and ensures that these processes are carried out under controlled conditions.

Where the results of processes cannot be fully verified through inspection and testing of the product, and where deficiencies may become apparent only during use, continuous monitoring is maintained through control of defined parameters, qualified personnel, and associated equipment. Special processes are prequalified to define their process capabilities. Records are maintained on all qualified processes, equipment, and personnel.

Compliance

Sample Electronics Company identifies and plans the prototype production, installation, and servicing of processes that affect quality and ensure that these processes are implemented under controlled conditions. Where special processes exist, continuous monitoring takes place through control of qualified processes and personnel.

TABLE QSM-2 Examples of Special Processes for Electronic Systems

- Plating, protective coating
- Electrostatic discharge sensitivity
- Automatic test equipment
- Magnetic particle inspection
- Eddy current inspection
- Ultrasonic inspection
- Radiographic (x-ray) inspection
- Wiring and crimping
- Soldering
- High-reliability soldering
- Corrosion control processes

Table QSM-2 is a sample of special processes at SEC. Process records are maintained.

Processes are performed by qualified personnel to criteria identified. Equipment is maintained to ensure continuous process capability. Periodic auditing and process verification is conducted at defined intervals, with records of these audits/verifications maintained.

4.10 Inspection and Testing

Purpose

Inspection and test requirements are to be addressed in all aspects of production, from receipt of material through shipment of completed product. Inspection and test records shall give evidence that: incoming material has been verified as conforming to specified requirements; material is controlled to prevent inappropriate release; product is subjected to in-process inspection and testing; nonconforming product is controlled; and all required inspections and tests have been accomplished before product is dispatched.

Compliance

SEC's procedures provide a documented process by which all aspects of inspection and testing of materials and products are conducted. Methods are provided for receiving inspection; urgent release of material; in-process inspection and testing; final inspection and testing; identification of nonconforming product; and generation and maintenance of records.

4.11 Control of Inspection, Measuring, and Test Equipment

Purpose

Sample Electronics Company (SEC) has established and maintains documented procedures to control, calibrate, and maintain inspection, measuring, and test equipment (including test software) used to demonstrate the conformance of product to the specified requirements. SEC's system is in compliance with ANSI/NCSL Z540-1-1994, *American National Standard for Calibration—Calibration Laboratories and Measuring and Test Equipment—General Requirements.*

Compliance

Sample Electronics Company uses comprehensive systems and procedures for measuring and test equipment calibration control, inspection tooling control, certification of test equipment/processes, and related issues. These documented processes ensure that the selection, application, and condition of equipment (including software) are evaluated and controlled for intended use.

4.12 Inspection and Test Status

Purpose

The identification of inspection and test status is maintained throughout product development, manufacturing, installation, and servicing so that only product which meets specification requirements is used, dispatched, or installed.

Compliance

Sample Electronics Company (SEC) procedures provide a documented process by which the inspection and test status of product is recorded throughout the product life cycle.

4.13 Control of Nonconforming Product

Purpose

Control of nonconforming product is maintained to avoid inadvertent use or installation of product which does not conform to specified requirements. Review of nonconforming product is required prior to acceptance whenever repair (with or without concession), rework, regard, or scrap disposition has been made.

If a customer desires to be part of the control of nonconforming product, it must be defined in the contract or purchase order.

Compliance

Sample Electronics Company (SEC) procedures provide documented processes by which nonconforming material is handled, reviewed, and dispositioned with responsibility for review and authority for disposition well-defined. The procedures also provide for identification and reinspection of such material and notification of functions concerned.

4.14 Corrective and Preventive Action

Purpose

Sample Electronics Company (SEC) has established and maintains a documented system for implementing corrective and preventive action to the degree appropriate for the prevention of actual or potential nonconformances.

Compliance

Sample Electronics Company procedures provide documented processes for corrective action, recurrence prevention, and preventive actions throughout the product life cycle, including the handling of customer complaints. Appropriate documentation is provided in all cases and relevant information is submitted for management review.

4.15 Handling, Storage, Packaging, Preservation, and Delivery

Purpose

Sample Electronics Company (SEC) uses methods which prevent damage to or deterioration of the product from receipt of material through shipment of completed product. Products are packaged and preserved in accordance with specified requirements to protect the quality of the product to its destination.

Compliance

Documented procedures are used for handling, storage, packaging, preservation, segregation, and delivery of product. Compliance is monitored in accordance with established procedures.

All products (raw material, supplies, component parts, assemblies, end item shipments, customer property, and packaging material) are protected during handling and storage to prevent damage, loss, deterioration, degradation, loss of identity, and loss of inspection/test status. Products are stored in designated stock areas pending use or delivery.

Before acceptance into stock, product is verified to ensure it is appropriately identified. This requirement includes product which is being returned to a controlled stock area. Products being returned to a controlled stock area are verified to ensure conformance to the requirements of the part number under which it is to be stocked. Product is issued in accordance with documented procedures.

An electrostatic discharge control program has been implemented in order to preserve the functional integrity and reliability of those products which are susceptible to damage from electrostatic discharge.

Stored product is inspected periodically for deterioration or damage in accordance with written procedures. These procedures include requirements for disposition of products which are subject to deterioration from prolonged storage (such as age-sensitive materials). Product having special storage requirements or limited shelf life is appropriately labeled.

Packing, packaging, and marking processes are controlled to ensure conformance to specified requirements. The quality of the product is protected after final inspection and test, and protection extends to include delivery to destination.

4.16 Control of Quality Records

Purpose

Sample Electronics Company (SEC) ensures that the quality records demonstrate achievement of the required product quality and the effective operation of the quality management system.

Compliance

Quality records (mostly on electronic media) are maintained to demonstrate conformance to specified requirements and the proper operation of the quality system. Records are identified in the governing procedures. The functional organization performing the work maintain the records, including collection, indexing, accessing, filing, storage, and disposition to the extent appropriate for the product's degree of completion. All records are treated in such a manner as to

TABLE QSM-3 Quality Records (Sample Table)

Records	Purpose	Process owner	Location/retention
Product specification	Defines requirements	R&D manager	Product file/5 years
Process specification	Defines manufacturing processes	Manufacturing manager	On-line database/10 years after last order
Validation	Identifies problems; shows results	R&D manager	On-line database/5 years
Engineering change notices (ECNs)	Document product changes	Document control	On-line document control database/10 years
Manufacturing work instructions	Define manufacturing process	Document control	On-line database/10 years after last order

prevent damage, loss, or deterioration. Retention times are established and recorded in corporate, divisional, and departmental schedules, as appropriate. When required by contract, records are available for review on customer request. Table QSM-3 is a sample summary of records kept by SEC.

4.17 Internal Quality Audits

Purpose

Sample Electronics Company (SEC) has established and maintains procedures ensuring that an internal audit system is in place to assess the effectiveness of the quality management system. There are two levels of audit: internal quality audits at the company level and, at the departmental level, internal audits in which each department assesses its compliance to the SEC quality management system.

Compliance

Internal Quality Audit (IQA) plans and conducts internal quality audits to objectively evaluate compliance with implementing documents and to determine the effectiveness of the quality system. Audits done by individual departments will be shared with the IQA function. The annual audit plan is developed on the basis of the status and importance of the activity to be audited. The audits are conducted by personnel independent of those having direct responsibility for the activity being audited. Management of audited organizations provides timely corrective action for deficiencies. IQA follows up to ensure that corrective actions are accomplished and effective. The management review process uses reports of internal quality audits to assess the effectiveness of the quality management system.

4.18 Training

Purpose

Sample Electronics Company (SEC) ensures that training needs are recognized, documented, and met for all personnel performing activities affecting quality.

Compliance

Management identifies and provides for the training needs of all personnel performing activities affecting quality. Education, training, and experience are considered in assigning tasks.

An effective training program is maintained by:

1. Identifying training requirements of all personnel
2. Identifying technical, procedural, or certification training, as required
3. Providing required formal and informal training in a reasonable period of time and assuring completion of the training
4. Developing procedures, certifications, and course outlines for all in-house training
5. Printing and making available on the local area network a listing of all training courses available

Records are maintained by the functional departments for their personnel and by the training department as part of training records.

4.19 Servicing

Purpose

Sample Electronics Company (SEC) has an established policy and procedures for servicing (logistics support and repair) and customer servicing (marketing). Separate logistics plans identify logistics support and repair and modification activities. Marketing procedures are established that determine specific customer requirements, communicate with customers and the company, and ensure the capability exists to meet customer requirements with new customers.

Compliance

It is Sample Electronics Company policy to develop, maintain, and follow company or contract-directed procedures for field distribution of product support and ensuring that verification and reporting requirements are met. Customer service of delivered product is handled by both the logistics and marketing personnel who maintain close contact with customers. Metrics are collected on delivered products so that product improvements can be made, and on customer service so that it too can be continuously improved.

4.20 Statistical Techniques

Purpose

The need for statistical techniques is required for establishing, controlling, and verifying process capability and product characteristics. It is identified for all programs. Sample Electronics Company (SEC) has established processes for the implementation and application of statistical process control/variability reduction (SPC/VR). During the planning for development/production processes, the SPC/VR techniques to be used are defined. Process capability is calculated for each production process.

Compliance

Sample Electronics Company evaluates the exact statistical processes techniques for individual programs during the development phase and documents the results in the SPC manual appendix and quality assurance program plan, as required. Sample Electronics Company implements the application of statistical techniques for development, production, repair, and software programs through documented procedures and manuals. Statistical process control is in place on all production and repair programs. In addition, software metrics are kept as part of our continuous improvement of the software development process.

Appendix: Procedures/Processes—Compliance to ISO 9001

4.1	Management Responsibility	A-1	Quality Policy
4.2	Quality Policy	A-2	Integrated Product/Process Development

	A-3	Program/Project Management
	A-4	Organization Manual
4.3 Contract Review	C-1	Contract Administration Guide
4.4 Design Control	D-1	Engineering Administration Process
	D-2	Systems Engineering Process
	D-3	Product Design Process
	D-4	Hardware Design Process
	D-5	Software Process Manual
	D-6	System Integration/Verification/Validation Process
	D-7	Configuration Management Manual
	D-8	Reliability and Maintainability Process
	D-9	Product Support Manual
	D-10	Drawing Requirements Manual
	D-11	Specification Requirements Manual
	D-12	Printed Wiring Board Manual
	D-13	Wafer Design Manual
	D-14	System Safety Manual
4.5 Document and Data Control	L-1	Engineering Data System
	L-2	Work Instruction Manual
	L-3	Backup of Computer Generated Design Data
	L-4	Prototype Engineering Procedure
	L-5	Document Control
	L-6	Document Changes
	L-5	Documents in Electronic Formats
	L-6	Data Requirements Manual
4.6 Purchasing	P-1	Purchased Material Identification, Codes, and Controls
	P-2	Control of Engineering Documentation

Section	Procedure	Title
	P-3	Evaluation of Subcontractors/Suppliers
	P-4	Approved Source List
	P-5	Purchase Order System
	P-6	Engineering Subcontract Management
	P-7	Requirements for Supplier Partnerships
	P-8	Electronic Data Interchanges
	P-9	Blanket Purchase Orders
	P-10	Source Selection Process
	P-11	Case File Documentation
	P-12	Requesting Quotations
	P-13	Material Estimating
	P-14	Negotiation Plans/Summaries
	P-15	Rejections of Purchased Product
	P-16	Source Inspection
	P-17	Inspection of Purchased Material
	P-18	Certification of Special Processes at Suppliers
	P-19	Direct Shipment of Purchased Material to Customers
	P-20	Metrics/Customer Survey Procedure
4.7 Control of Customer Supplied Product	CS-1	Property Management
	CS-2	Accountability for Property; Company/Customer
	CS-3	Physical Inventories
	CS-4	Customer Furnished Property
	CS-6	Control and Movement of Customer Furnished Property
4.8 Product Identification and Traceability	I-1	Serial Numbers for Products
	I-2	Microcircuit Control and Processing
	I-3	Wafer Control and Processing
	I-4	Change Processing for Identification
4.9 Process Control	PC-1	Planning

	PC-2	Scheduling
	PC-3	Production Control
	PC-4	Work Instructions
	PC-5	Tooling Manual
	PC-6	Advanced Quality Techniques
	PC-7	Processing Equipment Control
	PC-8	Test Equipment Instruction Manual
	PC-9	Handling Electrostatic Discharge Sensitive Parts
	PC-10	Certifying/Qualifying Special Processes
	PC-11	Controlling Special Processes
	PC-12	Document Control in Production
	PC-13	Shop Control Procedures
	PC-14	Engineering Prototype Work in Manufacturing
	PC-15	Line Proofing
	PC-16	Workmanship Program
	PC-17	Metrics for Manufacturing
4.10 Inspection and Testing	IT-1	Control of Bulk Chemicals
	IT-2	Receipt and Acceptance of Purchased Material
	IT-3	Use of Components Prior to Acceptance Test Completion
	IT-4	Specific Process Specifications
	IT-5	Process Control Surveillance
	IT-6	Certification of Test Equipment
	IT-7	Inspection of Purchased Material
	IT-8	Quality Assurance Data Systems
	IT-9	Inspection Instructions Manual
4.11 Control of Inspection, Measuring, and Test Equipment	C-1	Factory Test Equipment
	C-2	Operating Instruction Specifications
	C-3	Program Media for Test Equipment
	C-4	Test Equipment at Suppliers
	C-5	Calibration Control System

		C-6	Tool Proofing
4.12	Inspection and Test Status	ITS-1	Operation Completion Records
		ITS-2	Indicating Inspection Status
		ITS-3	Shipping Inspection Documentation
4.13	Control of Nonconforming Product	NC-1	Nonconforming Material Corrective Action and Disposition
		NC-2	Segregation of Nonconforming Material
		NC-3	Remarking Nonconforming Product
		NC-4	Customer Controlled Nonconforming Product System
4.14	Corrective and Preventive Action	CP-1	Failure Reporting, Corrective Action
		CP-2	Reliability Corrective Action
		CP-3	Corrective Action Process
		CP-4	Preventive Action Process
		CP-5	Customer Corrective Action Requests
		CP-6	Functional Test Corrective Action
4.15	Handling, Storage, Packaging, Preservation, and Delivery	HP-1	General Process Specifications
		HP-2	Packaging, Handling, Storage, Packaging, Preservation Process
		HP-3	Shipping Process Manual
		HP-4	Packaging Requirements Manual
		HP-5	Surveillance of Stored Items
		HP-6	In-Transit Protection of Material
		HP-7	Control of Special Handling
		HP-8	Shelf Life Control
		HP-9	Parts Protection Manuals
		HP-10	Transportation and Services
4.16	Control of Quality Records	QR-1	Records Retention and Destruction
		QR-2	Electronic Control of Records

4.17	Internal Quality Audits	A-1	SEC Audit Program
		A-2	Audit Process Improvement
		A-3	Functional Audit Responsibilities
4.18	Training	T-1	Personnel Development Manual
		T-2	Certification/Qualification Programs
		T-3	Professional Training
		T-4	Advanced Quality Techniques Training
		T-5	Training Catalog
		T-6	Inspection Training Program
		T-7	Control of Training Records
4.19	Servicing	S-1	Customer Survey Process
		S-2	Marketing Process
		S-3	Logistics Support Process Manual
		S-4	Modification and Repair Manual
		S-5	Product Deficiency Reports from Customers
		S-6	Technical Publications
4.20	Statistical Techniques	SPC-1	Statistical Process Control Manual
		SPC-2	Variability Reduction Process
		SPC-3	Process Capability Process
		SPC-4	SPC at Subcontractors/Suppliers
		SPC-5	Relationship of DOE with SPC/VR

Appendix

C

The USAF R&M 2000 Variability Reduction Process

NOTE: No longer available through the U.S. government.

The USAF R & M 2000

VARIABILITY REDUCTION PROCESS

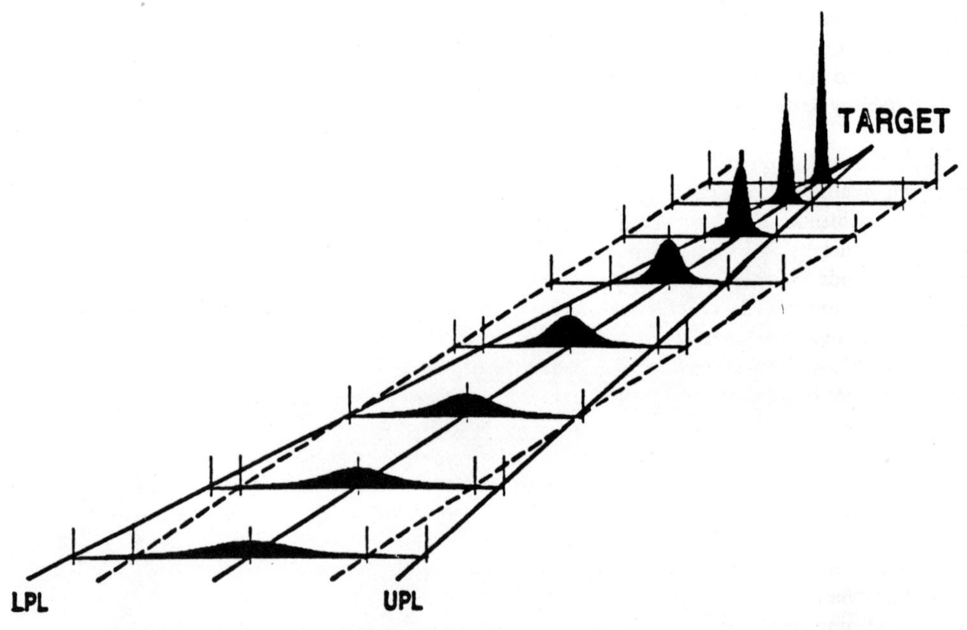

by the Special Assistant for Reliability and Maintainability
United States Air Force

EXECUTIVE SUMMARY

The Variability Reduction Process (VRP) is one of the building blocks of the USAF R&M 2000 process, a set of practices proven effective and efficient in achieving more reliable, maintainable, and producible weapon systems. VRP is an essential part of the Air Force's approach to the DOD Total Quality Management (TQM) initiative. VRP's principal focus is the acquisition process—the point of maximum leverage.

This guidebook describes the USAF R&M 2000 VRP—its benefits, techniques, and implementation. VRP draws on both Japanese quality experience and the more recent experience of U.S. industry in implementing new quality methods. The central emphasis of the new quality methods is continuous reduction of variability at each stage of the design and manufacturing cycle—thus the term Variability Reduction Process.

VRP uses modern quality techniques such as teamwork, quality function deployment, and design of experiments. An essential theme of this guidebook is that the VRP tools do not replace the existing methods by which weapon systems are designed and produced; they modify and supplement them in important ways. For example, the concept of the loss function is related to operating and support (O&S) costs; quality function deployment is closely tied to systems engineering. Explaining the tools of VRP in the context of system design and production is one of the objectives of this guidebook.

Improving combat capability is central to Air Force objectives. Improving combat capability, however, is becoming increasingly difficult in the face of constrained manpower and fiscal resources. VRP is an important component of the solution to this dilemma because it offers increased reliability and maintainability while simultaneously reducing the development time and cost of Air Force systems. The substantial improvements in combat capability available through improved R&M are well documented. The importance of reduced development time and cost should be obvious to all.

Accomplishing the purpose of VRP requires the achievement of three basic objectives:

1. Robust Designs: Designs that are insensitive to variations in the manufacturing process and environments

2. Capable Manufacturing Processes: Processes that produce uniform, defect-free products

3. Continuous Improvement: Improvement in all processes throughout the life of the system.

Because VRP focuses on design and manufacturing processes, not the product, it is equally applicable to single-item, low-rate, and high-volume production. The VRP tools include design of experiments (DOE), the loss function, and problem analysis techniques, which are used throughout most of the acquisition process; quality function deployment (QFD) and parameter design, which are applied principally in the design and development phases; and on-line statistical process control (SPC), used to detect and reduce variation during production.

There are also three organizational aspects of VRP. The first is teamwork because the essential idea is that better designs result from bringing together a cross-functional team of people early in the design phase rather than "throwing the design over the wall" to manufacturing. Teamwork is the essential element of concurrent or simultaneous engineering. It is also essential to the conduct of QFD, DOE, problem solving, and other continuous improvement activities.

The second organizational component is management involvement, the preeminent principle of both the R&M 2000 Process and VRP. Top management (in both government and industry) must be committed to deploying VRP throughout the organization. They must emphasize excellence and focus on user satisfaction, employee involvement, and supplier participation. Education and training in the use of VRP tools must be provided regularly and intensively to realize continuous improvement and to achieve the full potential of all organization personnel.

Motivation, the third component, is a key principle of the R&M 2000 Process that is specifically important to VRP. Source-selection ground rule incentives, and critical evaluation of contracto progress can provide industry with motivation t achieve the objectives of VRP.

This guidebook is designed to help those in volved in the Air Force acquisition process and t stimulate additional interest in VRP techniques The guidebook includes an introduction to VRP the VRP approach, an overview of the VRP tools techniques, VRP in the acquisition process, an VRP in the organization. The reader with limite time will want to focus on the basic guidebook; th appendices offer more comprehensive treatment the historical development of VRP, examples o VRP language that could be included in a statemen of work, a VRP review checklist, and related supplementary material. Together with the USAF R&M 2000 Process pamphlet (AFP 800-7), this document provides a reference for implementing VRP.

TABLE OF CONTENTS

CHAPTER I
Introduction

A. Variability Reduction
B. Target Value...
C. How Does VRP Relate to TQM?

CHAPTER II
Variability Reduction Approach and Concepts

A. Background: Conformance to Specification
B. The VRP Approach

CHAPTER III
Overview of VRP Tools and Techniques

A. Introduction
B. Descriptions of Tools/Techniques

CHAPTER IV
Deployment of R&M 2000 VRP in the Acquisition Process

A. Introduction
B. Source Selection
C. Incentives for Implementing VRP
D. Contract Monitoring and Evaluating Progress
E. Summary

CHAPTER V
R&M 2000 VRP in the Organization

A. Management
B. Education and Training

APPENDICES

Appendix

A. Historical Development of VRP
B. Statement of Work Sample Text
C. Useful Military Reference Documents
D. Bibliography
E. Acronyms
F. Barriers to Total Quality Management in the Department of Defense
G. VRP Review Checklist

LIST OF FIGURES

Figure

I-1 DOD TQM Implementation
I-2 Elements of TQM Implementation
I-3 Life-Cycle Application of VRP Techniques
II-1 Conformance to Specification
II-2 Variability Reduction Approach
II-3 Process Capability Relationships
II-4 Process Capability Index
II-5 Comparison of U.S./Japanese Transmission Companies
III-1 SPC Control Chart
III-2 Loss Function
III-3 Quality Function for Pyrotechnic Burn Time Levels (Target Values)
III-4 Problem Analyses Techniques
III-5 IR Maverick Inner Gimbal Ishikawa Diagram
III-6 QFD House of Quality
III-7 Product Planning
III-8 Teamwork
IV-1 Cultural Changes Needed
V-1 VRP Education and Training Structure
A-1 Evolution of Quality Management Awareness
A-2 Comparison of Traditional Way and New Way (PDCA Cycle)

LIST OF FIGURES (Continued)

Figure

A-3 Progress in Quality Assurance Methods
F-1 Percent of Responses by Category

LIST OF TABLES

Table

1 VRP Tools

Chapter I
Introduction

This chapter discusses variability reduction and its relationship to USAF R&M 2000 VRP objectives, presents the concept of target value, and explains how R&M 2000 VRP relates to TQM.

A. VARIABILITY REDUCTION

Systems fail for many reasons. Some components, like tires, wear out. *The major cause of eventual failures, however, is variability in the original manufacturing processes.* Variability results from changes in the conditions in which items are produced, including differences in raw materials, machines and their operators, and other aspects of the manufacturing environment. When variability increases, the product's physical properties or performance degrade, and the number of defects increases. The significance that variability has on a product's reliability and quality depends on the criticality of the manufacturing process and part characteristics. Defects result not only from limitations on parts and materials but more particularly from design.

Design methods based on modern quality techniques minimize the significance of variability. There are several ways to reduce variability and decrease defects. Traditionally, the approach has been to tighten design tolerances and increase inspections. Costs climb as scrap and rework increase; productivity drops. A better method is to reduce the variability by improving both the design and production processes. One good way to do this is to eliminate the causes of variability through statistical techniques. A better way is to design and develop robust products that are insensitive to the causes of variability in the first place. Reducing variability through statistical techniques and product design is aptly named the Variability Reduction Process (VRP).

The purpose of VRP is to provide a proven set of practices and techniques that yield more reliable and defect-free products at lower cost. It is a structured, disciplined design and manufacturing approach aimed at meeting (and exceeding) customer expectations and improving the development and manufacturing process while minimizing time and cost of acquisition.

To accomplish this, three objectives need to be achieved:

- Robust Designs: Designs that are insensitive to the manufacturing process, customer use, and the environment. A key element of achieving robust designs is action early in the design phase to ensure that robustness is incorporated up front with inputs from all functions involved in the product life cycle.

- Capable Manufacturing Processes: Processes that produce uniform, defect-free products that function well over their intended lifetime. This can only be achieved when the critical manufacturing parameters are known and the causes of variability are eliminated or minimized.

- Continuous Improvement: Improvement in design and manufacturing processes throughout the life of the system. For VRP to succeed, management from the top down must adopt new attitudes about reliability and quality and must become directly involved in continuously improving the design and manufacturing processes. They must implement programs to foster improvement, tear down the barriers that inhibit change, instill teamwork, establish goals for improvement, and provide education and training for successful implementation.

USAF R&M 2000 VRP is one of the USAF R&M 2000 Process building blocks, a set of reliability and maintainability (R&M) practices that have proven to be effective and efficient in achieving more reliable, defect-free weapon systems that do the job they were designed to do. VRP provides increased reliability and maintainability while simultaneously reducing the time and cost of developing Air Force systems. This guidebook presents the essential tools and techniques of VRP.

B. TARGET VALUE

The single most important concept behind VRP is the modern design and manufacturing practice of reducing variability around the target value. The key phrase is target value: target values are those that are the best for satisfactory performance of the product in the user's operational environment. Emphasizing the target value and reducing variability around it are fundamentally different from emphasizing conformance to requirements in the sense of meeting tolerances.

Reducing variability in this manner shortens development time, reduces costs of design and redesign, and creates a better performing, more reliable product. A reliable product is one that meets operational requirements and performs its functions over its design life. The key recognition is that the operational requirements can be the basis for key target values. They in turn determine design and production target values. An operational requirement for bombing accuracy would require target values of position coordinates, with accuracy being determined by the variation from the target values. Other examples of target values are a design value for gain in a traveling wave tube, or a manufacturing parameter for controlling surface flatness.

By minimizing variability around the target value, fewer defects are released into the field, which increases reliability and consequently combat capability. Producing fewer defects also lowers cost, thereby providing more combat capability for the same expenditure. The process is equally effective in achieving the bottom line for the Air Force and for industry. For the Air Force, combat capability is improved, and for industry, market share and profit are increased.

C. HOW DOES VRP RELATE TO TQM?

TQM is the DOD strategy for continuously improving performance in all functional areas and at every level. Meeting customer expectations in the minimum time and at the lowest cost is the objective. While implementing TQM involves all of the activities of an organization, commitment of top management is the key motivator. Implementing TQM includes the four steps shown in Figure I-1.

FIGURE I-1
DOD TQM IMPLEMENTATION

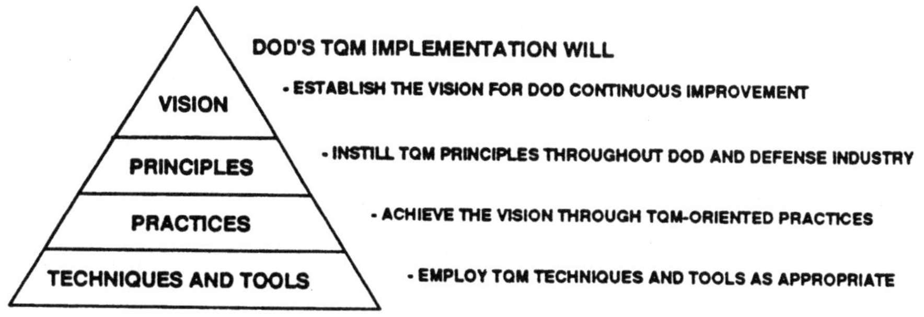
VISION
PRINCIPLES
PRACTICES
TECHNIQUES AND TOOLS
DOD'S TQM IMPLEMENTATION WILL
- ESTABLISH THE VISION FOR DOD CONTINUOUS IMPROVEMENT
- INSTILL TQM PRINCIPLES THROUGHOUT DOD AND DEFENSE INDUSTRY
- ACHIEVE THE VISION THROUGH TQM-ORIENTED PRACTICES
- EMPLOY TQM TECHNIQUES AND TOOLS AS APPROPRIATE

R&M 2000 VRP is an Air Force initiative that promotes variability reduction during design and manufacture. As seen in Figure I-2, VRP-related tools/techniques (discussed in Chapter III) are an integral part of TQM implementation.

Many of these techniques are best used by a cross-functional team. Teamwork ensures multidisciplinary collaboration and will help facilitate efficient, earlier, and faster resolution of problems. Cooperation and communication are keys to the success of all team efforts. Cross-functional teamwork differs from traditional matrix management in that team members participate throughout the life of the project, as opposed to on-call service in matrix management.

The order of the VRP-related tools and techniques in Figure I-2 reflects how they are applied through the life cycle of a product. Teamwork, problem analysis, and process improvement are ubiquitous. Historically, teamwork and statistical process control are the first ones introduced in the organization. The pattern typically starts with statistical process control (applicable in the main to manufacturing) and then moves back earlier and earlier in the life cycle. Figure I-3 shows where various VRP techniques might be applied during a program life cycle.

Implementing VRP and TQM requires the leadership and continuous support of management at all levels. It requires a commitment to teamwork, continuous improvement, process orientation, and supplier participation.

FIGURE I-2
ELEMENTS OF TQM IMPLEMENTATION

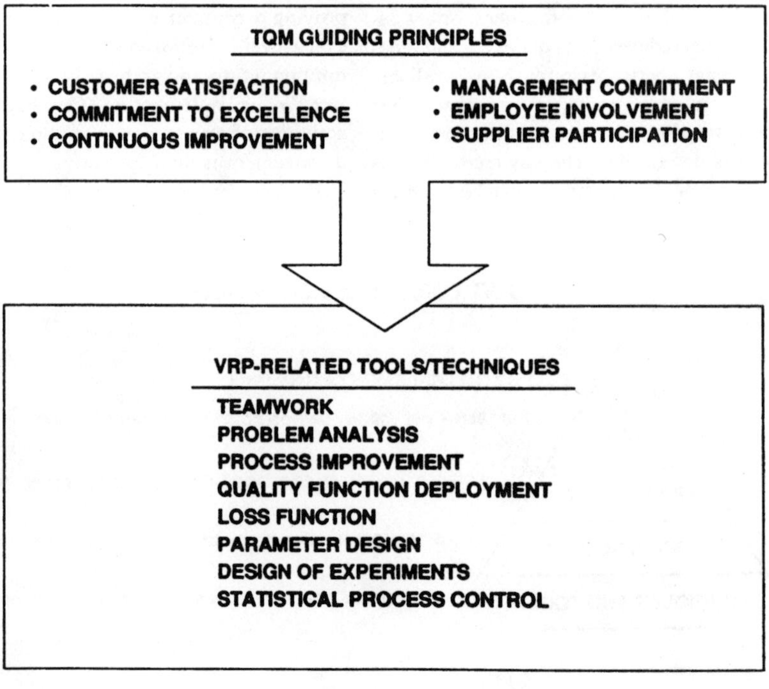

TQM GUIDING PRINCIPLES

- CUSTOMER SATISFACTION
- COMMITMENT TO EXCELLENCE
- CONTINUOUS IMPROVEMENT
- MANAGEMENT COMMITMENT
- EMPLOYEE INVOLVEMENT
- SUPPLIER PARTICIPATION

VRP-RELATED TOOLS/TECHNIQUES

TEAMWORK
PROBLEM ANALYSIS
PROCESS IMPROVEMENT
QUALITY FUNCTION DEPLOYMENT
LOSS FUNCTION
PARAMETER DESIGN
DESIGN OF EXPERIMENTS
STATISTICAL PROCESS CONTROL

FIGURE I-3
LIFE-CYCLE APPLICATION OF VRP TECHNIQUES

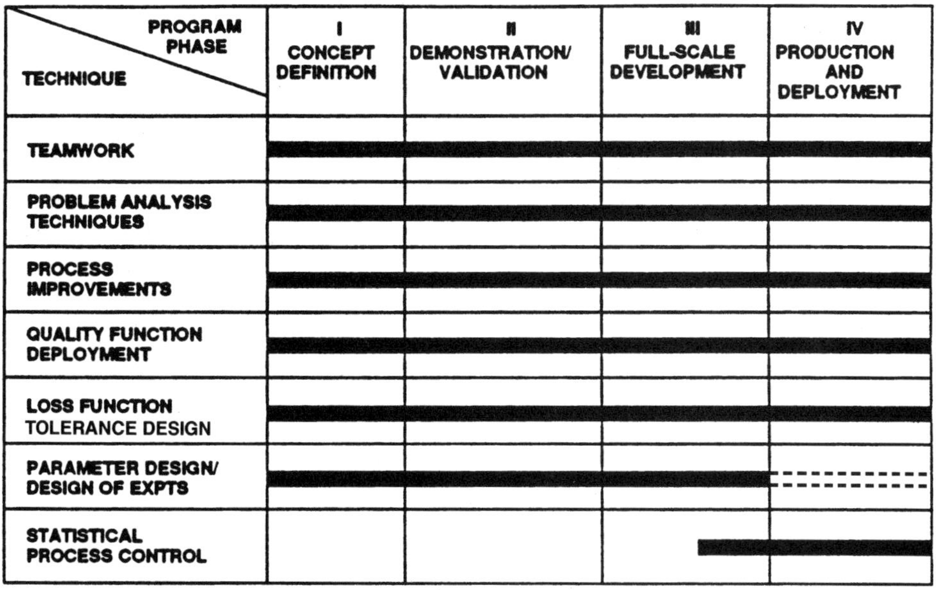

TECHNIQUE \ PROGRAM PHASE	I CONCEPT DEFINITION	II DEMONSTRATION/ VALIDATION	III FULL-SCALE DEVELOPMENT	IV PRODUCTION AND DEPLOYMENT
TEAMWORK				
PROBLEM ANALYSIS TECHNIQUES				
PROCESS IMPROVEMENTS				
QUALITY FUNCTION DEPLOYMENT				
LOSS FUNCTION TOLERANCE DESIGN				
PARAMETER DESIGN/ DESIGN OF EXPTS				
STATISTICAL PROCESS CONTROL				

Chapter II
Variability Reduction Approach and Concepts

This chapter describes the philosophy and the general approach necessary to understand what is needed to implement variability reduction. This approach defines the foundation for variability reduction activities and relates directly to all three R&M 2000 VRP objectives described earlier.

A. BACKGROUND: CONFORMANCE TO SPECIFICATION

In the past, it was common government and industry practice to measure a product's acceptability by its conformance to specification limits.[1] The goal was to produce products with operating variables/tolerances that always fell within prescribed lower and upper specification limits (LSL and USL). These measurements generally were established by engineers. Figure II-1 illustrates this concept. Product characteristics falling within the limits are considered acceptable. This concept is a barrier to improving systems and equipment because it fosters an attitude that meeting the specification limits is all that is required and that product quality is a "go/no-go" matter.

The concept of conforming to specification limits inevitably leads to increased costs and waste because some items will conform and others will not and also because not all the products within specification limits will work equally well. Further, according to this concept, manufactured parts and assemblies must be inspected to determine which can be accepted and which must be rejected. Rejects are scrapped or reworked. In this situation, after-the-fact inspection techniques prevail rather than process control or up-front quality design practices. Reducing the specification limits (tighter tolerances) and making more inspections are common approaches to "improved" quality. However, such solutions only treat the symptoms, increase scrap and rework, and raise cost; they do not resolve the problems.

FIGURE II-1
CONFORMANCE TO SPECIFICATION

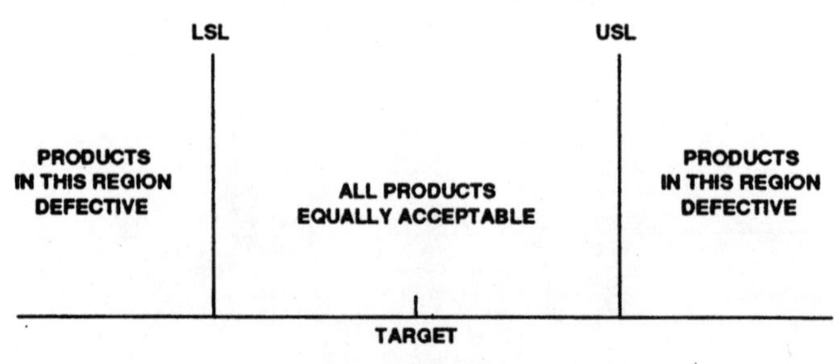

PRODUCT CHARACTERISTIC

LSL = LOWER SPEC LIMIT
USL = UPPER SPEC LIMIT

[1] Military standards and specifications defining quality processes for both hardware and software are in transition from the traditional approach to modern methods. This should not be a surprise since implementing modern quality, in DOD or elsewere, requires fundamental institutional change. To a large extent, military standards and specifications reflect the DOD institution. Current specifications and standards have a strong residual emphasis on conformance to specifications (i.e., inspecting in quality).

Of equal importance, even if the product char- teristics fall within the lower and upper specifi- tion limits, the customer is forced to accept a oduct that may not work well or last as long as eded in the field. There is, however, another proach.

. THE VRP APPROACH

Variability reduction seeks performance at a rget value that is well within any prescribed andards or specification limits.[2] Figure II-2 shows e difference between VRP and the specification ethod. Reducing the variations of performance ariability reduction) around the target value will sult in a more uniform, more defect-free product ith greater consistency of performance and with igher reliability. These variations related to the rget value are known as process limits, since they are associated with process control. The questions are: How is this variation measured, and why does higher reliability result when compared to operating just within specification limits?

The process variation around the target value can be measured by a process capability index, C_p, as shown in Figure II-3. The index C_p is the ratio of specification range (LSL to USL) to 6σ (the process range). It assumes that the process mean (the central tendency of the process) lies on the target value—a perfectly centered process as shown in Figure II-3. Examples of defect rates corresponding to C_p values are listed in the table included in Figure II-3. Note that building to spec (C_p=1.0) results in 2,700 defective parts per million (ppm). In Japan, a C_p of 1.33 (63 ppm) is normally the minimum acceptable level, with higher values often required. C_p values greater

FIGURE II-2
VARIABILITY REDUCTION APPROACH

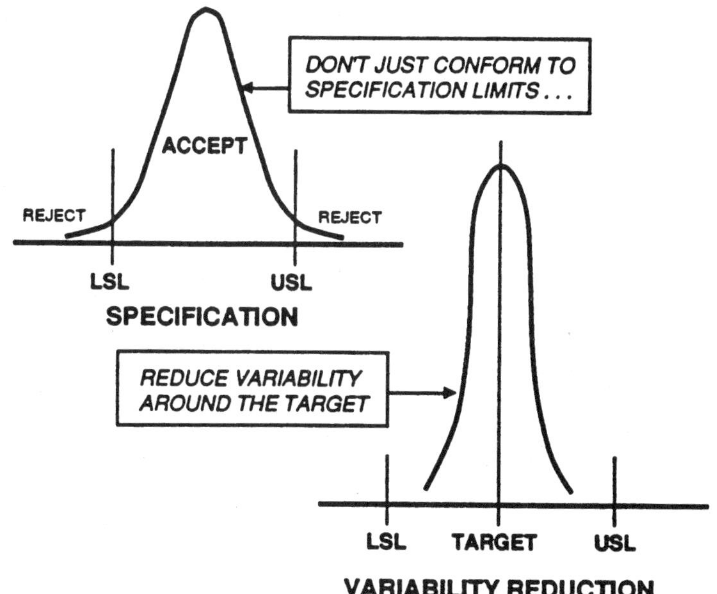

than 4 are achievable in practice. This small a defect rate, equivalent to 12 standard deviations away form the mean, cannot be found in a table of normal probabilities. It should be no surprise that when defect rates reach these levels (parts per billion), there is little point in hiring quality control inspectors to check if products conform; inspecting in quality becomes a thing of the past. Leading U.S. companies are aiming for comparable values.

A slightly different measure, C_{pk}, is commonly used when the process is not centered on the target value. See Figure II-4. C_{pk} is defined by

$$C_{pk} = \frac{(\text{process mean - nearer spec limit})}{3\sigma}$$

where

σ = standard deviation (of the measured characteristic in the manufacturing process).

FIGURE II-3
PROCESS CAPABILITY RELATIONSHIPS

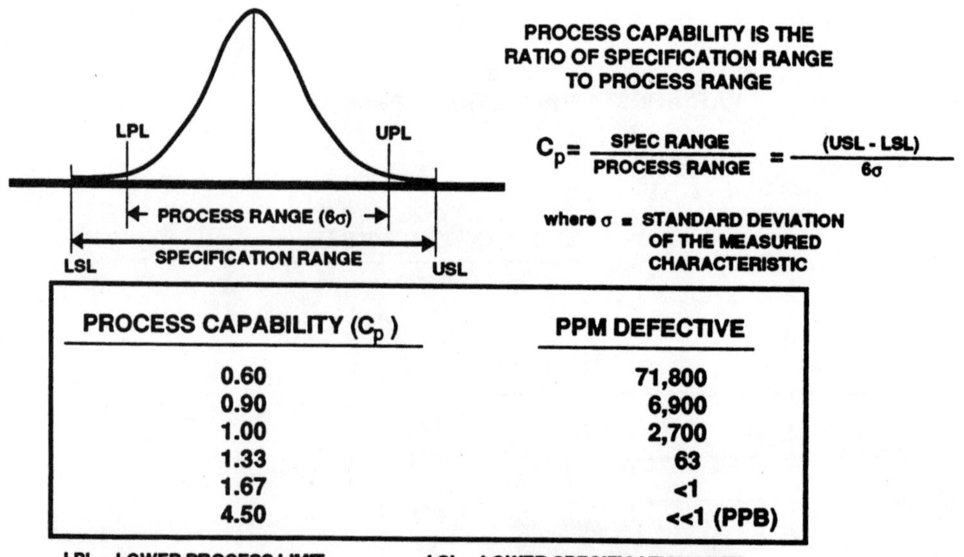
LPL
UPL
PROCESS RANGE (6σ)
SPECIFICATION RANGE
LSL
USL
PROCESS CAPABILITY IS THE RATIO OF SPECIFICATION RANGE TO PROCESS RANGE
Cp= SPEC RANGE / PROCESS RANGE = (USL - LSL) / 6σ
where σ = STANDARD DEVIATION OF THE MEASURED CHARACTERISTIC

PROCESS CAPABILITY (C_p)	PPM DEFECTIVE
0.60	71,800
0.90	6,900
1.00	2,700
1.33	63
1.67	<1
4.50	<<1 (PPB)

LPL = LOWER PROCESS LIMIT
UPL = UPPER PROCESS LIMIT
LSL = LOWER SPECIFICATION LIMIT
USL = UPPER SPECIFICATION LIMIT

C_p is referred to as a measure of process potential and is the most commonly used process capability index. C_{pk} is called a measure of process performance since it considers both variability and the location of the mean.

Improving process capability actually does two things. First, it reduces cost by reducing scrap, rework, and inspection—as just discussed. More importantly, it improves performance over time. A simple mechanical example is increased bearing life, from reduced vibration, when concentricity variations are reduced. As a second example, closer adherence to target values for electronic circuit board dimensions decreases future likelihood of delamination, bridging between adjacent traces, and failure of solder connections. Further, the improved predictability of physical, chemical, and electrical characteristics when variation in material composition is reduced has profound effects on reliability for both electronic and mechanical systems. Because of this tie between quality and reliability, VRP is an important component of the R&M 2000 Process.

An example of the value of variability reduction is found in the automotive industry. An American car manufacturer found that its U.S. supplier could not meet the demand for transmissions. A Japanese company became a second supplier. Specifications and blueprints for the transmission given to the Japanese-managed plant were identical to those used by the U.S. supplier. The transmissions delivered by both plants conformed to specifications, but the Japanese transmissions were superior in operation. Investigation revealed that the Japanese product had a finer finish, fewer chips and burrs, and less variation in important product characteristics because it had been made with each of its critical dimensions closer to the target value. While the U.S.

FIGURE II-4
PROCESS CAPABILITY INDEX
(PROCESS NOT CENTERED ON TARGET VALUE)

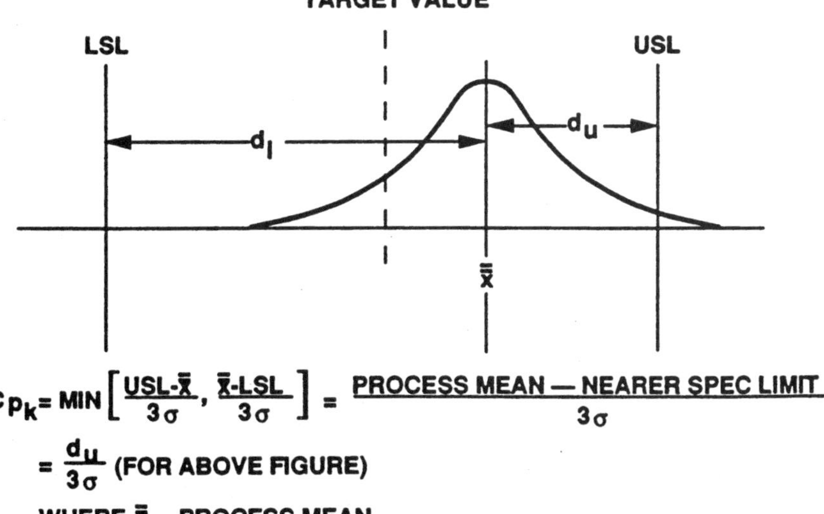

$$C_{p_k} = \text{MIN}\left[\frac{USL-\bar{x}}{3\sigma}, \frac{\bar{x}-LSL}{3\sigma}\right] = \frac{\text{PROCESS MEAN — NEARER SPEC LIMIT}}{3\sigma}$$

$$= \frac{d_u}{3\sigma} \text{ (FOR ABOVE FIGURE)}$$

WHERE $\bar{x}$ = PROCESS MEAN

plant was working to make every piece conform to drawing, the Japanese manufacturer was working to make each piece exactly the same.

The variability in the Japanese units was only 27 percent of the spread between upper and lower specification limits (equivalent to a C_p of 3.7 if the mean were centered on the target value), whereas the U.S. distribution curve was over 70 percent of the limit range ($C_p \simeq 1.4$). See Figure II-5. The resulting warranty costs for the Japanese units were significantly less. This U.S. car manufacturer now specifies target values for product characteristi and uses processes that produce minimum var ation from these values, as opposed to just emph sizing specifications.

Dr. W. Edwards Deming has likened the var ability situation to a comparison of the Londo Symphony Orchestra and his hometown orchestr playing a symphony from the same score. H stated, "Same music; same specifications. Not mistake. Both perfect. But listen to the difference Just listen to the difference."[3]

FIGURE II-5
COMPARISON OF U.S./JAPANESE TRANSMISSION COMPANIES

U.S. COMPANY

$$C_p = \frac{\text{SPEC RANGE}}{\text{PROCESS RANGE}} = \frac{1}{.7} = 1.43 \Rightarrow$$ 18 DEFECTS PER MILLION

JAPANESE COMPANY

$$C_p = \frac{\text{SPEC RANGE}}{\text{PROCESS RANGE}} = \frac{1}{.27} = 3.7 \Rightarrow$$ <<1 DEFECT PER BILLION

Chapter III
Overview of VRP Tools and Techniques

A. INTRODUCTION

VRP is more than just variability reduction. It includes understanding requirements and incorporating the requirements into a robust design that can be produced by capable processes. This chapter presents an overview of the most important VRP tools and the essential concept of teamwork. These tools have been demonstrated to succeed by competitive, world-class companies.

VRP uses a variety of tools and techniques to achieve robust design, capable processes, and continuous improvement. VRP:

- Identifies and controls the parameters important to the customer, process, and product
- Reduces the effort (and cost) of over-specification and control of unimportant parameters.

Table 1 shows where the most significant tools and techniques are used in relation to R&M 2000 VRP objectives. Statistical process control (SPC), the loss function, design of experiments, and parameter design are engineering methods applied at the detailed (i.e., circuit board, component, assembly) level. Quality function deployment and team-

TABLE 1
VRP TOOLS

OOLS/TECHNIQUES	R&M 2000 VRP OBJECTIVES		
	ROBUST DESIGN	CAPABLE PROCESSES	CONTINUOUS IMPROVEMENT
TEAMWORK	X	X	X
PROBLEM ANALYSIS TECHNIQUES	X	X	X
PROCESS IMPROVEMENT		X	X
QUALITY FUNCTION DEPLOYMENT	X		X
LOSS FUNCTION	X	X	X
PARAMETER DESIGN	X		X
DESIGN OF EXPERIMENTS	X	X	X
STATISTICAL PROCESS CONTROL		X	X

work are methods at the systems engineering level. The VRP problem analysis techniques apply over the entire process. SPC is probably the most familiar of the tools and has been used by many organizations.

B. DESCRIPTIONS OF TOOLS/TECHNIQUES

Statistical process control (SPC) is used to monitor the critical variables of a manufacturing process so that the causes of process variations can be identified and controlled. Capable manufacturing processes can be achieved only when the critical parameters are known and the causes of variations are eliminated or minimized. A wide variety of simple-to-apply SPC techniques is available, including control charts (Figure III-1), histograms, and run charts. SPC allows the operator to observe the process and distinguish patterns of random and abnormal variations. It assists the operator in making decisions such as adjusting or shutting down the process before defects are produced. Then the worker can isolate and remove the causes of abnormal variation, which may involve the use of other VRP tools or techniques. When the abnormal variations are removed from the process, the process is said to be under statistical control. The emphasis on operator and worker is deliberate. SPC is primarily an operator's tool, although the assistance of a statistician is valuable in analyzing results.

The loss function ("nominal the best" case shown in Figure III-2) was created by Dr. Genichi Taguchi and is a methodology for evaluating loss due to variability. The power of the loss function and the associated theory is that it reflects costs due to variation and lays a powerful foundation for decisionmaking. For the commercial world, where the loss function was born, "true cost" includes not only direct manufacturing costs and the cost of scrap and rework but also warranty returns, excess inventory and capital investment, customer dissatisfaction, and eventual loss of market share. However, in its most common interpretation,[4] the loss function measures the costs once the product is out the plant door. Thus it measures the same kinds of costs that the Air Force would interpret to be within operating and support (O&S). If that were the end of it, the loss function would add little to the discipline of O&S estimation. But the loss function has important benefits not provided by other existing methods.

- The first benefit is conceptual. As discussed in Chapter II, the traditional notion that loss only occurs when a product falls outside the specification limits, the traditional go/no-go approach, is too narrow a viewpoint. The loss function connects variability to cost.

- The second benefit is eminently practical. Whereas attempting to estimate quality-driven O&S costs at the component level using traditional O&S models would be difficult at best, the loss function is an easily implemented and practical engineering figure of merit.

Figure III-2 illustrates the case where nominal is best. However, as Figure III-3 illustrates, the loss function can be one-sided when variation in one direction from the target value produces negative results, while variation in the other direction has a positive impact. Figure III-3 illustrates the quality loss function for pyrotechnic burn time of a tracer cartridge. In this case, longer burn time for a tracer cartridge is better. Parameter design experiments led to reduction of the loss function from $2.00 for the old process to $0.81 for a new process, a savings of 60 percent. The arithmetic is useful in the same way that O&S estimates are; it is not the absolute values that count but the deltas. The loss function is an effective tool to help understand the payback of variability reduction so that the payback can be traded off with cost of implementation.

FIGURE III-1
SPC CONTROL CHART

PART	*PROCESS*		**TARGET: 95 FT•LBS**							
OPERATOR	*MACHINE*		**SPEC LIMITS: ±20 FT•LBS**							
DATE										
TIME										
MEASUREMENT	1	102	106	104	89	95	100	96	107	92
	2	94	97	91	94	105	95	85	103	100
	3	97	102	105	92	103	92	104	100	93
	4	106	107	98	96	92	93	90	105	90
AVERAGE, $\bar{X}$		100	103	99	93	99	95	94	104	94

115
UCL →
$\bar{\bar{X}}$ 95
LCL →
75
$\bar{\bar{X}}$

- **UCL & LCL, UPPER AND LOWER CONTROL LIMIT (CALCULATED)**
- **$\bar{\bar{X}}$ MEAN OF AVERAGES**

FIGURE III-2
LOSS FUNCTION

NOMINAL THE BEST

$$L = \begin{cases} k\,(y-m)^2 \\ k\,(\sigma^2 + (\mu - m)^2) \end{cases} \qquad k = \frac{A_o}{\Delta_o^2}$$

- m = **DESIGN NOMINAL OR TARGET VALUE**
- A_o = **DOLLAR LOSS AT THE TOLERANCE LIMIT**
- Δ_o = **HALF THE MANUFACTURING TOLERANCE**
- σ = **PROCESS STANDARD DEVIATION**
- μ = **PROCESS MEAN**

IF PROCESS CAPABILITY (Cp) IS KNOWN, PROCESS IS SYMMETRICAL AND CENTERED, AND (+) AND (-) TOLERANCE EQUAL, THEN THE AVERAGE DOLLAR LOSS PER UNIT IS

$$L = \frac{A_o}{9 \times Cp^2}$$

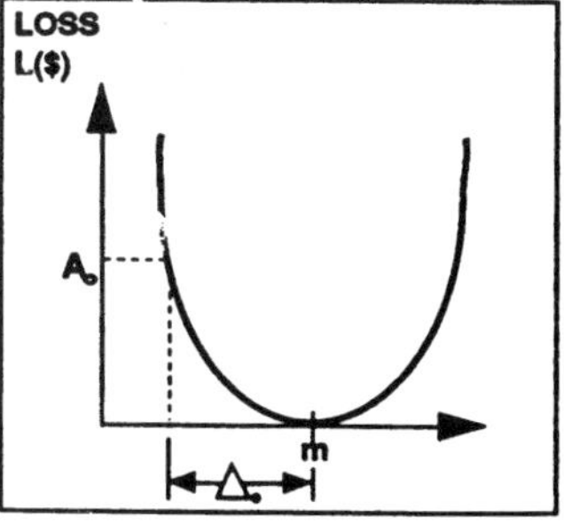

FIGURE III-3
QUALITY FUNCTION FOR PYROTECHNIC BURN TIME LEVELS (TARGET VALUES)

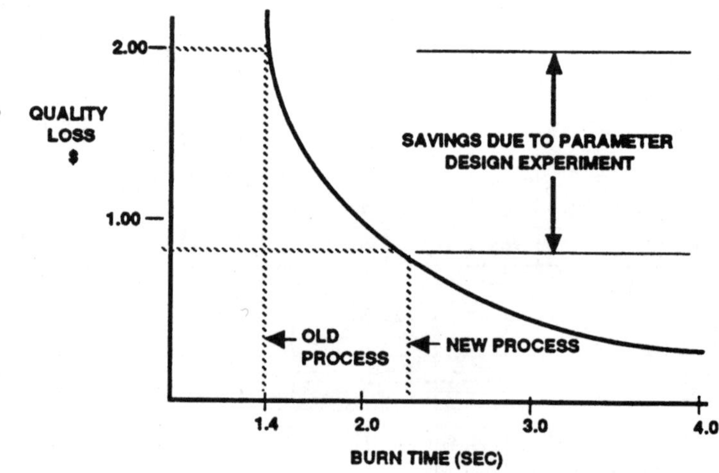

SOURCE: AEROJET GENERAL

Design of experiments (DOE) is another powerful tool. Its advantage to a systems acquisition program is that it will accelerate the rate at which product designs and manufacturing processes improve and, hence, result in less rejects and shorter flow times.

Many parameters influence the variability of a manufacturing process, but not all are equal in terms of their importance to variability reduction. There is no point in spending time and money controlling a parameter that does not make a difference. DOE can be used both to identify the critical process parameters and to establish their optimal levels (target values).

DOE works by measuring the effects that different inputs have on a process. This is done by varying input values during a series of carefully selected experiments, collecting the data, and analyzing the results. An input may be varied over a range of values, such as an oven's curing temperature, or conditionally, such as the decision to add or withhold a curing additive. The actual design of experiments rests on established statistical techniques. These techniques include full-factorial designs, partial-factorial designs such as orthogonal arrays, response surface analysis, two-level designs, three-level designs, etc., all of which are basically aimed at striking a balance between the information you need and the cost to obtain it. DOE is used in parameter design, as will be discussed next.[5]

Parameter design is a technique for achieving robust designs by selecting product and process parameters that minimize variability and achieve better performance over a wide range of operating conditions and environments. During parameter design, a set of parameters is identified (through brainstorming or the problem analysis techniques later described) that engineers and/or technicians

[5] DOE also can be used to reduce the cost and time of developmental tests (e.g., wind tunnel, structural, etc.).

believe could affect the process variability. Then a series of experiments (DOE) is conducted to quantify the effects of the parameters on the desired part characteristics. The results will identify parameter settings that will minimize variability and improve the product and increase yield. Parameter design provides a systematic approach to product/process design, as opposed to an intuitive approach.

Parameter design quickly shows that what everybody thought was the most important problem is not and may not be important at all, which means that money was being spent, perhaps for years, controlling the wrong thing. Frequently, parameter design will uncover problems that are a complete surprise.

Parameter design is the key for achieving enhanced reliability, maintainability, and producibility while simultaneously reducing cost.

Problem analysis techniques, such as Pareto analysis and Ishikawa diagrams, can be used for analyzing difficulties throughout the product life cycle (Figure III-4). These and other techniques are simple but effective methods for prioritizing, understanding, and analyzing problems in devel-

FIGURE III-4
PROBLEM ANALYSES TECHNIQUES

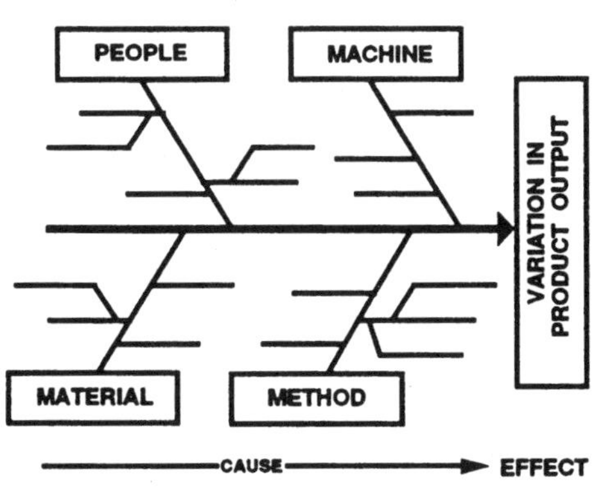

ISHIKAWA DIAGRAM

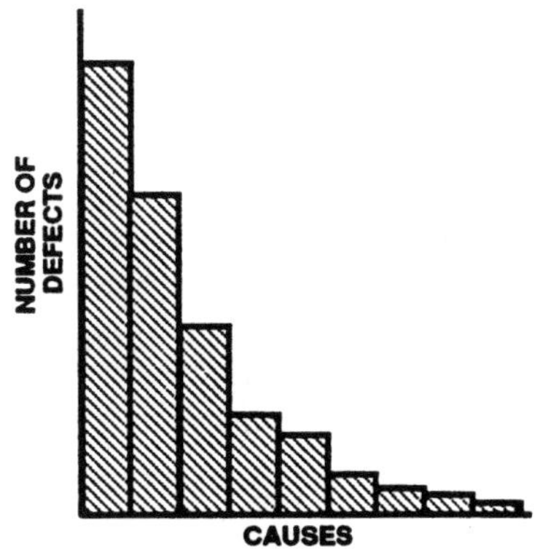

PARETO ANALYSIS

opment and production. Figure III-5 illustrates how the Ishikawa diagram was used by engineers at Hughes Missile Systems to depict the relationships between various types of defects and the nature of the operations performed in the fabrication of the IR Maverick Inner Gimbal. This chart helped identify possible causes of scrap at each operation, which led to operational changes that eventually resulted in increased yields.

Process improvement is the ongoing organizational commitment to continually apply the VRP and other tools and techniques to improve the process and thus improve the product. Defect-free products are the goal, and the organization continually strives to improve all facets of design and production to either reach that goal or to maintain that level of performance. The organization uses the PDCA (Plan-Do-Check-Act) cycle[6], together with the VRP and other tools, to provide a structured methodology for improving all processes.

The six concepts discussed thus far—SPC, the loss function, DOE, parameter design, problem analysis techniques, and process improvement—are highly focused tools. Quality function deployment and teamwork have a much broader reach.

Quality function deployment (QFD) is a structured procedure for translating the "voice of the customer" (i.e., voice of the airman) into design characteristics and target values that then are disseminated horizontally through the organization's product planning, engineering, manufacturing, quality assurance, cost, and service departments. It

FIGURE III-5
IR MAVERICK INNER GIMBAL ISHIKAWA DIAGRAM

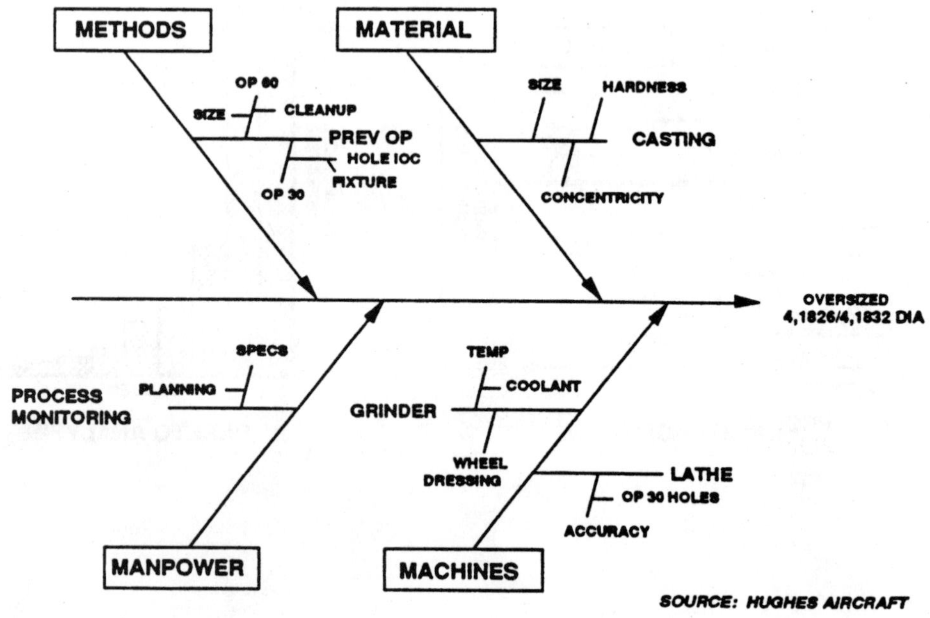

SOURCE: HUGHES AIRCRAFT

emphasizes early involvement in both product and process design decisions by people in all pertinent functional areas, and therefore it is a natural adjunct of, and a catalyst for, teamwork.

QFD employs a set of matrices, frequently referred to as the "house of quality" (Figure III-6). They provide a methodology for systematically deploying the customer's requirements throughout the product development and delivery process. The first matrix in the series is called the product planning matrix. It minimizes the chance of starting the design process with incomplete or erroneous requirements. Figure III-7 depicts the product planning matrix for an automotive turbocharger. The left-hand column is a list of the customer's needs, i.e., what features must be provided so that the product satisfies the customer. The horizontal items are the product requirements, i.e., how the customers' needs will be met in terms of engineering parameters. The matrix entries shown by the symbols indicate the degree to which the customer needs are met by the product requirements as determined by the QFD team.

The triangular roof of the matrix documents the degree of interaction between the various product requirements. The column on the far right shows how well the current product and competing products satisfy customer needs. Ratings of the product's importance are shown in the bottom row. Importance is classified in five ratings from 1 to 5, with 5 being most important.

FIGURE III-6
QFD HOUSE OF QUALITY

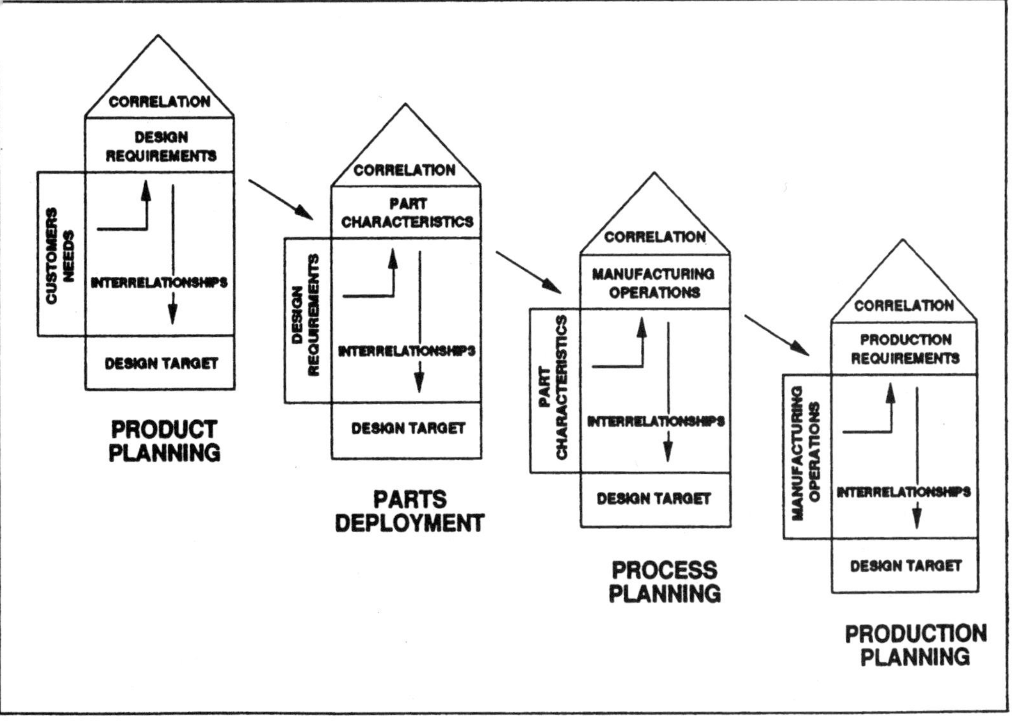

FIGURE III-7
PRODUCT PLANNING

T-X TURBOCHARGER

RELIABLE PRODUCT

CUSTOMER NEEDS \ PRODUCT REQUIREMENTS		TURBINE PERFORMANCE	COMPRESSOR PERF	LUBE SYSTEM INTEGRITY	VIBRATION RESISTANCE	ACTUATOR DURABILITY	C.H.R.A. BALANCE	REGULATOR CALIBRATION	OPERATING NOISE	BEARING DURABILITY	COKING RESISTANCE	MOUNTING INTEGRITY		CUSTOMER SATISFACTION / PRODUCTS COMPARED: T		A	B	C
INCREASED POWER	C	**O**	**O**			**O**		**O**		O		O		3		3	3	3
FAST RESPONSE	A	O	O			**O**		O		O		△		4		4	4	3
SMOOTH RESPONSE	B	**O**	O							O		O		4		4	5	3
LOW-END RESPONSE	B	**O**	**O**			**O**		**O**		O		△		3		3	3	3
FUEL ECONOMY	B	**O**	**O**			**O**		O				△		3		3	3	3
NO EXHAUST LEAKS	C				**O**							**O**		4		4	4	4
NO OIL LEAKS	B			**O**	O						O			2		1	2	4
SAFE PRODUCT	A			O	**O**									3		3	3	3
NO BREAKDOWNS	A	△	O	**O**	**O**	**O**	O	**O**	O	**O**	**O**	O		3		4	2	4
NO TURBO NOISE	B	O	O		O		**O**		**O**	**O**		**O**		2		4	3	5
NO EMISSIONS	A				**O**	**O**		**O**				**O**		4		4	4	4
IMPORTANCE		5	5	3	5	5	1	5	1	3	1	5						

SOURCE: ALLIED SIGNAL

Key to Rankings

Customer Needs: Relative importance to the customer—A (highest), B, or C.
Customer Satisfaction: Degree of customer satisfaction—1 - 5 (highest). The T is the company's product, and A, B, and C are their competitors.
Importance: Relative priority for action—1 - 5 (highest)
The Circles and Triangles in the body of the figure represent the degree of relationship between the customer needs and the product requirements— **O** (strongest), O (medium), △ (light).

QFD enhances systems engineering: it starts earlier (before a set of requirements exists) than systems engineering typically does; it provides a systematic method for enabling cross-functional teamwork. The demonstrated benefits of QFD in the commercial sector include much shorter development times (reduction by more than 50 percent); fewer engineering changes; fewer production start-up problems; and less scrap, rework, and repair. There is every reason to believe it will produce marked benefits for Air Force weapon system acquisition. Tools like the problem analysis techniques fix problems. QFD prevents them.

Teamwork is a means by which many functional areas and disciplines work cooperatively to improve products, processes, and services. Teamwork, despite how it might sound, is not a cheerleading exercise. It brings together multifunctional management teams, interdisciplinary design teams, integrated product development teams, process improvement teams, and employee involvement groups with the specific aim of combining their knowledge to improve quality and cut costs. An important application of the team approach is in product and process design, when representatives of all disciplines and functional areas are brought together at the start of a project. Teamwork is also essential to problem analysis, as discussed earlier.

Coaching, coordinating, and integrating the contributions of the members of the team will require hard work on the part of management. The returns are worth the effort: the interacting team design approach has proven to be superior to the sequential design practice often used in the past. Other terms with very nearly the same meaning are concurrent engineering and simultaneous engineering. Teamwork is crucial to VRP because concurrent engineering or simultaneous engineering cannot function effectively without it. Teamwork played a critical part in the development of the prototype X-31 fighter, an operational capability demonstrator. Because the effort involved many different organizations, including U.S. and West German Defense Departments as well as industry organizations (Rockwell and Messerschmitt Bolkow Blohm), strong working relationships and a concurrent engineering approach had to be developed and used. As a result, significant labor reductions were achieved as shown in Figure III-8. Additionally, no special fixtures were required for the X-31.

FIGURE III-8
TEAMWORK

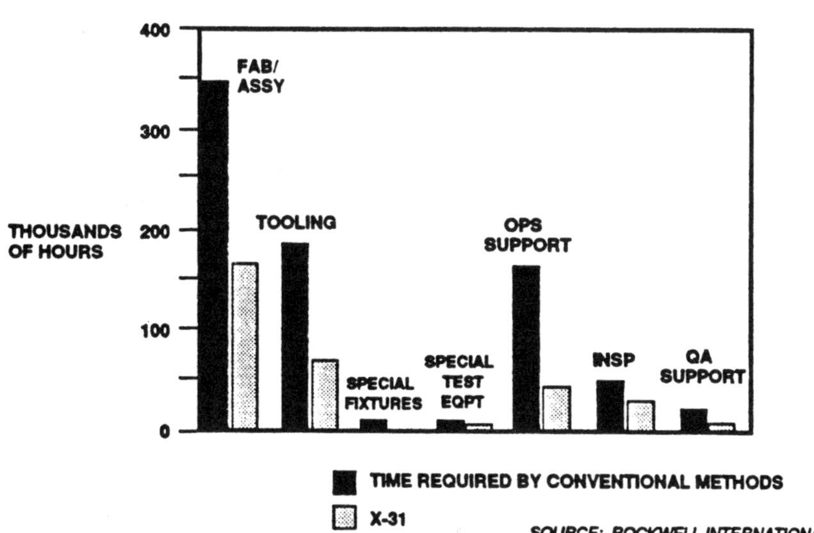

SOURCE: ROCKWELL INTERNATIONAL

Chapter IV

DEPLOYMENT OF R&M 2000 VRP IN THE ACQUISITION PROCESS

The previous chapters described the VRP concepts, objectives, and tools. This chapter discusses how to deploy R&M 2000 VRP in the acquisition process. The essentials are source selection procedures, incentives, and critical evaluation of variability reduction progress during the execution of a program.

A. INTRODUCTION

One of the central principles of the R&M 2000 Process is motivation. The motivation principle, applied to VRP, means communicating to industry the Air Force emphasis on variability reduction to foster contractor commitment. Key motivational methods are

1. Source Selection
   - Emphasize variability reduction in both evaluation criteria and general considerations to determine a company's ability to provide producible, reliable, and maintainable products
   - Select a contractor who has implemented/is implementing variability reduction practices such as those described in this guide
   - Select suppliers who demonstrate capable manufacturing processes
2. Incentives
   - Reward successful application of variability reduction practices
   - Reward continuous improvement
   - Establish long-term business relationships for qualified suppliers
3. Program Evaluation
   - Evaluate progress in VRP implementation
   - Publicize benefits accrued through VRP practices.

VRP utilizes a variety of tools to achieve more reliable, yet less costly, systems. An essential element is that steps are taken early in the design phase to ensure that quality is incorporated "up front" rather than attempting to inspect it into the product after production.

A number of VRP tools are described in this guidebook. *However, the Government should not specify the tools to be used in a particular contract, since there may be several applicable tools, or the contractor may choose to modify them to better suit organizational capabilities and program needs.* No one technique is a panacea for all problems, and efforts should be made to use the best of available technology in a logical manner. Best results will be obtained from an approach emphasizing robust design, capable processes, and continuous improvement. It is intended that VRP be a prominent part of the contractor's producibility, R&M, and company-wide TQM programs. But it is important that the procuring agency, Government or contractor, avoid specifying a rigid, by-the-book plan for VRP implementation.

B. SOURCE SELECTION

1. Overall Purpose

The objective is to select contractors who will apply R&M 2000 VRP throughout the program life cycle (requirements analysis, design, transition to production, manufacturing, and operations) to develop robust designs that are produced by capable processes and complemented by continuous improvement. For VRP to be successful, it should not be program dependent, but it needs to be deployed throughout the corporation. To accomplish this, contractors and their suppliers of necessity will

- Practice top-level commitment to VRP and TQM principles
- Involve personnel at all organizational levels and departments
- Apply proven, cost-effective VRP tools/techniques in an orderly manner
- Implement continuous improvement.

This section discusses three opportunities to stress VRP in the source selection process and, where appropriate, provides suggested language. The opportunities are the Commerce Business Daily, Source Selection Plan, and Request for Proposal. Because each program is unique, the approaches and language are starting points and will always require careful thought and tailoring.

2. Commerce Business Daily (CBD) Synopsis

a. Purpose: To bring attention to variability reduction in the initial screening of offerors.

b. CBD Requirements: Recommend the following statement: "Potential offerors shall include in their statement of corporate capabilities a description of their variability reduction strategy. Include a discussion of corporate policies and practices that support developing robust products and capable manufacturing processes and their continuous improvement."

3. Source Selection Plan (SSP)

a. Purpose: The SSP is prepared for the approval of the Source Selection Authority (SSA) for initiating and conducting the source selection. The purpose of the following text is to assure that the source selection process places priority on engineering and manufacturing processes and practices that achieve VRP objectives, as evidenced by in-place programs and past variability reduction results.

b. Air Force Program Office Responsibilities: Recommend the following: "Variability reduction will be an integral part of the design-development-manufacturing cycle for this program."

4. Request for Proposal (RFP)—Instruction to Offerors (ITO)

a. Purpose: To alert offerors to the emphasis on VRP.

b. Suggested Text by Section:

(1) Executive Summary

"This section shall include a brief overview of the contractor's management organization and approach to total quality as a means to achieve higher reliability and lower cost."

(2) Technical Proposal

"This section shall contain information that clearly demonstrates the offeror's approach and capability to implement variability reduction tools/techniques throughout the scope of the contract. As appropriate, discussions in this section should include items such as

"(a) How VRP practices will contribute to the achievement of reliability, maintainability, supportability, and producibility (RMS&P) goals.

"(b) Details of the proposed engineering effort to integrate design, manufacturing, test, and operations early in the program.

"(c) Application of available process control tools/techniques, such as statistical process control or Poka-Yoke (mistakeproofing), to minimize variation in product quality characteristics during the production phase of the program.

"(d) How robust design engineering procedures in the development phase will identify critical product and process characteristics that must be measured and controlled during production. This item should include a description of the approach to measuring and controlling process variability.

"(e) How personnel are trained and educated in variability reduction tools/techniques."

(3) Management Proposal

"This section should clearly demonstrate the offeror's top-level management commitment to VRP implementation. It should indicate what corporate policy has been developed and deployed and the degree to which it is practiced. It should demonstrate the contractor's ability to produce robust designs with capable manufacturing processes, using suppliers who share the same commitment."

(4) Subcontractors

"This section shall demonstrate how the offeror's subcontracts will be awarded using variability reduction as a key measure of merit."

5. RFP—Statement of Work (SOW)

Sample language for this section of the RFP is contained in Appendix B.

6. RFP—Evaluation Factors For Award

a. Evaluation Criteria:

For programs involving significant development, include "Design for Reliability, Maintainability, Supportability, and Producibility" in the evaluation factors. For engineering development of systems consisting of proven off-the-shelf products, use "Reliability, Maintainability, Supportability, and Producibility." In both cases, require that the role of variability reduction be shown. But since variability reduction is a means to reliability and maintainability, it would be counterproductive to place separate emphasis on it in the form of a specific criterion (or area, item, or factor). Instead, since VRP will normally cross two or more evaluation areas (e.g., reliability, maintainability, and producibility; engineering and integration), it will be an important consideration from the standpoint of an assessment criterion.[7] In particular, VRP is an important consideration when assessing soundness of technical approach and past performance. The following are suggested VRP criteria:

Specific criteria: none recommended

Assessment criteria: within the context of the solicitation's assessment criteria, consider the following:

- Management policy of commitment to excellence, continuous improvement, and process orientation
- An understanding of robust design requirements and how to achieve them

[7] AFR 70-15, Formal Source Selection for Major Acquisitions, distinguishes between specific criteria that relate to program characteristics and assessment criteria that relate to the offeror's proposal and capabilities. As indicated previously, VRP should not be program dependent but needs to be deployed throughout the organization (i.e., related to the offeror's capability).

- Use of product engineering and manufacturing teams with documented procedures
- An understanding of the most critical product and process characteristics that will need to be controlled during the production phase
- Strategy for developing participative supplier relationship
- Plans for flowdown of quality engineering requirements to major subcontractors and vendors of critical items
- An approach to the measurement and control of process variability to ensure capable manufacturing processes
- Management, employee, and supplier education and training program
- Past success with variability reduction.

b. General considerations: Within the context of the overall program and solicitation, consider pre-award surveys and evaluation of past variability reduction performance.

C. INCENTIVES FOR IMPLEMENTING VRP

1. Purpose

To provide incentives to the contractor for investing in the implementation of variability reduction and continuous improvement.

2. Need for Incentives

The contractor needs some form of economic incentive to invest in the continued implementation of VRP. However, unlike the commercial sector, where VRP-like activities are driven by higher profit margins, DOD contracting practices may unintentionally act to discourage extended use of VRP. Such a situation may occur where the cost incentive structure of a nonfixed-price contract limits the contractor's return on continuous improvement.[8] Another example is where the contractor's achievement of high reliability limits future revenue from spares or other support contracts. This simply says that VRP incentives will need to be carefully thought out.

3. Considerations for Contract Incentives

First, VRP contract incentives cannot stand alone from the rest of the contract incentive structure and also will need to vary depending on the stage of development.

Second, contract incentives presuppose some "measurement" that becomes the basis for determining if or how much of an award is made. Particularly in early engineering effort, it is the process part of VRP that is crucial—and process is inherently difficult, and subjective, to measure.

With these considerations, there is no reason that the cost incentive structure of a cost-plus-incentive-fee production contract cannot incentivize VRP—if it is clear that the contractor is sophisticated enough to understand that under VRP low cost and high reliability are synonyms rather than alternatives. Similarly (and with the same caveat), the profit opportunity from a fixed-price production contract should provide its own incentives. Sometimes an award fee with a fixed-price production contract can provide the initial motivation for a contractor to implement VRP.

For engineering contracts consider an award fee. Award fees would be given when

a. The contractor exhibits a thorough and successful VRP integration throughout the functional disciplines of the organization, and independent reviews of the contractor activities show that VRP technologies have been successfully used to arrive at robust design characteristics.

b. The contractor exhibits a complete understanding of the VRP, both at the contractor's

[8] Even under fixed-price contracts, unexpectedly high profits can bring unfavorable scrutiny.

as well as supplier's facilities, resulting in reduced variability of characteristics that are clearly linked to long-term performance, reliability, or supportability.

c. The contractor realizes cost reductions through continuous application of VRP, as measured (where possible on early hardware or existing manufacturing processes) by an increase in the capability index. This applies to both the contractor and suppliers.

4. Design Changes

When engineering changes are prepared by the contractor to baselined designs to make the design more robust, i.e., less sensitive to variation in manufacturing or operating conditions, existing value engineering incentive provisions may be appropriate as a contractor motivator. When the Government owns/controls drawings during the production phase, the Government can easily become the biggest stumbling block because the Government is now in the approval loop for design changes. Delays greater than six months have not been uncommon in practice. Consider a contract provision of the form where failure to disapprove an engineering change within xxx days constitutes approval.

D. CONTRACT MONITORING AND EVALUATING PROGRESS

The experience of commercial industry has shown that to determine how well VRP is being implemented in a given program, periodic reviews will be needed. There are three levels of reviews:

- VRP review of the contractor by the Government
- Internal review by contractor top management (head of profit center) and/or program manager
- VRP review of suppliers by the contractor.

The purpose of these reviews is to determine how well the system for implementing VRP is working and to point out areas for improvement. The emphasis of the review is on the process, in contrast to a conventional quality review, which is product oriented. The VRP review, if performed properly, can benefit both the contractor and the supplier. The VRP review could be part of a corporate quality audit.

The VRP checklist in Appendix G can be used as a starting point by Government, contractor, or supplier to evaluate progress in VRP implementation. Each organization needs to create its own checklist to best suit its program needs.

E. SUMMARY

Through source selection procedures, incentives, and critical evaluation of VRP progress, the VRP objectives of robust design, capable manufacturing processes, and continuous improvement can be made a part of an Air Force program.

The task will not be easy. A major change in mindset is required to get top management committed and the organization working together as cooperative teams. Some of the cultural changes needed are shown in Figure IV-1. Accomplishing this change will take training, work, and perseverance.

FIGURE IV-1
CULTURAL CHANGES NEEDED

FROM	TO
BOTTOM LINE EMPHASIS	QUALITY FIRST
MEET SPECIFICATION	CONTINUOUS IMPROVEMENT
GET PRODUCT OUT	SATISFY CUSTOMER
FOCUS ON PRODUCT	FOCUS ON PROCESS
SHORT-TERM OBJECTIVES	LONG-TERM VIEW
ASSIGNED QUALITY RESPONSIBILITY	MANAGEMENT-LED IMPROVEMENT
INSPECTION ORIENTATION	PREVENTION ORIENTATION
PEOPLE ARE COST BURDENS	PEOPLE ARE ASSETS
SEQUENTIAL ENGINEERING	TEAMWORK
MINIMUM-COST SUPPLIERS	QUALITY PARTNER SUPPLIERS
COMPARTMENTALIZED ACTIVITIES	COOPERATIVE TEAM EFFORTS
MANAGEMENT BY EDICT	EMPLOYEE PARTICIPATION

Chapter V

R&M 2000 VRP in the Organization

The expectations behind VRP are fundamentally different from those of traditional quality methods. Deployment in the organization will require hard work for several reasons.

- There will be a natural tendency to force the new methods to fit into the traditional mold, the "we already do that" reaction. As an example, expect the C_p concept to be misunderstood as simply tighter tolerances.

- Because the new concepts are outside traditional experience, initially they will appear unrealistic. The idea that quality levels can be achieved without inspections is a case in point. The idea of continuous improvement is another.

- Responsibilities and organizational structure will change. The quasi-adversarial role of quality assurance goes out. Teamwork comes in.

Because VRP represents a fundamental change, management's commitment is crucial to getting the organization to adopt VRP techniques and tools. Education and training to create a new set of expectations and to provide skills with the new techniques are key elements of the deployment process. This chapter first discusses management and then education and training.[9]

A. MANAGEMENT

There are a number of readable references available that credibly treat the management issues; further, the issues are basically the same for commercial and DOD-related industry.[10] Rather than repeat what is well covered elsewhere, the list below summarizes the major factors.

1. Management Commitment and Policy

- Management involvement and leadership
  - Continually stress importance of VRP
  - Facilitate employee participation
  - Show visible evidence of involvement
  - Emphasize that the next person in the process is a customer
  - Reward good participation in teamwork
- VRP review
  - Conduct annual review of VRP progress in each functional area

2. Education and Training

- Set up in-house instruction
- Provide external courses and seminars
- Establish on-the-job team-based training
- Obtain consultation by specialists

3. Planning

- Develop VRP implementation plan
- Set goals and objectives with schedules
- Establish agreed measures of variability reduction
- Develop data for process improvements

4. Organization

- Establish product development/process improvement teams
- Define cross-functional relationships for team building

[9] In addition, Appendix F summarizes the results of a recent extensive research project by the Air Force Institute of Technology into the implementation barriers that can be expected.

[10] See, for example, Lance A. Ealey, Quality by Design: Taguchi Methods and U.S. Industry, ASI Press, 1988; Philip Crosby, Quality is Free, New American Library, 1979; Mary Walton, The Deming Management Method, Dodd, Mead and Company, 1986.

- Set up internal review mechanisms, including presidential audits or reviews
- Set up a top-level management quality council, to which teams report

5. Team Projects

- Conduct team training courses
- Develop team facilitators for all VRP tools
- Establish teams to demonstrate VRP tools/techniques
- Generate successful demonstration cases and share results (This is important for the obvious reason that success breeds success.)

6. Employee Participation

- Promote team approach (particularly by allocating resources)
- Facilitate cross-functional dialogue
- Develop and support a suggestion program[11]

7. Supplier Involvement

- Conduct VRP seminars
- Participate in planning and design activities
- Implement a supplier qualification program to reward high-quality suppliers who are committed to VRP

8. Incentives

- Establish incentives for continuous improvement. This applies to both employees and to suppliers.

B. EDUCATION AND TRAINING

1. Importance

Education and training are keys to realizing the potential of both employees and supplier personnel. Professor Ishikawa has stated, "Quality control begins with education and ends with education. To promote QC with participation by all, QC education must be given to all employees, from the president to assembly line workers." According to L. P. Sullivan of the American Supplier Institute, the single most important factor contributing to Japanese successes in world markets is their system for educating and training all employees on a continuing basis.

A combination of education and hands-on training is the quickest method to implement VRP. Since the change in thought process associated with VRP requires leadership from top management, it is vital that the upper levels of management are trained and educated first. However, every person in the organization must be trained. Education and training will be needed recurrently throughout an employee's tenure in the organization. Not only must workers receive training in the use of basic problem-solving tools, but a vigorous education program must continuously keep them abreast of new developments in variability reduction.

From an acquisition program manager's standpoint, the investment by contractors in education and training is both an important and a simple way to judge the seriousness of their commitment to VRP.

2. Types of Training

In Japan, well-defined programs are available for each job level from the company directors, president, and senior management down to the factory worker. The Union of Japanese Scientists and Engineers (JUSE) has played a significant role in the education and training of management for quality. Courses are promoted throughout the year at all levels from the board of directors to the factory worker.

In the United States, the American Society for Quality Control (ASQC) has been prominent in advancing the theory and practice of quality control and management. Also, several training or-

[11] Few organizations do not already have a suggestion program. However, the typical suggestion program—characterized by low program visibility, many review layers with a predisposition to disapprove, slow response and feedback—will not work well for VRP. Envisioned here are the opposite characteristics: fast feedback (from hours to a few days), few review layers (and restrictions on who is permitted to say no), and high visibility. Some companies have implemented on-the-spot approval and reward.

ganizations offer a wide variety of courses on related subjects that cover both quality awareness and the tools necessary to implement VRP. Many universities also offer short courses in statistical methods. The training required in acquiring knowledge of VRP techniques, for the most part, comprises concentrated programs lasting from a few days to several weeks, generally not semester-long courses. For larger organizations, it may be more cost-effective to have in-house training programs. Courses are available to train the instructor or facilitator required for such an activity. Other commercially available training aids include video and audio cassettes, home-study courses, and workbooks. Seminars on special topics, including presentations by successful practitioners, also are useful for supplementing the training courses.

As an example of company-wide training, Xerox undertook a top-to-bottom recasting of its corporate culture with heavy emphasis on training. Virtually all 100,000 Xerox employees completed at least 40 hours of schooling in "Leadership through Quality," a problem-solving system that teaches workers to think of everyone as a customer to please.

Figure V-1 illustrates the matrix of education and training courses that could be prescribed for any organization. Three categories of training are identified: VRP awareness, robust design, and process control. For top management, strong emphasis should be placed on instruction in the area of VRP awareness; an overview of robust design and process control is also required. At the other extreme, supervisors and workers on the production floor should receive on-the-job training in process control techniques.

Methods involving interdisciplinary teams (e.g., design of experiments and QFD) are best taught by having a team train together. Dr. Taguchi advocates the case-study approach facilitated by a resident consultant, including interactions with man-

FIGURE V-1
VRP EDUCATION AND TRAINING STRUCTURE

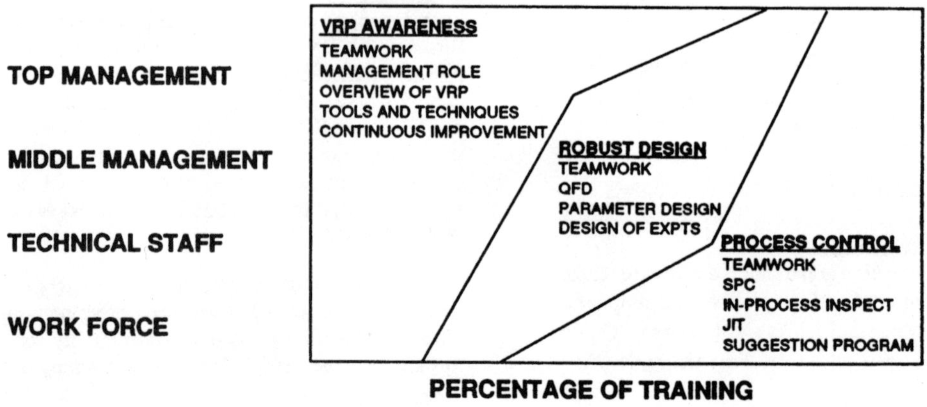

TOP MANAGEMENT
MIDDLE MANAGEMENT
TECHNICAL STAFF
WORK FORCE

VRP AWARENESS
TEAMWORK
MANAGEMENT ROLE
OVERVIEW OF VRP
TOOLS AND TECHNIQUES
CONTINUOUS IMPROVEMENT

ROBUST DESIGN
TEAMWORK
QFD
PARAMETER DESIGN
DESIGN OF EXPTS

PROCESS CONTROL
TEAMWORK
SPC
IN-PROCESS INSPECT
JIT
SUGGESTION PROGRAM

PERCENTAGE OF TRAINING

agement. Seminars featuring case-study results are a good way to exchange experience and ideas. For example, the American Supplier Institute holds an annual Taguchi Methods Symposium and presents a Taguchi medal for the best paper. U.S. companies are beginning to conduct their own symposia where case studies are presented by the teams who worked the actual cases.

3. Supplier Education

From the standpoint of the relationship between prime contractors and suppliers, the goal of R&M 2000 VRP is to eliminate the need for inspecting incoming supplier products. Since this is a significant departure from the present way of doing business, education will be needed for suppliers just as it is needed for prime contractors. Suppliers, particularly small suppliers, will need assistance. The prime contractor will have to take an active role in seeing that its suppliers are familiar with VRP techniques. One way to promote the use of VRP tools/techniques and relay the message of commitment to VRP is to sponsor seminars on quality awareness and VRP. The prime contractor also will need to visit its suppliers, conduct VRP audits, and provide guidance on implementation. Technical experts also can be assigned to work with suppliers on the use of VRP tools/techniques. Another device is to have supplier personnel participate in the prime contractor's in-house training programs. Probably the most effective way is to establish a qualified supplier program to establish a means of assuring a supplier's quality and recognizing quality suppliers who are committed to VRP.

Appendix A

HISTORICAL DEVELOPMENT OF VRP

This appendix traces the historical development of VRP by discussing the evolution of quality management from the turn of the century to the present. The evolution and implementation of quality assurance methods are discussed, followed by success stories of companies that have adopted these practices.

A. EVOLUTION OF TQM AND VRP

In the past, the operations of U.S. industry were largely governed by methods with roots in the work of Fredrick W. Taylor. The Taylor method, a major step forward for its time (late 19th century), is still used in the Western World. It emphasizes management by specialists who formulate technical and work standards and workers who follow the standards established by management. Taylor turned work planning over to various specialists, leaving the foremen and workers only the job of executing plans prepared by someone else. Workers had virtually no say as to how the work was done.

Such a separation of planning from execution dealt a crippling blow to the concept of craftsmanship. Since jobs were broken down into simple repetitive tasks, monetary piecework incentives evolved as a way to motivate workers. Such an incentive scheme was established to eliminate the worker as a source of variation and free engineers and other professionals to work on improving methods, materials, and machines necessary for increased efficiency.

The Taylor method was initially good for producing large quantities of the same goods. However, as workers' economic and educational levels increased, their dissatisfaction increased. Workers became increasingly unhappy with doing simple repetitive tasks where money was the only reward and management allowed little if any input. Absenteeism and turnover increased, and work quality suffered. Furthermore, as the level of worker education increased, much potential for variability reduction from workers' ideas was lost.

In an attempt to maintain quality, various quality assurance and control departments run by nonworker specialists were formed. The main activity here was inspection and testing—separating the good from the bad. However, this "after-the-fact" approach did not address the problem at its source. Scrap, rework, and redesign were the results. Moreover, such a philosophy paid no attention to continuous variation from target. Sure, in football, a ball kicked anywhere between the goalposts scores the same as one going dead center; however, the analogy does not carry over into product performance. A product made just within tolerance is not much better than one a little bit outside; yet in the "goalpost" philosophy, still prevalent in the United States today, the former passes inspection and the latter does not.

Of course, such prime reliance on inspection and test was fundamentally unsound. However, it was not a significant handicap if all competitors used it. In fact, despite the deficiencies inherent in this approach to quality improvement, until about 20 years ago, U.S. goods were well regarded as far as quality was concerned. Thus, in the early postwar period, U.S. companies considered Japanese competition to be an issue of price rather than quality.

However, during the 1960s and 1970s, a number of Japanese manufacturers greatly increased their share of the U.S. market—primarily through superior quality. In fact, more recent research by the Strategic Planning Institute (SPI), using its extensive PIMS (Profit Impact of Market Strategy) data base, shows not only that increasing market share is important in increasing return on investment but also that changes in relative quality have a far more potent effect on market share than do changes in price.

In this light, some U.S. firms have been seriously looking at more fundamental approaches to quality improvement. One such approach is the concept of VRP. The adaptation of the more modern approaches that make up VRP has been stimulated by Japanese developments during the past 30 years. These developments, in turn, originated in the teachings of Sir Ronald Fisher and his school over 60 years ago as well as several key Americans just after World War II. (See Figure A-1.)

Other approaches such as just in time (JIT), flexible manufacturing systems (FMS), and computer-integrated manufacturing (CIM) developed concurrently.

In the late 1940s, General MacArthur's command recruited Dr. W. Edwards Deming to help the Japanese prepare for the 1951 census. The Japanese took note of his statistical control background, and in 1950 they invited Deming to lecture to the Union of Japanese Scientists and Engineers (JUSE) on statistical control and to top-level business leaders on the importance of statistical quality control (SQC) as a management tool. SQC was devised by Dr. W. A. Shewhart of Bell Laboratories, the father of modern SQC. He introduced the Shewhart Cycle, also known as the "PDCA Cycle" for "Plan, Do, Check, Act," which has

FIGURE A-1
EVOLUTION OF QUALITY MANAGEMENT AWARENESS

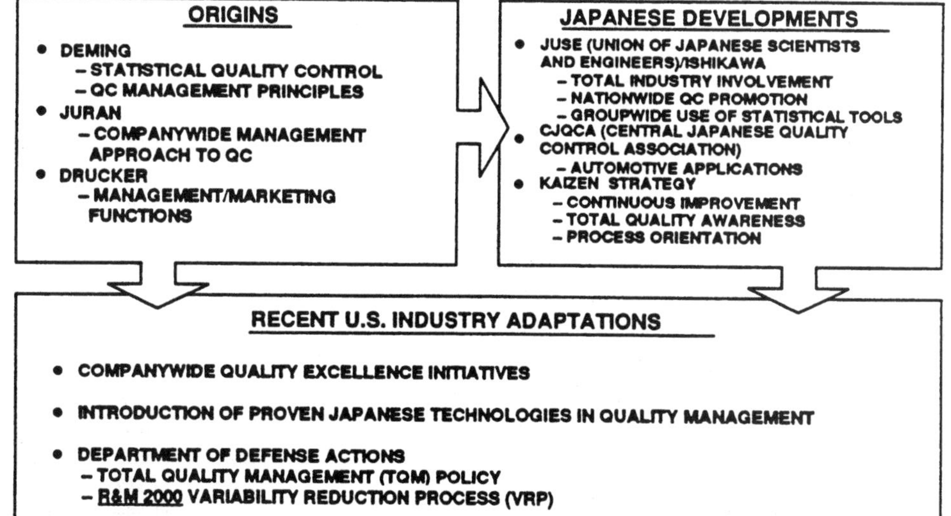

become one of the most widely used tools for continuous improvement. PDCA is depicted in Figure A-2.

Dr. J. M. Juran's significant contribution began in 1954 when, at the invitation of JUSE, he conducted seminars for top- and middle-level managers on their role in promoting quality control (QC) methods. He brought to top management's attention the fact that QC was a management tool. Dr. Peter Drucker worked as a consultant with Japanese top management on treating management as a complete entity and on defining management/marketing functions. In the 1960s, Dr. Armand V. Feigenbaum introduced the concept of total quality systems involving cross-functional management. He also emphasized the role of workers in quality improvement.

JUSE was instrumental in promoting total industry involvement and QC throughout Japan. Dr. Kaoru Ishikawa played a key role in developing the QC movement and QC circles in Japan. The Central Japanese Quality Control Association (CJQCA) of Nagoya, under Prof. Fukuhara, made the development of quality function deployment (QFD) a major project. QFD was applied in Mitsubishi's Kobe shipyards and subsequently used and enhanced by Toyota.

A significant step in promoting quality awareness in Japan was the establishment of the Deming Prizes (one category for application and one for research and education) by JUSE. The prizes have been awarded every year since 1951 and serve both to recognize QC achievements and to stimulate QC implementation. The awards are given on prime-time TV. More recently, an overseas category was established for the Deming Prize, and in 1989 Florida Power and Light Company became the first non-Japanese company to receive this prize.

In the United States, industry made isolated efforts to emulate the Japanese, but many companies missed the importance of management's involvement. Beginning in the late 1970s, Japan's competitive standing had reached such proportions that U.S. industry became aware of the power of total quality control and a strategy of continuous improvement called Kaizen. Since then, a growing

FIGURE A-2
COMPARISON OF TRADITIONAL WAY AND NEW WAY
(PDCA CYCLE)

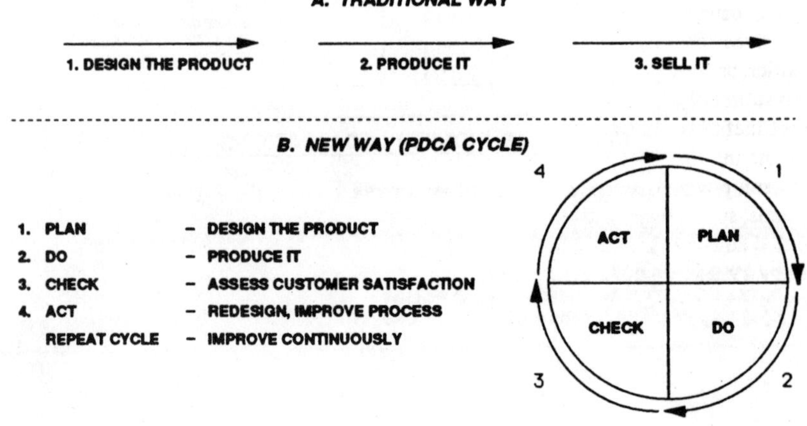
A. TRADITIONAL WAY
1. DESIGN THE PRODUCT
2. PRODUCE IT
3. SELL IT
B. NEW WAY (PDCA CYCLE)
1. PLAN - DESIGN THE PRODUCT
2. DO - PRODUCE IT
3. CHECK - ASSESS CUSTOMER SATISFACTION
4. ACT - REDESIGN, IMPROVE PROCESS
REPEAT CYCLE - IMPROVE CONTINUOUSLY
4
1
ACT
PLAN
CHECK
DO
3
2

commercial sector of U. S. industry has implemented company-wide quality excellence initiatives. Some examples are the Ford Motor Company, IBM, and Xerox. At the same time, many management consultant organizations have been established to help U. S. industry develop quality awareness and implement quality management. In 1988, the U. S. Government created the Malcolm Baldrige National Quality Awards, the American version of the Deming Prizes.

In 1988, the DOD, recognizing that drastic changes were needed, announced that it would give top priority to the DOD Total Quality Management (TQM) effort as the vehicle for attaining continuous quality improvement and as a major strategy to meet the President's productivity objectives. In August 1988, DOD issued the TQM Master Plan, delineating DOD's approach to implementing TQM objectives.

In October 1987, the Air Force Office of the Special Assistant for Reliability and Maintainability (R&M) issued the USAF R&M 2000 Process (now AF Pamphlet 800-7). Its aim was to achieve increased combat capability through good R&M practices. One of the building blocks of the R&M 2000 Process is the R&M 2000 Variability Reduction Process (VRP), which delineates an approach to achieving improved combat capability with higher reliability and lower cost. VRP objectives are to achieve robust designs, using capable processes, with a focus on continuous improvement. These objectives can be implemented throughout the acquisition process. Thus, VRP represents the acquisition subset of TQM. In July 1988, the Air Force Vice Chief of Staff issued a policy letter to all commands stating that VRP was to be applied to acquisition and postproduction support of all Air Force systems, subsystems, and equipment. This letter directed all commands to achieve the policy objectives by 1993.

B. EVOLUTION OF QUALITY ASSURANCE METHODS

The type of quality assurance (QA) that predominates in many U. S. production activities emphasizes verification of the product after it is produced. If defects occur, then quality assurance activities center on defining the cause of defects, developing corrective actions, and obtaining data to verify the results.

Verification (i.e., the process of inspecting or testing for conformance) is product oriented and does nothing to improve the basic quality of the product. As can be seen from Figure A-3, QA was 100 percent verification (i.e., inspection) oriented when Deming first went to Japan. The next step involving the use of statistical process control (SPC) (i.e., quality control during production) started about this time. SPC is process oriented and has the potential to provide a better product at a lower cost than does inspection.

The third improvement started around 1960 and focused on process and product design optimization. It involved the use of several tools. Initially it involved the concept of parameter design. This idea, as developed by Dr. Genichi Taguchi, directly addressed VRP by determining the combination of product and process parameter values that were least sensitive to environmental conditions and noise factors (i.e., most robust at minimum cost). Design of experiments methodology was used to find the experimental results that minimized variability and loss. In this respect, the earlier work establishing the principles of DOE by Sir R.A. Fisher and later extended by others such as Dr. George E.P. Box were important. Parameter design has been practiced for more than 30 years in Japan. It is generally acknowledged to have produced dramatic improvements in quality.

In the 1970s, the use of QFD further improved process and product design optimization. This approach, which emphasizes building quality into the product, offers the best potential for quality at lower cost. It assures that customers' needs drive the product's design and production process. The right-hand chart of Figure A-3 illustrates this concept. Although process control is shown to decrease to virtually zero on the left-hand chart, it should not be interpreted that SPC has been abandoned. On the contrary, SPC is used to control the production process but is no longer the dominant method used to ensure quality excellence.

C. SUCCESS STORIES

The following are examples of successful companies that have adopted quality-oriented principles of management. Higher quality has been accompanied by higher productivity and lower costs. All companies share common characteristics—the commitment and participation by top management. Such participation is one of the main ideas of Dr. W. Edwards Deming. Higher reliability, quality, and productivity can be achieved with labor and management in the United States if they make a commitment to a new approach. The Air Force can implement such an approach by using VRP.

- Ford Motor Company
  - Quality consciousness made an integral part of corporate culture
  - Company mission defined to stress continuous improvement of product and services to meet customers' needs and expectations
  - Techniques used: team approach, statistical process control, and design of experiments, including traditional and Taguchi methods
  - Results
    - Increased market share and profits
    - Recovery from large corporate deficits to highest profitability among U.S. automakers
    - 65-percent reduction in customers' reports of "things gone wrong" for both cars and trucks (from 1980 through 1988)
    - 35-percent increase in owner satisfaction (from 1982 through 1988)
- Xerox
  - Leadership Through Quality Program
  - Accomplishments/Benefits
    - Manufacturing costs down by 20 percent (1982-86)
    - Time to bring new products to market reduced by up to 60 percent

FIGURE A-3
PROGRESS IN QUALITY ASSURANCE METHODS

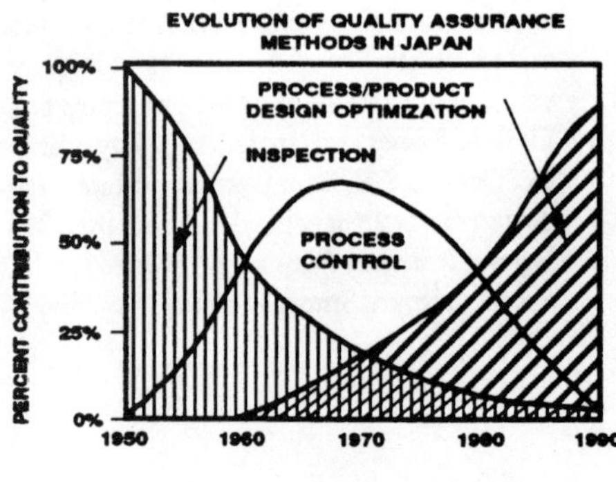

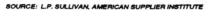

SOURCE: L.P. SULLIVAN, AMERICAN SUPPLIER INSTITUTE

SOURCE: USAF/LE-RD

- Revenue produced per employee up by 20 percent
- Reprographics market share regained from Japanese

- Westinghouse Electric Corporation, Commercial Nuclear Fuel Division (CNFD)
  - Used Total Quality Approach
    - Customer orientation
    - Human resource excellence
    - Product/process leadership
    - Management leadership
  - Results (between 1984 and 1987)
    - Manufacturing yield of once-through fuel rod increased from less than 50 percent to 87 percent (reduced scrap, rework, and manufacturing cycle time)
  - Cowinner of 1988 Malcolm Baldrige National Quality Award

- Yokogawa-Hewlett Packard
  - Total Quality Control Program
    - Objective: Reduce hardware failure rates 90 percent in 10 years
  - Accomplishments/Benefits (1977-84)
    - Failure rate down 79 percent
    - Manufacturing cost down 42 percent
    - Productivity up 120 percent
    - Market share up 19 percent
    - Profit up 244 percent

- Tennant Company
  - Quality Program Emphasis: Do things right the first time
    - Crosby principles introduced
  - Cost of assuring quality

1980		1988	
<u>17% of Sales</u>		<u>2.5% of Sales</u>	
Failure	50%	Failure	15%
Appraisal	35%	Appraisal	35%
Prevention	15%	Prevention	50%

- John Deere Company
  - Manufacturing Excellence Program
    - JIT, Design for Mfg., Group Technology, Employee Participation
  - Cellular Manufacturing Results for Ottumwa Works
    - Inventory down 60 percent
    - Required floor space down 27 percent
    - Required machine tools down 31 percent
    - Direct labor down 11 percent.

Appendix B
STATEMENT OF WORK SAMPLE TEXT

A. Purpose: To ensure that R&M 2000 VRP is integrated into program activities.

B. Example Text:

(i) Objective
In order to meet the reliability and cost goals established for the ____ program, the contractor shall emphasize the use of variability reduction. Variability reduction tools and techniques shall be used to achieve high reliability and low cost by means of defect reduction and defect avoidance.

(ii) Systems Engineering (Systems Integration)
The contractor shall describe an approach for implementing teamwork and robust design, which emphasizes designing quality into the product from the beginning of the design and development cycle. Emphasis will be placed on continuous improvement, with the use of variability reduction tools and techniques to minimize variation about a target value. Engineering drawings shall have target values indicated, in addition to their specification limits.

(iii) Engineering Management Plan
The contractor shall plan for the conduct and management of a fully integrated engineering effort, using variability reduction tools and techniques.

(iv) Quality Assurance
Through quality engineering participation in the development effort, the contractor shall identify the most critical quality characteristics for each configuration item. These characteristics shall be derived from an analysis of development test results, design of experiments or design optimization studies, integrity program, results of hazard analysis, fault tree analysis, failure modes and effects criticality analysis, single-point failure analysis, sneak circuit analysis, etc. This information should be compiled into an integrated quality structure from which specific material or process characteristics, whose variation must be controlled, are identified. The quality structure shall include product and process characteristics for both "make" or "buy" items, as well as critical characteristics of purchased items or material. Appropriate quality engineering tasking shall be flowed down to major subcontractors or vendors of critical items.

(vi) Manufacturing
The contractor shall develop and qualify manufacturing processes in parallel with hardware development. During production, the contractor shall ensure that the manufacturing processes are controlled. A primary goal of manufacturing is to minimize variability of the processes that produce the critical functional characteristics so as to produce hardware conforming to design target values. The degree of manufacturing conformance to design target values can be quantified through the use of terms such as "process capability index (C_p)."

The contractor shall determine the capability of manufacturing processes to control variability for critical quality characteristics. Process capability shall be measured, and there shall be prompt producibility feedback to design or manufacturing planning for any process found to be incapable of achieving variability reduction around given target values. The status of process capability studies shall be summarized and briefed to the program office at design and program reviews.

(vii) Quality Management Review
The contractor shall conduct a monthly quality Program Management Review to provide a critique of the contractor's variability reduction program status for the Government program office. The contractor shall conduct an in-depth quality review each year. Such a review will be headed by the Program Manager or top management of the organization's profit center.

(viii) Subcontractor
The contractor shall implement a plan for selecting subcontractors as part of the concurrent engineering teams. These organizations shall implement quality awareness methods and variability reduction tools and techniques.

(ix) Education and Training
The ____ Program shall include a plan for training all personnel (contractor and supplier) in variability reduction tools and techniques. The plan shall indicate the extent and level of the training or education activities and the extent to which this training will be offered to the subcontractor team.

Appendix C
USEFUL MILITARY REFERENCE DOCUMENTS

Dept. of the Air Force, "The USAF R&M 2000 Process"
HQ USAF/LE-RD, Washington, DC
Air Force Pamphlet 800-7, January 1989

"Total Quality Management Master Plan"
Department of Defense
August 1988

DOD 4245.7M Transition from Development to Production
September 1985 (Willoughby) Navy Quality Effort

NAVSO P-6071 Best Practices - How to Avoid Surprises in the World's
Most Complicated Technical Process - The Transition from Development to Production
(Dept. of Navy), March 1986

Total Quality Management Implementation: Selected Readings, TN 89-17
Navy Personnel Research & Development Center
San Diego, CA
April 1989

Appendix D
BIBLIOGRAPHY

TOTAL QUALITY

Bartus, K. M.
"Quality Improvement for the Defense Industry"
MIT MS Thesis, May 1988.

Crosby, Philip.
"Quality Without Tears - The Art of Hassle-Free Management"
ASQC Quality Press, 1984.

Deming, W. Edwards.
"Out of the Crisis"
MIT-CAES, 1986.

Drucker, Peter F.
"The Practice of Management"
Harper & Row, original edition 1954, reprint 1986.

Feigenbaum, A. V.
"Total Quality Control"
ASQC Quality Press, Third Edition, 1983.

Imai, Masaaki.
"Kaizen - The Key to Japan's Competitive Success"
Random House Business Division, 1986.

Ishikawa, Kaoru.
"Quality Control Circles at Work - Cases from Japan's Manufacturing and Service Sectors"
Asian Productivity Organization, 1984.

Ishikawa, Kaoru.
"What is Total Quality? The Japanese Way"
Prentice-Hall, Inc., 1985.

Iura, Toru, White, Herbert M., and Forrest, Lester.
"Applying Quality Management Techniques to the ALS Program"
Proceedings of the Space Systems Productivity and Manufacturing Conference-V sponsored by the U.S. Air Force Space Division and The Aerospace Corporation, August 16-17, 1988.

Juran, J. M.
"Managerial Breakthrough"
McGraw-Hill, 1964.

Elliott, J. G.
"Statistical Methods and Applications"
Allied-Signal, Inc. Automotive Sector, 1987.

Schmidt, S. R., and Lannsby, R. G.
"Understanding Industrial Designed Experiments"
CQG Ltd., Longmont, CO, March 1989.

Sullivan, L. P.
"The Power of Taguchi Methods to Impact Change in U.S. Companies"
Target, pp. 18-22, Summer 1987.

"Taguchi Methods - Quality Engineering"
Executive Briefing Book
American Supplier Institute, 1988.

Taguchi, Genichi.
"Introduction to Quality Engineering - Designing Quality into Products and Processes"
Asian Productivity Association, 1986.

STATISTICAL PROCESS CONTROL

American Society for Quality Control, Automation Division
"Statistical Process Control Manual"
ASQC, 1986.

AT&T
"Statistical Quality Control Handbook," Select Code 700-444
AT&T Technologies, Indianapolis, Indiana, 1956.

Grant, E. L., and Leavenworth, R. S.
"Statistical Process Control" (5th Ed.)
McGraw Hill, New York, 1979.

Ishikawa, Kaoru.
"Guide to Quality Control"
Asian Productivity Organization, 1982.

Navy Personnel R&D Center
"Measurement of Work Processes Using Statistical Process Control: Instructor's Manual," NPRDC TN 87-17, March 1987.

Shewhart, Walter A.
"Statistical Method - From the Viewpoint of Quality Control"
The Graduate School, Department of Agriculture, 1939.

CONTROL METHODS

Shingo, Shigeo.
"Zero Quality Control: Source Inspection and the Poka-Yoke System"
Productivity Press, 1986.

MISCELLANEOUS

"Application Guidelines 1991- Malcolm Baldrige National Quality Award"
United States Department of Commerce, National Bureau of Standards, 1991.

Deming Prize Committee
"The Deming Prize Guide - For Overseas Companies"
Union of Japanese Scientists and Engineers, 1986.

Schonberger, Richard.
"World Class Manufacturing - The Lessons of Simplicity Applied"
The Free Press, Macmillan, Inc., 1986.

Appendix E
ACRONYMS

AFIT	Air Force Institute of Technology
AFLC	Air Force Logistic Command
AFP	Air Force Pamphlet
ASI	American Supplier Institute
ASQC	American Society for Quality Control
CAE	Computer-Aided Engineering
CBD	Commerce Business Daily
CDRL	Contract Data Requirements List
CIM	Computer-Integrated Manufacturing
CJQCA	Central Japanese Quality Control Association
CNFD	Commercial Nuclear Fuel Division
C_p	Process Capability Index
C_{pk}	Process Capability Shift Index
DOD	Department of Defense
DOE	Design of Experiments
FAR	Federal Acquisition Regulations
FMEA	Failure Modes and Effect Analysis
FMS	Flexible Manufacturing System
FTA	Fault Tree Analysis
IBM	International Business Machines
ITO	Instruction to Offerors
JIT	Just in Time
JUSE	Union of Japanese Scientists and Engineers
LPL	Lower Process Limit
LSL	Lower Specification Limit
MTBF	Mean Time Between Failure
O&S	Operations And Support
PDCA	Plan-Do-Check-Act
PIMS	Profit Impact of Market Strategy
PPB	Parts per Billion
PPM	Parts per Million
QC	Quality Control
QFD	Quality Function Deployment
R&D	Research and Development
R&M	Reliability and Maintainability
RFP	Request for Proposals
RMS&P	Reliability, Maintainability, Supportability, and Producibility
SOW	Statement of Work
SPC	Statistical Process Control
SPI	Strategic Planning Institute
SQC	Statistical Quality Control

SSA	Source Selection Authority
SSP	Source Selection Plan
TPM	Total Productive Maintenance
TQM	Total Quality Management
UPL	Upper Process Limit
USAF	United States Air Force
USL	Upper Specification Limit
VRP	Variability Reduction Process

Appendix F
BARRIERS TO TOTAL QUALITY MANAGEMENT IN THE DEPARTMENT OF DEFENSE

Understanding the perceived barriers to implementing TQM has relevance to VRP because VRP supports TQM. Major H. Rumsey and Lt Col P. Miller of the Air Force Institute of Technology (AFIT) conducted a survey of perceived barriers to Total Quality Management (TQM) and published their results in the proceedings of the 1990 Annual Reliability and Maintainability Symposium. The respondents to the study were from all organizational levels of AFLC. The survey was administered during a two-day introduction to TQM and in conjunction with a discussion of Deming's 14 points.[12] The authors state that "students were asked to identify the most formidable barriers to TQM in their organizations."

The responses from the students indicated that they accepted the instructors' belief that formidable barriers to TQM existed. Major Rumsey and Lt Col Miller grouped the responses into four categories. These categories are listed below, and the percent of responses in each category is shown graphically in Figure F-1.

1. Relationships between managers and subordinates

2. Philosophy, policy, and procedures

3. Manpower and training

4. Miscellaneous

The responses in the first category indicate that managers, workers, and the relationships between them could be a barrier to TQM. In particular, the responses indicate acceptance of the status quo (by both managers and workers) and a lack of commu-

FIGURE F-1
PERCENT OF RESPONSES BY CATEGORY

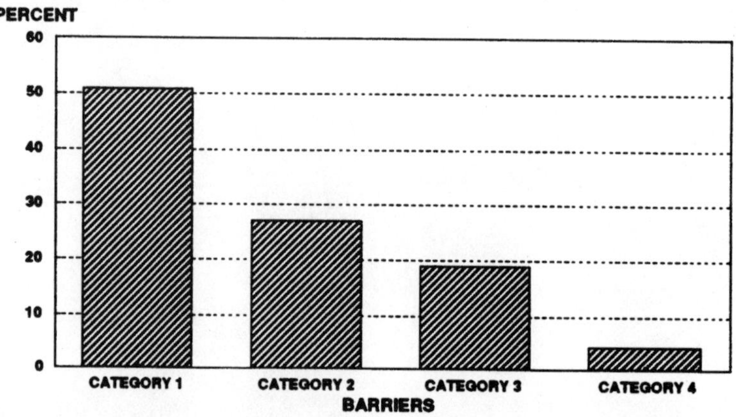

[12] One of Deming's 14 points is that organizational barriers to quality must be overcome. He was referring to barriers between departments within a corporation. However, in their presentation of Deming's 14 points, the instructors did not limit the discussion of barriers to the original context.

nication could interfere with TQM. Subordinates thought that lack of support from management could oppose the implementation of TQM. Managers thought that resistance and lack of motivation from their subordinates could impede the implementation of TQM. Both managers and subordinates indicated that a lack of motivation, trust, and cooperation could obstruct TQM.

The second category includes responses that are related to policies and procedures. One response was that promotion, hiring, and firing policies could present a formidable barrier to TQM. Another was concerned with production quotas. Some of the respondents also felt that an emphasis on quantity would conflict with TQM. Lack of management support, availability of equipment, and over-regulation also were considered potential problems.

The third category comprises responses from students who thought a lack of available manpower and training could be a barrier to TQM. Major Rumsey and Lt Col Miller concluded that the respondents did not understand TQM since it is generally held that TQM increases efficiency and that it does not require more manpower.

The fourth category comprises less than 4 percent of the responses. These responses included concern for inadequate budgets, lack of process controls, union resistance, and awarding contracts to the lowest bidder.

Appendix G
VRP REVIEW CHECKLIST

A. TOP-LEVEL COMMITMENT TO VRP AND TQM

- Management Involvement and Leadership
  - Encouragement and promotion of VRP concepts and techniques
  - Evidence of management involvement, such as biweekly meetings on VRP implementation progress
  - Resources allocated by management to obtain improvement
  - Top management attendance at VRP-oriented education and training classes
  - VRP made key factor in employee recognition
  - Corporate streamlining and assignment of process ownership

- Corporate Policy with Regard to Management and Quality/Reliability
  - Clear statement of policy and plans for excellence and continuous improvement
  - Extent that policy is deployed through the organization

- Education and Training in Defect Prevention
  - Existence of a permanent VRP education and training program
  - Focus of program (type and extent)

- Strategic Planning
  - Extent to which variability reduction awareness and continuous improvement are integrated into the corporate planning process
  - Evidence of employee, customer, and user involvement in the planning process
  - Plans for promoting variability reduction awareness in the future
  - Use of reviews within the company
  - Use of policy deployment process

B. INVOLVEMENT OF ALL ORGANIZATIONAL FUNCTIONS AND LEVELS

- Organization and Administration
  - Organizational structure for implementing VRP
  - Utilization of staff and consultants
  - Utilization of teamwork and natural group activities

- Employee Knowledge Level
  - Robust design techniques
  - Capable manufacturing processes
  - Variability reduction tools
  - Statistical tools
  - Problem-solving tools

- Teamwork
  - Use of interdisciplinary product development teams and process improvement teams
  - Multifunctional teams to obtain cooperation and communications between separate departments

- Supplier Involvement
  - Cooperation and communications with suppliers
  - Involvement of suppliers in the robust design process
  - System of supplier selection—emphasis

on supplier selection that is not determined solely by price
- Use of statistical process control
- Education and training—type and extent

C. APPLICATION OF ROBUST DESIGN AND CAPABLE PROCESS TECHNIQUES

- Implementation of Robust Design Practices
  - Use of cross-functional design teams
  - Methods for intergroup communications and integration
  - Use of computer-aided engineering (CAE)
  - Use of tools/techniques for variability reduction
- Customer (Voice of the Airman) Requirements and Analysis
  - Type of quality-related data obtained from customers and how the data are used
  - Use of QFD or analogous method to deploy the voice of the customer throughout the organization
- Design of Experiments (DOE)
  - Use of DOE in parameter design and on-line process control
  - DOE methodology employed
  - Impact of DOE on variability reduction
- Loss Function
  - Use of loss function or other economic measures for continuous improvement
- Capable Production and Repair Processes
  - Application of statistical process control (SPC) for critical product characteristics and process parameters
  - Use of statistical data to measure variability reduction
  - Use of control charts to demonstrate process control and capability

D. EVIDENCE OF CONTINUOUS IMPROVEMENT

- Results of Continuous Improvement Efforts
  - Defect reduction
  - Process capability improvement
  - Cost reduction and reliability gains
- Future Plans For VRP Improvement
  - Ensuring continuity of continuous improvement process
  - Utilization of new techniques or modification of existing techniques
  - Implementation of employee suggestions
- Education and Training in VRP Techniques
  - Percentage of employees trained, by organizational categories
  - Types of courses offered
  - Education and training for suppliers
- Incentives and Recognition
  - Encouragement of suggestions from employees for reducing defects
  - Recognition procedures for employee/team efforts
  - Incentives for supplier improvement efforts

Index

ABC (activity-based costing), 119–120
Acceptable quality levels (AQLs), 198, 450
Accreditation, 289, 291, 403
Action limits, 312
Active devices, 5
Activity-based costing (ABC), 119–120
Actual cycle time, 122
Advanced quality concepts, 154–198
Advanced quality management, 15–16, 132
Advanced quality management systems, 134
Advanced quality technology, 154–198
Aesthetics, 33
Affinity diagrams, 161–162
AIAG (Automotive Industry Action Group), 184–185, 189, 251–252
AIAG QS-000, 184, 250–252, 269, 271
Air Force, U. S., 196, 198, 250
Airlines, 13
Akao, Yoji, 156, 158
Alpha Industries, 121–123
American Management Association, 61
American National Standards Institute (ANSI), 19, 37, 40–41, 52, 88, 181, 188–189, 198, 204–209, 255, 406–407, 414
American Productivity and Quality Center, 489
American Society for Quality (ASQ), 3, 19, 37, 52, 88, 132, 174, 176, 188, 204–210, 291–292, 414, 512
American Supplier Institute (ASI), 156–157, 174, 176
Ames Rubber Corporation, 509
Analog circuits, 5–6
Analysis systems, 118
ANSI (*see* American National Standards Institute)
ANSI/ASQC Q9000, 206–209, 266, 291
ANSI/ASQC Q9001, 52, 88, 203, 205–207, 244
ANSI/ASQC Q9002, 206, 208, 414
ANSI/ASQC Q9003, 206, 208
ANSI/ASQC Q9004, 206, 209
ANSI/EIA 554-1988, 189
ANSI/EIA 555-8, 36
ANSI/EIA 557-1989, 189
ANSI/EIA 584-91, 198
ANSI/EIA 599, 406–407
ANSI/IPC PC-90, 181
ANSI/NCSL Z540, 239
AOQL (average outgoing quality limit), 198
Applications, 16–25
Appraisals, performance, 335
Appreciation of system, 302–308
AQLs (acceptable quality levels), 198, 450
Armstrong World Industries, 509
ASI (American Supplier Institute), 156–157, 174, 176
ASQ (*see* American Society for Quality)
Assessments, 118, 199
  and certification, 397–402
  conformity, 19
  first-party, 398–399
  maturity, 384
  on-site, 402–403
  of product compliance, 414–421
  quality, 397–409
  second-party, 399–400
  self-, 269, 272–277, 398–399, 439–443, 555–566
  third-party, 400–402
Assignable causes, 179–180, 194, 309–310
AT&T, 168, 509, 544–545
ATE (automated test equipment), 89
Attribute data, 177–178
Audits, 278
  first-party, 256
  functional configuration, 40, 224, 418
  implementation, 267, 289
  internal quality, 140, 244–248
  ISO 9000 checklist, 281–289
  physical configuration, 40, 48, 224, 418
  reassessment, 268
  quality system, 402, 404
  second-party, 256–257
  third-party, 257
Authority, 213, 263
Autocratic management, 1
Automated test equipment (ATE), 89
Automated testing, 89, 91–92, 420
Automobiles, 3, 6, 144, 156, 250–252, 486, 499–500
Automotive Industry Action Group (AIAG), 184–185, 189, 251–252
Availability, 54, 56

Average outgoing quality limit (AOQL), 198

Balance of power, 448–449
Baldrige, Malcolm, 512
Baldrige Award, 21, 23, 70, 109, 131, 133, 135–136, 163, 261, 383, 451, 455–456, 463, 469, 497–498, 507–566
Banks, 13
Bardeen, John, 7
Baseline data, 372–373
Baselines, 405
Basic quality management systems, 133–134
Belgium, 255
Bell Laboratories, 7, 167–168
Benchmarking, 13, 22–23, 504
  benefits of, 457–460
  continuous improvement, 502–503
  data analysis, 490–492
  defined, 445–447
  ethical data gathering, 487–490
  goals, 495–497
  history of, 448–451
  objections to, 460
  process, 461–473
  process understanding, 473–483
  steps, 463–473
  teams, 492–494
  types of, 462–463
  uses of, 451–461
  vital signs, 483–487
  world class, 497–502
Benchmarking Exchange, 451, 457, 492–493
Best value, 66–68, 390
Beta testing, 60
Big “Q,” 17, 21–22, 24, 28–30, 32, 37, 50, 63, 132–133, 142, 144, 191–192
Block diagrams, 8, 52, 83
Body language, 411
Boeing Company, the, 179, 358
Bottlenecks, 124
Brainstorming, 374–375
Britain, 255
Brittain, Walter H., 7
Burn-ins, 92, 232
Business results, 521, 536–537, 553
Buy-in, 471

CAD (computer-aided design), 396
Cadillac, 509
Calibration, 239
CAM (computer-aided manufacturing), 396
Campbell Soup Company, 396
Canon Corporation, 504
Capability indices, 186–188
Capacitors, 5
Capacity principles, 105–106
Carnegie Mellon University, 248, 473
CASE (Coordinated Aerospace Supplier Evaluation), 400, 404
Catalogs, supplier, 388
Cause-and-effect diagrams, 374–375
CE (Conformit, Europ,ene) mark, 253–254
CENELEC, 205, 255–256
Center for Quality Management, 456
CEOs (*see* Chief executive officers)
Certification, 257–259
  and assessments, 397–402
  and inspection, 416–418
  preparation for, 269–292
  process, 233–234, 406–407
  product, 252–256
  supplier, 431–436
  supplier audit, 291
  task scheduling for, 265–269
Change processes, TQM, 366–377
Changes, 3–4, 7–8, 23–25, 224, 226, 339, 357–358
Characteristics, performance, 80–88
Characterization efforts, 92, 94
Charts, 49, 178–179, 181, 188, 194, 312, 372–375, 379, 422–423
Checksheets, 373–374
Chemical analyses, 417
Chief executive officers (CEOs), 9, 13–14, 20, 117, 143, 174, 199, 212–213, 244, 248–249, 256, 259, 261, 265
China, 4
Chips, 2, 6–7, 121–122
Chrysler Corporation, 250–252
CIDs (commercial item descriptions), 52
Circuits:
  analog/digital, 5–6
  integrated, 5, 7–8, 39, 54, 173
Citizenship, 516
Closed-loop communication, 411–412, 431
Coaching, 327–328
Comfort, 500–501
Commerce Department (*see* Department of Commerce)
Commercial item descriptions (CIDs), 52
Commercial off-the-shelf (COTS) items, 60, 66, 68–70, 91
Commitment, 146
Common cause strategy, 310–311
Common causes, 179–180, 194, 308
Communication, 146–147
Communication loops, 410–414, 431
Companywide quality control (CWQC), 32, 341
Competition, 13, 25
Competitive benchmarking, 462–463
Complexity of products, 2–3

Compliance, 256–269, 414–421
Components, electronic, 35–37, 42–44, 52, 54, 60
Computer-aided design (CAD), 396
Computer-aided manufacturing (CAM), 396
Computers, 3, 7, 174, 391
Concept exploration phase, 56, 58
Concurrent engineering, 8, 77, 156
Conductors, 5
Conformance, 33, 47–51
Conformity assessment, 19
Consensus decisions, 369–370
Continuity, 92
Continuous improvement plans, 472
Continuous project improvement (CPI), 11–13, 95–96, 120, 122, 126, 265, 305–306, 313, 330–331, 355–356, 368–373, 502–503, 514, 521, 540
Contract process, 79–80
Contract reviews, 138, 220
Contracts, 17, 62–92
Control charts, 178–179, 181, 188, 194, 312
Conway, William, 75, 143, 355
Coordinated Aerospace Supplier Evaluation (CASE), 400, 404
Core business competencies, 388
Corning Telecommunications, 509, 545–546
Corporate responsibility, 516
Corrective action, 140, 241–242, 267
Costs:
  activity-based, 119–120
  of advanced quality techniques, 155
  registration, 278
  and world class manufacturing, 95, 192
COTS (commercial off-the-shelf) items, 60, 66, 68–70, 91
Counseling, 327–328
CPI (*see* Continuous project improvement)
Critical product specification, 47
Critical tolerances, 49
Crosby, Phil, 355
Crystals, 6–7
Culture, TQM, 357–366
Customer service, 61
Customer-driven quality, 513
Customer-specific requirements, 252
Customer-supplied items, 139, 230
Customers:
  feedback from, 118
  focus on, 4, 10–11, 24, 37, 521–522, 538–540, 553
  inquiries by, 396
  interfaces with, 70–73
  loyal, 306
Customers: (*Cont.*)
  and TQM, 358–359
  voice of, 159, 162, 196–197
CWQC (companywide quality control), 32, 341
Cycle time, 122–123
Cycle time reduction, 124–125

Data:
  attribute/variable, 177–178
  baseline, 372–373
  for benchmarking, 469–470, 481–483
  collection/analysis of, 313, 325–326
Data analyses, 490–492
Data control, 139, 224, 227
Data gathering, ethical, 487–490
DCMC (Defense Contract Management Command), 135, 291
De Forest, Lee, 6
Decision trees, 179, 182
Defects, prevention/detection of, 15–16, 414–416
Defense Contract Management Command (DCMC), 135, 291
Defense Department (*see* Department of Defense)
Defense Electronics Supply Center (DESC), 235–236, 405
Defense Logistics Agency, 135, 291
Defense Standardization Program, 80
Defense Supply Center, Columbus (DSCC), 405, 407
Definitions, 1–9
Delegated inspection, 432
Delivery, 140, 242–243, 424–425
Demilitarization phase, 59
Deming, W. Edwards, 14, 20–21, 66–67, 117, 133, 138, 179, 191, 212–213, 250, 256–257, 295–348, 353–355, 358, 360–362, 378, 381, 389, 508, 544, 559–560, 562, 566
Deming Prize, 21, 28, 109, 297, 301, 316, 341–347, 355, 456, 498, 508
Democratic management, 1
Department of Commerce, 255, 462, 512
Department of Defense (DOD), 7, 32, 44, 58–59, 68, 79, 135, 168, 203, 234–235, 291, 351, 353, 462
Deployment phase, 58–59
DESC (Defense Electronics Supply Center), 235–236, 405
Design, 8–9, 33, 35–47, 118
Design changes, 224, 226
Design control, 138–139, 220–226
Design of equipment (DOE), 164, 167–168, 174
Design input, 222
Design leverage, 100–101

Design output, 222
Design planning, 222–223
Design principles, 104, 106
Design process, 8–9
Design quality, 33, 35–47, 514–515
Design review, 224–225
Design validation, 224
Design verification, 224, 226
Designs, robust, 165, 167–176
Destructive parts analyses (DPAs), 419
Detail specifications, 84
Detection of defects, 15–16
Deutsch, J. M., 203
Digital circuits, 5–6
Digital Equipment Corporation, 289
Diodes, 5
Disposal phase, 59
Distribution phase, 58–59
Dixon, J. R., 119
Document control, 139, 224, 227–228
Documentation, 52–55, 123–124, 214–215, 268
DOD (*see* Department of Defense)
DOD 4120, 83
DOD 5000, 33, 56, 75, 291
DOE (design of experiments), 164, 167–168, 174
Dormant, 255
Dow-Corning, 467–468
DPAs (destructive parts analyses), 419
Drawings, 52
Drucker, Peter, 315
Drug use, 3
Dryden Research Center, 451
DSCC (Defense Supply Center, Columbus), 405, 407
Dun & Bradstreet, 261, 389
DuPont, 259, 261
Durability, 33

Eastman Chemical, 509
Economic loss, minimization of, 311–312
Economy of multiples, 97
Education, 327–328
EIA (*see* Electronic Industries Association)
EIA 554-A, 36, 198
EIA 599-1, 236
EIA 681, 406–407
Electromagnetic interference (EMI), 222
Electronic age, 7
Electronic components, 35–37, 42–44, 52, 54, 60
Electronic Industries Association (EIA), 36, 89, 132, 189, 198, 205, 234, 236, 292, 394, 406–408, 462
Electronic systems, 35–37, 42, 52, 54, 83
Electronics:
  active/passive devices, 5
  analog/digital, 5–6
  building blocks, 5
  change for, 7–8
  conformance, 47–51
  defined, 4–5
  design, 35–47
  evolution of, 6–7
  fitness for use, 51–62
  processes, 8–9
  schematic diagrams, 8
Electrostatic discharge (ESD), 45, 222
EMI (electromagnetic interference), 222
Employee participation/development, 328–329, 514
Employees, and TQM, 362–366
Employer-employee loyalty, 3
Empowerment, 12, 96, 144–148, 363–365
EN (European Norms), 205
Engineering development phase, 58
Environment stress screening (ESS), 92
Environmental tests, 92–93
Environmentalism, 3
Equipment control/maintenance, 139
ESD (electrostatic discharge), 45, 222
ESS (environment stress screening), 92
Ethical data gathering, 487–490
Ethnic nationalism, 3
EU (*see* European Union)
European Committee for Electrotechnical Standardization (CENELEC), 205, 255–256
European Community (*see* European Union)
European Norms (EN), 205
European Union (EU), 19, 204–205, 253–256, 291
Evaluation, 88–94
Expansion of quality, 4
Experimental designs, 164, 167–168, 174
Extrinsic motivation, 319–320

Failure modes, 89, 91
Fast response, 514
FCAs (functional configuration audits), 40, 224, 418
Fear, 320
Features, 33
Federal Acquisition Streamlining Act (1994), 64
Federal Express, 451, 509
Feedback, 118, 411–414, 431
Feigenbaum, Armand V., 355
Ferner, Ron, 396
Fielding phase, 58–59
Filters, 411
Final assembly, 477, 479

First-party assessments, 398–399
First-party audits, 256
Fishbone diagrams, 374–376
Fisher, R. A., 171
Fitness for use, 33, 51–62, 92, 94
Fleeting events, 309
Florida Power and Light, 28, 508
Flow diagrams, 299, 306–308
Flowcharts, 372
Flowing down requirements, 410–414
Ford Motor Company, 62, 156, 168, 250–252, 298
Fortune 500, 469
France, 255
Fuji Film, 168
Functional benchmarking, 463
Functional configuration audits (FCAs), 40, 224, 418
Future, long-range view of, 515

Gainsharing, 383
Gantt charts, 379
GAO (General Accounting Office), 141, 383, 554
GAP analyses, 239, 261, 263, 266–267, 470
Garvin, David A., 32–35, 51, 92
Gates, Bill, 4
Gauges, 188
General Accounting Office (GAO), 141, 383, 554
General Electric, 255, 381
General Motors Corporation, 145, 156, 177, 250–252, 298
General principles, 103–104
General Services Administration (GSA), 83
Germany, 101, 255
GIQLP (Government and Industry Quality Liaison Panel), 131–136, 155
Globalization, 4
Globe Metallurgical, 509
GOAL/QPC conference, 156–158, 161
Goals, 30, 199, 333–334, 495–497
Godfrey, A. Blanton, 198
Government, quality policy of, 17–18
Government and Industry Quality Liaison Panel (GIQLP), 131–136, 155
Granite Rock Company, 509
Grant, E. L., 179, 189
Graphs, linear, 171–172
Greif, Michel, 150
GSA (General Services Administration), 83
GTE, 509
Gyrna, Frank M., 29, 50, 61, 189

Hagan, John T., 24
Handbooks, supplier, 388
Handling, 140, 242–243
Harry, M. J., 190–191
Hertz, Harry S., 510
Hewlett-Packard Company, 191, 289, 368
Hot lots, 124–125
House of Quality, 71, 81, 157–158, 163–165, 358
Hudiburg, John J., 28
Human resource development, 520, 529–530, 552
Human resources principles, 104–106
Hybert, Peter, 64, 66, 70–71, 74–75, 77, 79

IBC (International Benchmarking Clearinghouse), 451, 489
IBM, 391, 509, 546–547
ICs (integrated circuits), 5, 7–8, 39, 54, 173
Ideal cycle time, 122–123
IEC (International Electrotechnical Commission), 204–205, 252, 255
IECQ (International Electrotechnical Commission Quality) System, 89, 255–256
IEEE (Institute of Electrical and Electronics Engineers), 39, 52
IEEE P1220, 39, 52, 221
Imagineering, 148
Implementation:
  ISO 9000 standards, 226, 228–231, 236, 238–243, 245, 248–251, 259, 264, 267–268
  systematic paths for, 181–182
Implementation audits, 267, 289
Implementation plans, 265–266, 472–473
Implementation principles, 378–380
Implementation processes, TQM, 376–380
Important versus urgent, 339–340
Improvement (*see* Continuous project improvement)
Incentive pay, 335–336
India, 4
Inductors, 5
Industry standards, 89
Industry Week, 109–117, 154, 462, 469, 497–498
Information:
  and analysis, 518, 520, 525–526, 552
  availability of, 120
  principles, 105–106
Information revolution, 4
Innovation, 305–306
Inspection, 50
  and certification, 416–418
  conventional, 27–28
  delegated, 432
  ISO 9000 standards, 139, 236–240
  source, 421
Institute of Electrical and Electronics Engineers (IEEE), 39, 52

Institute for Interconnecting and Packaging Electronic Circuits (IPC), 40–41, 89, 181, 188–189, 394, 409
Integrated circuits (ICs), 5, 7–8, 39, 54, 173
Integrated process development, 141, 173
Integrated product development (IPD), 8, 72, 77, 141, 173
Integrated product teams (IPTs), 100
Integration of quality, 4
Integrity, 338
Intel Corporation, 7, 89, 289
Interactions, 329
Interdependencies, 302
Interfaces with customers, 70–73
Internal benchmarking, 462
Internal quality audits, 140, 244–248
International Benchmarking Clearinghouse (IBC), 451, 489
International Electrotechnical Commission (IEC), 204–205, 252, 255
International Electrotechnical Commission Quality (IECQ) System, 89, 255–256
International Organization for Standardization (ISO), 10, 20, 138, 204–205, 210, 252, 255
International Trade Association, 253, 255
Internet, 4, 13, 70, 292, 382, 388, 392, 396, 462
Intrinsic motivation, 319–320
Inventory turns, 95
Investments, 118
IPC (*see* Institute for Interconnecting and Packaging Electronic Circuits)
IPC D-330, 41
IPD (integrated product development), 8, 72, 77, 141, 173
IPTs (integrated product teams), 100
Ishikawa, Kaoru, 137, 355, 374
Ishigawa diagrams, 374
ISO (*see* International Organization for Standardization)
ISO 8402, 50, 136, 209, 211, 215
ISO 9000 series, 203–292, 388, 397–399, 401–404, 409, 412–414, 435–437, 463, 473, 509–511
ISO 9000, 10, 18–20, 56, 60, 66, 117, 131–133, 135–136, 144, 150, 199, 203–207, 209, 211, 281–289
ISO 9001, 22, 37, 75, 84, 89, 123, 145, 162, 204, 206–253, 263, 269, 280, 392, 406, 418–419, 510
ISO 9002, 204, 206–209, 211, 215, 219, 224, 235, 252, 259, 269, 280, 406
ISO 9003, 204, 206–209, 211, 215, 219, 252, 269, 280
ISO 9004, 204, 206, 209, 211, 249
ISO 10011, 211, 248
ISO 10012, 211, 239
ISO 10013, 211, 215
ISO 10014, 211
ISO 10015, 211
ISO 10016, 211
ISO 209000, 413
Item specifications, 84
ITT, 156

Japan, 18, 20–21, 28, 32, 49, 95–96, 98–101, 132, 137, 142, 145, 156–158, 167–168, 176–177, 248, 296–299, 314, 353–355, 450, 456, 504, 507–508, 521
JESD 46, 236
JIT (just-in-time) manufacturing, 97–99, 120, 181, 237, 418
Juran, Joseph M., 29, 32, 50–51, 61, 82–83, 189, 198, 355, 360
Just-in-time (JIT) manufacturing, 97–99, 120, 181, 237, 418

Kearns, David T., 446
Kerwin, Robert E., 198
Key processes, 195
Key suppliers, 196
Kilby, Jack, 7
King, B., 158
Kiser, Kenneth J., 351, 362
Knowledge, 314–318, 320, 326
Kochran, Thomas, 144
Kotter, John P., 361–362

L. L. Bean, 456
LANs (local-area networks), 70, 266
Lawler, Edward, 362, 382–383
Lead times, 95
Leadership, 4, 10–12
  and Baldrige award, 513, 518, 523–524, 551–552
  Deming on, 337–338
  and empowerment, 146
  and management, 142–144, 300, 361–362
  and WABM, 120
Learning, 340–341, 514
Leavenworth, R. S., 179, 189
Life cycle, 56–60
Line proofing, 47–48
Linear graphs, 171–172
Lischefska, John, 396
Little "q," 17, 27–30
Local-area networks (LANs), 70, 266
Logic diagrams, 8, 52
Loss function, 171–172

Machine capability studies, 186
Maintainability, 54, 56
Malcolm Baldrige National Quality Award (*see* Baldrige Award)
Malcolm Baldrige National Quality Improvement Act (1987), 512
Management:
advanced quality, 15–16
autocratic/democratic, 1
commitment by, 265
common practices, 331–337
defined, 1, 300
of electronic systems, 9–10
by fact, 515
getting job done, 329–330
and leadership, 142–144, 300, 361–362
and nonconforming material, 14
as prediction, 317–318
principles, 10–14, 103–108, 321–324
process, 520–521, 532–535, 552–553
reactive, 332–333
responsibility, 138, 210, 212–213, 263, 306
self-, 142, 338–339, 365–366
and supervision, 300
and TQM, 359–362
of transformation, 101–102
workforce activity-based, 120–121
Management by objectives (MBO), 305, 336–337
Management representatives, 213, 263
Management reviews, 213, 263, 267
Manuals, 52, 214–219, 266
Manufacturing, world class, 32–33, 92–93, 95–126, 436–443, 463
Manufacturing development phase, 58
Manufacturing process, 45–46
Manufacturing process surveys, 402
Marconi, Guglielmo, 6
Marketing principles, 105–106
Marlow Industries, 509, 547–548
Maslow, Abraham, 381
Material specifications, 84
Matrix of Matrices, 157–158, 160
Maturity assessment, TQM, 384
MBO (management by objectives), 305, 336–337
McClaskey two-circle model, 559
McKinsey 7-S model, 366
Mean time between failure (MTBF), 54, 56
Mean time to repair (MTTR), 54, 56
Measurement principles, 378–380
Measurement processes, TQM, 376–380
Measurement systems, 421–431
Measurement systems analysis (MSA), 185
Metrics, 148–153, 313
to achieve world class, 153–154
for conformance, 51
in contracting process, 79–80
Metrics, (*Cont.*)
for design, 45, 47
for fitness for use, 60–62
process, 397
Microprocessors, 2, 6–7
Microsoft, 4, 61, 289, 391
MIL-H 38534, 235
MIL-I 45208, 245
MIL-PRF 19500, 405
MIL-PRF 38534, 405
MIL-PRF 38535, 405
MIL-Q 9858A, 245
MIL-STD 883, 235
MIL-STD 973, 221
MIL-STD 45662, 238–239
Miles per gallon, 486
Milliken & Company, 381, 509
Models, use of, 43–44, 356–357, 366, 559
Modifications, 58
Motivation, 118
Motorola, 3, 13, 15, 189–192, 289, 450, 509, 548–549
MSA (measurement systems analysis), 185
MTBF (mean time between failure), 54, 56
MTTR (mean time to repair), 54, 56
Multiples, economy of, 97

NASA (National Aeronautics and Space Administration), 6, 59, 150, 451
National Aeronautics and Space Administration (NASA), 6, 59, 150, 451
National Institute of Standards and Technology (NIST), 23, 510, 512, 555, 558, 566
Navy, U. S., 7, 42
NCR Corporation, 184
New product lead times, 95
New quality, 27–62, 132
New United Motor Manufacturing, 298
Nippon Denso Company, 168
NIST (*see* National Institute of Standards and Technology)
Noise, 165, 170–171
Non-value-added, 74–75
Nonconforming material, 14, 140, 240–241
Nondevelopmental items, 54, 60, 66, 68–70, 91
Numerical goals, 333–334

OEMs (original equipment manufacturers), 10, 68–69, 187, 208, 255
On-site assessments, 402–403
One quality system, 132–140

Open architecture, 64
Open systems architecture (OSA), 64
Open-loop communication, 410
Operating principles, 321–324
Operational definitions, 316–317
Operational support, 58
Operations principles, 104, 106
Optimization, 299, 305–306, 310, 325–327
Order-to-delivery lead times, 95
Organizational interfaces, 222
Organizational responsibilities, 118
Organizational systems, 357, 365–366, 368
Original equipment manufacturers (OEMs), 10, 68–69, 187, 208, 255
Orthogonal arrays, 171–172
OSA (open systems architecture), 64
Overlapping production, 98
Overspecification, 82

Packaging, 140, 242–243
Paperwork, unnecessary, 332
Paradoxes, 339
Parameter design, 170–171
Pareto, Vilfredo, 373
Pareto charts, 373–375
Partnering, supplier, 359, 392–396
Partnerships, development of, 515–516
Parts counts, 97
Passive devices, 5
PATs (process action teams), 369–370, 378
Pay for performance, 335–336, 383
PCAs (physical configuration audits), 40, 48, 224, 418
PCBs (printed circuit boards), 184
PDSA (*see* Plan-do-study-act)
Perceived quality, 33
Perceptual mapping charts, 422–423
Performance, 33
Performance appraisals/rankings, 335
Performance characteristics, 80–88
Performance classes, 40–41
Performance specifications, 52, 54, 63–64, 70, 80–88, 405
Performance verification, 91–92, 94
Performance-based contracting, 17, 62–92
Performance-based pay, 335–336, 383
Performance-based quality, 28
Personal computers, 391
Peters, Tom, 351–352
Physical configuration audits (PCAs), 40, 48, 224, 418
Plan-do-study-act (PDSA), 20, 118, 310, 313, 315–316, 318, 326, 348, 544, 559–560, 562, 566
Planning, 118, 199
  design, 222–223
  quality, 214, 217, 219
  strategic, 520, 527–528, 552
Plans:
  continuous improvement, 472
  implementation, 265–266, 472–473
  project, 74–76
  strategic, 265
  strategy action, 379–380
Poirier, Charles C., 144
Porter, Michael, 357
Power, 363
Preassessment audits, 268
Prediction, management as, 317–318
Preservation, 140, 242–243
Prevention of defects, 15–16, 514–515
Preventive action, 140, 241–242
Principles:
  capacity, 105–106
  design, 104, 106
  general, 103–104
  human resource, 104–106
  implementation, 378–380
  information, 105–106
  management, 10–14, 103–108
  marketing, 105–106
  measurement, 378–380
  operating, 321–324
  operations, 104, 106
  process improvement, 105–106
  promotion, 105–106
  quality, 105–106
Printed circuit boards (PCBs), 184
Problems, handling of, 304
Procedure upgrades, 267
Process action teams (PATs), 369–370, 378
Process capability, 132, 185–187
Process certification, 233–234, 406–407
Process control, 139, 231–236
Process design lead times, 95
Process flow diagrams, 475–478
Process improvement principles, 105–106
Process management, 520–521, 532–535, 552–553
Process maps, 372
Process metrics, 397
Process specifications, 49, 64, 84
Process surveys, 407–409
Process vital signs, 484–486
Processes, 11
  benchmarking, 473–474
  change, 366–377
  contract, 79–80
  design, 8–9
  electronic, 8–9
  implementation, 376–380
  ISO 9000 standards, 136–137, 266
  key, 195
  management, 520–521, 532–535, 552–553
  manufacturing, 45–46
  measurement, 376–380

Processes (*Cont.*)
new, 266
propriety, 8–9
special, 407
strategic change, 366–367
systems engineering, 37–40, 45–46, 57
and TQM, 368–369
Product certification, 252–256
Product compliance, 414–421
Product design, 118
Product identification/traceability, 139, 230–231
Product specifications, 47, 64, 83–84, 188
Product vital signs, 484–486
Production:
focus on, 96–97
overlapping, 98
Production phase, 58–59
Productivity, 95
Products, complexity of, 2–3
Profound knowledge, 320
Program definition phase, 58
Promotion principles, 105–106
Propriety processes, 8–9
Prototypes, testing of, 42–43
Psychology, 318–320, 326–327
Purchasing, 139, 227–229, 438

QCs (quality circles), 314–315
QFD (*see* Quality function deployment)
QML 38534-20, 235
QMLs (qualified manufacturers lists), 89, 235–236, 404–407
QPLs (qualified product lists), 234–235, 404–405
QS 9000, 398
Qualification tests, 92
Qualified manufactuers lists (QMLs), 89, 235–236, 404–407
Qualified product lists (QPLs), 234–235, 404–405
Qualifying agencies, 405
Quality, 95
challenges/changes, 2–4, 23–25
of conformance, 33, 47–51
customer-driven, 513
defined, 1–2, 24, 27–62, 510
design, 33, 35–47
evolution of, 2, 27–28
of fitness for use, 33, 51–62
integration of, 4
little "q" versus big "Q," 17, 27–30, 32
new, 27–62, 132
one system, 132–140
perceived, 33
performance-based, 28
world class, 507–511
Quality assessments, 397–409
Quality circles (QCs), 314–315
Quality control, 32, 50–51, 341, 355
Quality engineering, 170
Quality function deployment (QFD), 8, 41–42, 51, 64, 71–73, 79, 81, 132, 156–166, 173, 221, 403
Quality Function Deployment Institute, 164
Quality image, 51
Quality manuals, 214–219, 266, *Appendix B*
Quality planning, 214, 217, 219
Quality policy, 212, 263
Quality principles, 105–106
Quality Progress, 174, 188
Quality records, control of, 140, 243–245
Quality system audits, 402, 404
Quality system surveys, 402–404
Quality systems, 138, 214–219, 402–404
Quality-oriented organizations, 30–31

R & D (research and development), 451
RAB (Registrar Accreditation Board), 289, 291, 403
Radio, 6
Radio Manufacturers' Association, 205
Railroads, 3
Rand Corporation, 42
Random causes, 179–180, 194, 308
Rankings, performance, 335
Rating systems, 421–431
Rayner, Steven, 145–146
Reactive management, 332–333
Reagan, Ronald, 448, 512
Recognition, TQM, 380–383
Records, 140, 243–245
Redesign, 147
Redlines, 429
Registrar Accreditation Board (RAB), 289, 291, 403
Registrars, selection of, 271, 278–280
Registration, 257–259, 262, 269–292
Reimann, Curt W., 509
Reinforcement, 147
Reliability, 7, 33, 42–45, 54, 56
Reliability engineering, 42
Renewal, 147
Repair, 8, 54, 61
Reputation, 51
Requests for proposals (RFPs), 67–68
Research and development (R & D), 451
Resistors, 5
Resources, 213, 263
Respect, 338
Responsibilities, organizational, 118
Responsibility:
assignment of, 424
corporate, 516

Responsibility (*Cont.*)
management, 138, 210, 212–213, 263, 306
personal, 340
and TQM, 360
Responsibility centers, 97–100
Results orientation, 516
Reviews:
contract, 138, 220
design, 224–225
management, 213, 263, 267
specification, 220
RFPs (requests for proposals), 67–68
Risk reduction, 15–16
Risk reduction phase, 58
Ritz-Carlton Hotel Company, 509
Roadmap for Quality in the 21st Century, 131, 133, 135, *Appendix A*
Robust designs, 165, 167–176
Robustness, 165, 168–169
Rockwell International Corporation, 425
Rome (ancient), 30
Rubinstein, Saul, 144
Ruggedized equipment, 91
Run-ins, 232

SAC (supplier audit certification), 291
Sashkin, Marshall, 351, 362
Saturn, 144–145
Schematics, 8, 52, 54, 83
Scholte, Peter, 369
Schonberger, Richard J., 95–108, 120, 122–123, 144–145, 148, 462, 497–498
Scope, 9–16
Screens, 232
Second-party assessments, 399–400
Second-party audits, 256–257
Sector-specific requirements, 252
Securities and Exchange Commission, 389
SEI (Software Engineering Institute), 248, 269, 473
Selection, supplier, 388–392
Self-assessment, 269, 272–277, 398–399, 439–443, 555–566
Self-declaration, 268–269, 280
Self-evaluation, 269, 272–277
Self-managing, 142, 338–339, 365–366
Semiconductors, 6–7, 15, 39, 89, 121–122
Senge, Peter, 302–303
Service contracts, 61
Serviceability, 33
Servicing, 61, 140, 248–249
Shewhart, Walter A., 49, 176, 309, 311–312, 353–355
Shockley, William, 7
Shores, A. Richard, 118–119
Sigma, 190–191
Signal-to-noise ratio, 170–171
Simplicity, 96, 101
Simultaneous engineering, 77
Single Process Initiative (SPI), 135
Six-sigma product quality, 189–192
SLA criteria, 265, 291
Sloan, Daniel, 148
Slogans, 358
Slowdowns for problems, 98–99
SMDs (standard military drawings), 235
Software Engineering Institute (SEI), 248, 269, 473
Software specifications, 84
Solectron Corporation, 509, 549
Solicitations, 67–68
Solid state, 7
Sony Corporation, 15, 193–194
SOPs (standard operating procedures), 309
Source inspection, 421
Source surveillance, 421
Soviet Union, 448–449
Space shuttle, 6
SPC (*see* Statistical process control)
Special cause strategy, 311
Special causes, 179–180, 194, 309–311
Special processes, 407
Specification reviews, 220
Specifications:
detail, 84
item, 84
material, 84
performance, 52, 54, 63–64, 70, 80–88, 405
process, 49, 64, 84
product, 47, 64, 83–84, 188
software, 84
system, 84
SPI (Single Process Initiative), 135
SPI (supplier performance index), 425, 428–431
SRIP (supplier rating and incentive program), 425–431
Standard military drawings, 235
Standard operating procedures (SOPs), 309
Standard & Poor 500 Index, 555
Standards, 10
industry, 89
ISO 9000 series, 19–20, 138, 203–205
voluntary, 17
Statistical process control (SPC), 18, 32, 41, 49–50, 132, 173, 176–189, 192, 234, 250, 252, 312, 353–354, 368, 397, 404, 432–433, 435–436
Statistical techniques, 140, 250–251
Steering committees, 266
Storage, 140, 242–243
Strategic change process, 366–367

Strategic information analyses, 448
Strategic planning, 520, 527–528, 552
Strategic plans, 265
Strategy action plans, 379–380
Strategy descriptions, 378–379
Subcontractors, 21–22, 227–229
Sullivan, L. P., 248
Supervision, 300
Supplier audit certification (SAC), 291
Supplier partnering, 359, 392–396
Supplier performance index (SPI), 425, 428–431
Supplier rating and incentive program (SRIP), 425–431
Supplier Research Group, 121
Suppliers, 10–11, 13–14, 21–22, 48, 99, 227–229, 237, 358–359, 387–388
  certification, 431–436
  flowing down requirements, 410–414
  key, 196
  measurement/rating systems, 421–431
  partnerships/teaming, 392–396
  product compliance, 414–421
  quality assessments, 397–409
  selection, 388–392
  world class, 436–443
Supply chain management, 22
Supportability, 51
Surveys, 402–404, 407–409
Suttler, Goodlow, 456
System approach, 11–12
System design, 170
System specifications, 84
Systematic paths for implementation, 181–182
Systems:
  advanced quality management, 134
  analysis, 118
  appreciation for, 302–308, 325
  basic quality management, 133–134
  defined, 302–303
  electronic, 35–37, 42, 52, 54, 83
  measurement, 421–431
  one quality, 132–140
  organizational, 357, 365–366, 368
  quality, 138, 214–219, 402–404
  rating, 421–431
Systems engineering process, 37–40, 45–46, 57

Tabak, Lawrence, 143
Taguchi, Genichi, 18, 161, 164–165, 167–169, 176
Taguchi Method, 45, 132, 169–174, 176, 192, 222
Target benchmarks, 471–472
Target values, 168–169
Teams, 118, 327
  benchmarking, 492–494
  as integrated approach, 72, 75–76, 173–174
  integrated product, 100
  ISO 9000 standards, 259–260
  process action, 369–370, 378
  and suppliers, 388–389, 392–396
  and TQM, 140–142, 365–367, 369–370, 378
  and WABM, 120–121
Technical interchange meetings (TIMs), 413
Technical interfaces, 222
Technical performance measurements (TPMs), 149
Technicomp, 163
Technological support, growth in, 3
Technology review boards (TRBs), 405–406
Telecommunications Industry Association (TIA), 205
Teleconferencing, 392
Television, 5–6
Temperature, 92
Templates, for specification development, 85–88
Test proofing, 92, 94
Test status, 139, 239–240
Testing, 52, 54, 69, 94, 139
  automated, 89, 91–92, 420
  beta, 60
  environmental, 92–93
  impact on quality assurance, 88–92
  ISO 9000 standards, 236–239
  of prototypes, 42–43
  qualification, 30
Texas Instruments, 7, 191, 194, 509, 550
Theory, importance of, 314–315
Third-party assessments, 400–402
Third-party audits, 257
TIA (Telecommunications Industry Association), 205
TIMs (technical interchange meetings), 413
Tokarz, Steven J., 144
Tolerance design, 171
Tolerances, 49
Toll-free telephone numbers, 61
Toshiba Corporation, 89–90
Total quality control, 355
Total Quality Management (TQM), 12–13, 18, 21, 29–30, 32, 61–62, 74, 98, 103, 120, 133, 135, 140–142, 144–145, 148, 265, 289, 298, 313, 554
  additional information, 383–385
  background, 352–355
  change processes, 366–377

Total Quality Management (TQM), (*Cont.*)
  culture, 357–366
  defined, 351–352
  implementation processes, 376–380
  measurement processes, 376–380
  model, 356–383
  recognition, 380–383
Total Quality Metrics, 133
Toyota Motor Company, 156, 168, 296–298, 508
TPMs (technical performance measurements), 149
Traditional organizations, 30–31
Training, 101, 120, 140, 247–249, 266, 268, 327–328
Transformation, management of, 101–102
Transistors, 5, 7
TRBs (technology review boards), 405–406
Tree diagrams, 161–162, 375, 377
Troubleshooting, 8, 54, 56–60
Trust, 320

UL (Underwriters Laboratories), 204, 255–256
Underspecification, 82
Underwriters Laboratories (UL), 204, 255–256
Union of Japanese Scientists and Engineers, 28, 297, 354
Unique quality management systems, 134
United Auto Workers, 145
United Parcel Service, 451
United States Postal Service, 451
University of Pennsylvania, 7
Urgent versus important, 339–340
USAF R + M 2000 Variability, Reduction Process, *Appendix C*
User friendly, 7
User manuals, 52

Vacuum tubes, 6–7
Validation:
  design, 224
  and suppliers, 418–419
  and verification, 39–40, 418–419
Value, best, 66–68, 390
Value-added, 74–75
Value-added work, 74–78
Values, 3, 118
Variability, 165, 170–172, 195
Variability reduction (VR), 18, 41, 49–50, 96, 132, 173, 192–198, 250, 397, 404, 432–433, 435, *Appendix C*
Variable data, 178
Varian's Chromatography Systems, 181, 183–184
Variation, 308–314, 325–326
Velocity of change, 4
Vendors (*see* Suppliers)
Verification, 48
  benchmarking, 473
  design, 224, 226
  ISO 9000 standards, 229, 237
  performance, 91–92, 94
  and suppliers, 418–419
  and validation, 39–40, 418–419
Vertical integration, 96
Visual aids, 124
Vital signs, 483–487
Voice of customer. 159, 162, 196–197
Voluntary standards, 17

WABM (workforce activity-based management), 120–121
Wainwright Industries, 509
Wallace, Thomas, 145
Wallace Company, 509
WalMart, 381
Walton, Mary, 354
Warranties, 61
WCM (*see* World class manufacturing)
Wealth, distribution of, 3
Weibull probability, 61–62
Westinghouse Electric Corporation, 509
Whirlpool Corporation, 289
WIP (work in process), 95, 121, 123
Work in process (WIP), 95, 121, 123
Worker involvement, 328–329, 514
Workforce, 4, 144–148
Workforce activity-based management (WABM), 120–121
Working definitions, 317
World class benchmarking, 463, 497–502
World class factories, 95
World class manufacturing (WCM), 32–33, 92–93, 95–126, 192, 436–443, 463
World class quality, 507–511

Xerox Corporation, 168, 194, 232–233, 248–249, 289, 446, 450–451, 456, 504, 509

Yield, improvement of, 125–126

Zuckerman, Amy, 253
Zytec Corporation, 509, 550–551

ABOUT THE EDITOR

Marsha Ludwig-Becker is an Operations Project Manager, The Boeing Company, and has experience in managing defense, aerospace, and commercial programs for NASA, the DOD, and foreign governments. Her team developed a nationally based supplier rating system that has been adopted widely by industry; her work in new quality management policies has been published in *Quality Digest,* and *International Standards Desk Reference*, *Your Passport to World Markets*. As an instructor and consultant with the Technology Training Corporation in Torrance, California, and Systems Management and Development Corporation in Springfield, Virginia, Ms. Ludwig-Becker conducts seminars and consults on the new quality management system.